Target Tracking with Random Finite Sets

Weihua Wu · Hemin Sun · Mao Zheng ·
Weiping Huang

Target Tracking with Random Finite Sets

国防工業出版社
National Defense Industry Press

Springer

Weihua Wu
Air Force Early Warning Academy
Wuhan, Hubei, China

Hemin Sun
Air Force Early Warning Academy
Wuhan, Hubei, China

Mao Zheng
Air Force Early Warning Academy
Wuhan, Hubei, China

Weiping Huang
Air Force Early Warning Academy
Wuhan, Hubei, China

ISBN 978-981-19-9814-0 ISBN 978-981-19-9815-7 (eBook)
https://doi.org/10.1007/978-981-19-9815-7

Jointly published with National Defense Industry Press
The print edition is not for sale in China (Mainland). Customers from China (Mainland) please order the print book from: National Defense Industry Press.

This Springer imprint is published by the registered company Springer Nature Singapore Pte Ltd.
The registered company address is: 152 Beach Road, #21-01/04 Gateway East, Singapore 189721, Singapore

Preface

In the process of research on target tracking and information fusion, the authors were deeply impressed by the great influence of the random finite set (RFS) theory. As a scientific top-down method, the RFS theory provides a unified theoretical description framework and solution for target detection, tracking and identification, situation assessment, sensor management and other problems involved in information fusion, differing from the traditional methods which decompose these problems into independent sub-problems to be solved separately.

The RFS theory, started at the end of the twentieth century, was created and pushed forward by Ronald Mahler et al. However, due to its abstraction and complexity, the RFS theory was not highly regarded by the academic community for a time. With the pioneering work of B-N Vo, such as the sequential Monte Carlo (SMC) and Gaussian mixture (GM) implementations of probability hypothesis density (PHD) filter, the realizations of the RFS theory were opened up. The above implementation methods were adopted with terms and symbols commonly used in the field of target tracking and information fusion, which greatly promotes the development of the RFS theory. Based on these, the implementation methods of cardinalized probability hypothesis density (CPHD) and multi-Bernoulli (MB) filter were proposed in a few years. In particular, in 2013, the labeled RFS filters such as the generalized labeled multi-Bernoulli (GLMB) filter were put forward, which made the RFS theory more and more perfect, and attracted widespread attention from famous domestic and foreign scholars in the field of target tracking and information fusion, resulting in an explosion of research results. To some extent, the RFS theory has become a new development direction of target tracking and information fusion. However, currently, there are only a few books that introduce this theory in a systematic manner—two books written by Mahler were published. As mentioned earlier, these books mainly focused on the theory, which can be difficult for beginners to understand. How can we popularize the RFS theory in a way that can be easily accepted by more people and promote the vigorous development of RFS theory? That is the purpose of this book.

This book adopts terms and symbols that are friendly to the academic world of target tracking and data fusion, with a focus on the systematic introduction of the specific implementations of the RFS theory in the target tracking field. It covers most of the current research results in this field: PHD, CPHD, MB, labeled MB (LMB), GLMB, δ-GLMB and marginalized δ-GLMB (Mδ-GLMB) filters. These filters are the latest technologies in the target tracking field and have provided new ideas and effective solutions for target tracking. After a systematic introduction to the above filters, their extensions and popular applications are described in detail, including maneuvering target tracking, target tracking for Doppler radars, track-before-detect of dim targets, target tracking with non-standard measurements and distributed multi-sensor target tracking. This book is well organized with systematic and comprehensive contents keeping up with popular development of advanced technology, making it suitable for beginners, graduate students, engineering and technical personnel in the field. According to the authors' learning experience, the authors suggest that readers can read this book first, and then study Mahler's original work, so that the books can complement each other and improve the learning efficiency of readers.

Upon the publication of this book, the author recollected that the draft had been revised numerous times since the end of 2015 in order to continuously absorb the latest research results. Although the book is completed, the author is deeply gratified by the fact that there are still more to learn. In addition, the authors have a lot to thank for. Thanks to the pioneering work of Ronald Mahler, B-N Vo and others, the authors have been able to feed on and draw upon their extensive work in writing; thanks to the National Natural Science Foundation of China (No. 61601510), the Young Talent Support Project of China Association for Science and Technology (No. 18-JCJQ-QT-008) and National Defense Industry Press for their funding and strong support; thanks to the editors of National Defense Industry Press for their hard work; thanks to the reviewers as well as the experts and scholars who care about the publication of this book.

This book provides a systematic and comprehensive introduction to the application of the RFS theory in the field of target tracking, and its application in the field of higher-level information fusion also has exciting and broad prospects. The authors sincerely hope that this book can be used to promote the further development of the RFS theory. Although we have made the greatest efforts to write this book, inadequacies are hard to avoid due to the limit of our knowledge, so we welcome any criticisms and corrections from readers.

Guiyang, China
October 2022

Weihua Wu

Contents

Abbreviations

CBMeMBer	Cardinality balanced MeMBer
CI	Covariance intersection
C-K	Chapman–Kolmogorov
CKF	Cubature Kalman filter
CPEP	Circular position error probability
CPHD	Cardinalized PHD
CRLB	Cramer–Rao lower bound
DA	Data association
DBZ	Doppler blind zone
DLI	Distinct label indicator
EAP	Expected a posterior
EK-CPHD	Extended Kalman CPHD
EKF	Extended Kalman filter
EK-PHD	Extended Kalman PHD
EMD	Exponential mixture density
ET	Extended target
FISST	Finite set statistics
GCI	Generalized covariance intersection
GGIW	Gamma Gaussian inverse Wishart
GIF	Generalized indicator function
GIW	Gaussian inverse Wishart
GLMB	Generalized labeled multi-Bernoulli
GM	Gaussian mixture
GMTI	Ground moving target indicator
GSF	Gaussian sum filter
IID	Independent and identically distributed
IIDC	IID cluster
IMM	Interacting multiple model
IPDA	Integrated PDA
ITS	Integrated track splitting
IW	Inverse Wishart

JIPDA	Joint IPDA
JM	Jump Markov
JPDA	Joint PDA
KDE	Kernel density estimation
KF	Kalman filter
KLA	Kullback–Leibler average
KLD	Kullback–Leibler divergence
LG	Linear Gaussian
LGJM	Linear Gaussian jump Markov
LGM	Linear Gaussian multi-target
LMB	Labeled multi-Bernoulli
MAP	Maximum a posterior
MB	Multi-Bernoulli
MCMC	Markov chain Monte Carlo
Mδ-GLMB	Marginalized δ-GLMB
MDV	Minimum detectable velocity
MeMBer	Multi-target multi-Bernoulli
MHT	Multiple hypothesis tracking
MM	Multiple model
MoE	Multi-object exponential
MS	Multi-sensor
MTT	Multi-target tracking
NWGM	Normalized weighted geometric mean
OSPA	Optimal sub-pattern assignment
PDA	Probabilistic data association
PDF	Probability density function
PF	Particle filter
PGF	Probability generating function
PGFl	Probability generating functional
PHD	Probability hypothesis density
PMF	Probability mass function
RD	Renyi divergence
RFS	Random finite set
SMC	Sequential Monte Carlo
SNR	Signal-to-noise ratio
TBD	Track-before-detect
UK-CPHD	Unscented Kalman CPHD
UKF	Unscented Kalman filter
UK-PHD	Unscented Kalman PHD
UT	Unscented transformation
δ-GLMB	δ-Generalized labeled multi-Bernoulli

Chapter 1
Introduction

1.1 Basic Concepts of Target Tracking and Random Finite Sets

Target tracking is a process carried out to estimate target states based on sensor measurements. A target is usually an object of interest, e.g. a vehicle, a ship, an aircraft, or a missile. The single-target refers to one target while multi-target refers to more-than-one targets. The target state refers to the unknown but interest information about the target. The typical target state includes the location and speed in the Cartesian coordinate system and it may also include other target characteristics such as identity, attribute, amplitude, size, shape or similarity. To estimate the unknown state of a target, sensor measurements are needed, which typically include time stamp, range, azimuth, elevation, Doppler, and amplitude information. The scope of measurement differs from sensors to sensors. For an active radar, the measurement mainly includes time stamp, range and azimuth; for a three-dimensional radar, elevation is also included; for an airborne Doppler radar, Doppler (or radial speed) information is also contained; and for the passive sensors such as IR and electronic support measurement (ESM) sensors, angle information such as azimuth and elevation is usually measured. Sensor measurements may originate from targets of interest, or from targets of no interest or clutters. Even if a sensor measurement come from a target of interest, the target may be detected with certain errors, or the target even missed due to limited observing capability of a sensor. Besides, sensor measurements may be collected by a single sensor or multiple homogeneous or heterogeneous sensors.

Generally speaking, the Bayesian filter is often used in target tracking. The mainstream Bayesian filter includes the Kalman filter (KF), extended KF (EKF), unscented KF (UKF), cubature KF (CKF), Gaussian sum filter (GSF) and particle filter (PF). The Bayesian filter usually involves "two models" and "two steps". The two models are the motion model (also known as dynamic model) and the measurement model, both of which are collectively referred to as the state-space model. The motion model describes the evolution of the target state over time. Common motion models

W. Wu et al., *Target Tracking with Random Finite Sets*,
https://doi.org/10.1007/978-981-19-9815-7_1

include constant velocity (CV), constant acceleration (CA), and coordinate turning (CT). The measurement model describes the linear or nonlinear relationship between the measurement and the target state. The two steps are prediction and update. The prediction step utilizes the motion model to predict the target state, while the update step utilizes the collected measurement to update the target state according to the measurement model.

The above filters are mainly used to reduce the impact of noise and, generally, only applied to ideal conditions (e.g. a single target is always present and there is no clutter). In fact, due to the limitation of sensor detection capabilities, in addition to noise interference, missed detection of targets occurs from time to time, and is inevitably accompanied by interference from non-target items (e.g. clutter) and other targets. The real challenge of target tracking is multi-target tracking (MTT) in clutter environment. Multi-target tracking refers to the estimation of unknown and time-varying number of targets and their tracks based on sensor observations. While the terms "multi-target tracking" and "multi-target filtering" are often interchangeable, there are still subtle differences between the two. Multi-target filtering involves estimation of unknown and time-varying number of targets and their independent states based on sensor observations; but for multi-target tracking, target tracks (track labels are required in the actual multi-target tracking system to distinguish different targets) are also of interest [1]. Therefore, multi-target tracking is essentially multi-target filtering that can provide target track estimation. Strictly speaking, multi-target tracking should be referred to as multi-target tracking filter.

Compared with the (single-target) Bayesian filtering algorithm, the main difficulty of multi-target tracking is the further increase of uncertainty. In addition to the uncertainty caused by the measurement noise and missed detection of targets, there is also the association uncertainty caused by the corresponding relationship between the measurement, clutter and each target. To overcome the association uncertainty, data association (DA) is usually carried out before Bayesian filtering in multi-target tracking to determine whether a measurement is from a clutter or to which target the measurement belongs. Through data association, multi-target tracking is thus decomposed into multiple single-target tracking problems. Famous DA-based MTT algorithms mainly are the joint probabilistic data association (JPDA) and multiple hypothesis tracking (MHT). However, the DA-based MTT problem is a non-deterministic polynomial (NP) problem, which requires a lot of computation and has the problem of combination explosion. In addition, it is difficult for these algorithms to give satisfactory results when multiple targets are close to each other and behaviors such as newborn, spawning, merging and death of targets are considered. Essentially, in these algorithms, the target state is modeled as a random variable (or random vector). Because of the birth[1] and death processes of each target, the number and states of moving targets are time-varying and unknown, hence it is difficult to model the finite and time-varying targets and measurements mentioned above with random variables. In order to track multiple targets with time-varying numbers, a complete and practical multi-target tracking should include track management (such

[1] Target birth includes target newborn and target spawning.

as track initiation and track termination) for behaviors such as target birth, merging and death, in addition to data association. Therefore, in a certain sense, the DA-based MTT algorithms adopt a divide-and-conquer, bottom-up approach.

In recent years, a type of tracking algorithm based on random finite sets (RFS) has emerged and attracted great attention. The RFS approach provides a multi-target Bayesian formula for the problem of multi-target filtering/tracking, where a set of target states (referred to as the multi-target state) is regarded as a finite set. Both random variables and RFS are random, except that the number of elements (referred to as the "cardinality") in the RFS is random and out of order. In particular, random variables take values according to a certain probability distribution in a certain space, while the RFS is a set. Unlike the traditional concept of a set, the number of elements in the RFS is uncertain, namely, the cardinality of the set is a random variable, and every element of the set is also random, which means it may or may not exist. If it does exist, its value follows a certain distribution. Therefore, an RFS is a set-valued random variable, which is the generalization of the concept of the random variable in probability theory. In fact, an RFS is actually a set in which elements and their number are also random variables. A random variable is used for solving random point functions while the RFS is used for solving random set-valued functions. The RFS theory is a generalization of point variable (vector) statistics to "set variable" statistics (finite set statistics). The RFS theory is also known as the point process theory, or more accurately, simple point process theory.[2]

In conclusion, different from the DA-based tracking algorithms, the RFS-based tracking algorithm models the multi-target state and the multi-target measurement as RFSs, and naturally incorporates the mechanism for track initiation and track termination. Hence, it is a top-down scientific approach and can realize the simultaneous estimation of the number of targets and their states. In addition to the MTT application, it also provides a unified theoretical framework and solution for target detection, tracking and identification, situation assessment, multi-sensor (MS) data fusion and sensor management [2].

Due to the systematic and scientific features of the RFS theory, it has developed into the 4th generation filter (Here, the 2nd, 3rd, and 4th generation filters refer to the cardinalized probability hypothesis density (CPHD) [3], multi-Bernoulli (MB) [4, 5], and generalized labeled multi-Bernoulli (GLMB) filters, respectively; see the following for details) in just over 10 years since the proposal of the 1st generation probability hypothesis density (PHD) [6, 7] filter, and has rapidly penetrated into various applications in the tracking field, showing its tenacious vitality. On top of the detailed introduction of the four generations of filters, this book also respectively introduces the main extensions and applications of each filter, including maneuvering target tracking, target tracking with Doppler radars, track-before-detect (TBD) for dim targets, target tracking with non-standard measurements, and distributed multi-sensor target tracking.

[2] A set for simple finite point process does not allow repeated elements, and only contains a limited number of elements.

1.2 Research Status of Target Tracking

Target tracking covers a wide range of contents. From the perspective of the number of targets, it can be classified into single target tracking and multi-target tracking; from the perspective of motion models, it contains constant velocity, constant acceleration and coordinate turning models; from the perspective of targets' environment, it can be classified into target tracking with clutter and target tracking without clutter; from the perspective of the number of sensors, it can be classified into single-sensor target tracking and multi-sensor target tracking; from the perspective of sensor property, it contains active sensors and passive sensors; and from the perspective of the spatial dimensions, it can be classified into two-dimensional target tracking and three-dimensional target tracking. The main process of target tracking generally includes data pre-processing, track initiation, filtering/tracking, and track termination. A variety of implementation methods are available for each process, especially for filtering/tracking. In addition, the integration of target tracking and detection, recognition, sensor management and decision-making is also the current research focus. In recent years, the emerging RFS approach provides a unified theoretical framework for the development of integration, and it is developing vigorously.

The following will briefly introduce the research status of single-target tracking and classical multi-target tracking [8–13], and then focus on the development status of RFS-based multi-target tracking [14, 15].

1.2.1 Single Target Tracking

The mainstream single target tracking filters mainly include: KF [16, 17], EKF [18, 19], converted measurement KF (CMKF) [20, 21], UKF [22, 23], CKF [24, 25], and PF [26, 27]. The KF is the optimal solution for linear Gaussian (LG) systems, which has made important contributions to the development of filtering theory, and the subsequent advanced filters are derived and developed from it to some extent. However, almost all practical systems are nonlinear and non-Gaussian. For example, measurements such as slant distance, azimuth and Doppler in the tracking problem are nonlinear functions of unknown states. Generally, the optimal solution for non-linear filtering can not be obtained. The EKF had once become the "standard solution" for nonlinear systems. It obtains the sub-optimal solution by linearizing the nonlinear system. However, the Jacobian matrix needs to be solved in the EKF method, which limits its scope of application. Compared with first-order EKF, the performance improvement for high-order EKF is small, but its computational complexity increases greatly. The converted measurement KF (CMKF), as the name implies, is to reconstruct the linear measurement equation after the measurement is converted, and to derive the corresponding covariance of converted measurement errors, and then execute the KF. It has been proved that the performance (estimation accuracy, robustness, consistency, etc.) of the CMKF is better than that of the EKF for nonlinear measurement systems [20, 21]. Based on the unscented transformation (UT) [28], the UKF approximates the statistical properties of random variables with a finite number of parameters. Unlike the EKF, which approximates the nonlinear dynamic and/or nonlinear measurement models by linearization, the UKF approximates the probability density function (PDF) of the state vector. The UKF is widely used because it does not need to derive and calculate the complex Jacobian matrix or higher order Hessian matrix, but its estimation can reach the 2nd order accuracy. The CKF is a new filtering algorithm based on the principles of cubature numerical integration. It has the advantages of high numerical accuracy and strong robustness, and its estimation can reach the 3rd order accuracy for the nonlinear Gaussian system. All the above algorithms assume that the PDF approximately complies with Gaussian or Gaussian mixture (GM) distribution. For (nonlinear) non-Gaussian systems, the PF method, which is also known as sequential Monte Carlo (SMC) [29] method, is required. A large number of improved methods have emerged, which have promoted the significant development of the PF algorithm, but its computational amount is significantly increased, compared with the aforementioned filtering algorithms.

The above filtering algorithms are mainly for single target non-maneuvering models. When a target maneuvers, it will lead to the mismatch between the motion model used in the filtering algorithms and the actual motion model of the target, causing the filter to diverge. For maneuvering target tracking, the algorithms adopted can be classified into tracking algorithms with maneuvering detection and adaptive tracking algorithms. The former category mainly includes tunable white noise model [30], variable-dimensional filtering [31], and input estimation algorithm [32]. However, these algorithms have problems such as detection delay and detection reliability; and the latter category mainly includes the first-order temporal correlated noise model [33], current statistics (CS) model, Singer model, Jerk model [34], multiple model (MM), and interacting MM (IMM) [35]. The IMM is a suboptimal filter with good cost-effectiveness ratio. In addition, it has strong extensibility and can be easily combined with other algorithms, such as IMM probabilistic data association (PDA) [36], IMM-JPDA [37], and IMM-MHT [38]. Due to its excellent performance, the IMM has gradually become the mainstream for maneuvering target tracking.

1.2.2 Classical Multi-target Tracking

Due to the complexity of the actual environment and since the tracking performance is affected by factors such as clutter, sensor performance (such as missed detection) and multiple targets, it is difficult for pure filtering algorithms to achieve effective target tracking. A classic tracker generally consists of two steps: data association (DA) and filtering. Filtering is meaningful only if it is based on correct data association. Data association refers to the process of determining the one-to-one correspondence between measurements and targets. Through DA processing, multi-target tracking is decomposed into multiple single-target tracking tasks. Therefore, data association, as the core of DA-based MTT algorithms, is mainly classified into two categories:

maximum likelihood data association and Bayesian data association. The former includes the track splitting method [39], the joint maximum likelihood method, and 0–1 integer programming method. They are usually achieved by batch processing, and their basic ideas are to maximize the likelihood function. The latter includes PDA [40], global nearest neighbor (GNN) [41], S-dimensional (S-D) assignment [42], integrated PDA (IPDA) [43], joint probabilistic data association (JPDA) [44], joint IPDA (JIPDA) [45], integrated track splitting (ITS) [46, 47], belief propagation (BP) method [48, 49] based on sum-product algorithm [50, 51], Markov Chain Monte Carlo (MCMC) data association [52, 53], and MHT [54]. The advantage of these methods is that they have recursive forms and can obtain the state estimation in real time. As mentioned earlier, the DA-based MTT problem is an NP problem, which requires a lot of computation and has the problem of combination explosion. In tracking algorithms, 60–90% of the computation time is consumed by data association [11], among which, the maximum likelihood data association is usually larger than the Bayesian data association. Therefore, one of the important research contents of classic MTT algorithms is to reduce the calculation amount and improve the real-time performance. For example, the m-best S-D assignment algorithm [55] and K-best MHT method [56] constrain the number of hypotheses by limiting the optimal m (or K) number of association hypotheses.

1.2.3 RFS-Based Multi-target Tracking

The first to use the RFS theory to systematically process multi-sensor multi-target filtering is Mahler's FISST (FInite Set STatistics) [14, 15]. The FISST is a systematic and unified method for multi-sensor multi-target detection, tracking and information fusion. It can realize the Bayesian unification of detection, classification, tracking, decision-making, sensor management, group target processing, expert system theories (fuzzy logic, DS theory, etc.), and performance evaluation for multi-platform, multi-source, multi-evidence, multi-target, and multi-group problems. It has the following advantages: it is an explicit, comprehensive and unified statistical model based on multi-sensor multi-target systems; it can combine the two independent purposes of multi-target tracking, namely target detection and state estimation, into a single, seamless, Bayesian optimal step; it serves as a rich soil for cultivating new methods of multi-source multi-target tracking and information fusion, and has promoted new multi-target tracking algorithms, such as the PHD, CPHD and MB filters. Although these new algorithms do not require data association between measurements and tracks, their tracking performance (in terms of tracking accuracy and execution efficiency) is comparable to or even better than that of conventional multi-target tracking algorithms [2].

The core of Mahler's RFS theory is multi-target Bayesian filter [57]. This filter, similar to single target Bayesian filter, is also composed of prediction and update. Although the form is elegant and simple, the RFS-based optimal multi-target Bayesian filter is not practical due to the combinatorial nature of multi-target density and the multiple integration of infinite-dimensional multi-target state space. To this end, Mahler derived a variety of principle approximations, such as the PHD, CPHD, and MB filters. Specifically, the PHD filter is obtained by the first-order moment approximation of the optimal multi-target Bayesian filter. Based on the classic point process theory, another derivation method of the PHD filter is presented in [58]. Then, the Poisson hypothesis condition for the number of targets is further relaxed, and the CPHD filter, the second-order moment approximation of the optimal multi-target filter, is derived. It not only propagates the intensity, but also propagates the cardinality distribution (the probability distribution of the number of targets). At the cost of increasing the computation amount, both of its filtering accuracy and estimation accuracy for number of targets are better than those of the PHD filter. In addition, Mahler also proposed the MB filter, also known as the multi-target multi-Bernoulli filter (MeMBer) [14, 59–61]. Different from the PHD and CPHD filters, which approximate the first- and second-order moments of the optimal filter, the MB filter is the probability density approximation of the optimal filter.

Despite the approximate processing, there are still complex operations such as multiple integrals in recursive expressions of the above three filters, resulting in no analytical solutions under general conditions. To this end, Vo developed an SMC implementation (also known as particle implementation) method [62] for the PHD filter under general conditions, denoted as the SMC-PHD filter. Then, under the linear Gaussian condition, the analytic form of the PHD recursion was derived, and the Gaussian mixture (GM) implementation of the PHD filter was developed, denoted as the GM-PHD filter [63]. By using linearization and unscented transformation, the closed-form recursive formula applicable only to linear models can be extended to moderate nonlinear models. Based on this, Vo et al. developed two CPHD implementation methods [64], namely the SMC-CPHD and GM-CPHD filters. Vo et al. also found that the MeMBer filter proposed by Mahler was biased in estimating the number of targets. Therefore, a cardinality balanced MeMBer (CBMeMBer) filter was proposed, and two corresponding versions were also developed, namely the SMC-CBMeMBer [4] and GM-CBMeMBer [5] filters. The above approximation algorithms have been proved to have fine convergence properties [65–68]. Nevertheless, these filters are not multi-target trackers in principle and cannot provide track labels [62]. In order to solve the problem of target track output, the concept of labeled RFS was introduced in [69, 70], in which the first GLMB analytical implementation of multi-target Bayesian filter was derived by using the conjugacy of the GLMB family (with respect to the standard measurement model), and the GLMB and δ-GLMB filters (also referred to as Vo-Vo filters) were put forward. Simulation results showed that the δ-GLMB filter is superior to the approximations of multi-target Bayesian filter. However, the δ-GLMB filter requires a large amount of computation. Inspired by the PHD and CPHD filters, the labeled multi-Bernoulli (LMB) [71] and marginalized δ-GLMB (Mδ-GLMB) [72] filters were developed respectively. Among them, the LMB filter only matches the first-order statistical moment of the δ-GLMB posterior, while the Mδ-GLMB filter matches the first-order moment of the δ-GLMB posterior and cardinality distribution. In addition to these filters, multi-target trackers incorporating the label in the target state also include the particle marginal Metropolis-Hasting tracker [73]. The monographs [14, 15] and Refs. [74, 75] have provided systematic and comprehensive introductions to this field. It is worth pointing out that, under the harsh conditions of high false alarm rate, high missed detection rate and high uncertainty of measurement source (because targets are very close to each other), Ref. [76] first verified that, the GLMB filter was able to track multiple targets with a peak value of 1 million on ordinary commercial computers. The work above has laid a solid foundation for the research of RFS-based tracking algorithms.

In short, as the RFS-based tracking algorithm becomes more and more mature, its scope of application becomes more and more extensive, such as sonar image tracking [77, 78], audio signal tracking [79], video tracking [80, 81], robot simultaneous localization and mapping (SLAM) [82, 83], traffic surveillance [84], ground moving target indication (GMTI) tracking [85], track-before-detect [86], multi-station passive radar tracking [87], angle-only tracking [88], multiple input multiple output (MIMO) radar tracking [89], sensor networks and distributed estimation [90–92], as well as target tracking with non-standard measurement models, such as tracking of unresolved targets (or tracking with merging measurement) [93, 94], extended target (ET) tracking [95], and group target tracking [96, 97].

1.2.3.1 Probability Hypothesis Density Filter

The PHD filter recursively propagates the posterior intensity (first-order moment) of the multi-target state set, and projects the posterior probability density of the multi-target state set onto the single-target state space with "minimum loss" [6]. In this way, the PHD filter only needs to implement recursion in single-target state space, which significantly reduces the complexity of computation. However, calculating the posterior probability by Bayes rule requires the integration of the product of prior and likelihood functions to obtain the normalization constant. Practically, it is still very difficult to implement multi-dimensional integration, and there is generally no closed analytic form available under nonlinear non-Gaussian conditions. Therefore, Mahler and Vo et al. studied the implementation of the PHD filter, and presented the SMC-PHD filter [62] for nonlinear non-Gaussian conditions and the GM-PHD filter [63] for linear Gaussian conditions.

In order to improve the effectiveness of particle implementation in the SMC-PHD filter, an auxiliary random variable was introduced in [98] to incorporate the measurement information into the importance function, and the auxiliary particle implementation of the PHD filter was proposed as well. In [99], the authors used the Gaussian mixture expression to approximate the importance function and predicted density function through the unscented information filter, and then put forward the Gaussian mixture unscented SMC-PHD filter. Reference [100] called the Rao-Blackwellisation to achieve a more effective SMC implementation for some specific formal models. In order to avoid the degradation and not be limited to the Gaussian system, considering that any curve can be represented by splines, the spline PHD (SPHD) filter was proposed in [101]. However, this algorithm is only suitable for tracking a few valuable targets under severe conditions because of its high complexity. In addition, considering that the statistical characteristics of noise are usually unknown in most of the actual systems, the robustness of a filter is very important. Taking into account the unknown characteristics of nonlinear models and uncertain noise, the H_∞ filter is more robust than the Kalman filter in terms of model errors and noise uncertainty. Therefore, based on the H_∞ filter, a new GM implementation for the PHD recursion was proposed in [102]. In order to solve the problem of close target occlusion in multi-target tracking, considering the "one-to-one correspondence" assumption of measurement-target in the GM-PHD filter, a competitive GM-PHD method is proposed in [103] by calling a re-normalization scheme to reset the weight assigned to each target in the GM-PHD recursion. In [104, 105], the PHD filter is applied to joint detection, tracking and classification (JDTC) of multiple targets. The goal of JDTC is to simultaneously estimate the time-varying number, kinematic states, and class labels of targets.

State extraction is a necessary step for PHD filters. For the SMC-PHD filter, state extraction is usually carried out according to the spatial distribution of particles and by k-mean method [106, 107] or clustering technology based on the finite mixture model (FMM) [108]. Reference [107] compared the state extraction performances of k-mean clustering and the expectation maximization (EM)-based FMM, and the results indicated that compared with the EM method, k-mean algorithm significantly reduced the complexity of computation. In fact, the EM is a deterministic method and is not suitable for estimating parameters of complex multi-mode distribution. Therefore, the random method of Markov Chain Monte Carlo (MCMC) was used for FMM parameter estimation in [108]. In addition to the spatial distribution of particles, the weight information of particles can also be used to better extract the states of close or near targets [109, 110]. Different from the SMC-PHD filter, the GM-PHD filter can easily extract state estimation without clustering (which requires a lot of computation and may lead to inaccurate estimation), but it is still limited to the linear Gaussian system. To this end, a hybrid GM/SMC implementation method of the PHD filter was proposed in [111]. Reference [112] proposed a new particle implementation algorithm for the PHD filter based on the GM-PHD filter, which can not only extract target states without clustering technology, but also be applicable to highly nonlinear non-Gaussian models.

Standard PHD filters cannot provide the track information of targets [62]. One way to solve this problem is to use estimation-track association [106, 113]. The idea is to take the multi-target state estimation output by the PHD filter as the "measurement" input of another data association-based multi-target tracker, and then implement the estimation-track association through the tracker to obtain each target track; another approach is to use the PHD filter as a clutter filter to eliminate the clutter in the measurement set that are unlikely to originate from targets, and then input the remaining measurements into the tracker [106]. It can be considered a gate operation at the global level, which eliminates most clutter measurements. Both methods have reduced the number of measurements used for data association, nevertheless, the track information is output by the tracker, and the PHD filter itself does not utilize the track information of targets. Besides the estimation-track association method, another method called labeling is also commonly used for track output. It has be used for the GM-PHD [114] and SMC-PHD [106, 107, 115] filters. The labeling method can effectively provide track information. However, when targets are passing across or close to each other, the labeling method is prone to misjudgment of targets. For this reason, in [114], the estimation-track association method was used for state association when targets were passing across or close to each other, and the labeling method was used to carry out state association when targets were far away. Additionally, [116] combined the PHD filter with multi-frame association to further reduce the false association rate.

1.2.3.2 Cardinalized Probability Hypothesis Density Filter

In the PHD filter, since the mean and variance of the Possion distribution are equal, when the number of targets is high, it is easy to cause strong fluctuations in the estimation of the number of targets under the condition of missed detections or high false alarm density, which makes the estimation unreliable and leads to the missing measurement problem [64]. In response to this problem, Ref. [117] pointed out that for the PHD filter, not only the propagation of the first-order multi-target moment, but also the propagation of the higher-order number of targets is required. Based on this, a CPHD filter with the cardinality distribution is proposed in [3] using FISST tools such as the probability generating functional (PGFl) or functional derivative. The CPHD is a general version of the PHD. It not only propagates the posterior intensity of the multi-target state set, but also propagates the posterior cardinality distribution of the set at the same time. Its recursion formula is more complex and has the cubic complexity, but it still has better performance than the JPDA with the non-polynomial complexity [64]. Different from [3], Ref. [118] obtained a new derivation of the PHD and CPHD filters by directly performing the Kullback–Leibler divergence (KLD) minimization on predicted and posterior multi-target densities. Due to the introduction of cardinality distribution, the CPHD improves the accuracy and stability of the estimation of the number of multiple targets and their states, but the response speed of the CPHD to target birth and target death is not as fast as that of the PHD [64]. In addition, although the update formula of the overall CPHD cardinality distribution is accurate, when targets are missed, it will exhibit a local singular behavior—the "spooky effect" [119], i.e., the weight of the PHD will shift from the missed part to the detected part, no matter how far apart the two parts are, resulting in a significant underestimation of the number of local targets in the vicinity of the missed measurements [119]. To this end, the surveillance area was divided into different regions in [119], and then the CPHD was applied to each region in turn. However, the clutter density needs to be modified after the division, which in turn may increase the uncertainty of the cardinality estimation. Reference [120] reduced the influence of weight drift and estimation error through a dynamic reweighting scheme.

Unlike the PHD filter, in the standard CPHD filter [64], the prediction step does not include the target spawning model. Although the CPHD filter for birth targets can be used to solve the target spawning issue, it is obviously more appropriate to use a spawning model that can consider specific issues. Taking resident space object (RSO) as an example, the natural and artificial earth orbiting satellites are composed of space shuttles, scrapped payloads, and debris. If there is no spawning model, the best option is to use diffused birth regions; however, this requires a large birth area to cover the corresponding spatial volume. To improve the performance of the CPHD filter when tracking the spawning RSO, a spawning model with accurate description of the physical process of generating a birth RSO is proposed in [121], which can obtain better accuracy and faster confidence time for birth targets. In [122], Poisson or Bernoulli spawning models are incorporated into the CPHD filter, while in [123], the CPHD prediction equation suitable for any spawning process is derived through partial Bell polynomials [124]. Moreover, for the three specific models (Poisson, zero-inflated Poisson, and Bernoulli models), a GM-CPHD filter applicable to spawning targets can be obtained without additional approximations. Other literatures on CPHD filters considering spawning targets also include [125].

1.2.3.3 Multi-Bernoulli Filter

Unlike the PHD (CPHD), which recursively propagating the moment (and cardinality distribution) of the posterior multi-target density, the Bernoulli RFS models the target track [14] through two parameter pairs, the probability of target existence and the PDF when the target exists, which has some similarity with the IPDA filter [43] that simultaneously estimates the existence probability of a single target and its state. In [126], the Bernoulli RFS was used to derive the optimal Bayesian solution to the single-target detection and tracking problem, and the Bernoulli filter, also known as the joint target detection and tracking (JoTT) filter [14, 127–131], was obtained. The JoTT here refers to the joint estimation of the number of targets and their states from sensor measurements.

The JoTT filter in the multi-target background is called the multi-Bernoulli (MB) filter. As the name implies, the MB RFS is the union of multiple Bernoulli RFSs. In essence, the MB filter propagates the parameters for approximating the MB distribution of the posterior multi-target RFS density. The multi-target multi-Bernoulli (MeMBer) update equation proposed in [14] has a significant bias (overestimation) in the estimation of the number of targets, and this bias disappears only when the detection probability is 1 [4]. For this reason, literature [4] deduced the analytical expression of the "cardinality bias" in the MeMBer filter, calculated the updated existence probability by using the accurate probability generating functional, eliminated the cardinality bias problem by correcting the measurement updated track parameters, and gave an unbiased cardinality balanced MeMBer (CBMeMBer) filter, as well as two implementation methods: SMC-CBMeMBer and GM-CBMeMBer. However, in order to calculate the effective spatial PDF, the filter makes strict assumptions about the target detection probability. For this reason, Ref. [132] eliminated this bias by introducing a fake Bernoulli target without any strict assumption, and proposed an improved MeMBer filter. Reference [59] balanced the posterior cardinality distribution by multiplying the probability of missed detection. The new algorithm has better performance than the CBMeMBer method. However, the approximation of the missed detection probability in the new algorithm is based on the assumption that the tracks are separable. In order to solve the problem of close targets, Ref. [133] extended [134] and presented a principled, highly efficient approximation method to find the MB distribution that minimizes the KLD with the full RFS distribution. To overcome the problem that the parameters such as clutter intensity, detection probability and sensor field-of-view (FoV) need to be known a priori in the MB filter, Ref. [60, 61] respectively proposed the MeMBer filters under the conditions of unknown detection probability and clutter intensity, as well as unknown non-uniform clutter intensity and sensor FoV.

For approximation algorithms such as the PHD, CPHD and CBMeMBer, the GM-CPHD algorithm has the best performance in the linear Gaussian case, while the GM-CBMeMBer has the similar performance as the GM-PHD, not showing advantages. However, under highly nonlinear non-Gaussian conditions, the MB filter should be a better option. Unlike the particle realization of the PHD/CPHD filter, which needs particle clustering to extract target states, requiring a large amount of computation and being unreliable, the SMC-CBMeMBer filter, however, does not require additional clustering operation, and can directly extract the multi-target state estimation. Under the condition of high signal-to-noise ratio (SNR), the SMC-CBMeMBer filter not only has lower computation but also has better performance compared with the SMC-PHD/CPHD filter. Additionally, the MB filter can also provide the existence probability of targets. As a result, the GM-CPHD filter has the best performance under linear Gaussian conditions, while the SMC-CBMeMBer filter has obvious advantages under nonlinear non-Gaussian conditions.

In terms of the amount of computation, Ref. [135] has verified the real-time performance of the RFS algorithm using actual data. In general, the complexity of algorithms such as the PHD, CPHD, and MeMBer are $\mathcal{O}(mn)$, $\mathcal{O}(m^3n)$, and $\mathcal{O}(mn)$, respectively, where m and n represent the number of measurements and targets respectively. In other words, the MeMBer and PHD filters have the same linear complexity, while the CPHD filter has a higher cubic complexity. By reducing the cardinality of the measurement set, the calculation amount of the algorithm can be reduced. In [136, 137], the computational amount was reduced without any significant performance loss by incorporating the ellipsoid gate technique used in conventional tracking algorithms. In addition, Ref. [138] proposed a CPHD filter with linear complexity based on a relatively simple clutter model.

1.2.3.4 Labeled Random Set Filters

It should be noted that the filters mentioned above are not multi-target trackers in nature, because the target states are unresolved, which is one of the reasons why the RFS framework was once criticized: the algorithms derived from the RFS framework cannot obtain target labels. Besides, they are all approximate filters, and even assuming special observation models, such as the standard point target observation model, they are not closed-form solutions of the optimal Bayesian filter. Therefore, in [69], the concept of labeled RFS is introduced to solve the problem of target track and its uniqueness, and a new RFS distribution class called GLMB distribution was proposed. The GLMB distribution is conjugated with respect to the multi-target observation likelihood, and is closed with respect to the multi-target transition kernel under the multi-target Chapman-Kolmogorov (C-K) equation, thus providing an analytical solution for the multi-target inference and filtering problems, namely the δ-GLMB filter, which exploits the conjugation of the GLMB family to accurately forward propagate the (labeled) multi-target filtering density over time. It is an exact closed-form solution of multi-target Bayesian recursion, yielding the joint estimation of states and labels (or tracks) in the presence of clutter, missed detection, and association uncertainties, and is the first tractable RFS-based multi-target tracking filter that can generate track estimation in a principled way, refuting the view that the the RFS methods cannot generate track estimates. Reference [139] further extended the GLMB filter to be suitable for spawning target conditions, Refs. [140, 141] extended it to multi-frame sliding window processing, and Ref. [142] extended it to be suitable for related multi-target systems. In view of the complex and lengthy derivation process in [69], the probability generating functional (PGFl) method was proposed in [143] to provide a simplified derivation of the GLMB filter, and another tractable multi-target tracker, namely labeled multi-Bernoulli mixture (LMBM) filter, with an accurate closed form was derived using the PGFl method. The LMBM filter may be more practical because the LMB mixture is simpler to calculate than the GLMB distribution.

Nevertheless, the specific implementation of the δ-GLMB filter was not given in [69] and [143]. Therefore, an efficient and highly parallel implementation of the δ-GLMB filter was given in [70], which complements the theoretical contribution of [69] with practical algorithms. Specifically speaking, each iteration of the δ-GLMB filter involves multi-target predicted density and filtering density, both of which are weighted sums of multi-object exponentials (MoE). Despite these weighted sums have a closed form, the number of component terms in the posterior grows super-exponentially over time due to the explicit data association in the δ-GLMB filter. It is obviously infeasible to adopt the pruning strategy, which first exhaustively calculates all items of the multi-target density and then discards the secondary components. Hence, a pruning strategy that does not need to exhaustively calculate all components of the multi-target density is proposed in [70], in which the multi-target predicted density and filtering density are pruned using the K shortest path and ranked assignment algorithms [144] respectively. Meanwhile, the relatively low-computational PHD filter derived from the same RFS framework is used as a look ahead strategy to significantly reduce the number of calls of the K shortest path and ranked assignment algorithms.

The two-step implementation proposed in [70] is intuitive and highly parallel, specifically, in the prediction step, pruning is realized by solving two different K shortest path problems, one for the existing tracks and the other for the birth tracks; in the update step, pruning is achieved by solving the ranked assignment problem for each predicted δ-GLMB component. However, the two-step implementation is structurally inefficient, because the pruning of the predicted and updated δ-GLMB components is carried out separately, so a large proportion of the predicted components may produce updated components with negligible weights. As a result, a large amount of computation is wasted on solving a large number of ranked assignment problems, each of which has at least a cubic complexity with the number of measurements. Therefore, Refs. [145, 146] pruned the GLMB filtering density by combining the prediction and update steps into a single step, and using a random Gibbs sampler with a linear complexity with the number of measurements and an exponential convergence rate based on the Markov chain Monte Carlo (MCMC) [147] method. There is no need to discard samples during the burn-in phase in the pruning application, so it is not necessary to wait for samples from a stable distribution. The stochastic solution has two advantages over the deterministic ranked assignment (in the order of non-increasing weights) for pruning strategies: first, it eliminates the unnecessary computations caused by component sorting and reduces the cubic complexity with the number of measurements to the linear complexity; second, it automatically adjusts the number of generated significant components by using the statistical characteristics of component weights, thus resulting in a more efficient implementation of the GLMB filter, significantly improving the running speed without affecting the filtering performance. It should be pointed out that the recommended Gibbs sampler also provides an effective solution to the data association problem or the more general ranked assignment problem. In conclusion, the new implementation method is an online multi-target tracker, which has a linear complexity with the number of measurements and a quadratic complexity with the number of hypothetical tracks. It can be applied to complex scenarios such as nonlinear dynamic and measurement models, non-uniform survival probability, sensor field of view, and clutter intensity. Since it was proposed, the δ-GLMB filter has been rapidly promoted and applied [83, 86, 94, 148–151], which shows that the GLMB filter is a general model with excellent performance.

In addition to using the above acceleration strategies, some scholars have sought cheaper approximations of the δ-GLMB filter to improve the performance, the most famous of which are the Mδ-GLMB [72] and LMB filters [71]. As mentioned earlier, the performance improvement of the δ-GLMB filter is obtained at the cost of higher computational complexity, which mainly comes from data association. For some applications, such as multi-sensor tracking or distributed estimation, it is not feasible to apply the δ-GLMB filter due to limited computational resources. Inspired by Mahler's independent and identically distributed cluster (IIDC) approximation in the CPHD filter, Ref. [152] derives a special tractable GLMB class: marginal δ-GLMB (Mδ-GLMB) density, which can be used to define a principle approximation of the δ-GLMB density representing the true posterior of the multi-target Bayesian filter. Since the δ-GLMB density can be used to optimally approximate any labeled multi-object density [152], the Mδ-GLMB density provides a tractable multi-target density approximation to a general labeled RFS density that captures statistical correlations between targets. Especially, it matches the cardinality distribution and the first order moment (probability hypothesis density, PHD) of labeled multi-target distribution of interest (such as the true δ-GLMB density), and minimizes the Kullback-Leibler divergence (KLD) on tractable GLMB densities (such as the Mδ-GLMB density). Based on this, Ref. [72] proposes the Mδ-GLMB filter, as it can be interpreted as performing marginalization on the data association histories generated by the δ-GLMB filter. Therefore, the Mδ-GLMB filter is computationally cheaper than the δ-GLMB filter, while preserving the key statistics of the multi-target posterior, in particular, it is easier to develop efficient tractable multi-sensor trackers based on Mδ-GLMB filter.

Another efficient approximation to the δ-GLMB filter is given in [71], namely, the labeled multi-Bernoulli (LMB) filter, which uses the δ-GLMB update step in each iteration. However, it approximates the δ-GLMB posterior from each update step using the LMB distribution, to reduce the computational complexity. The LMB filter, a generalization of the MB filter, inherits the advantages of the MB filter concerning particle implementation and state estimation, and also advantages of the δ-GLMB filter. It delivers a more accurate update approximation than the MB filter by calling the conjugate prior form of the labeled RFS, without the cardinality bias problem. In addition, it can also output the target tracks (labels), with performance significantly better than performances of the PHD, CPHD and MB filters [153], and comparable to that of the δ-GLMB filter. Besides, it gets rid of the limitation that the MB filter is only suitable for high signal-to-noise ratio (low clutter and high detection probability) conditions. In conclusion, the LMB filter can formally estimate the tracks with unbiased posterior cardinality distribution even in difficult scenarios such as low detection probability and high false alarm. This filter has been used in the environment perception system of the autonomous vehicle with multiple sensors (radar, lidar, and video sensor) [154], which demonstrates its real-time performance and robustness. Reference [155] improves the real-time performance of the LMB filter by further approximation.

To synthesize the advantages of the LMB and δ-GLMB filters: the low complexity of the LMB filter and the accuracy of the δ-GLMB filter, Ref. [156] proposes an adaptive LMB (ALMB) filter, which automatically switches between the LMB and δ-GLMB filters based on the KLD [157] and entropy [158]. Aiming at the problem that most methods usually discard partial or all of statistical correlations in order to reduce the amount of calculation, Ref. [159] proposes an improved labeled multi-object (LMO) density approximation by adaptively decomposing the LMO density into the densities of several independent subsets according to the analysis of the true statistical correlation between target states. Considering that the labeled Poisson RFS and labeled IIDC RFS are special cases of the GLMB RFS, Ref. [160] derives the labeled PHD (LPHD) and labeled CPHD (LCPHD) filters based on the GLMB filter. When the targets approach each other for a long time and then separate, a mixed label problem will occur [161]. In such case, the label-switching improvement (LSI) method that is not interested in label information [134] can be applied. The basic idea of this approach is as follows: the label can be regarded as an auxiliary variable when only the inference of the label-free target set is interested. At this time, it is equivalent to generate an additional degree of freedom, which helps improve the approximation of the posterior PDF, so that any labeled posterior PDF can be selected, as long as the corresponding unlabeled posterior PDF remains unchanged.

1.2.3.5 Track Initiation and Parameter Adaptation

Track initiation is a relatively simple yet important part of multi-target tracking. If a track can be initialized in a statistically consistent manner, it can avoid the divergence of the filter, improve measurement-track correlation, reduce the number of false tracks, and help track management in multi-target tracking. In the context of random finite sets, the birth distribution or intensity of targets plays a similar role to track initiation of conventional trackers. Specifically, for the PHD/CPHD filters, the track initiation process is handled by the birth intensity, while for the MB/LMB/GLMB filters, track initiation is handled by the birth distribution. In most studies, the birth distribution or intensity is always assumed to be known a priori, but this is too strict for practical applications. In fact, a target may appear randomly at any position at any time. Therefore, when a target appears in an area where the predefined birth distribution or intensity is not covered, the standard RFS filters will completely "blind" to the presence of the target. A natural approach is to model the target birth distribution or intensity as a uniform density. It is, however, obviously inefficient and will lead to a higher incidence of short-life false tracks and a longer confidence time of true tracks, as a proper approximation of desired uniform density needs a large number of Gaussian components. To solve this issue, Refs. [109, 162–164] proposed an adaptive estimation approach of the birth intensity for PHD/CPHD filters. In [165], birth targets are assumed to be only appeared in the limited space around the measurements, and then newborn particles are extracted from the Gaussian mixture centered on the measurement components. Reference [166] proposed a GM-PHD filter with the ability to estimate the birth intensity, based on the assumption that each measurement may originate from a birth target and the birth target can always be detected. Reference [94] assumes that there is no missed detection in the first two frames, and then constructs the birth density with consecutive frames. According to the Bernoulli RFS concept, Ref. [167] strictly derived the statistics PDF (namely, the posterior existence probability and related state distribution of newborn tracks) of posterior RFS of birth targets, and proposed a brand-new approach for newborn track detection and state estimation using data-driven importance sampling and tree structure measurement sequences. References [71, 168] presented the measurement driven birth (MDB) models for the LMB and GLMB filters, respectively. The models adapt to the target birth process based on measurement data, thereby eliminating the dependence on the prior knowledge concerning birth target distribution.

Although the above algorithms can adaptively estimate the newborn targets to a certain extent, most of estimation algorithms for the birth intensity/distribution only consider the measurable components (such as positions) of the target state, and the unmeasurable parts (such as velocities) are regarded as prior information [166] or modeled as a simple distribution (such as a zero-mean Gaussian distribution), which is not efficient enough to represent the initial state of newborn targets. For example, Ref. [163] extracts newborn particles with equal weights to approximate the target birth intensity according to the measurement information, but the velocity component is set to 0. By contrast, Ref. [169] proposes a simple solution to make the intensity cover the complete state-space, but the computational cost is higher. Reference [162] divides the state space into a measurable space and an unmeasurable space, and then approximates them with a Gaussian distribution and a uniform distribution, respectively. Considering that the single-point track initiation algorithm can quickly locate the position of the newborn targets, and the two-point differencing track initiation algorithm can be used to estimate the target speed, Ref. [170] proposes an adaptive estimation approach of the target birth intensity that integrates the single-point and two-point differencing track initiation algorithms.

In addition to track initiation, the knowledge of clutter density and detection probability is also very important in multi-target Bayesian tracking. In a large number of multi-target tracking studies, it is usually assumed that the clutter density and detection probability are known to be time-invariant or at least uniform. In practical applications, however, these parameters, especially the clutter distribution, are often unknown and non-uniform, and their values are time-varying as the environment changes. Therefore, Refs. [171, 172] propose an online estimation approach of the clutter density, but they assume that the clutter background does not change too fast compared with the measurement update rate. Hence, when the clutter intensity changes rapidly, and the numbers of clutters and target measurements are at the same level, the performance of the approach may be deteriorated severely. Reference [173] proposed an approach for jointly estimating unknown clutter and detection probability, but the performance of this algorithm is inferior to that of the conventional CPHD filter with accurately known clutter density. To improve the performance, Ref. [174] first uses the joint estimation method of unknown clutter and detection probability to estimate the clutter density, then uses the conventional CPHD filter to estimate the PHD and cardinality distribution of the true targets, and proposes a bootstrap filtering method, which has almost the same performance as the GM-CPHD filter that matches the clutter intensity. Based on the RFS, Ref. [175] proposes an improved multi-Bernoulli filter suitable for nonlinear dynamic and measurement models and unknown clutter rates. This approach incorporates the amplitude information into the state space and measurement space to improve the discrimination between true targets and clutter.

1.2.3.6 Application in Maneuvering Target Tracking

There are many methods for dealing with maneuvering targets. Classical maneuvering target tracking algorithms include Singer model, current statistics model and multiple model approaches. The multiple model approach, which detects the maneuver and identifies the corresponding model, has been proven to be highly effective. The multiple model approach includes the generalized pseudo-Bayesian (GPB) [176] algorithm. The GPB algorithm of order n (GPB n) requires N_μ^n filters, where N_μ is the number of models. In this method, a limited number of filters work in parallel, which assume that the target model follows a certain model in the model set of the tracker. Considering the trade-off between the complexity and the performance, the interactive multiple model (IMM) approach has been shown to be the most effective among the known multiple model approaches, and has gradually become the mainstream algorithm for maneuvering target tracking. The IMM estimator, with performance similar to that of the 2nd order GPB (GPB_2), needs only N_μ parallel filters, thus it has the same computational complexity as the GPB_1, with significantly reduced complexity. In addition, the IMM estimator does not need to make maneuver detection decisions like the variable state dimension (VSD) filter, but performs soft switching between models based on the updated model probabilities.

However, the above-mentioned classical maneuvering target tracking algorithms mainly aim at the single-target case. For multiple maneuvering target tracking, it involves jointly estimating the number of targets and their states at each time under complex conditions such as noise, clutter, target maneuvering, data association and detection uncertainty. Therefore, this problem is extremely challenging both in theory and practice. Under the linear Gaussian condition, the closed form solution of the PHD recursion in the jump Markov (JM) system is derived in [177, 178], where a multiple model PHD is proposed and its GM implementation (GM-MM-PHD) is given. This algorithm has a better performance than IMM-JPDA with less computation. Reference [179] applied the GM-MM-PHD filter to jointly detect and track multiple maneuvering targets in clutter. However, the model set used assumes that the model of each target is identical at each moment. Therefore, Ref. [180] gives a variable structure GM-MM-PHD (VSGM-MM-PHD) filter, to determine the model set used by different targets at different times by calling the likely-model set (LMS). The above-mentioned GM-MM-PHD is non-interactive [181]. Reference [181] solves the problem of how to incorporate the "interactive" IMM into the PHD algorithm. Without any assumption of target's dynamics, the PHD filter under the linear JM system is derived from the perspective of the IMM. Reference [182] approximates the multiple model prior probability density function by recursively using the best-fitting Gaussian (BFG) distribution at each step, so that the multiple model estimation of the linear JM system is transformed into the single-model estimation of the LG system, and then the GM-PHD filter is applied to the approximate LG system. Compared with the existing non-interactive multiple model GM-PHD filter, this algorithm enjoys two advantages. First, the estimation is more accurate, because the performance of the BFG approximation is very close to that of IMM estimator. Second, the computational cost is lower, because the resulting filter is a single-model estimator. Reference [183] proposed the GM-MM-CPHD filter to improve the accuracy. However, the amount of computation is remarkably increased and the computational complexity is $\mathcal{O}(m^3nr^2)^9$, where m and n represent the numbers of measurements and targets, respectively, and r is the number of models. According to the characteristics of the time-lapse cellular microscopy imaging system, Ref. [184] combines the CPHD filter [173, 174] for adaptative estimation of parameters such as clutter rate and detection probability with the multiple model method, to track multiple maneuvering targets under the condition of unknown and time-varying clutter rate and detection probability.

For the nonlinear non-Gaussian problem, Ref. [185] extends the results derived from [178] to a mildly nonlinear JM system via the unscented transformation. Reference [186] gives a PHD filter for the nonlinear JM system based on the virtual linear fractional transform model framework. Reference [187] proposes a multiple model spline PHD (MM-SPHD) filter, which is a multiple model extension of the spline PHD (SPHD) filter. The MM implementation of this method is similar to that of the MM-PHD filter in [188]. References [188, 189] propose the SMC-MM-PHD algorithm, which can handle highly nonlinear non-Gaussian models. However, due to the need for particle clustering calculation, the estimation of the number of targets in this method is not accurate enough and requires high computational cost.

It should be noted that, in view of the fact that different targets should have independent motion models, Ref. [190] points out that most existing methods assume that multiple targets must maneuver with the same model, hence solves the problem of how to correctly extend the JM model to multi-target systems, derives the JM versions of the multi-sensor multi-target Bayesian filter and its approximate filters (PHD and CPHD filters), and compares the derived results with different MM methods of existing PHD filter [183, 188, 189], thus clearly pointing out that only the PHD filter for the JM system in [177, 178, 185] is the unique MM-PHD filter with rigorous theory.

Under the framework of (multi-)Bernoulli filters, Ref. [191] presents a multiple model extension of the joint target detection and tracking (JoTT) filter, along with its SMC and GM implementations. Based on the JM system, Refs. [192, 193] propose a multiple model CBMeMBer (MM-CBMeMBer) filter and its SMC implementation (SMC-MM-CBMeMBer) and GM implementation (GM-MM-CBMeMBer). Reference [194] introduces the particle labeling technology in the SMC-MM-CBMeMBer algorithm to obtain target tracks, with performance better than that of the SMC-MM-PHD and SMC-MM-CPHD filters. Besides, Ref. [195] proposes a multiple model LMB (MM-LMB) filter based on the LMB filter and the jump Markov system.

These filters, however, are only approximate solutions of the multi-target Bayesian filter for maneuvering targets. Therefore, based on the GLMB filter, Ref. [196] proposes an analytical solution of the multi-target Bayesian filter for maneuvering targets using the JM system. In addition, Ref. [197] also proposes an adaptive GLMB filter for dealing with maneuvering targets with time-varying dynamics and gives its GM implementation.

1.2.3.7 Application in Target Tracking with Doppler Radars

The Doppler radar measures the Doppler, slant range, azimuth (and elevation) of a target, among which the Doppler measurement is very important. In the traditional multi-target tracking algorithms, the Doppler information acts as an additional source of measurement differentiation, which can improve the convergence speed of track initiation [198] and track confirmation [199], as well as the correctness of data association [200], etc. The improvement of tracking performance by Doppler information has always been paid attention to. Reference [200] points out that the Doppler measurement can significantly improve the tracking accuracy, but does not specify whether this is due to the improvement of data association or more measurement information. Reference [201] comprehensively compares the abilities to improve filtering accuracy from position measurement, Doppler measurement, and amplitude SNR measurement. From the point of view of information theory, more measurement information is helpful to improve the filtering accuracy. However, the research in [202] shows that the improvement ability is limited. Reference [203] compares different nonlinear filters such as EKF, UKF and PF, and the results show that different filters have almost similar performances. In addition, the nonlinear Doppler measurement can be approximately processed as a linear measurement by using the measuring angle [204] or the estimated direction cosine [205] of the target relative to the platform.

The above methods generally assume that the measurement errors of Doppler and slant range are statistically uncorrelated [198–200, 206]. However, for some waveforms, there may be a correlation between the Doppler measurement and the slant range [207]. When there is a correlation between the two, ignoring the correlation will lose a certain performance, while the rational use of the correlation can improve the performance. Reference [207] quantitatively analyzes the performance improvement after considering the correlation factor under the condition of a simplified 1-dimensional coordinate system and linear measurement function. However, the model is relatively simple. In order to be close to reality, Ref. [208] takes into consideration the correlation between Doppler and slant range, and proposes a sequential converted measurement filtering method with Doppler measurement based on the earth-centered earth-fixed (ECEF) coordinate system, which can be applied to the airborne platform with time-varying attitudes.

In terms of track initiation, unlike conventional radars that use the two-point difference method [198, 209–211], the Doppler radar mostly uses the single-point initiation method [198, 203]. Reference [203] uses the slant range and azimuth measurements and their associated covariance to obtain the two-dimensional (2D) position estimation and the corresponding covariance, while the 2D velocity estimates are set to 0 and a large prior covariance is set based on the maximum possible target velocity. Then, the target state is updated with the Doppler measurement (that is a function of the target position and velocity) based on the linear minimum mean square error (LMMSE) estimator. The single-point initiation algorithm in [198] is similar to that in [203], except that the method of [212] is used to obtain the corresponding position measurement and its covariance, and the Doppler measurement is used to update the velocity estimate through the LMMSE estimator, with cross-covariance between the position and velocity components set to zero. However, the radial velocity measurement only includes the target velocity component along the radar line of sight, and cannot measure the vertical velocity component. Therefore, the aforementioned track initiation algorithms will produce a biased estimate of the velocity. For this reason, Ref. [213] corrects two mistakes in [198] and presents a modified track initiation algorithm based on the heading-parameterized multiple model (HPMM) method, which comprehensively uses the positive and negative characteristics of Doppler measurements, the prior knowledge of the maximum speed and the minimum detectable velocity (MDV) of a target.

In Doppler radar target tracking, an unavoidable practical problem is the influence of Doppler blind zone (DBZ). The DBZ stems from the physical limitation of the sensor. When the amplitude of the target's radial velocity is below a certain threshold, namely, the MDV, the sensor fails to detect the target [214, 215]. The width of the DBZ is usually determined by the MDV. Generally speaking, the width is a constant sensor parameter in the case of a single-station side-looking radar. However, Ref. [216] has proved that the width is no longer constant in a separate configuration of the receiving station and the transmitting station.

The target occlusion caused by the DBZ will lead to a series of missed detections, significantly deteriorating the tracking performance and seriously affecting the tracking accuracy and track consistency. Unlike the conventional missed detection, which is represented by specifying a detection probability $P_D < 1$ in tracking algorithms, the missed detection caused by the DBZ may provide "useful" dynamic information about the target, e.g., the target seems to be moving in the blind zone [215]. To address the challenge caused by the DBZ, a natural choice is to call the particle filter, since the hard constraint in the state space and nonlinear measurement space are easily incorporated into this framework [217]. In [218], an interactive multiple model particle filter is proposed to track a maneuvering target hidden in the DBZ. To reduce the computational complexity of the particle filter, Ref. [219] gives an analytical method under the DBZ constraint based on the Gaussian mixture approximation of the conditional density of the target state. Reference [220] developed an improved particle filter, in which the importance distribution utilizes the noise-related Doppler blind (NRDB) filtering algorithm in [219]. Another Gaussian mixture tracking algorithm similar to that in [219] was proposed by Koch [214, 215], which introduced a fake measurement to represent the missed detection, and lowered its detection probability value in the DBZ by constructing suitable state-dependent detection probability. However, the resulting Gaussian mixture approximation may have negative weights, which are prone to cause numerical instability.

The above-mentioned methods consider the missed detection problem caused by the DBZ by modifying the measurement model. In addition, some methods take this problem into account by modifying the motion model. Reference [204] proposes to solve this problem with a variable structure IMM estimator that includes an additional stopping model. The stopping model is added to the model set when the target state enters the DBZ and is removed once the state leaves the DBZ. Considering that the maximum deceleration of the target is always limited in practical scenarios, and it is impossible to switch from high speed to stopping model immediately, Ref. [221] develops a corresponding processing method by using the state-dependent model transition probability.

However, the above algorithms fail to address the data association problem, so they cannot be directly used for multi-target tracking in the clutter environment. For this reason, Ref. [222] proposes a two-dummy assignment method, in which one dummy measurement represents the conventional missed detection corresponding to $P_D < 1$, and the other additional dummy measurement represents the missed detection caused by DBZ. To address the problem of intermittent tracks caused by the DBZ, Ref. [198] proposes a track segment association (TSA) method by associating tracks from the same target at different times (old or terminated tracks and new or young tracks) based on the two-dimensional (2D) assignment method. Reference

[206] further combines the above method with the state-dependent model transition probability in [221].

The above-mentioned MTT algorithms involve the data association between measurements and targets. Due to the combination characteristics, the corresponding calculation is large. As an alternative to traditional data association-based algorithms, RFS-based multi-target tracking algorithms have significant advantages. For example, the PHD and CPHD filters are only executed in a single-target state space, avoiding the complicated data association. Under the RFS framework, Ref. [223] studies the problem of using Doppler information for track initiation and clutter suppression based on the GM-PHD filter [63], showing that the Doppler information can significantly improve the multi-target tracking performance. References [224, 225] respectively use the GM-PHD and δ-GLMB filters to study the problem of tracking multiple targets using only Doppler measurement in passive multi-station radar systems.

When considering target tracking in the presence of the DBZ, the DBZ effect is modeled by the target state-dependent detection probability [215]. References [226] and [85, 227] respectively apply the GM-PHD and GM-CPHD filters to GMTI tracking in the DBZ by modeling the detection probability as a function of the target state, and obtain a more stable estimation of the number of targets. However, these references do not provide rigorous derivation and specific implementation steps, and only use the MDV knowledge related to the DBZ, without utilizing the Doppler information. For this reason, Ref. [228] derives the GM-PHD update equation in the presence of the DBZ, by substituting the MDV-incorporated detection probability model into the standard GM-PHD update equation, and proposes a GM-PHD-based multi-target tracking algorithm with the MDV in the presence of the DBZ (GM-PHD-D&MDV). The algorithm makes full use of the MDV and Doppler information, thus effectively improving the tracking performance. In addition, the network consisting of multiple airborne Doppler radars may be affected by the systematical error, and is prone to cause the problem of "target splitting", which means that the measurements of the same target from multiple radars form multiple targets in a unified situation due to the existence of the systematical error. To solve this problem, Ref. [229] proposes an augmented state GM-PHD multi-target tracking algorithm with registration errors for Doppler radars.

1.2.3.8 Application in Tracking Before Detection of Small Targets

In target tracking, the tracker usually deals with plot measurements (data that exceed a detection threshold). In terms of storage and computational requirements, it is efficient to compress the observation data into a limited set of plots. However, the method may not be suitable for the low signal-to-noise ratio condition, because the information loss caused by the detection process becomes crucial, prone to result in a low detection probability and a large number of false alarms. Therefore, it is necessary to make full use of all the information contained in the original observation data (such as amplitude or energy information) to improve the tracking performance. In fact, the target echo amplitude is usually stronger than that from false alarms. Thereby, the amplitude information is a valuable source of information to determine whether the measurement comes from a false alarm or from a true target. Therefore, more accurate target likelihood and false alarm likelihood can be obtained by incorporating the amplitude information, thus improving the multi-target state estimation. For conventional tracking filters (such as the PDA [230] and MHT [231]), it has been proved that the target amplitude characteristics can improve the performance of data association and obtain better target tracking performance at low signal-to-noise ratio. In addition, as a sought-after incoherent energy accumulation technology in recent years, track-before-detection (TBD) jointly processes a number of consecutive frames of measurements without threshold processing (or low-threshold processing), to form a candidate track set indexed by a certain metric (typically related to the likelihood function or signal strength). The detection is triggered only when the metric exceeds a given threshold, and the track estimation is returned along with the detection decision.

Starting from [232], many TBD techniques have been proposed, mainly including TBD methods based on Hough transform (HT) [233], particle filter [234], dynamic programming (DP) [235], and maximum likelihood [236]. Please refer to the review [237] for details. The TBD method using target amplitude information usually assumes that the SNR of a target is known. Reference [238] uses the Fisher information and Cramer-Rao lower bound (CRLB), indicating that a large amount of measurement data is required to reliably estimate the SNR of Rayleigh distributed targets. Therefore, it is unlikely to obtain a reliable estimate of SNR in practice. Hence, Ref. [238] constructs an amplitude likelihood, and obtains an analytical solution of the Rayleigh target likelihood by marginalizing the likelihood on a certain range of probable values, which provides a possible method for scenarios where the SNR is unknown.

The above algorithms are mainly aimed at the single-target scenario. For the multi-target situation, Refs. [234, 239] propose a PF-based multi-target TBD solution; Refs. [240, 241] propose a DP-TBD-based MTT algorithm by concatenating multiple independent target state vectors into the multi-target state and then searching track estimates in the augmented state-space. Although the implementation is intuitive in principle, there are two main problems. First, it involves a high-dimensional maximization solution, which is computationally infeasible under the multi-target condition. Second, it is difficult to be applied to the tracking of an unknown number of targets. In view of this, Ref. [241] develops a sub-optimal algorithm based on the multi-path Viterbi algorithm [242] to solve the high-dimensional optimization problem. Although the amount of calculation is reduced to a certain extent, the tracking performance will be seriously affected by the interference of adjacent targets, in which case a strong target may obscure a weak one. References [243, 244] propose a brand-new two-stage architecture for multi-frame detection in radar systems. The architecture consists of a detector and plot-extractor (DPE) and a TBD processor. The DPE here extracts the set of candidate detections (or plots) from original measurements via a processing chain (including clustering, constant false alarm filtering, threshold operation, etc.), just like the conventional DPE, but the difference lies in that it uses a lower threshold value to obtain more candidate plots. The TBD processor uses the spatial-temporal correlation of candidate plots between different frames to jointly process multi-frame measurements and confirm reliable plots. Because the TBD processor is just an additional module placed between the DPE and the subsequent tracking modules, this architecture has the following significant advantages. no modification of any hardware or DPE's operating mode is required. In addition, different from what Refs. [240, 241] describe, this method directly operates on the plot list, requiring no spatial discretization. On this basis, Ref. [245] further applies the successive track cancellation (STC) in [240, 241] to the aforementioned TBD processor, thus enhancing the performance under the condition of multiple close-range targets. Reference [246] carries out test with measured data in heavy (sea) clutter environment obtained by nautical and shore-based radars, and the results show that the two-stage architecture is very effective for sea clutter suppression.

With the rise of the RFS theory, the researches of the TBD technology under this frame have also gained extensive attention. For the single-target scenario, Ref. [247] presents the optimal solution of the TBD smoother based on the Bernoulli filter. By using the actual MIMO radar data, Ref. [248] verifies the practicability of the TBD method based on the Bernoulli filter. Reference [130] summarizes the theory of the Bernoulli filter, and the implementation and application of different measurement models (e.g., TBD measurement model, standard point target measurement model, and non-standard measurement model introduced hereinafter). For more complicated multi-target scenarios, Ref. [238] explains how to incorporate amplitude information into a general multi-target Bayesian filter and its computationally feasible approximation (i.e., PHD/CPHD filters), and verifies that the performances of the PHD and CPHD filters incorporating amplitude information are significantly improved compared with those using only position measurements.

Essentially, in the TBD solution, the key is to obtain the observation likelihood function, which is a conjugate prior for some specific multi-target distributions. For example, for generic observation models (GOM), such as TBD and non-standard measurement models, the GLMB density is not necessarily a conjugate prior, which means that the posterior multi-target density is no longer a GLMB. Therefore, in the applications involving the GOM, multi-target densities are generally not be processed numerically. A simple strategy to significantly reduce the numerical complexity is to assume that the measurement likelihood has a separable form, in which case Poisson, IIDC, MB, and GLMB are conjugate with respect to the measurement likelihood.

Based on the assumption that observation regions affected by independent targets do not overlap with each other (in this case, the measurement likelihood is separable), Ref. [249] derives analytical characteristics of posterior distributions corresponding to different prior distributions (Poisson, IIDC, MB). Among them, the MB-based RFS method is the Bayesian optimal method for multi-target filtering on TBD data, which has been successfully applied to the TBD [250] and computer vision field [80, 81]. This method, however, is essentially multi-target filtering, rather than multi-target tracking. Drawing on the conclusion of [69, 249], Ref. [251] presents the first labeled RFS solution for the multi-target TBD problem when targets do not overlap (that is, the targets are not too close to each other in the measurement space). Specifically, this literature uses the GLMB distribution family in [69] to model the multi-target state, and the separable measurement likelihood function in [249] to model the measurement data, wherein the separability of the measurement likelihood function ensures that the GLMB distribution family is closed under the Bayesian recursion. Reference [152] proposes to use the Mδ-GLMB density to approximate product-labeled multi-object (P-LMO) density, which is the product of the joint existence probability of the label set and the joint probability density of the states conditioned on the corresponding labels. Based on this, Ref. [252] proposes a generalized LMB filter for the GOM, which is a principle approximation of the P-LMO filter. It not only inherits the advantages of multi-Bernoulli filter [249], with the intuitive mathematical structure of a MB RFS, but also has the accuracy of the P-LMO filter with less computational burden. For likelihood functions with specific separable forms, Refs. [253, 254] respectively apply the GLMB and LMB filters to solve multi-target visual tracking. Reference [255] focuses on point target tracking in infrared focal plane arrays, and proposes a TBD method based on the MM-LMB filter.

Although the separable approximation simplifies the development of the corresponding filters, it often leads to biased estimates when the separability assumption is violated (such as in the case of close-range targets). For the multi-target TBD problem of close-range targets, it can be attributed to a more general "superposition measurement" model. The so-called superposition measurement refers to a function of the measurement as the sum of the contributions of the existing targets in the surveillance area. For example, in the problems such as target tracking with unresolved or merged measurements, multi-target TBD of close-range targets, etc., the sensor output is a function of the sum of the contributions from individual sources. To address the long-time crossover of targets, Ref. [256] gives the SMC implementation of the LMB filter for the generalized TBD measurement model based on the MCMC method [257] and the label-switching improvement (LSI) algorithm. To alleviate the degradation of the sampling of the SMC method in high-dimensional space, Ref. [86] constructs the LMB and GLMB densities according to the superpositional approximate CPHD (SA-CPHD) filter [258], then uses these densities to design effective proposal distributions for RFS-based multi-target particle filter, and proposes an effective multi-target sampling strategy.

In many practical applications, additional information about the environment and/or targets is often obtained and can be described by the dynamic constraint of targets. For example, in the ground target tracking, the additional information may be a polygon representing the constraints of the road network. Similarly, a polygon might also be used to describe the strait/river and port areas in nautical applications. Reference [259] takes into account the application of the δ-GLMB filter to terrestrial and/or nautical multi-target TBD problems with additional information. Specifically, the additional information about the surveillance area is modeled with the constraint on target dynamics. According to the available set of state constraints, the generalized likelihood function is derived, and the constraint is imposed in the update step of the δ-GLMB filter. Attributed to the non-linearity of describing the additional information constraints, the SMC method is typically used to approximate the posterior density [260]. According to simulation results, the proposed constraint δ-GLMB filter is a closed form solution of the multi-target Bayesian filter when targets do not overlap in the measurement space, and is an optimal approximate solution for close-range targets from the perspective of minimizing the KLD with respect to the posterior densities.

1.2.3.9 Application in Non-standard Target Tracking

Most trackers adopt the so-called standard measurement model, also known as the point target model, which assumes that each target produces at most one measurement at a given time, and that each measurement is produced by at most one target. This model simplifies the development of multi-target trackers, yet may be a non-realistic representation of the actual measurement process under some conditions. For example, in the case of high sensor resolution, large target size, close distance between sensors and targets, multipath effect, etc., multiple resolution units of a sensor are occupied by one target, so that a sensor may observe multiple measurements of a single target, or even observe the extension of the target. In general, a target that generates multiple measurements at a given time is often referred to as an extended target [261]. For the convenience of description, the size, shape, and orientation of the extended target are collectively referred to as "extension" hereinafter, and the conventional target's position and speed are referred to as "kinetic state" (or "dynamic"). Examples of extended target tracking include tracking aircrafts and ships in close proximity with ground or marine radars (e.g., high-resolution X-band radars), and tracking vehicles and pedestrians with cameras, light detection and ranging (LIDAR), and red-green-bue-depth (RGB-D) sensors [262]. On the other hand, since most radar and sonar systems divide the measurement space into discrete detection units, the sensor with low resolution may fail to generate detection for each target once multiple targets fall into the same unit due to close proximity. When that happens, measurements of multiple targets will be merged together and unresolved to the sensor, in which case the target is known as a merged measurement target or an unresolved target [94]. Another example where merged measurements occur is in computer vision, where detection algorithms often generate merged measurements for groups of targets that are close to or occluded by each other in an image. In these cases, sensors often produce fewer measurements than the number of surviving targets. If the tracking algorithm assumes that each target generates a measurement independent of each other, it may often infer that some targets have disappeared, and at this time, these measurements have actually been merged.

The above measurement models pose challenges to multi-target tracking algorithms as they violate important assumptions of standard measurement models. In this case, a non-standard measurement model needs to be used. The non-standard measurement model relaxes the above assumptions and can handle a more general measurement process, however, this usually comes at the cost of increased computation.

In non-standard target tracking, the field of extended target tracking is becoming more active. Extended targets are usually modeled as targets with a certain spatial extension. To make full use of all available information and achieve accurate estimation, both the measurement model that can represent the target extension and the algorithm that can handle more complex track-measurement correlation problems are required. Typically, the extended target measurement model requires two models, one for modeling the number of measurements produced per target and the other one for modeling the spatial distribution of measurements. The two models depend strongly on the sensor characteristics and the type of target being tracked. For example, in radar tracking, some targets may generate many separate detections since they have many scattering points. However, it is also possible for some targets that most of the energy fails to be reflected back to the receiver, resulting in very little or no detection. In general, when the target is far away from the sensor, the detection of the target is often characterized as a point cluster, manifested as an unrecognizable geometric structure. In these cases, the commonly used extended target measurement model is the non-homogeneous Poisson (PPP) [261, 263], where at each moment the number of measurements is modeled using a Poisson distribution, while the spatial distribution of measurements is simply assumed to be distributed around the center of the target.

In order to better estimate the extension of the extended target, it can be realized by assuming that the extension is a certain parametric shape, and then estimating the corresponding parameters based on the spatial distribution of the measurements. For the spatial distribution, the commonly used model is the random matrix model [264, 265], which assumes that the measurements are Gaussian distributed around the target's center of mass. More specifically, the model assumes that the target is elliptical-shaped, and the multi-dimensional Gaussian parameters can be characterized by a random covariance matrix with inverse Wishart (IW) distribution. Therefore, it is also known as the Gaussian inverse Wishart (GIW) method, which can estimate the target extension online without specifying a prior. Reference [266] integrates the model in [264] into the framework of the probabilistic multiple hypothesis tracking (PMHT). In addition to the random matrix model, the other is the random hypersurface model [267]. Under the assumption of single target, Ref. [268] compares the random matrix and random hypersurface models. For other methods of estimating the target extension, please refer to [269, 270].

RFS-based multi-target filters have been applied to extended object tracking. Reference [271] first proposes to use the CPHD filter to solve the single extended target tracking problem under the condition of detection uncertainty and clutter. Reference [272] proposes a Bernoulli filter suitable for extended targets, yet it is limited to at most one target in clutter. For tracking multiple extended targets, Ref. [273] derives the extended target PHD filter in detail based on the PPP extended target model [261, 263]. On this basis, Ref. [95] presents a Gaussian mixture implementation of the extended target PHD filter. In multiple extended target tracking, partitioning the measurement set is an important step. The optimal filter needs to deal with all possible partitions of the measurement set, which is obviously not feasible. In fact, it is not necessary to consider all possible partitions of the measurement set, instead, it is sufficient to consider a subset of the partitions, as long as the subset contains the most probable partitions [95]. This can be done by calculating the Mahalanobis distance between measurements and grouping the measurements within a certain threshold to limit the number of partitions considered. Reference [274] corrects the mistake concerning the proof of the uniqueness of distance partitioning in [95]. In [275], aiming at the target tracking problem of sky-wave over-the-horizon radar (OTHR), a general multi-detection PHD (MD-PHD) filter is derived based on the PHD filter, and its GM implementation is given. Moreover, the proposed MD-PHD filter can be converted into the extended target PHD (ET-PHD) and multi-sensor PHD (MS-PHD) filters through a reasonable approximation, and even can be generalized to the multi-sensor multi-detection scenario. However, the proposed MD-PHD filter is generally only suitable for tracking a few high-value targets under challenging conditions due to its high computational complexity.

It should be noted that the above algorithms only estimate the dynamic properties of target's centroid, ignoring the estimation of the target extension. Inspired by the GIW method used by conventional extended target tracking algorithms to estimate the target extension, a large number of GIW methods have emerged under the RFS framework. By using a symmetric positive definite random matrix to represent the target ellipse shape, Ref. [276] developed a PHD filter based on the GIW model (GIW-PHD) to simultaneously estimate the dynamics and extension of an unknown number of extended targets in the presence of clutter and missed measurements. Reference [277] verifies the proposed algorithm using measured data of X-band marine radars. Reference [278] derives the corresponding CPHD filter for extended targets, but does not provide a specific implementation. To deal with high-density clutter and close-range extended targets, Ref. [279] further modifies the GIW model so that it can estimate the target measurement ratio. The algorithm regards the ratio parameter of the Poisson PDF (which represents the average number of measurements generated by a target) as a random variable, models its distribution as a gamma PDF, and incorporates the modified GIW model into the CPHD filter. Therefore, the algorithm is called the Gamma Gaussian inverse Wishart CPHD (GGIW-CPHD) filter [279]. Reference [280] presents a Poisson multi-Bernoulli mixture (PMBM) conjugate prior for extended targets, which is very similar to the point target PMBM [281]. It allows one set of targets to be decomposed into two disjoint subsets: targets that have been detected and targets that have not been detected. Besides, the PMBM has yielded computationally efficient algorithms [281]. Based on the random matrix model, Ref. [282] proposes a Gamma Gaussian inverse Wishart (GGIW) implementation of the PMBM filter for tracking multiple extended targets, and gives the update and prediction of parameters of the GGIW-PMBM density. For measurements with Poisson distribution in number and Gaussian distribution in space, the GGIW density is a conjugate prior of a single extended target, while for existing targets, birth targets and clutter measurements with Poisson distribution in number, the PMBM density is a multi-target conjugate prior. Specifically, the Poisson part of the GGIW-PMBM multi-target density represents the distribution of missed targets, while the multi-Bernoulli mixture part of the GGIW-PMBM multi-target density represents the distribution of targets that were detected at least once. In addition, the GGIW distribution has also been applied to the labeled RFS filters. References [148, 283] model the multi-target state as the GLMB RFS, in which the extended target is modeled using the GGIW distribution. At the same time, a cheap version of the proposed algorithm is given based on LMB filter, and the proposed GGIW-GLMB and GGIW-LMB algorithms are verified with the actual data set obtained from lidar radar sensors in autonomous driving applications. Reference [284] integrates the Gaussian process (GP) measurement model proposed in [270] into the extended target LMB filter. Compared with [148], the GP measurement model allows simultaneous tracking of multiple extended targets with different shapes. Furthermore, by using the random hypersurface model, Refs. [269, 285, 286] provide an estimation method for extended targets with elliptical, rectangular or more general shapes. Reference [287] gives a clear definition of the extended target tracking problem and discusses its boundaries with other types of target tracking, with a detailed overview of the current researches on extended target tracking.

In addition to extended target tracking, target tracking with merged measurements is another important research content in non-standard target tracking. Currently, a lot of multi-target Bayesian tracking algorithms have been proposed to deal with merged measurements. Among them, one approach is the joint probability data association (JPDA), which was first applied to the sensor model with limited resolution [288]. This method calculates the merging probability of two targets with grid-based model, and then incorporates it into the JPDA update. Therefore, the technique is limited to the case where at most two targets are merged at a certain moment. References [289, 290] apply this resolution model to multiple model JPDA. Reference [291] defines a more general resolution model and extend the above method to deal with any number of merged targets. This method combines the pairwise merging probability, thus providing a tractable approximation of the probabilities of different resolved events, which are then incorporated into the JPDA update. Introducing the resolution model into the JPDA has been shown to improve performance, however, a fundamental limitation of these methods is that the JPDA implicitly assumes that the number of targets is known and fixed. Reference [292] proposes a method based on the multiple hypothesis tracking (MHT), which utilizes a "two-target" resolution model to maintain the tracks in the presence of merged measurements. Besides, a host of data association techniques for merged measurements have been developed, such as the PDA [293], Markov chain monte carlo (MCMC) [294], probabilistic MHT (PMHT) [295], linear multitarget integrated PDA (LM-IPDA) [296] as well as integrated track splitting (ITS) [297]. These are useful techniques, however, they can only deal with a handful of targets, with the exception of the LM-IPDA, which sacrifices performance for complexity. Furthermore, it is difficult for these algorithms to establish any conclusions on Bayesian optimality in the sense of minimizing the posterior Bayesian risk.

The RFS theory provides a new technical means for target tracking with merged measurements. Mahler first applies the PHD filter to the unresolved target tracking [93]. Based on the concept of the continuous number of targets, Mahler derives the measurement update equation of the corresponding PHD filter for the unresolved target model. Although it is accurate in theory, it is difficult to deal with in calculation. Reference [117] studies the performance of the PHD filter in tracking multiple unresolved targets with multiple sensors. Reference [298] derives the CPHD filter for extended targets and unresolved targets. The derivation, however, relies on a strict assumption that extended targets and unresolved targets cannot be too close relative to the sensor resolution and the clutter density cannot be too high. References [86, 94] generalize the GLMB filter to a sensor model including merged measurements, thus making it suitable for a wider range of practical applications.

1.2.3.10 Application in Multi-sensor Fusion

The multi-sensor multi-target tracking system can achieve performance improvement in robustness, space–time coverage, ambiguity, spatial resolution, and system reliability. The rapid development of wireless sensor network technology makes it possible to call a large number of low-cost and low-energy sensors and communication devices with sensing, communication and processing capabilities in the monitoring area to perform a distributed target tracking. The main goal of multi-sensor distributed networking is to combine information from different independent nodes (usually with limited observability) to provide a more complete situation image in a scalable, flexible, and reliable way by using appropriate information fusion steps, so that it can achieve maximum agreement with all information obtained from multi-sensor measurements.

In order to maximize the multi-sensor performance, the structure and algorithm of multi-target trackers need to be redesigned considering the following issues: (1) limited by energy consumption, a single node has limited sensing and communication capabilities and limited data transmission capacity; (2) the implementation should be performed in a distributed manner without coordination with a central node and in a scalable manner with the sensor scale; (3) for each node, the degree of correlation between its own information and the information received from other nodes is unknown.

Depending on the processing structure, data fusion can typically be classified into the centralized fusion and the distributed fusion. The scalability requirement precludes the centralized fusion. Compared to the centralized fusion architecture, distributed multi-target tracking has become increasingly important due to its advantages such as lower communication cost, scalability, flexibility, robustness, and fault tolerance. In a distributed architecture, independent nodes need to work without central fusion nodes or without knowledge of information flow in the network.

The scalability with relative to the network scale and lack of knowledge of fusion centers and network topology need to adopt the consensus method [299, 300] to achieve information fusion on the whole network by iterating local fusion steps between adjacent nodes. The consensus method has become a powerful tool for distributed computing on the network. Besides, one of the prominent problems of distributed multi-sensor multi-target tracking lies in that due to the unknown level of correlation between estimates from different sensors, the computational cost of common information between nodes is unacceptable, which may make the optimal solution for distributed multi-sensor multi-target tracking infeasible in many practical applications and needs to resort to robust suboptimal information fusion rules. In fact, the problems of consensus and "data incest" (which leads to double counting of information) require the use of the generalized covariance intersection (GCI) [301, 302] method to fuse the multi-target densities calculated by different nodes of the network. The GCI is a generalization of covariance intersection (CI) [303], which only utilizes the mean and covariance and is thus limited to Gaussian posteriors. The advantage of the GCI is that it can sub-optimally fuse Gaussian and non-Gaussian multi-target distributions from different sensors whose correlations are completely unknown. The GCI fusion is also known as the Chernoff fusion rule [304], Kullback-Leibler average (KLA) rule [92, 305], exponential mixture density (EMD) fusion rule [301]. It is essentially a geometric mean operation.

Ultimately, due to the use of data link transmission, the limited processing-communication capability precludes the execution of complex multi-target tracking algorithms in network nodes, requiring the multi-target information to be expressed as "stingy" as possible.

The FISST mathematical tool can be used to derive the conceptual solutions and their principal approximations in the multi-sensor multi-target tracking problem. Despite the RFS provides a theoretical framework [15] for multi-sensor fusion, its specific implementation is still challenging. For example, the extended multi-sensor versions of the standard PHD and CPHD filters have several drawbacks: higher computational complexity, dependence on sensor update order, and numerical instability. References [306, 307] first derive the extended multi-sensor version of the PHD filter in the two-sensor case, detailing the PHD algorithm with only two targets and two sensors. Reference [308] proposes to implement a multi-sensor PHD filter by repeatedly applying the two-sensor PHD filter. Reference [309] further generalizes the filtering equation to the case involving any number of sensors, but its exact implementation is almost infeasible. Reference [310] studies the asymptotic performance of the multi-sensor PHD filter in the case of static multiple targets when the number of sensors tends to an infinite limit. Reference [311] derives the update equations of multi-sensor PHD (MS-PHD) and multi-sensor CPHD (MS-CPHD) filters. The update equation of the multi-sensor PHD/CPHD filter is of the form similar to that of single-sensor PHD/CPHD filter for extended targets [273]. Similar to the fact that the update equation for extended targets requires partitioning of the single-sensor measurement set, its exact implementation involves the summation on all partitions of measurements from different sensors. This requires that all sensor measurements are partitioned into disjoint subsets. A subset of divisions is disjoint and constitutes all sensor measurements, each subset including at most one measurement per sensor, corresponding to measurements produced by all sensors for a potential target. Solving for all partitions and subsets is impractical, and its exact update equation can only be applied in case of a small number of targets. Reference [311] proposes a computationally tractable two-step greedy partitioning method based on the GM implementation that associates possible subsets of measurements with independent target densities from the predicted PHD function. To lower the combination complexity of the multi-sensor PHD filter, Ref. [312] approximates the multi-sensor update equation when different sensor fields of view have limited overlap, and derives a simplified version of the update equation.

In the specific implementation, a simple way is to use multi-sensor sequential update method [313]. The iterated-corrector PHD filter proposed in [307] processes the information of different sensors in a sequential manner, however, the method is severely affected by the processing order of the sensors [314]. Reference [315] develops a computationally inexpensive iterated-corrector strategy for the multi-sensor PHD/CPHD filter. Besides, the problem of the dependence on the sensor order can be mitigated to some extent using approximate product multi-sensor PHD and CPHD filters [316]. Although the final result is independent of the sensor order, Ref. [317] shows that the SMC implementation of the approximate product multi-sensor PHD filter is unstable and the problem worsens as the number of sensors increases.

In addition to being based on the PHD/CPHD filter, Refs. [318, 319] study distributed detection and tracking on Doppler-shifted sensor networks using the Bernoulli filter. Reference [320] derives a multi-sensor multi-Bernoulli (MSMB) filter for multi-target tracking. The exact implementation of the MSMB update step is computationally intractable. For this reason, an efficient approximation implementation is proposed by using the greedy measurement partitioning mechanism. Compared with the multi-sensor CPHD filter, the MSMB filter boasts improved accuracy and lower computational cost under both linear Gaussian and nonlinear models, especially in the case of low detection probability.

The above algorithms are mainly suitable for the centralized multi-sensor fusion structure. As mentioned earlier, the distributed fusion structure is generally used in large-scale multi-sensor fusion. Reference [303] proves the Chernoff fusion is inherently immune to repeated computation of information in the single target case, thus validating its application in the distributed configuration. Although the challenges of multi-target tracking are more complicated in the distributed structure, the concept of the multi-target probability density in the RFS representation enables the consistency of distributed state estimation in the single-target system to be directly applicable in the multi-target system [91, 92, 321]. Based on the Chernoff fusion, Ref. [321] extends the method in [303] to the multi-target scenario and proposes a consensus PHD/CPHD filter. References [322, 323] put forward the concept of "partial consensus", which compares the difference and comparative advantages between the (currently mainstream) geometric average and the (simple but often overlooked) arithmetic average. References [301, 324, 325] study the robust fusion when the correlation is unknown in practical situations. Specifically, Ref. [301] first generalized the covariance intersection in the context of multi-target fusion. On this basis, Ref. [324] derives the explicit expressions for the GCI fusion of PHD functions and cardinality distributions. According to these conclusions, Ref. [91] proposes a SMC implementation for the GCI fusion of PHD filters. By utilizing the consensus algorithm and the KLD concept of the PDFs to be fused, Refs. [92, 326] propose a consensus GM-CPHD filter, which provides a fully distributed, scalable and computationally effective solution to the distributed multi-target tracking problem on a heterogeneous node network with sensing, communication and processing capabilities. Compared with the PHD/CPHD filter [14, 63, 64, 119], the multi-Bernoulli filter [4, 68, 249] is an extension of the Bernoulli filter [130], which is more useful in the problem requiring particle implementation or independent existence probabilities of targets.

As the advantages of labeled RFS filters become increasingly prominent, it makes sense to generalize them to a distributed setting. Aiming at the problem that the complexity of multi-sensor multi-target tracking increases super-exponentially with the number of sensors, Refs. [327, 328] give an efficient implementation of the multi-sensor GLMB filter by calling the joint GLMB prediction and update step based on the Gibbs sampling technique. Its complexity is proportional to the product of the number of measurements per sensor, and proportional to the square of the number of hypothetical targets. Reference [149] proposes a new algorithm for multi-sensor multi-target tracking based on the Mδ-GLMB density class. The complexity of the algorithm is linearly scalable with respect to the number of sensors. Reference [329] proposes a fully distributed multi-target estimation method for sensor networks based on the labeled RFS, and derives a closed form solution of the GCI fusion using Mδ-GLMB and LMB posteriors, thus developing two consensus tracking filters, namely consensus Mδ-GLMB tracking filter and consensus LMB tracking filter. The labeled RFS-based filter avoids the so-called spooky effect and can output target tracks. However, the above methods are based on the assumption that different sensors share the same label space: not only must the label spaces of different sensors be the same, and the same elements of different sensors have the same physical meaning, or actually represent the same target. The assumption is difficult to hold in practice. Reference [330] refers it as the "label space mismatch" problem, which means that the same realizations sampled from label spaces of different sensors do not necessarily have the same meaning.

One way to overcome the label mismatch is to perform the GCI fusion using an unlabeled version of the GLMB distribution family. References [331, 332] first recommend a robust strategy for distributed fusion with labeled set posteriors, where the labeled set posteriors are transformed into the corresponding unlabeled versions, and then, the mathematical expression of the unlabeled version of the common labeled set distribution in the GLMB family is derived, and it is proved that they all belong to the same (unlabeled) RFS family, which is called the generalized multi-Bernoulli (GMB) family. Finally, the GCI fusion is performed on the unlabeled posteriors. However, the GCI fusion of two (G)MB posterior distributions no longer has an exact closed-form expression. To solve this problem, inspired by the extension of PHD and CPHD filters to LMB and Mδ-GLMB filters, two effective approximations for the GCI fusion of GMB families are derived, one of which is to approximate each GMB posterior by using the MB distribution that matches the first order moment (PHD), wherein this approximate distribution is also called the first-order approximation of GMB (FO-GMB) [331]; and the other of which is the second-order approximation of GMB (SO-GMB) [333], which preserves the PHD and cardinality distribution. This provides the basis for distributed fusion of the GLMB filter family, including the GLMB, δ-GLMB, Mδ-GLMB and LMB filters.

In multi-sensor data fusion problems, sensor registration is a prerequisite for successful fusion. The traditional association-based methods address data association and sensor registration problems separately. That is, the traditional sensor registration algorithm first obtains the association relationship through a classical association method, and then uses the registration algorithm to estimate the sensor bias and target state with the measurements from the same target according to the estimated association result. However, they actually affect each other, data association affects sensor registration, and sensor registration in turn directly affects data association. In other words, sensor registration requires the correct association data, and data with sensor bias will lead to wrong associations [334]. By taking into account the nonlinear and non-Gaussian condition, Ref. [335] gives the registration results of three heterogeneous yet synchronous sensors tracking multiple targets, based on the SMC-PHD filter on the target state-space augmented with bias vectors. This method can jointly estimate the number of targets and states as well as sensor biases without data association. With the RFS framework, Ref. [336] presents a joint registration and tracking method in a more general form. The algorithm can also be applied to the environment in which the asynchronous sensors may have missed detections and dynamic targets randomly appear and disappear, in addition to requiring no prior knowledge of multi-sensor measurement-target association. For non-cooperative targets, Ref. [337] presents a Bayesian estimation method for parameter vectors including sensor biases. This method is very general. In addition to the correction of sensor bias, it can also be applied to the correction of any parameter in the target dynamic model and sensor measurement model, including the process noise level, environmental characteristics (clutter properties, propagation loss) or sensor parameters (gain, sensor bias, detection probability), etc. However, the method adopts the batch processing. To develop a recursive version of the registration technique, Ref. [338] studies the sequential Bayesian estimation problem of the composite state composed of both multi-target dynamic states and multi-sensor biases. Although these filters do not require prior information about the association between targets and measurements, and can jointly estimate the number of targets, states, and sensor biases, they inherit the shortcomings of the SMC-PHD filter, such as the high computational complexity and the unreliability of clustering operation. To overcome the above problems, Ref. [339] applies the GM-PHD filter to solve multi-sensor, multi-target tracking with registration errors. In addition to the problem of multi-sensor registration, out of sequence measurement (OoSM) is also a practical problem that needs to be considered in multi-sensor fusion. Reference [340] extends the GM-PHD filter to making it suitable for the OoSM problem and gives the corresponding closed form recursive solution.

1.2.3.11 Application in Sensor Management

Sensor management based on the RFS framework mainly involves sensor control and sensor selection applications. In the context of multi-target tracking, sensor control usually needs to solve two main problems: multi-target filtering and sequential decision-making. The purpose of sensor control is to find the optimal control command from the allowable control command set to guide the sensor to obtain the maximum observability of an unknown number of targets, so as to obtain the most accurate estimation of the number of targets and their states. Generally, different control commands guide the sensor to different new states, thus resulting in different measurement sets. Each resulting measurement set contains information different from other sets, and the obtained information can be analyzed by a decision-making process (such as maximizing a objective function) to determine the optimal control command. The complexity of this problem mainly arises from the uncertainties in the state space and measurement space. It needs to be solved by stochastic control theory, wherein the number of targets may vary with time. In the Bayesian filtering paradigm, the control command affects the measurement, and the likelihood function only appears in the update step of the Bayesian filtering scheme. Therefore, the most common method is to first define a criterion based on which the quality of updated multi-target densities caused by different control commands is compared with each other, and then select the optimal control command, in the hope that the optimal updated density can be obtained after the prior (predicted) density is updated by using sensor measurements.

In terms of sensor selection, since the bandwidth and energy of sensor networks are limited, it is expensive to directly use all the information of sensor nodes for target detection and tracking. Therefore, sensor selection is required. At this point, the problem of sensor selection becomes the selection of sensor nodes that maximize the observability using limited computational and communication resources. Generally speaking, sensor selection also consists of two parts: a multi-target filtering process and an optimal decision-making method. Therefore, the problem of sensor selection is essentially similar to that of sensor control, which is also a sequential decision-making process under random uncertainties. These uncertainties originate from the multi-target tracking process or from the influence of selecting different sensor nodes. For the decision-making process, the most common method is to select sensor nodes through the optimization of the objective function. In fact, the objective function in sensor selection corresponds to the criterion in sensor control.

It can be seen from the above that the selection of objective functions is very critical in the application of sensor management. Objective functions can be roughly divided into two categories: the task-driven objective function and the information-driven objective function. In the former category, the objective function is described as a cost function, which often depends on performance metrics such as variances of state and cardinality estimates or other distribution-dependent metrics. Commonly used task-driven objective functions mainly include: posterior expected number of targets (PENT) [341, 342], and posterior expected error of cardinality and states (PEECS) [150, 343, 344]. In the latter category, the objective function is usually a reward function, which is directly related to the information content of the multi-target distribution. In information theory and statistical analysis, the similarity/difference between random variables can be measured by the information-theoretic divergences such as the Renyi divergence [345]. Similarly, for random finite sets, the corresponding information-driven objective functions [346] have also been developed. The common ones are Renyi divergence (RD) [342, 347, 348], Cauchy-Schwarz divergence [349, 350].

1.2.3.12 Performance Evaluation Metrics for Multi-target Tracking

The performance evaluation of tracking algorithms is very important in the design, parameter optimization, and comparison of tracking systems. In single-target tracking, the posterior CRLB (PCRLB) is one of the commonly used performance evaluation metrics. Reference [351] gives a recursive PCRLB to evaluate the performance of nonlinear filters, and Ref. [352] presents a recursive form of the single target error bound in the presence of single-sensor missed detections, which is, however, only suitable for clutter-free conditions. References [353–355] generalize it to the scenario where both clutter and missed detections are present. These PCRLBs, however, can hardly be applied to multi-target tracking problems, as they only consider the estimation error of a single target state and do not consider the estimation error of the number of targets. Reference [356] generalizes the results of [352] to the single-sensor multi-target case, but it is limited to a more stringent condition of clutter-free and no missed detection. It should be noted that due to the strict constraint on the sensor's observation model, the number of targets can actually be completely determined by the number of measurements, and thus, the error bound in [356] does not actually include the detection error caused by the uncertainty of the number of targets. For this reason, Ref. [357] gives a single sensor multi-target error bound when clutter and missed detections exist simultaneously.

Currently, the most commonly used metric for multi-target estimation performance evaluation of RFS filters is the optimal sub-pattern assignment (OSPA). It is originated from the Hausdorff metric, which is, however, disadvantaged by being insensitive to the difference between the cardinality (number of elements in a set) of true state set and the cardinality of estimated state set. Based on the Wasserstein distance, Hoffman and Mahler recommend a new optimal mass transfer (OMAT) metric [358]. Compared with the Hausdorff metric, the OMAT is more sensitive to cardinality differences and has a physically intuitive explanation when the cardinalities of two target states are the same. However, when the cardinalities are different, it has no physically consistent explanation and suffers from many other serious limitations. For this reason, Ref. [359] proposes the OSPA measure, which incorporating the cardinality error and state error and has a natural physical explanation even if two sets have different cardinalities. Reference [360] further incorporates the label error. Aiming at the ambiguous target identity (namely, the problem of mixed labeling) caused by cross targets, Ref. [361] proposes a label uncertainty metric with clear physical explanations (such as label probability and label error) based on the concept of the labeled RFS. Reference [362] further incorporates the quality information and proposes a so-called quality-based OSPA (Q-OSPA), which provides a more accurate performance metric of multi-target estimation algorithms. If the estimated quality is not available, the Q-OSPA degenerates to the OSPA by assigning equal quality to the estimates. The OSPA is mainly aimed at the state filtering error at each time step. To compare the similarity of the tracks over a period of time, Ref. [363] proposes the $\mathrm{OSPA}^{(2)}$ measure.

It should be noted that, however, the application of the OSPA shall be built on the precondition that the true multi-target state is available. To a certain extent, the OSPA corresponds to the single-target tracking performance metric: normalized estimation error square (NEES). The NEES is often used for the consistency check of trackers, and the prerequisite for its use is also to obtain the true target state. In other words, the NEES cannot be computed without the true target state. In this case, the normalized innovation square (NIS) is generally used instead. It determines whether the residual between the predicted measurement and the true measurement is within the uncertainty range given by the innovation covariance matrix of the Kalman filter. Therefore, convergence detectors such as the NIS are used to detect whether the hypothetical motion model or measurement model deviates significantly from the actual behaviors of the tracked targets or sensors. Inspired by this, Ref. [364] extends the single-target generalized NIS to the multi-target generalized NIS (MGNIS), and derives the MGNIS for the PHD and CPHD filters. In addition, Ref. [365] derives the MGNIS for the GM implementation of the δ-GLMB filter.

1.2.3.13 Other Relevant Issues

The aforementioned single-target or multi-target filters all estimate the target state online. Compared with the filtering, the smoothing can significantly improve the estimation accuracy by using time-lag data to delay the decision time [356]. For the single-target case, many types of smoothers have been proposed. In the case of linear Gaussian dynamics and measurement models, the Kalman smoother [367] provides an exact analytical solution; for the linear Gaussian mixture model, Ref. [366] proposes a closed forward and backward smoother; for nonlinear models, many smoothers have appeared based on the SMC method. For the case that there is at most one target yet its existence is uncertain, Refs. [131, 366] give a smoother based on the Bernoulli target model. Compared with the single-target case, multi-target smoothing is more complicated due to the time-varying number of targets, the uncertainty of the measurement source, and the presence of clutter measurement and missed detections. Under the RFS framework, approximate multi-target smoothing algorithms have been proposed including the PHD smoother [368–370], CPHD smoother [371], and MB smoother [372]. To estimate the target tracks, Ref. [373] gives the analytical form of multi-target forward and backward smoothing recursion based on the GLMB multi-target model, and proves that the GLMB family is also closed for the standard multi-target system model (including the standard multi-target motion model and the standard measurement model) under the backward smoothing operation.

As a novel multi-target tracking algorithm, the RFS-based tracking algorithm actually has a certain connection with the traditional mature algorithms. Reference [374] shows that the random vector framework and the random set framework are closely related in mathematics. Under ideal conditions, the equivalence of the two frameworks can be derived by making certain assumptions on detection and clutter statistics. Reference [281] points out that although (unlabeled) RFS framework avoids the need to explicitly model data association, data association actually exist implicitly in the full Bayesian filter. Moreover, by approximating the discrete distribution of the implicit data association, a track-oriented marginal Bernoulli/Poisson filter and a measurement-oriented marginal Bernoulli/Poisson filter are obtained, the former being very similar to the JITS and the JPDA. Therefore, the JITS and JIPDA algorithms and their extensions can be derived from the RFS framework after appropriate modifications. Reference [117] shows that the IPDA can be derived from the RFS formulation. In fact, the Bernoulli filter degrades into the IPDA filter under the assumption that the sensor field of view is uniform, the clutter is uniformly distributed, and the Gaussian mixture posterior densities can be merged into a single Gaussian component [129]. Reference [85] proves that in the single-target case, the GM-CPHD is equivalent to the MHT that uses the sequential likelihood ratio test for track extraction. Reference [375] discretize the PHD surface into infinitesimal "boxes", predict and update the probability that each box contains a target, and give the physical space interpretation of the PHD/CPHD filters from another intuitive and easy-to-understand perspective, avoiding sophisticated RFS knowledge and facilitating the engineers and technicians to understand and master. Reference [376] discriminates the concept of measurement-track association under the RFS framework and the MHT framework. Reference [377] compares two methods using the Gaussian mixture: the PHD and the ITS. Reference [378] compares the GM-CPHD and MHT algorithms. Reference [379] combines the MHT and the GM-CPHD to track multiple targets.

1.3 Overview of Chapters and Contents

The book mainly introduces the emerging RFS-based multi-target tracking algorithms, including the probability hypothesis density (PHD) filter, cardinalized PHD (CPHD) filter, multi-Bernoulli filter, labeled RFS filters and the applications of these algorithms. The overall structure of the book is shown in Fig. 1.1, and the specific contents are organized as follows.

The book is divided into 12 chapters. This chapter is the Introduction. It gives the basic concepts of the random finite sets (RFS) and target tracking, introduces the current research and development status of target tracking, and provides readers with an overview of this book.

Chapter 2 introduces the mainstream filtering algorithms commonly used in single target tracking. In addition to the classical Kalman filter, extended Kalman filter, and unscented Kalman filter, it also includes the frontier cubature Kalman filter, Gaussian sum filter (GSF), and particle filter. These contents are not only the foundation of classical multi-target tracking algorithms, but also the foundation of the RFS-based multi-target tracking algorithms. Especially, the GSF and particle filter are respectively the foundation of Gaussian mixture (GM) and sequential Monte Carlo (SMC) implementations of the RFS-based multi-target tracking algorithms.

Chapter 3 describes the basic knowledge of the RFSs. The chapter introduces the statistical description, intensity, cardinality distribution, and main types of the RFSs, multi-target system models (including the multi-target motion model and the multi-target measurement model), cornerstone (such as multi-target Bayesian

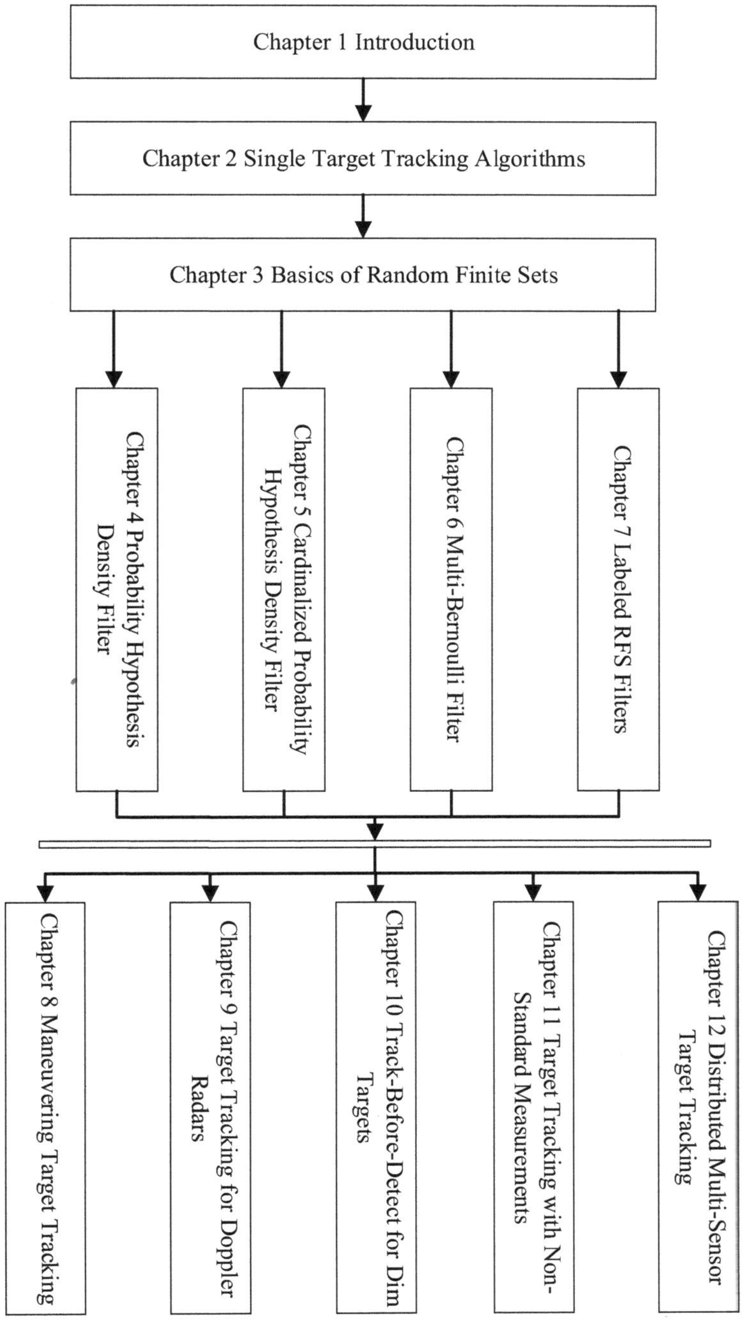

Fig. 1.1 Overall framework and chapter arrangement of the whole book

filtering, multi-target Bayesian recursion and its particle implementation, i.e., particle multi-target filter) of subsequent filters, multi-target formal modeling paradigm, and multi-target tracking performance evaluation metrics.

Chapter 4 introduces the PHD filter. First, the PHD recursive formula is given, based on which two implementation methods, namely the SMC-PHD filter and the GM-PHD filter, are given respectively. In addition, the extensions of the GM-PHD filter are also introduced.

Chapter 5 introduces the CPHD filter. Similar to Chap. 4, the CPHD recursive formula is given first, and then two implementation methods, namely the SMC-CPHD and GM-CPHD filters, and the extensions of the GM-CPHD filter are given.

Chapter 6 introduces the multi-Bernoulli (MB) filter. The chapter focuses on the CBMeMBer filter, with two implementation methods provided, namely, the SMC-CBMeMBer and GM-CBMeMBer filters.

Chapter 7 introduces the labeled RFS filters. Four filters are introduced respectively, namely, the GLMB filter, δ-GLMB filter, LMB filter, and marginal δ-GLMB (Mδ-GLMB) filter, and the corresponding implementation methods are given.

Subsequent chapters are extensions and specific applications of the above filters.

Chapter 8 introduces maneuvering target tracking. Based on the jump Markov system, the chapter takes the PHD, CBMeMBer, and GLMB filters as examples to describe their multiple model versions accordingly.

Chapter 9 introduces target tracking with Doppler radars. To make full use of the Doppler measurement, a GM-CPHD filter with Doppler measurement is given. Then, a more complicated GM-PHD filter in the presence of Doppler blind zone is further introduced. Finally, the augmented state GM-PHD filter with registration errors for Doppler radar networks is presented.

Chapter 10 introduces track-before-detect for dim targets. First, the multi-target TBD measurement model is introduced, then the analytical characteristics of the multi-target posteriors under different priors are provided, and finally the corresponding TBD methods based on the MB and Mδ-GLMB filters are described respectively.

Chapter 11 introduces target tracking with non-standard measurements, including extended target tracking and target tracking with merged measurements. First, a relatively simple extended target GM-PHD filter without considering the estimation of target extension is presented. Then, based on the GGIW distribution, an extended target tracking method for estimating target extension is introduced. Finally, target tracking with merged measurements is given.

Chapter 12 introduces distributed multi-sensor target tracking. First, the distributed multi-target tracking problem is described. Then, the relatively simple distributed single target filtering and fusion is described. Finally, based on the relevant conclusions of multi-target density fusion, the SMC and GM implementations of distributed fusion of RFS filters are introduced respectively.

Chapter 2
Single Target Tracking Algorithms

2.1 Introduction

Target tracking refers to the process during which the number and states of targets are jointly estimated based on uncertain measurements. Measurement uncertainty includes noise pollution, clutter interference, detection uncertainty, and data association uncertainty. Among them, detection uncertainty is mainly caused by the sensor's detection capability, which cannot secure 100% detection for a target and is usually characterized by a detection probability, whereas the data association uncertainty refers to that the measurement-target corresponding relation cannot be exactly determined by the tracker.

For single target tracking, the measurement uncertainty usually only considers the uncertainty of noise pollution. In other words, there is no clutter interference, the target detection probability is 1, and there is no data association problem. At this time, the number of targets is naturally known a priori, and the main focus is on the target state. The target state usually includes the position, velocity, identity (trajectory), attributes, and other information related to the target of interest. Therefore, single target tracking is also called single-target filtering, which is mainly dedicated to improving the estimation accuracy with certain filtering algorithms.

The selection of filtering algorithms mainly depends on system models and posterior distributions (or posterior densities). The so-called system models include the (target) dynamical model and the (sensor) measurement model. Generally, system models can be divided into linear Gaussian model, nonlinear Gaussian model, and nonlinear non-Gaussian model according to the linearity of system models and Gaussianity of posterior distributions. It is well established that the Kalman filter (KF) is the optimal solution for linear Gaussian models. For nonlinear Gaussian models, mainstream solutions include the extended Kalman filter (EKF), unscented Kalman filter (UKF) and cubature Kalman filter (CKF); Generally, there is usually no analytical solution for nonlinear non-Gaussian models due to the complexity of models and distributions, but the Gaussian sum filter (GSF) and particle filter (PF) can be used to seek approximate solutions. The GSF (an extension of Kalman filter) and

W. Wu et al., *Target Tracking with Random Finite Sets*,
https://doi.org/10.1007/978-981-19-9815-7_2

PF approximate arbitrary nonlinear density respectively through Gaussian mixture distribution (weighted sum of Gaussian density) and random samples (particles).

To some extent, the above algorithms are derived from the (single-target) Bayesian recursion equation, indicating that the Bayesian recursion is the theoretical basis of each filtering algorithm. Therefore, the chapter first introduces the Bayesian recursion and then the KF, EKF, UKF, CKF, GSF, and PF on that basis. It should be noted that this chapter also lays foundation for various RFS tracking algorithms introduced hereinafter. In particular, the GSF and PF filters are highly comparable with the Gaussian mixture (GM) and sequential Monte Carlo (SMC) implementations of RFS-based tracking algorithms.

2.2 Bayesian Recursion

The Bayesian recursion is based on the target dynamic model (target motion model) and the sensor observation model. The target dynamic model can be described by a discrete-time model or a continuous-time stochastic difference equation. Here we only consider the discrete-time model. In general, it is assumed that the target state $\boldsymbol{x}_k$ evolves according to the following state transition equation

$$\boldsymbol{x}_k = f_{k|k-1}(\boldsymbol{x}_{k-1}, \boldsymbol{v}_{k-1}) \tag{2.2.1}$$

where $f_{k|k-1}(\cdot, \cdot)$ represents a nonlinear state transform and $\boldsymbol{v}_{k-1}$ represents the process noise. Also, the equation can be described with Markov transition density $\phi_{k|k-1}(\cdot|\cdot)$[1]

$$\phi_{k|k-1}(\boldsymbol{x}_k|\boldsymbol{x}_{k-1}) \tag{2.2.2}$$

The above equation represents the probability density of transition from state $\boldsymbol{x}_{k-1}$ at time $k-1$ to state $\boldsymbol{x}_k$ at time k. It should be noted that for each $\boldsymbol{x} \in \mathbb{X}$, $\phi_{k|k-1}(\cdot|\boldsymbol{x})$ represents the probability density in state space $\mathbb{X}$. Commonly used dynamic models include constant speed (CV), constant acceleration (CA), and coordinate turning (CT). Please refer to the overview [380] for details.

At time k, state $\boldsymbol{x}_k$ generates measurement z_k according to the following measurement equation

$$z_k = h_k(\boldsymbol{x}_k, \boldsymbol{n}_k) \tag{2.2.3}$$

where $z_k = h_k(\cdot, \cdot)$ represents a nonlinear measurement transformation and $\boldsymbol{n}_k$ denotes the measurement noise. The measurement equation can also be described by the following likelihood function $g_k(\cdot|\cdot)$

[1] For the sake of description, no distinction is made between random variables and their implementations in the book.

$$g_k(\boldsymbol{z}_k|\boldsymbol{x}_k) \tag{2.2.4}$$

The above equation represents the probability density when receiving a measurement $z_k \in \mathbb{Z}$ given state $\boldsymbol{x}_k$. Let $z_{l:k} = (z_l, \ldots, z_k)$ represent the measurement history from time l to k. Similarly, $\boldsymbol{x}_{l:k} = (\boldsymbol{x}_l, \ldots, \boldsymbol{x}_k)$ represents the state history from time l to k. Assume that the probability density of the measurement history $z_{1:k}$ conditioned on $\boldsymbol{x}_{1:k} = (\boldsymbol{x}_1, \ldots, \boldsymbol{x}_k)$ has a separable form, that is,

$$g_{1:k}(\boldsymbol{z}_{1:k}|\boldsymbol{x}_{1:k}) = g_k(\boldsymbol{z}_k|\boldsymbol{x}_k)g_{k-1}(\boldsymbol{z}_{k-1}|\boldsymbol{x}_{k-1}) \ldots g_1(\boldsymbol{z}_1|\boldsymbol{x}_1) \tag{2.2.5}$$

The posterior density $p_{0:k}(\boldsymbol{x}_{0:k}|\boldsymbol{z}_{1:k})$ encapsulates all the information about the state history up to time k. For any time $k \geq 1$, started from the initial prior $p_0 \triangleq p_0(\boldsymbol{x}_0|z_0) = p_0(\boldsymbol{x}_0)$ (z_0 represents no measurement), the posterior density can be recursively calculated with the following Bayesian recursion

$$p_{0:k}(\boldsymbol{x}_{0:k}|\boldsymbol{z}_{1:k}) \propto g_k(\boldsymbol{z}_k|\boldsymbol{x}_k)\phi_{k|k-1}(\boldsymbol{x}_k|\boldsymbol{x}_{k-1})p_{0:k-1}(\boldsymbol{x}_{0:k-1}|\boldsymbol{z}_{1:k-1}) \tag{2.2.6}$$

Filtering density (or updated density) $p_k(\boldsymbol{x}_k|\boldsymbol{z}_{1:k})$ is the marginal density of posterior density $p_{0:k}(\boldsymbol{x}_{0:k}|\boldsymbol{z}_{1:k})$, representing the probability density of the state $\boldsymbol{x}_k$ at time k given measurement history $z_{1:k}$. In the target tracking problem, we are mainly interested in the filtering density, which is also referred to as the posterior density in the following.

According to Bayesian point of view, the filtering density at time k can be calculated recursively from the initial density p_0 through the Bayesian equation. Specifically, it is composed of the prediction step expressed by Chapman-Kolmogorov (C-K) equation and Bayesian update step

$$p_{k|k-1}(\boldsymbol{x}_k|\boldsymbol{z}_{1:k-1}) = \int \phi_{k|k-1}(\boldsymbol{x}_k|\boldsymbol{x})p_{k-1}(\boldsymbol{x}|\boldsymbol{z}_{1:k-1})d\boldsymbol{x} \tag{2.2.7}$$

$$p_k(\boldsymbol{x}_k|\boldsymbol{z}_{1:k}) = \frac{g_k(\boldsymbol{z}_k|\boldsymbol{x}_k)p_{k|k-1}(\boldsymbol{x}_k|\boldsymbol{z}_{1:k-1})}{\int g_k(\boldsymbol{z}_k|\boldsymbol{x})p_{k|k-1}(\boldsymbol{x}|\boldsymbol{z}_{1:k-1})d\boldsymbol{x}} \tag{2.2.8}$$

where $p_{k|k-1}(\boldsymbol{x}_k|\boldsymbol{z}_{1:k-1})$ is called the predicted density. The posterior density $p_k(\boldsymbol{x}_k|\boldsymbol{z}_{1:k})$ encapsulates all the information of state $\boldsymbol{x}_k$ at time k, and the state estimation at this moment can be obtained with the minimum mean squared error (MMSE) or maximum a posterior (MAP) criterion.[2] Another marginal density of posterior density is the smooth density $p_{k|k+l}(\boldsymbol{x}_k|\boldsymbol{z}_{1:k+l})$, $l > 0$, which represents the probability density of state $\boldsymbol{x}_k$ at time k when measurement history $z_{1:k+l}$ is given.

---

[2] These criteria may not be applicable to multi-target scenarios.

2.3 Conjugate Priors

It can be seen from (2.2.8) that in the Bayesian recursion, the calculation of the posterior requires the integration of the product of the prior and the likelihood function to obtain the normalization constant of the denominator. As the solution of the integral is highly difficult, it becomes a key research topic in the Bayesian method. Generally, there are two solutions: one is to use a cheap and powerful computer to numerically approximate the integral (such as the particle filter introduced in Sect. 2.9); the other method is to use paired likelihood function and prior distribution that allow the normalization constant to have a tractable analytical form. For a given likelihood function, if a posterior and a prior belong to the same family, the posterior and the prior are called conjugate distributions, and the prior is called the conjugate prior. For example, the Gaussian family is conjugate with respect to the Gaussian likelihood function. Other well-known likelihood-prior combinations in exponential family include Binomial-Beta, Poisson-Gamma and Gamma-Gamma models. These prior distribution families play an important role in the Bayesian inference. Due to the integration in the normalization constant, the calculation of the posterior is generally difficult to deal with, which is more prominent in the nonparametric inference, where the posterior may be very complex. Because of the curse of dimensionality and the inherent combinatorial nature of the problem, computing the posterior of a non-conjugate prior is not tractable. By contrast, the conjugate prior is convenient for algebraic processing by providing a closed form of the posterior, thus can avoid tricky problem of numerical integration. In addition, the posterior that has the same functional form as the prior usually inherits the desirable properties that are crucial for analysis and interpretation.

2.4 Kalman Filter

The Kalman filter (KF) is a closed form solution of the Bayesian recursion under linear Gaussian models. Specifically, the dynamic model and the measurement model are both linear transformations, and the corresponding process noise and measurement noise are additive Gaussian noises, i.e.,

$$\boldsymbol{x}_k = \boldsymbol{F}_{k|k-1}\boldsymbol{x}_{k-1} + \boldsymbol{v}_{k-1} \tag{2.4.1}$$

$$\boldsymbol{z}_k = \boldsymbol{H}_k\boldsymbol{x}_k + \boldsymbol{n}_k \tag{2.4.2}$$

where $\boldsymbol{F}_{k|k-1}$ denotes the state transition matrix, $\boldsymbol{H}_k$ represents the measurement matrix, and $\boldsymbol{v}_{k-1}$ and $\boldsymbol{n}_k$ are independent zero-mean Gaussian noises, with the corresponding covariances recorded as $\boldsymbol{Q}_{k-1}$ and $\boldsymbol{R}_k$. At this time, the transition density and the likelihood function can be written respectively as

$$\phi_{k|k-1}(\boldsymbol{x}_k|\boldsymbol{x}_{k-1}) = \mathcal{N}(\boldsymbol{x}_k; \boldsymbol{F}_{k|k-1}\boldsymbol{x}_{k-1}, \boldsymbol{Q}_{k-1}) \tag{2.4.3}$$

$$g_k(\boldsymbol{z}_k|\boldsymbol{x}_k) = \mathcal{N}(\boldsymbol{z}_k; \boldsymbol{H}_k\boldsymbol{x}_k, \boldsymbol{R}_k) \tag{2.4.4}$$

where $\mathcal{N}(\cdot; \boldsymbol{m}, \boldsymbol{P})$ represents Gaussian density with mean $\boldsymbol{m}$ and covariance $\boldsymbol{P}$.

Based on the above assumption, when the initial prior is assumed as a Gaussian distribution $p_0 = \mathcal{N}(\cdot; \boldsymbol{m}_0, \boldsymbol{P}_0)$, all subsequent filtering densities are in the Gaussian form. Specifically, if the filtering density at time $k-1$ is in the Gaussian form

$$p_{k-1}(\boldsymbol{x}_{k-1}|\boldsymbol{z}_{1:k-1}) = \mathcal{N}(\boldsymbol{x}_{k-1}; \boldsymbol{m}_{k-1}, \boldsymbol{P}_{k-1}) \tag{2.4.5}$$

then the predicted density at time k is also in the Gaussian form, given by

$$p_{k|k-1}(\boldsymbol{x}_k|\boldsymbol{z}_{1:k-1}) = \mathcal{N}(\boldsymbol{x}_k; \boldsymbol{m}_{k|k-1}, \boldsymbol{P}_{k|k-1}) \tag{2.4.6}$$

where

$$\boldsymbol{m}_{k|k-1} = \boldsymbol{F}_{k|k-1}\boldsymbol{m}_{k-1} \tag{2.4.7}$$

$$\boldsymbol{P}_{k|k-1} = \boldsymbol{F}_{k|k-1}\boldsymbol{P}_{k-1}\boldsymbol{F}_{k|k-1}^{\mathrm{T}} + \boldsymbol{Q}_{k-1} \tag{2.4.8}$$

At this time, the filtering (updated) density at time k is also in the Gaussian form

$$p_k(\boldsymbol{x}_k|\boldsymbol{z}_{1:k}) = \mathcal{N}(\boldsymbol{x}_k; \boldsymbol{m}_k(\boldsymbol{z}_k), \boldsymbol{P}_k) \tag{2.4.9}$$

where

$$\boldsymbol{m}_k(\boldsymbol{z}_k) = \boldsymbol{m}_{k|k-1} + \boldsymbol{G}_k(\boldsymbol{z}_k - \boldsymbol{H}_k\boldsymbol{m}_{k|k-1}) \tag{2.4.10}$$

$$\boldsymbol{P}_k = (\boldsymbol{I} - \boldsymbol{G}_k\boldsymbol{H}_k)\boldsymbol{P}_{k|k-1} \tag{2.4.11}$$

$$\boldsymbol{G}_k = \boldsymbol{P}_{k|k-1}\boldsymbol{H}_k^{\mathrm{T}}\boldsymbol{S}_k^{-1} \tag{2.4.12}$$

$$\boldsymbol{S}_k = \boldsymbol{H}_k\boldsymbol{P}_{k|k-1}\boldsymbol{H}_k^{\mathrm{T}} + \boldsymbol{R}_k \tag{2.4.13}$$

In the above equations, $\boldsymbol{I}$ is the identity matrix, residual $\boldsymbol{z}_k - \boldsymbol{H}_k\boldsymbol{m}_{k|k-1}$ is also referred as the innovation, and matrices $\boldsymbol{G}_k$ and $\boldsymbol{S}_k$ respectively represent the Kalman gain and the innovation covariance.

In many real tracking scenarios (such as angle only tracking, radar tracking and video tracking), the dynamic model and/or the measurement model are usually nonlinear, and the process noise and/or the measurement noise may also be non-additive and non-Gaussian. In these cases, a closed form solution is generally unavailable and the Kalman filter cannot be used. To solve this problem, a wealth of approximate filtering algorithms are developed, including the extended Kalman filter (EKF), unscented Kalman filter (UKF), cubature Kalman filter (CKF), Gaussian sum filter (GSF) and particle filter (PF).

2.5 Extended Kalman Filter

The extended Kalman filter (EKF) uses the linearization technique to transform the nonlinear filtering problem into an approximate linear filtering problem, and then applies the KF filter to solve. Hence, it is a sub-optimal solution. The commonly used linearization technique is the Taylor series expansion method, which can be divided into first-order EKF and higher-order (such as second-order) EKF according to the number of retained Taylor series items. Here we only introduce the first-order EKF. In short, the standard EKF filter is linearized based on Taylor series expansion and is a first-order approximate KF.

Consider the following discrete time nonlinear system model

$$\boldsymbol{x}_k = f_{k|k-1}(\boldsymbol{x}_{k-1}) + \boldsymbol{v}_{k-1} \tag{2.5.1}$$

$$z_k = h_k(\boldsymbol{x}_k) + \boldsymbol{n}_k \tag{2.5.2}$$

In the model, $f_{k|k-1}(\cdot)$ is the nonlinear state transform, $h_k(\cdot)$ is the nonlinear measurement transform, $\boldsymbol{v}_{k-1}$ and $\boldsymbol{n}_k$ are independent zero-mean additive Gaussian noises, wherein the former and the latter are respectively the process noise and the measurement noise, with the corresponding covariances recorded as $\boldsymbol{Q}_{k-1}$ and $\boldsymbol{R}_k$ accordingly, and $\boldsymbol{v}_{k-1}$ and $\boldsymbol{n}_k$ are assumed to be independent of the initial state.

Based on the above assumptions, if the initial prior is a Gaussian distribution $p_0 = \mathcal{N}(\cdot; \boldsymbol{m}_0, \boldsymbol{P}_0)$, all subsequent filtering densities are in the (approximate) Gaussian form. Specifically, if the filtering density at time $k-1$ is in the following Gaussian form

$$p_{k-1}(\boldsymbol{x}_{k-1}|z_{1:k-1}) = \mathcal{N}(\boldsymbol{x}_{k-1}; \boldsymbol{m}_{k-1}, \boldsymbol{P}_{k-1}) \tag{2.5.3}$$

the predicted density at time k is also in the Gaussian form, given by

$$p_{k|k-1}(\boldsymbol{x}_k|z_{1:k-1}) \approx \mathcal{N}(\boldsymbol{x}_k; \boldsymbol{m}_{k|k-1}, \boldsymbol{P}_{k|k-1}) \tag{2.5.4}$$

where

$$\boldsymbol{m}_{k|k-1} = f_{k|k-1}(\boldsymbol{m}_{k-1}) \tag{2.5.5}$$

$$\boldsymbol{P}_{k|k-1} = \boldsymbol{F}_{k|k-1}\boldsymbol{P}_{k-1}\boldsymbol{F}_{k|k-1}^{\mathrm{T}} + \boldsymbol{Q}_{k-1} \tag{2.5.6}$$

In the above equation, $\boldsymbol{F}_{k|k-1}$ is the Jacobian matrix of $f_{k|k-1}(\cdot)$, calculated as follows

$$\boldsymbol{F}_{k|k-1} = \frac{\partial}{\partial \boldsymbol{x}} f_{k|k-1}(\boldsymbol{x})\big|_{\boldsymbol{x}=\boldsymbol{m}_{k-1}} \tag{2.5.7}$$

Besides, the filtering density at time k is also in the Gaussian form

$$p_k(\boldsymbol{x}_k|z_{1:k}) \approx \mathcal{N}(\boldsymbol{x}_k; \boldsymbol{m}_k(z_k), \boldsymbol{P}_k) \tag{2.5.8}$$

where

$$\boldsymbol{m}_k(z_k) = \boldsymbol{m}_{k|k-1} + \boldsymbol{G}_k(z_k - h_k(\boldsymbol{m}_{k|k-1})) \tag{2.5.9}$$

$$\boldsymbol{P}_k = (\boldsymbol{I} - \boldsymbol{G}_k\boldsymbol{H}_k)\boldsymbol{P}_{k|k-1} \tag{2.5.10}$$

$$\boldsymbol{G}_k = \boldsymbol{P}_{k|k-1}\boldsymbol{H}_k^{\mathrm{T}}\boldsymbol{S}_k^{-1} \tag{2.5.11}$$

$$\boldsymbol{S}_k = \boldsymbol{H}_k\boldsymbol{P}_{k|k-1}\boldsymbol{H}_k^{\mathrm{T}} + \boldsymbol{R}_k \tag{2.5.12}$$

In the above equations, the residual $z_k - h_k(\boldsymbol{m}_{k|k-1})$ is the innovation, matrices $\boldsymbol{G}_k$ and $\boldsymbol{S}_k$ are respectively the gain and the innovation covariance, and $\boldsymbol{H}_k$ is the Jacobian matrix of $h_k(\cdot)$, calculated by

$$\boldsymbol{H}_k = \frac{\partial}{\partial \boldsymbol{x}} h_k(\boldsymbol{x})\big|_{\boldsymbol{x}=\boldsymbol{m}_{k|k-1}} \tag{2.5.13}$$

2.6 Unscented Kalman Filter

The UKF filter uses the unscented transform (UT) to obtain the approximate solution for the nonlinear filtering problem. For this reason, the UT is introduced before the UKF filter.

The basic idea of the UT is to select a fixed number of deterministic σ points to accurately capture the mean and covariance of the original distribution of $\boldsymbol{x}$, and then use the σ points propagated through nonlinear transformations to estimate the mean and covariance of the transformed variables. It should be noted that the UT is essentially different from the Monte Carlo technique used in the particle filter, because these σ points are selected deterministically.

The UT is used to perform the Gaussian approximation of joint distribution of random variables $\boldsymbol{x} \in \mathbb{R}^N$ and $\boldsymbol{z} \in \mathbb{R}^M$, that is, it is assumed that joint variable $[\boldsymbol{x} \quad \boldsymbol{z}]^{\mathrm{T}}$ follows a Gaussian distribution

$$\begin{bmatrix} \boldsymbol{x} \\ \boldsymbol{z} \end{bmatrix} \sim \mathcal{N}\left(\begin{bmatrix} \boldsymbol{x} \\ \boldsymbol{z} \end{bmatrix}; \begin{bmatrix} \overline{\boldsymbol{x}} \\ \overline{\boldsymbol{z}} \end{bmatrix}, \begin{bmatrix} \boldsymbol{P}_x & \boldsymbol{P}_{xz} \\ \boldsymbol{P}_{zx} & \boldsymbol{P}_z \end{bmatrix}\right) \tag{2.6.1}$$

where $\boldsymbol{x}$ is a Gaussian variable with a known statistical property

$$\boldsymbol{x} \sim \mathcal{N}(\boldsymbol{x}; \overline{\boldsymbol{x}}, \boldsymbol{P}_x) \tag{2.6.2}$$

and $\boldsymbol{z}$ is its nonlinear transform, given by

$$\boldsymbol{z} = h(\boldsymbol{x}) \tag{2.6.3}$$

The goal now is to use the UT to solve $\overline{\boldsymbol{z}}$, $\boldsymbol{P}_z$ and $\boldsymbol{P}_{xz}$ ($\boldsymbol{P}_{zx} = \boldsymbol{P}_{xz}^{\mathrm{T}}$) in (2.6.1). The specific solution steps are as follows.

(1) Calculate $2N+1$ σ points according to matrix $\sqrt{(N+\lambda)\boldsymbol{P}_x}$ of size $N \times N$

$$\boldsymbol{x}^{(n)} = \begin{cases} \overline{\boldsymbol{x}}, & n = 0 \\ \overline{\boldsymbol{x}} + [\sqrt{(N+\lambda)\boldsymbol{P}_x}]_n, & n = 1, \ldots, N \\ \overline{\boldsymbol{x}} - [\sqrt{(N+\lambda)\boldsymbol{P}_x}]_{n-N}, & n = N+1, \ldots, 2N \end{cases}$$

and calculate relevant first order weight $w_{\mathrm{m}}^{(n)}$ and second order weight $w_{\mathrm{c}}^{(n)}$ as follows

$$w_{\mathrm{m}}^{(n)} = \begin{cases} \frac{\lambda}{N+\lambda}, & n = 0 \\ \frac{0.5}{N+\lambda}, & n = 1, 2, \ldots, 2N \end{cases}, \quad w_{\mathrm{c}}^{(n)} = \begin{cases} \frac{\lambda}{(N+\lambda)+(1-\alpha^2+\beta)}, & n = 0 \\ \frac{0.5}{N+\lambda}, & n = 1, 2, \ldots, 2N \end{cases}$$

where $\lambda = \alpha^2(N+\tau) - n$, α determines the spread of σ points and usually takes a small positive value (such as 0.001); τ is usually set as 0; β describes the distribution information of $\boldsymbol{x}$ (for the Gaussian case, the optimal value of β is 2); note that $[\sqrt{(N+\lambda)\boldsymbol{P}_x}]_n$ represents the n th column of square root matrix of matrix $(N+\lambda)\boldsymbol{P}_x$.

(2) Calculate the transformed value after nonlinear transformation for each σ point

$$\boldsymbol{z}^{(n)} = h(\boldsymbol{x}^{(n)}), \quad n = 0, 1, \ldots, 2N \tag{2.6.4}$$

(3) Calculate $\overline{\boldsymbol{z}}$, $\boldsymbol{P}_z$ and $\boldsymbol{P}_{xz}$ as follows

$$\overline{\boldsymbol{z}} = \sum_{n=0}^{2N} w_{\mathrm{m}}^{(n)} \boldsymbol{z}^{(n)} \tag{2.6.5}$$

$$\boldsymbol{P}_z = \sum_{n=0}^{2N} w_c^{(n)} (z^{(n)} - \overline{z})(z^{(n)} - \overline{z})^{\mathrm{T}} \tag{2.6.6}$$

$$\boldsymbol{P}_{xz} = \sum_{n=0}^{2N} w_c^{(n)} (\boldsymbol{x}^{(n)} - \overline{\boldsymbol{x}})(z^{(n)} - \overline{z})^{\mathrm{T}} \tag{2.6.7}$$

It can be seen from the above that the UT can be regarded as a function from $(h, \overline{\boldsymbol{x}}, \boldsymbol{P}_x)$ to $(\overline{z}, \boldsymbol{P}_z, \boldsymbol{P}_{xz})$, namely,

$$(\overline{z}, \boldsymbol{P}_z, \boldsymbol{P}_{xz}) = \mathrm{UT}(h, \overline{\boldsymbol{x}}, \boldsymbol{P}_x) \tag{2.6.8}$$

The UKF propagates the first- and second-order moments of predicted density and updated density through the deterministic sampling rule of the UT transform. For the non-linear problems given by (2.5.1) and (2.5.2), based on the UT, the prediction and update steps of the UKF are as follows.

Assume that the prior density at time $k-1$ is in the Gaussian form

$$p_{k-1}(\boldsymbol{x}_{k-1}|z_{1:k-1}) = \mathcal{N}(\boldsymbol{x}_{k-1}; \boldsymbol{m}_{k-1}, \boldsymbol{P}_{k-1}) \tag{2.6.9}$$

then the predicted density at time k is also in the Gaussian form

$$p_{k|k-1}(\boldsymbol{x}_k|z_{1:k-1}) \approx \mathcal{N}(\boldsymbol{x}_k; \boldsymbol{m}_{k|k-1}, \boldsymbol{P}_{k|k-1}) \tag{2.6.10}$$

where

$$(\boldsymbol{m}_{k|k-1}, \boldsymbol{P}'_{k|k-1}, \sim) = \mathrm{UT}(f_{k|k-1}(\cdot), \boldsymbol{m}_{k-1}, \boldsymbol{P}_{k-1}) \tag{2.6.11}$$

$$\boldsymbol{P}_{k|k-1} = \boldsymbol{P}'_{k|k-1} + \boldsymbol{Q}_{k-1} \tag{2.6.12}$$

Furthermore, the filtering density at time k is also in the Gaussian form

$$p_k(\boldsymbol{x}_k|z_{1:k}) \approx \mathcal{N}(\boldsymbol{x}_k; \boldsymbol{m}_k(z_k), \boldsymbol{P}_k) \tag{2.6.13}$$

where

$$\boldsymbol{m}_k(z_k) = \boldsymbol{m}_{k|k-1} + \boldsymbol{G}_k(z_k - \boldsymbol{m}_{z,k}) \tag{2.6.14}$$

$$\boldsymbol{P}_k = \boldsymbol{P}_{k|k-1} - \boldsymbol{G}_k \boldsymbol{S}_k \boldsymbol{G}_k^{\mathrm{T}} \tag{2.6.15}$$

$$(\boldsymbol{m}_{z,k}, \boldsymbol{P}_{z,k}, \boldsymbol{P}_{xz,k}) = \mathrm{UT}(h_k(\cdot), \boldsymbol{m}_{k|k-1}, \boldsymbol{P}_{k|k-1}) \tag{2.6.16}$$

$$G_k = P_{xz,k} S_k^{-1} \tag{2.6.17}$$

$$S_k = P_{z,k} + R_k \tag{2.6.18}$$

2.7 Cubature Kalman Filter

The CKF is similar to the UKF. Both filters use a set of weighted samples going through nonlinear transform to calculate the first- and second-order moments required by filtering, thus can avoid the linearization for nonlinear models and are suitable for any form of nonlinear models. However, there are essential differences between the two filters: the CKF uses even-numbered samples with the same weight, whereas the UKF uses odd-numbered samples with different weights. In a high-dimensional system, the weight of σ point for the UKF is prone to be negative, whereas the weight of the CKF is always positive. For this reason, the CKF has a better numerical stability and filtering accuracy than the UKF in the high-dimensional case. In a word, compared with the EKF, the UKF and other nonlinear filters, the CKF boasts a better nonlinear approximation performance, numerical accuracy and filtering stability. Furthermore, it also stands out for simple implementation, low computational workloads, and high filtering precision. Hence, the CKF has been widely used to address estimation problems in various fields since it was proposed.

The CKF solves the integral problem in the Bayesian filter with a set of equal-weighted cubature points, namely, it calculates the mean and covariance of random variables through a nonlinear transform based on the principle of cubature numerical integration [24]. Specifically, it uses M weighted cubature points $\{w_i, \boldsymbol{\xi}_i\}$ to approximate the following Gaussian weighted integral

$$\int_{\mathbb{R}^n} g(\boldsymbol{x})\mathcal{N}(\boldsymbol{x}; \boldsymbol{0}, \boldsymbol{I})d\boldsymbol{x} \approx \sum_{i=1}^{M} w_i g(\boldsymbol{\xi}_i) \tag{2.7.1}$$

where $\boldsymbol{\xi}_i$ and w_i are the ith cubature point and its weight, $\mathcal{N}(\boldsymbol{x}; \boldsymbol{0}, \boldsymbol{I})$ is a standard normal distribution with zero mean and unit covariance, $g(\boldsymbol{x})$ is a general nonlinear function, and n is the dimension of the state vector. For the 3-DOF (degrees of freedom) spherical-radial rule, the total number of cubature points is $M = 2n$, and

$$\boldsymbol{\xi}_i = \sqrt{n}[1]_i \tag{2.7.2}$$

$$w_i = 1/(2n), \quad i = 1, 2, \ldots, 2n \tag{2.7.3}$$

In (2.7.2), symbol [1] represents the completely symmetric set of points obtained by fully arranging the elements of the n-dimensional unit vector and changing the sign of the elements. For example, $[1] \in \mathbb{R}^2$ represents the following point set

$$\left\{ \begin{pmatrix} 1 \\ 0 \end{pmatrix}, \begin{pmatrix} 0 \\ 1 \end{pmatrix}, \begin{pmatrix} -1 \\ 0 \end{pmatrix}, \begin{pmatrix} 0 \\ -1 \end{pmatrix} \right\}$$

Furthermore, $[1]_i$ represents the ith point in point set [1].

Assume that $\boldsymbol{x} \in \mathbb{R}^n$ is a Gaussian variable with known statistical properties, obeying distribution $\boldsymbol{x} \sim \mathcal{N}(\boldsymbol{x}; \overline{\boldsymbol{x}}, \boldsymbol{P}_x)$, and z represents its nonlinear transform $z = h(\boldsymbol{x})$. According to the above principle of cubature numerical integration, $\overline{z}$, $\boldsymbol{P}_z$ and $\boldsymbol{P}_{xz}$ ($\boldsymbol{P}_{zx} = \boldsymbol{P}_{xz}^{\mathrm{T}}$) in (2.6.1) are solved with the cubature transform. The specific steps are as follows.

(1) Calculate $2n$ cubature points according to covariance $\sqrt{\boldsymbol{P}_x}$

$$\boldsymbol{x}^{(i)} = \overline{\boldsymbol{x}} + \sqrt{\boldsymbol{P}_x}\xi_i, \quad i = 1, 2, \ldots, 2n \tag{2.7.4}$$

(2) Calculate the cubature points after passing through the nonlinear measurement equation

$$z^{(i)} = h(\boldsymbol{x}^{(i)}), \quad i = 1, 2, \ldots, 2n \tag{2.7.5}$$

(3) Calculate $\overline{z}$, $\boldsymbol{P}_z$, and $\boldsymbol{P}_{xz}$ as follows

$$\overline{z} = \sum_{i=1}^{2n} w_i z^{(i)} \tag{2.7.6}$$

$$\boldsymbol{P}_z = \sum_{i=1}^{2n} w_i z^{(i)} (z^{(i)})^{\mathrm{T}} - \overline{z}\,\overline{z}^{\mathrm{T}} \tag{2.7.7}$$

$$\boldsymbol{P}_{xz} = \sum_{i=1}^{2n} w_i \boldsymbol{x}^{(i)} (z^{(i)})^{\mathrm{T}} - \overline{\boldsymbol{x}}\,\overline{z}^{\mathrm{T}} \tag{2.7.8}$$

It can be seen from the above that the cubature transform (CuT) can be deemed as a function from $(h, \overline{\boldsymbol{x}}, \boldsymbol{P}_x)$ to $(\overline{z}, \boldsymbol{P}_z, \boldsymbol{P}_{xz})$, namely

$$(\overline{z}, \boldsymbol{P}_z, \boldsymbol{P}_{xz}) = \mathrm{CuT}(h, \overline{\boldsymbol{x}}, \boldsymbol{P}_x) \tag{2.7.9}$$

The CKF uses the deterministic sampling rules of the cubature transform to propagate the first- and second-order moments of predicted density and updated density. Based on the cubature transform, for the nonlinear problems given by (2.5.1)–(2.5.2), the prediction and update steps of the CKF are the same as those of the UKF, except that function $\mathrm{UT}(\cdot)$ in (2.6.11) and (2.6.16) are replaced with $\mathrm{CuT}(\cdot)$, which is not repeated here.

In addition to the standard CKF, Ref. [24] also gives a more robust square root CKF (SR-CKF). Please refer to [24] if readers are interested.

2.8 Gaussian Sum Filter

The GSF, also known as the Gaussian mixture filter (GMF), can be regarded as an extension of the Kalman filter as it replaces Gaussian distribution in the Kalman filter with the Gaussian mixture distribution. Here we just briefly introduces the GSF. For a detail description, please refer to [381].

The GSF assumes that the Markov transition density and the likelihood function are both nonlinear, but such a nonlinear density can be approximated through weighted sum of the following Gaussian densities (namely, Gaussian mixture density)

$$\phi_{k|k-1}(\boldsymbol{x}|\boldsymbol{x}') \approx \sum_{i=1}^{I_{k-1}} \varpi_{k-1}^{(i)} \mathcal{N}(\boldsymbol{x}; \boldsymbol{F}_{k-1}^{(i)}\boldsymbol{x}', \boldsymbol{Q}_{k-1}^{(i)}) \tag{2.8.1}$$

$$g_k(\boldsymbol{z}|\boldsymbol{x}) \approx \sum_{j=1}^{J_k} \omega_k^{(j)} \mathcal{N}(\boldsymbol{z}; \boldsymbol{H}_k^{(j)}\boldsymbol{x}, \boldsymbol{R}_k^{(j)}) \tag{2.8.2}$$

where $\varpi_{k-1}^{(i)} \geq 0, \omega_k^{(j)} \geq 0, \sum_{i=1}^{I_{k-1}} \varpi_{k-1}^{(i)} = 1$ and $\sum_{j=1}^{J_k} \omega_k^{(j)} = 1$. In addition, the GSF assumes that posterior distributions $p_{k-1|k-1}(\boldsymbol{x}_{k-1}|\boldsymbol{z}_{1:k-1})$ and $p_{k|k-1}(\boldsymbol{x}_{k-1}|\boldsymbol{z}_{1:k-1})$ are both Gaussian mixture forms, namely,

$$p_{k-1}(\boldsymbol{x}_{k-1}|\boldsymbol{z}_{1:k-1}) = \sum_{n=1}^{N_{k-1}} w_{k-1}^{(n)} \mathcal{N}(\boldsymbol{x}_{k-1}; \boldsymbol{m}_{k-1}^{(n)}, \boldsymbol{P}_{k-1}^{(n)}) \tag{2.8.3}$$

$$p_{k|k-1}(\boldsymbol{x}_k|\boldsymbol{z}_{1:k-1}) = \sum_{n=1}^{N_{k|k-1}} w_{k|k-1}^{(n)} \mathcal{N}(\boldsymbol{x}_k; \boldsymbol{m}_{k|k-1}^{(n)}, \boldsymbol{P}_{k|k-1}^{(n)}) \tag{2.8.4}$$

Thus, the GSF propagates posterior component $(w_k^{(n)}, \boldsymbol{m}_k^{(n)}, \boldsymbol{P}_k^{(n)})$ over time according to the recursive Bayesian filtering equation, namely,

$$\begin{aligned}
&(w_0^{(n)}, \boldsymbol{m}_0^{(n)}, \boldsymbol{P}_0^{(n)})_{n=1,\ldots,N_0} \to (w_{1|0}^{(n)}, \boldsymbol{m}_{1|0}^{(n)}, \boldsymbol{P}_{1|0}^{(n)})_{n=1,\ldots,N_{1|0}} \\
&\quad \to (w_1^{(n)}, \boldsymbol{m}_1^{(n)}, \boldsymbol{P}_1^{(n)})_{n=1,\ldots,N_1} \to \cdots \to \\
&(w_{k-1}^{(n)}, \boldsymbol{m}_{k-1}^{(n)}, \boldsymbol{P}_{k-1}^{(n)})_{n=1,\ldots,N_{k-1}} \to (w_{k|k-1}^{(n)}, \boldsymbol{m}_{k|k-1}^{(n)}, \boldsymbol{P}_{k|k-1}^{(n)})_{n=1,\ldots,N_{k|k-1}} \\
&\quad \to (w_k^{(n)}, \boldsymbol{m}_k^{(n)}, \boldsymbol{P}_k^{(n)})_{n=1,\ldots,N_k} \to \cdots
\end{aligned} \tag{2.8.5}$$

- GSF prediction (time update)

According to (2.2.7) and Lemma 4 in Appendix A, the prediction of the GSF can be obtained as follows

$$
\begin{aligned}
p_{k|k-1}(\boldsymbol{x}_k|z_{1:k-1}) &= \int \phi_{k|k-1}(\boldsymbol{x}|\boldsymbol{x}')p_{k-1}(\boldsymbol{x}'|z_{1:k-1})d\boldsymbol{x}' \\
&= \sum_{i=1}^{I_{k-1}}\sum_{n=1}^{N_{k-1}} \varpi_{k-1}^{(i)} w_{k-1}^{(n)} \int \mathcal{N}(\boldsymbol{x}_k; \boldsymbol{F}_{k-1}^{(i)}\boldsymbol{x}', \boldsymbol{Q}_{k-1}^{(i)})\mathcal{N}(\boldsymbol{x}'; \boldsymbol{m}_{k-1}^{(n)}, \boldsymbol{P}_{k-1}^{(n)})d\boldsymbol{x}' \\
&= \sum_{i=1}^{I_{k-1}}\sum_{n=1}^{N_{k-1}} w_{k|k-1}^{(n,i)}\mathcal{N}(\boldsymbol{x}_k; \boldsymbol{m}_{k|k-1}^{(n,i)}, \boldsymbol{P}_{k|k-1}^{(n,i)}) = \sum_{n=1}^{N_{k|k-1}} w_{k|k-1}^{(n)}\mathcal{N}(\boldsymbol{x}_k; \boldsymbol{m}_{k|k-1}^{(n)}, \boldsymbol{P}_{k|k-1}^{(n)})
\end{aligned} \tag{2.8.6}
$$

where

$$
w_{k|k-1}^{(n,i)} = \varpi_{k-1}^{(i)} w_{k-1}^{(n)} \tag{2.8.7}
$$

$$
\boldsymbol{m}_{k|k-1}^{(n,i)} = \boldsymbol{F}_{k-1}^{(i)}\boldsymbol{m}_{k-1}^{(n)} \tag{2.8.8}
$$

$$
\boldsymbol{P}_{k|k-1}^{(n,i)} = \boldsymbol{F}_{k-1}^{(i)}\boldsymbol{P}_{k-1}^{(n)}(\boldsymbol{F}_{k-1}^{(i)})^{\mathrm{T}} + \boldsymbol{Q}_{k-1}^{(i)} \tag{2.8.9}
$$

Therefore, $p_{k|k-1}(\boldsymbol{x}_k|z_{1:k-1})$ represents a Gaussian mixture distribution with $N_{k|k-1} = I_{k-1}N_{k-1}$ components, indicating that Gaussian components of predicted distribution $p_{k|k-1}(\boldsymbol{x}_k|z_{1:k-1})$ will increase in combination over time, only when $I_{k-1} = 1$ (that is $\varpi_{k-1}^{1} = 1$), $N_{k|k-1} = N_{k-1}, w_{k|k-1}^{(n,1)} = w_{k-1}^{(n)}$.

- GSF update (measurement update)

Based on the predicted distribution (2.8.6), and according to correction Eq. (2.2.8) of the Bayesian filter and Lemma 5 in Appendix A, it can be concluded after derivation that the updated posterior distribution is also in the Gaussian mixture form

$$
\begin{aligned}
p_k(\boldsymbol{x}_k|z_{1:k}) &= \frac{g_k(z_k|\boldsymbol{x}_k)p_{k|k-1}(\boldsymbol{x}_k|z_{1:k-1})}{\int g_k(z_k|\boldsymbol{x})p_{k|k-1}(\boldsymbol{x}|z_{1:k-1})d\boldsymbol{x}} \\
&= c^{-1}\sum_{j=1}^{J_k}\sum_{n=1}^{N_{k|k-1}} \omega_k^{(j)} w_{k|k-1}^{(n)}\mathcal{N}(z_k; \boldsymbol{H}_k^{(j)}\boldsymbol{x}_k, \boldsymbol{R}_k^{(j)})\mathcal{N}(\boldsymbol{x}_k; \boldsymbol{m}_{k|k-1}^{(n)}, \boldsymbol{P}_{k|k-1}^{(n)}) \\
&= c^{-1}\sum_{j=1}^{J_k}\sum_{n=1}^{N_{k|k-1}} \omega_k^{(j)} w_{k|k-1}^{(n)}\mathcal{N}(z_k; \hat{z}_{k|k-1}^{(n,j)}, \boldsymbol{S}_k^{(n,j)})\mathcal{N}(\boldsymbol{x}_k; \boldsymbol{m}_{k|k}^{(n,j)}, \boldsymbol{P}_{k|k}^{(n,j)})
\end{aligned}
$$

$$
\begin{aligned}
&= \sum_{j=1}^{J_k} \sum_{n=1}^{N_{k|k-1}} w_k^{(n,j)} \mathcal{N}(\boldsymbol{x}_k; \boldsymbol{m}_{k|k}^{(n,j)}, \boldsymbol{P}_{k|k}^{(n,j)}) \\
&= \sum_{n=1}^{N_k} w_k^{(n)} \mathcal{N}(\boldsymbol{x}_k; \boldsymbol{m}_k^{(n)}, \boldsymbol{P}_k^{(n)})
\end{aligned}
\tag{2.8.10}
$$

where

$$w_k^{(n,j)} = c^{-1} \omega_k^{(j)} w_{k|k-1}^{(n)} \mathcal{N}(z_k; \hat{z}_{k|k-1}^{(n,j)}, \boldsymbol{S}_k^{(n,j)}) \tag{2.8.11}$$

$$\boldsymbol{m}_k^{(n,j)} = \boldsymbol{m}_{k|k-1}^{(n)} + \boldsymbol{G}_k^{(n,j)}(z_k - \hat{z}_{k|k-1}^{(n,j)}) \tag{2.8.12}$$

$$\boldsymbol{P}_k^{(n,j)} = \boldsymbol{P}_{k|k-1}^{(n)} - \boldsymbol{G}_k^{(n,j)} \boldsymbol{S}_k^{(n,j)} (\boldsymbol{G}_k^{(n,j)})^{\mathrm{T}} \tag{2.8.13}$$

$$\hat{z}_{k|k-1}^{(n,j)} = \boldsymbol{H}_k^{(j)} \boldsymbol{m}_{k|k-1}^{(n)} \tag{2.8.14}$$

$$\boldsymbol{S}_k^{(n,j)} = \boldsymbol{H}_k^{(j)} \boldsymbol{P}_{k|k-1}^{(n)} (\boldsymbol{H}_k^{(j)})^{\mathrm{T}} + \boldsymbol{R}_k^{(j)} \tag{2.8.15}$$

$$\boldsymbol{G}_k^{(n,j)} = \boldsymbol{P}_{k|k-1}^{(n)} (\boldsymbol{H}_k^{(j)})^{\mathrm{T}} (\boldsymbol{S}_k^{(n,j)})^{-1} \tag{2.8.16}$$

$$c = \sum_{j=1}^{J_k} \sum_{n=1}^{N_{k|k-1}} \omega_k^{(j)} w_{k|k-1}^{(n)} \mathcal{N}(z_k; \hat{z}_{k|k-1}^{(n,j)}, \boldsymbol{S}_k^{(n,j)}) \tag{2.8.17}$$

Obviously, the number $N_k = J_k N_{k|k-1}$ of Gaussian components in posterior distribution $p_k(\boldsymbol{x}_k|z_{1:k})$ will increase in combination over time, only when $J_k = 1$ (namely, $\omega_k^1 = 1$), $N_k = N_{k|k-1}$, $w_k^{(n,1)} = w_k^{(n)}$.

It can be seen from the above that for the more general case where the transition density and/or likelihood function are in the form of Gaussian mixture, the number of Gaussian components accurately representing the filtering density will increase exponentially over time. Therefore, Gaussian mixture pruning techniques (such as removing components with small weights and merging similar components) should be adopted to manage memory and computational workloads [382, 383].

- GSF state estimation

After the posterior distribution $p_k(\boldsymbol{x}_k|z_{1:k})$ is obtained, the expected a posterior (EAP) estimation of the state is calculated as

$$\hat{\boldsymbol{x}}_k^{\mathrm{EAP}} = \int \boldsymbol{x} \cdot p_k(\boldsymbol{x}_k|z_{1:k}) d\boldsymbol{x} = \sum_{n=1}^{N_k} w_k^{(n)} \boldsymbol{m}_k^{(n)} \tag{2.8.18}$$

However, when $p_k(\boldsymbol{x}_k|z_{1:k})$ has a significant multimodal form, the performance of the EAP estimator will be unsatisfactory. In this case, the MAP (maximum a posterior) estimator should be adopted, which needs to determine the position of the maximum value in Gaussian weighted sum. By using component removal and merging techniques, the MAP estimate can be approximated as

$$\hat{\boldsymbol{x}}_k^{\text{MAP}} \approx \boldsymbol{m}_k^{(n_*)} \tag{2.8.19}$$

where $\boldsymbol{m}_k^{(n_*)}$ is the mean of the Gaussian component corresponding to the maximum mixture coefficient $w_k^{(n_*)}$. If Gaussian mixture components are far apart from each other, such approximation is highly accurate.

2.9 Particle Filter

The particle filter (PF) is also known as the sequential Monte Carlo (SMC) method. Its basic principle is to use random samples (particles) to approximate the probability distribution of interest. It is a kind of numerical approximation solution for the Bayesian recursion. This method can be applied to nonlinear non-Gaussian dynamic and measurement models.

Let $\boldsymbol{x}^{(i)} \sim p(\boldsymbol{x})$ represent the independently and identically distributed (IID) samples extracted from probability density $p(\boldsymbol{x})$. For any function $h(\boldsymbol{x})$, its (limited) expectation can be approximated through empirical expectation with N IID samples $\{\boldsymbol{x}^{(i)}\}_{i=1}^N$, namely,

$$\int h(\boldsymbol{x})p(\boldsymbol{x})d\boldsymbol{x} \approx \frac{1}{N}\sum_{i=1}^{N} h(\boldsymbol{x}^{(i)}) \tag{2.9.1}$$

When N becomes infinite, the above empirical expectation is unbiased and tends to be the true expectation. Alongside that, the convergence rate is independent of the dimension of the integral, and it mainly depends on the number N of independent samples [384]. Thereby, samples $\{\boldsymbol{x}^{(i)}\}_{i=1}^N$ can be regarded as a point mass approximation of p, namely,

$$p(\boldsymbol{x}) \approx \frac{1}{N}\sum_{i=1}^{N} \delta_{\boldsymbol{x}^{(i)}}(\boldsymbol{x}) \tag{2.9.2}$$

where δ denotes the Dirac delta function.

In the Bayesian recursion, the normalization constant is typically difficult to calculate, so $p(\boldsymbol{x}) \propto \tilde{p}(\boldsymbol{x})$ is commonly considered. Since it is difficult to sample from density p, it is suggested to extract N IID samples $\{\boldsymbol{x}^{(i)}\}_{i=1}^{N}$ from a known density q, which is called the proposed density or importance density, and then corresponding weights are assigned to these samples, so that the weighted point mass approximation of p is obtained. More exactly, the (limited) expectation of any function h can be approximated through the following empirical expectation

$$\int h(\boldsymbol{x})p(\boldsymbol{x})d\boldsymbol{x} \approx \sum_{i=1}^{N} w^{(i)}h(\boldsymbol{x}^{(i)}) \tag{2.9.3}$$

where

$$\boldsymbol{x}^{(i)} \sim q(\boldsymbol{x}) \tag{2.9.4}$$

$$w^{(i)} = \tilde{w}^{(i)} \Big/ \sum\nolimits_{j=1}^{N} \tilde{w}^{(j)} \tag{2.9.5}$$

$$\tilde{w}^{(i)} = p(\boldsymbol{x}^{(i)})/q(\boldsymbol{x}^{(i)}) \tag{2.9.6}$$

In the above equations, $\tilde{w}(\boldsymbol{x}^{(i)})$ and $w^{(i)}$ are respectively referred to as the importance weight and the normalized importance weight. A desirable proposed distribution should allow all weights $\{w^{(i)}\}_{i=1}^{N}$ to be equal approximately. For the above importance sampling approximation, this empirical expectation is biased. However, as N tends to infinity, it almost certainly tends to be the true expectation. For this reason, weighted particles $\{w^{(i)}, \boldsymbol{x}^{(i)}\}_{i=1}^{N}$ can be deemed as a weighted point mass approximation of p, that is,

$$p(\boldsymbol{x}) \approx \sum_{i=1}^{N} w^{(i)}\delta_{\boldsymbol{x}^{(i)}}(\boldsymbol{x}) \tag{2.9.7}$$

It is assumed that posterior density $p_{0:k-1}$ at time $k-1$ is represented by weighted particle set $\{w_{k-1}^{(i)}, \boldsymbol{x}_{0:k-1}^{(i)}\}_{i=1}^{N}$, namely,

$$p_{0:k-1}(\boldsymbol{x}_{0:k-1}|\boldsymbol{z}_{1:k-1}) \approx \sum_{i=1}^{N} w_{k-1}^{(i)}\delta_{\boldsymbol{x}_{0:k-1}^{(i)}}(\boldsymbol{x}_{0:k-1}) \tag{2.9.8}$$

When a proposed density $q_k(\cdot|\boldsymbol{x}_{k-1}^{(i)}, \boldsymbol{z}_k)$ easy for sampling is specified, posterior density $p_{0:k}$ at time k can be represented by new weighted particle set $\{w_k^{(i)}, \boldsymbol{x}_{0:k}^{(i)}\}_{i=1}^{N}$, namely,

$$p_{0:k}(\boldsymbol{x}_{0:k}|\boldsymbol{z}_{1:k}) \approx \sum_{i=1}^{N} w_k^{(i)}\delta_{\boldsymbol{x}_{0:k}^{(i)}}(\boldsymbol{x}_{0:k}) \tag{2.9.9}$$

where

$$\boldsymbol{x}_{0:k}^{(i)} = (\boldsymbol{x}_{0:k-1}^{(i)}, \boldsymbol{x}_k^{(i)}) \tag{2.9.10}$$

$$\boldsymbol{x}_k^{(i)} \sim q_k(\cdot|\boldsymbol{x}_{k-1}^{(i)}, z_k) \tag{2.9.11}$$

$$w_k^{(i)} = \tilde{w}_k^{(i)} \Big/ \sum_{j=1}^{N} \tilde{w}_k^{(j)} \tag{2.9.12}$$

$$\tilde{w}_k^{(i)} = w_{k-1}^{(i)} \frac{g_k(z_k|\boldsymbol{x}_k^{(i)})\phi_{k|k-1}(\boldsymbol{x}_k^{(i)}|\boldsymbol{x}_{k-1}^{(i)})}{q_k(\boldsymbol{x}_k^{(i)}|\boldsymbol{x}_{k-1}^{(i)}, z_k)} \tag{2.9.13}$$

If only the filtering density is interested, only the samples at the latest time need to be saved, namely, the filtering density is represented by weighted samples $\{w_k^{(i)}, \boldsymbol{x}_k^{(i)}\}_{i=1}^N$.

A critical step in the particle filter is to recursively approximate the posterior density by sequentially using importance sampling, which is the well-known sequential importance sampling (SIS) method. However, the use of this method has the inevitable sample degeneracy or particle depletion problem: after several iterations, except for a few particles (with large weight), the weights of a large number of particles have been so small (approximately 0) that they can be ignored (that is, the variance of particle weights decreases with time). Therefore, a lot of calculations are wasted on particle updates that contribute little to the posterior. To overcome this problem, a better proposed density can be selected. For practical strategies regarding selection of the optimal proposed density and construction of a better proposed density, please refer to [385, 386]. Besides, a more general way is to adopt a re-sampling method, that is, re-sampling weighted particles $\{w_k^{(i)}, \boldsymbol{x}_k^{(i)}\}_{i=1}^N$ to generate more copies of particles with higher weights and eliminate those with lower weights. It should be noted that re-sampling alleviates the sample degradation, but also causes a new problem of sample impoverishment, which means, the diversity of samples is reduced as high-weighted samples are selected for several times. In this case, Markov chain Monte Carlo (MCMC) steps can be employed to make particles more diverse [53].

The complete steps of the particle filter focusing only on the filtering density are summarized below.

- Prediction
  - Given $\{w_{k-1}^{(i)}, \boldsymbol{x}_{k-1}^{(i)}\}_{i=1}^N$, for $i = 1, \ldots, N$, sample $\tilde{\boldsymbol{x}}_k^{(i)} \sim q_k(\cdot|\boldsymbol{x}_{k-1}^{(i)}, z_k)$ and calculate the predicted weight

$$w_{k|k-1}^{(i)} = w_{k-1}^{(i)} \frac{\phi_{k|k-1}(\boldsymbol{x}_k^{(i)}|\boldsymbol{x}_{k-1}^{(i)})}{q_k(\boldsymbol{x}_k^{(i)}|\boldsymbol{x}_{k-1}^{(i)}, z_k)} \tag{2.9.14}$$

- Update
  - Update weights

$$w_k^{(i)} = w_{k|k-1}^{(i)} g_k(z_k | \boldsymbol{x}_k^{(i)}), \quad i = 1, \ldots, N \tag{2.9.15}$$

  - Normalize weights

$$\tilde{w}_k^{(i)} = w_k^{(i)} / \sum_{j=1}^{N} w_k^{(j)}, \quad i = 1, \ldots, N \tag{2.9.16}$$

- Re-sample
  - Re-sample $\{\tilde{w}_k^{(i)}, \tilde{\boldsymbol{x}}_k^{(i)}\}_{i=1}^N$ to obtain $\{w_k^{(i)}, \boldsymbol{x}_k^{(i)}\}_{i=1}^N$.

There are many methods to implement re-sampling, such as the multinomial re-sampling, stratified re-sampling, and residual re-sampling. The selection of different strategies affects the amount of calculation and the quality of particle approximation [387, 388]. The effective stratified re-sampling delivers better statistical properties when compared with the multinomial re-sampling. Most re-sampling methods obtain $\{\tilde{\boldsymbol{x}}_k^{(i)}\}_{i=1}^N$ by copying $n_k^{(i)}$ copies for each particle $\tilde{\boldsymbol{x}}_k^{(i)}$ under constraint $\sum_{i=1}^N n_k^{(i)} = N$. A (random) re-sampling mechanism that satisfies condition $\mathrm{E}[n_k^{(i)}] = N\omega_k^{(i)}$ can be selected, where $\omega_k^{(i)} > 0$ and $\sum_{i=1}^N \omega_k^{(i)} = 1$ are a series of user-set weights. Typically, $\omega_k^{(i)} = \tilde{w}_k^{(i)}$ can be set, or $\omega_k^{(i)} \propto (\tilde{w}_k^{(i)})^\tau$ can also be selected, where $\tau \in (0, 1)$. Then, new weights are set to $w_k^{(i)} \propto \tilde{w}_k^{(i)} / \omega_k^{(i)}, \sum_{i=1}^N w_k^{(i)} = 1$.

A host of improved particle filtering algorithms have been proposed to enhance the performance. For state-space models of specific types, the Rao-Blackwellization technique can be incorporated into the particle filter [384]. Its core idea is to decompose the state vector into linear Gaussian and nonlinear Gaussian components. Then, the former is solved analytically by using the Kalman filter, while the latter is solved by the particle filter, thereby reducing the dimension of the problem and reducing the amount of calculation. Such an example is the mixture Kalman filter (MKF) [389]. Additionally, a continuous approximation of the posterior density can be obtained through the kernel smoothing, with examples of the convolution particle filter and the regularized particle filter [384]. Besides, other related methods include the Gaussian particle filter [390] and the Gaussian sum particle filter [391].

2.10 Summary

Following the complexity from easy to difficult and the development history from first to last, this chapter introduces the filtering algorithms widely used in single target tracking, such as the Bayesian filter, Kalman filter, extended Kalman filter, unscented Kalman filter, cubature Kalman filter, Gaussian sum filter, and particle filter. At the same time, this chapter also serves as an introduction to the more complex multi-target tracking algorithm based on random finite sets (RFS). As will be seen later, most of the single target filtering algorithms described in this chapter will have the corresponding RFS-based multi-target versions.

Chapter 3
Basics of Random Finite Sets

3.1 Introduction

As discussed in the previous chapters, important challenges faced by multi-target filtering/tracking include clutter, detection uncertainty, and data association uncertainty. Until now, three mainstream solutions have emerged for the multi-target tracking problem: multiple hypothesis tracking (MHT), joint probabilistic data association (JPDA), and the emerging random finite set (RFS). Thanks to a long time of development, the first two solutions are relatively mature with abundant reference materials. By contrast, materials about the RFS method are relatively lacking. For this reason, the book aims to detail the RFS-based target tracking algorithms.

Considering that the readers are relatively unfamiliar with the RFS, and at the same time, in order to facilitate the readers to get started quickly, the contents of this chapter are designed as follows: First, the preliminary knowledge of statistical descriptors, moments and common types of random finite sets is introduced. This part of the content will be frequently used later, which is the basis for understanding various RFS algorithms; Then, the multi-target system models and multi-target Bayesian recursion are introduced separately, and the multi-target formal modeling paradigm is given. These contents can be compared with the single target case. In particular, the multi-target Bayesian recursion serves as the core formula of each RFS algorithm in the following text; Finally, for the complex multi-target Bayesian recursion, a very general implementation method (i.e., the particle multi-target filter) is given, and the performance measure for multi-target tracking is introduced, which provides a specific evaluation criteria for weighing up the pros and cons of each algorithm. In addition, it is also a necessary element for higher-level sensor management.

W. Wu et al., *Target Tracking with Random Finite Sets*,
https://doi.org/10.1007/978-981-19-9815-7_3

3.2 RFS Statistics

This section aims to introduce in brief important concepts related to the RFS, such as statistical descriptors, first-order moment and and cardinality of the RFS.

3.2.1 *Statistical Descriptors of RFSs*

The RFS is a natural extension of the random vector (RV), which provides a natural way to deal with multi-sensor multi-target information fusion. The difference between the RFS and the RV is that the number of points is random, and the point itself is random and disordered. Generally speaking, an RFS is a random (spatial) point pattern, such as measurements on a radar screen. In fact, a RFS X is only a finite set-valued random variable, which can be fully described by a discrete probability distribution and a family of joint probability densities [392–394], wherein the discrete distribution characterizes the cardinality of RFS X, and for a given cardinality, the joint probability density characterizes the joint distribution of elements in RFS X.

The probability density function of an RFS is a natural generalization of the probability density function of a random vector. It is a non-negative function π on $\mathcal{F}(\mathbb{X})$ such that for any area $S \subseteq \mathbb{X}$,

$$P(X \subseteq S) = \int_{S} \pi(X)\delta X \tag{3.2.1}$$

The above equation establishes the relationship between the multi-target probability density $\pi(X)$ and the multi-target probability distribution. For any $X, \pi(X) \geq 0$ and the integral over the entire definition space $\mathbb{X}$ is 1, that is,

$$\int_{\mathbb{X}} \pi(X)\delta X = 1 \tag{3.2.2}$$

It should be noted that the integral in (3.2.1) and (3.2.2) is the set integral [6, 393] defined in (3.2.3), namely

$$\begin{aligned}\int \pi(X)\delta X &= \sum_{i=0}^{\infty} \frac{1}{i!} \int \pi(\{\boldsymbol{x}_1, \ldots, \boldsymbol{x}_i\}) d\boldsymbol{x}_1 \ldots d\boldsymbol{x}_i \\ &= \pi(\emptyset) + \sum_{i=1}^{\infty} \frac{1}{i!} \int \pi(\{\boldsymbol{x}_1, \ldots, \boldsymbol{x}_i\}) d\boldsymbol{x}_1 \ldots d\boldsymbol{x}_i \end{aligned} \tag{3.2.3}$$

The set integral sometimes takes the following form

$$\int \pi(X)\delta X = \sum_{i=0}^{\infty} \frac{1}{i!} \int \pi(\{\boldsymbol{x}_1, \ldots, \boldsymbol{x}_i\}) d(\boldsymbol{x}_1, \ldots, \boldsymbol{x}_i) \tag{3.2.4}$$

The book defaults that the set integral exists and is limited. Since $\pi(\emptyset)$ is a dimensionless probability, to make the summation operation in (3.2.3) well defined, it is required that the integral term of each summation in (3.2.3) must also be dimensionless. In this form, the dimension of $d\boldsymbol{x}_1 \ldots d\boldsymbol{x}_i$ is K^i, and thus the dimension of $\pi(\{\boldsymbol{x}_1, \ldots, \boldsymbol{x}_i\})$ is K^{-i}. According to the above analysis, although π is not a probability density, the function $\pi(X)K^{|X|}$ is a probability density, where $|X|$ represents the cardinality of RFS X (the number of elements in X).

The set representation $\pi(\{\boldsymbol{x}_1, \ldots, \boldsymbol{x}_n\})$ of a multi-target distribution can be expressed as a vector form $\pi(\boldsymbol{x}_1, \ldots, \boldsymbol{x}_n)$, and the relationship between the two representations is

$$\pi(\{\boldsymbol{x}_1, \ldots, \boldsymbol{x}_n\}) = n!\pi(\boldsymbol{x}_1, \ldots, \boldsymbol{x}_n) \tag{3.2.5}$$

This is because the probability of a finite set $\{\boldsymbol{x}_1, \ldots, \boldsymbol{x}_n\}$ is equally distributed over $n!$ possible vectors $(\boldsymbol{x}_{\sigma(1)}, \ldots, \boldsymbol{x}_{\sigma(n)})$, where σ is a permutation of numbers $1, \ldots, n$.

Another basic descriptor of the RFS is the probability generating functional (PGFl). The PGFl $G[\cdot]$ of X is defined as [392, 394]

$$G[h] \triangleq \mathrm{E}[h^X] \tag{3.2.6}$$

where $\mathrm{E}[\,\cdot\,]$ represents the expectation operator, h is any real-valued function on $\mathbb{X}$ that satisfies $0 \leq h(\boldsymbol{x}) \leq 1$, and h^X is the multi-object exponential (MoE), which is,

$$h^X \triangleq \prod_{\boldsymbol{x} \in X} h(\boldsymbol{x}) \tag{3.2.7}$$

and $h^{\emptyset} = 1$ is defined.

For the RFS, there is also a basic description: the belief mass function, in addition to the multi-target probability density and the PGFl. The belief mass function of an RFS is a natural generalization of the probability mass function (PMF) of a random vector. For a random vector $\boldsymbol{x}$ defined on space $\mathbb{X}$, its PMF is defined as

$$P(S) = P(\boldsymbol{x} \in S) \tag{3.2.8}$$

where S is any closed set on space $\mathbb{X}$. Similarly, for RFS X, its belief mass function is defined as

$$\beta(S) = P(X \subseteq S) = \int_S \pi(X)\delta X \tag{3.2.9}$$

The belief mass function is actually a special case of the PGFl, and the two have the following relationship

$$\beta(S) = \int_S \pi(X)\delta X = \int 1_S^X \pi(X)\delta X = G[1_S] \tag{3.2.10}$$

where $1_Y(X)$ is the generalized indicator function (GIF), also known as the inclusion function, defined as

$$1_Y(X) \triangleq \begin{cases} 1, \ if \ X \subseteq Y \\ 0, \ others \end{cases} \tag{3.2.11}$$

It is abbreviated as $1_Y(\{\,\boldsymbol{x}\}) \overset{\text{abbr.}}{=} 1_Y(\boldsymbol{x})$ when $X = \{\boldsymbol{x}\}$.

In addition, the belief mass function has the following relationship with the multi-target density function

$$\beta(S) = \int_S \pi(X)\delta X, \ \ \pi(X) = \frac{\partial \beta}{\partial X}(\emptyset) \tag{3.2.12}$$

Using the above-mentioned fundamental theorem of generalized calculus, the density of non-additive set function can be determined, and the relationship between set integrals and set derivatives can be established. Appendix B provides a strict definition of set derivatives from a more general perspective of functional derivatives.

In short, it can be seen from (3.2.10) and (3.2.12) that the three RFS statistical descriptors, namely, probability density function, PGFl, and belief mass function, actually contain the same information. In this sense, any one of these descriptors can be derived from any other one by using the multi-target calculus rule. However, although the three are equivalent in the mathematical sense, they each focus on different aspects of the multi-target problem.

The multi-target probability density function is the core of the multi-target Bayesian filter, which serves as the theoretical basis of multi-target detection, tracking, localization and recognition. The multi-target Bayesian filter propagates the multi-target probability density at different times, and the multi-target probability distribution at this time contains all the information regarding the number of targets and the states at the corresponding time.

The PGFl is similar to the integral transformation and are mainly used to transform complex mathematical problems into simple ones. Using the PGFl, it is convenient to deal with multi-target predictors, MB distributions and multi-target priorities [14].

The belief mass function is at the heart of multi-source, multi-target formal Bayesian modeling (see Sect. 3.6 for details). In particular, it plays a crucial role in the construction of the multi-target Markov density function based on the multi-target motion model and of the true multi-target likelihood function based on the multi-target measurement model. For example, given a multi-target measurement model, a specific expression of the belief mass function can be constructed, thus the multi-target likelihood function can be further obtained.

It should be noted that in addition to the above three RFS statistical descriptors, there are other descriptors, such as empty probability functional (or simply referred to as empty probability) [393-396] etc., which will not be detailed here.

3.2.2 Intensity and Cardinality of RFSs

In the tracking field, the intensity function [6, 62] is also known as the probability hypothesis density (PHD), a term first introduced in [396]. The intensity (PHD) v can be obtained from the multi-target density π, i.e., [6]

$$v(\boldsymbol{x}) = \int \pi(\{\boldsymbol{x}\} \cup X)\delta X \tag{3.2.13}$$

In addition, the PHD can also be obtained by differentiating the PGFl of the RFS. The integral of intensity v over any area S gives the expected number $\overline{n}$ of elements in RFS X in S, namely,

$$\overline{n} = \mathrm{E}\left[|X \cap S|\right] = \int_S v(\boldsymbol{x})d\boldsymbol{x} \tag{3.2.14}$$

In other words, for a given point $\boldsymbol{x}$, intensity $v(\boldsymbol{x})$ is the density of the expected number of targets per unit volume at $\boldsymbol{x}$.

The PHD is actually the first order moment of the RFS, similar to the expectation of a random vector. However, since the notion of addition does not exist for sets, the expectation of the RFS is meaningless. Thus, there is no multi-target expected posterior (EAP) version similar to the single-target EAP estimator. However, since the local maximum of intensity v is the point with the highest local concentration of the desired number of elements in S, it can be used to obtain an estimate of the elements in X. A simple multi-target estimator, the PHD filter (see Chap. 4 for details), can be obtained as follows: first round $\overline{n}$ to determine the number $\hat{n}$ of estimated states, and then select the $\hat{n}$ maxima of PHD v as the states to be estimated. In other words, let $\hat{n}$ be the integer closest to $\overline{n}$, and if the PHDs of $\hat{\boldsymbol{x}}_1, \ldots, \hat{\boldsymbol{x}}_{\hat{n}}$ are the $\hat{n}$ maxima of $v(\hat{\boldsymbol{x}}_1), \ldots, v(\hat{\boldsymbol{x}}_{\hat{n}})$, then $\hat{X} = \{\hat{\boldsymbol{x}}_1, \ldots, \hat{\boldsymbol{x}}_{\hat{n}}\}$ is the multi-target state estimation.

The cardinality of an RFS is also a very important concept. The cardinality $|X|$ of RFS X is a discrete random variable, whose probability distribution (namely, cardinality distribution) ρ is defined as

$$\rho(|X|) = \frac{1}{|X|!} \int_{\mathbb{X}^{|X|}} \pi(\{\boldsymbol{x}_1, \ldots, \boldsymbol{x}_{|X|}\}) d(\boldsymbol{x}_1, \ldots, \boldsymbol{x}_{|X|}) \tag{3.2.15}$$

The probability generating function (PGF) of cardinality $|X|$ is

$$G(z) = \sum_{j=0}^{\infty} \rho(j) z^j \tag{3.2.16}$$

It can be seen from the above equation that the cardinality distribution ρ is actually an inverse Z transform of PGF $G(\cdot)$, and $G(\cdot)$ can be obtained by substituting the constant function $h(\boldsymbol{x}) = z$ into the PGFl (3.2.6). It should be noted that the difference between PGF $G(\cdot)$ and PGFl $G[\cdot]$ is distinguished by parentheses and square brackets on the parameters, and Appendix C explains both from the perspective of integral transformation.

According to the cardinality distribution, another multi-target estimator, the CPHD filter (see Chap. 5 for details), can be obtained, which is basically the same as the previous PHD filter, but the estimation method of the number of states may be different. It can use EAP cardinality estimation $\hat{n} = \mathrm{E}[|X|]$ or MAP cardinality estimation $\hat{n} = \arg_n \max \rho(n)$ as the estimation of the number of targets, and then use the $\hat{n}$ highest local peaks of the PHD as the multi-target state estimation.

3.3 Main Categories of RFSs

The RFS can be divided into the labeled RFS and the label-free (or unlabeled) RFS according to whether it contains the label information. Main types of unlabeled RFSs include Poisson RFS, independent and identically distributed cluster (IIDC) RFS, Bernoulli RFS, and multi-Bernoulli (MB) RFS. For the labeled RFS, common types include labeled Poisson RFS, labeled IIDC RFS, labeled MB (LMB) RFS, generalized LMB (GLMB) RFS, and δ-GLMB RFS.

3.3.1 Poisson RFS

If cardinality $|X|$ of RFS X on $\mathbb{X}$ follows a Poisson distribution with mean $\langle v, 1\rangle$, and for any finite cardinality, the elements in X are independent and identically distributed (IID) according to the probability density $v(\cdot)/\langle v, 1\rangle$, then X is said to be a Poisson RFS [392, 394] with an intensity function v defined on $\mathbb{X}$, where $\langle f, g\rangle$ is the inner product, and $\langle f, g\rangle \triangleq \int f(\boldsymbol{x})g(\boldsymbol{x})d\boldsymbol{x}$ for any continuous functions, while $\langle f, g\rangle \triangleq \sum_{i=0}^{\infty} f(i)g(i)$ for discrete sequences. The Poisson RFS is a very important type of RFSs with unique properties, which can be fully described by intensity function v, that is, the probability density of a Poisson RFS can be explicitly expressed by v as [14]

$$\pi(X) = \exp(-\langle v, 1\rangle) \cdot v^X \tag{3.3.1}$$

The PGFl of a Poisson RFS with intensity function v is [14]

$$G[h] = \exp(\langle v, h - 1\rangle) \tag{3.3.2}$$

If the elements of the RFS are independently and identically distributed (IID) according to probability density v/N, and yet have an arbitrary cardinality distribution, such RFS is referred to as the IID cluster (IIDC) RFS. The Poisson RFS is a special case of the IIDC RFS with Poisson cardinality.

3.3.2 *IIDC RFS*

The elements contained in the IIDC RFS X are independent and identically distributed (IID) according to probability density p. The IIDC RFS can be fully characterized by probability density p and cardinality distribution ρ [392], i.e.,

$$\pi(\{\boldsymbol{x}_1, \ldots, \boldsymbol{x}_n\}) = n!\rho(n)\prod_{i=1}^{n} p(\boldsymbol{x}_i) \tag{3.3.3}$$

where $\pi(\emptyset) = \rho(0)$ is specified. Equation (3.3.3) can also be written in the following more concise form[1]

$$\pi(X) = |X|!\rho(|X|)p^X \tag{3.3.4}$$

The PGFl $G[\cdot]$ of the IIDC RFS is [14]

$$\begin{aligned} G[h] &= G(\langle p, h\rangle) = G(\langle v, h\rangle / \langle v, 1\rangle) \\ &= \sum_{n=0}^{\infty} (\langle v, h\rangle / \langle v, 1\rangle)^n \rho(n) \end{aligned} \tag{3.3.5}$$

The above equation employs the relation: $p = v/\langle v, 1\rangle$. In addition, there is a relationship between v and ρ, that is, $\sum_{n=0}^{\infty} n\rho(n) = \langle v, 1\rangle$. It should be noted that $G(\cdot)$ in (3.3.5) is the PGF defined by (3.2.16).

3.3.3 *Bernoulli RFS*

For a Bernoulli RFS defined on $\mathbb{X}$, the probability that it is an empty set is $1 - \varepsilon$, and the probability that it is a singleton (namely, a single element) is ε (i.e., the

[1] Throughout the book, the multi-target density is denoted by π, while the single-target density is denoted by p.

probability of existence), and the element obeys the distribution with probability density p (defined) on $\mathbb{X}$, that is, the cardinality distribution of the Bernoulli RFS is a Bernoulli distribution with parameter ε, and the probability density is [14]

$$\pi(X) = \begin{cases} 1-\varepsilon; & X=\emptyset \\ \varepsilon \cdot p(\boldsymbol{x}); & X=\{\boldsymbol{x}\} \end{cases} \tag{3.3.6}$$

The PGFl of the Bernoulli RFS is [14]

$$G[h] = 1-\varepsilon+\varepsilon\langle p, h\rangle \tag{3.3.7}$$

3.3.4 *Multi-Bernoulli RFS*

A multi-Bernoulli (MB) RFS is the union of multiple independent Bernoulli RFSs. Specifically, a MB RFS defined on $\mathbb{X}$ is the union of M independent Bernoulli RFSs $X^{(i)},\ i=1,\ldots,M$, i.e.,

$$X = \bigcup_{i=1}^{M} X^{(i)} \tag{3.3.8}$$

The probability density of the MB RFS X is [14]

$$\pi(\{\boldsymbol{x}_1,\ldots,\boldsymbol{x}_{|X|}\}) = \pi(\emptyset) \sum_{1\le i_1\ne\cdots\ne i_{|X|}\le M} \prod_{j=1}^{|X|} \frac{\varepsilon^{(i_j)} p^{(i_j)}(\boldsymbol{x}_j)}{1-\varepsilon^{(i_j)}} \tag{3.3.9}$$

where $\pi(\emptyset)=\prod_{j=1}^{M}(1-\varepsilon^{(j)})$, $\varepsilon^{(i)}\in(0,1)$ and $p^{(i)}$ respectively correspond to the probability of existence and probability density of Bernoulli RFS $X^{(i)}$. The cardinality of the MB RFS is a discrete MB random variable with parameter $\varepsilon^{(1)},\ldots,\varepsilon^{(M)}$, the mean of the cardinality is $\sum_{i=1}^{M}\varepsilon^{(i)}$, and the cardinality distribution is also in the MB form, which is [14]

$$\rho(|X|) = \pi(\emptyset) \sum_{1\le i_1\ne\cdots\ne i_{|X|}\le M} \prod_{j=1}^{|X|} \frac{\varepsilon^{(i_j)}}{1-\varepsilon^{(i_j)}} \tag{3.3.10}$$

The PHD of the MB RFS is

$$v(\boldsymbol{x}) = \sum_{i=1}^{M} \varepsilon^{(i)} p^{(i)}(\boldsymbol{x}) \tag{3.3.11}$$

Taking advantage of the independence of $X^{(i)}$, the PGFl of the MB RFS is [14]

$$G[h] = \prod_{i=1}^{M} (1 - \varepsilon^{(i)} + \varepsilon^{(i)} \langle p^{(i)}, h \rangle) \tag{3.3.12}$$

Equations (3.3.9) and (3.3.12) both indicate that the MB RFS can be completely described by the MB parameter set $\{(\varepsilon^{(i)}, p^{(i)})\}_{i=1}^{M}$. For convenience, the probability density in the form of (3.3.9) is abbreviated as $\pi = \{(\varepsilon^{(i)}, p^{(i)})\}_{i=1}^{M}$, and the multi-target density in the form of (3.3.9) or the PGFl in the form of (3.3.12) is also called the MB.

3.3.5 *Labeled RFSs*

3.3.5.1 Label and Labeled State

The reason why the RFS theory was once criticized in the tracking field is that it cannot provide identity (track) information for multi-target estimation, and the existence of this problem drives the development of labeled RFSs. To incorporate target tracks in the multi-target Bayesian filtering framework, targets can be uniquely identified by labels extracted from a discrete countable space $\mathbb{L} = \{\alpha_i : i \in \mathbb{N}\}$, wherein $\mathbb{N}$ represents a set of positive integers and α_i is different from each other. For target tracking, a better way of labeling is to distinguish targets through an ordered integer pair $\ell = (k, i)$, wherein k is the birth time and $i \in \mathbb{N}$ is the unique index to distinguish birth targets at the same time. Therefore, the label space of birth targets at time k is $\mathbb{L}_k = \{k\} \times \mathbb{N}$, and the label space of the targets at time k (including the label space of the targets born before time k) is denoted as $\mathbb{L}_{0:k}$. Since $\mathbb{L}_{0:k-1}$ and $\mathbb{L}_k$ are mutually exclusive, $\mathbb{L}_{0:k}$ can be constructed recursively by $\mathbb{L}_{0:k} = \mathbb{L}_{0:k-1} \cup \mathbb{L}_k$. The state of the target born at time k is $\mathbf{x} = (\boldsymbol{x}, \ell) \in \mathbb{X} \times \mathbb{L}_k$, composed of state $\boldsymbol{x}$ and label ℓ, which means that state $\boldsymbol{x}$ is augmented by label ℓ. Therefore, $\mathbf{x}$ is essentially an augmented state.

For the sake of simplicity, when there is no confusion, the timestamp subscript k of the label set is ignored: $\pi \triangleq \pi_k$, $\pi_+ \triangleq \pi_{k|k-1}$,$g \triangleq g_k$, and $\phi \triangleq \phi_{k|k-1}$ for short; let $\mathbb{L}$ represent the label space at the current moment, $\mathbb{B}$ the label space of birth targets at the next moment, and $\mathbb{L}_+ \triangleq \mathbb{L} \cup \mathbb{B}$ the label space at the next moment. It should be noted that $\mathbb{L}$ and $\mathbb{B}$ are mutually exclusive, thus, $\mathbb{L} \cap \mathbb{B} = \emptyset$.

3.3.5.2 Conventions About Related Symbols

To distinguish the labeled RFS from the unlabeled RFS, the following conventions are adopted: the track label is denoted by ℓ, the single-target state is denoted by lowercase letters, such as $\boldsymbol{x}$ and $\mathbf{x}$, while the multi-target state is denoted by uppercase letters, such as X and X. Besides, the unlabeled single-target state and multi-target state are respectively represented by italic $\boldsymbol{x}$ and X, whereas the corresponding labeled states are respectively represented by normal $\mathbf{x}$ and X. Thus, the multi-target state X at time k (including before time k) is a finite subset of the Cartesian product $\mathbb{X} \times \mathbb{L}_{0:k}$. In addition, the symbols of (single-target or multi-target) distribution/statistics corresponding to the labeled state are the same as the unlabeled version, yet can be distinguished according to the corresponding parameters. For example, for the unlabeled version, the single-target probability density, multi-target density, PHD (or multi-target intensity), single-target state transition density, single-target likelihood function, multi-target state transition density, multi-target likelihood function, and multi-target cardinality distribution are respectively expressed as $p(\boldsymbol{x})$, $\pi(X)$, $v(\boldsymbol{x})$, $\phi(\boldsymbol{x}|\boldsymbol{x}')$, $g(z|\boldsymbol{x})$, $\phi(X|X')$, $g(Z|X)$, and $\rho(|X|)$, whereas, the corresponding labeled versions are respectively represented by $p(\mathbf{x})$, $\pi(\mathrm{X})$, $v(\mathbf{x})$, $\phi(\mathbf{x}|\mathbf{x}')$, $g(z|\mathbf{x})$, $\phi\big(\mathrm{X}\big|\mathrm{X}'\big)$, $g(Z|\mathrm{X})$, and $\rho(|\mathrm{X}|)$, where, z represents the observation vector, and Z represents the observation set. Furthermore, various spaces are represented in blackboard bold, such as state space $\mathbb{X}$, measurement space $\mathbb{Z}$, label space $\mathbb{L}$, natural number space $\mathbb{N}$, model space $\mathbb{M}$, etc.

3.3.5.3 Labeled RFS and Related Formulas

Definition 1 Let $\mathcal{L} : \mathbb{X} \times \mathbb{L} \to \mathbb{L}$ be the label projection function $\mathcal{L}(\boldsymbol{x}, \ell) = \ell$, $\mathcal{L}(\mathrm{X}) = \{\mathcal{L}(\mathbf{x}) : \mathbf{x} \in \mathrm{X}\}$ denote the set of labels in X, and the labeled RFS with state space $\mathbb{X}$ and label space $\mathbb{L}$ is the RFS on $\mathbb{X} \times \mathbb{L}$ that satisfies the condition that each realization X has labels different from each other, namely,

$$|\mathcal{L}(\mathrm{X})| = |\mathrm{X}| \tag{3.3.13}$$

The above formula shows that if and only if X and its label set $\mathcal{L}(\mathrm{X}) = \{\mathcal{L}(\mathbf{x}) : \mathbf{x} \in \mathrm{X}\}$ have the same cardinality (namely, $\delta_{|\mathrm{X}|}(|\mathcal{L}(\mathrm{X})|) = 1$), the finite subset X in space $\mathbb{X} \times \mathbb{L}$ has mutually different labels, where $\delta_y(x)$ is the Dirac delta function defined by (2.9.2), and $\delta_Y(X)$ is the generalized Kronecker delta function, defined as

$$\delta_Y(X) \triangleq \begin{cases} 1, & \text{if } X = Y \\ 0, & \text{others} \end{cases} \tag{3.3.14}$$

The function $\Delta(\mathrm{X})$ is called the distinct label indicator (DLI), which is defined as

$$\Delta(\mathrm{X}) \triangleq \delta_{|\mathrm{X}|}(|\mathcal{L}(\mathrm{X})|) \tag{3.3.15}$$

That is, if the labels of X are different from each other, the value is 1, otherwise, the value is 0.

The unlabeled version of the labeled RFS is a projection from $\mathbb{X} \times \mathbb{L}$ to $\mathbb{X}$, which can be obtained by simply discarding the labels of the labeled RFS. In fact, the unlabeled version of the labeled RFS obeying distribution π obeys the following marginal distribution

$$\pi(\{\boldsymbol{x}_1, \ldots, \boldsymbol{x}_n\}) = \sum_{(\ell_1,\ldots,\ell_n)\in\mathbb{L}^n} \pi(\{\mathbf{x}_1, \ldots, \mathbf{x}_n\}) = \sum_{(\ell_1,\ldots,\ell_n)\in\mathbb{L}^n} \pi(\{(\boldsymbol{x}_1, \ell_1), \ldots, (\boldsymbol{x}_n, \ell_n)\}) \tag{3.3.16}$$

In the context of the labeled RFS, the set integral is also different from that of the unlabeled RFS defined in (3.2.3), and needs to be modified appropriately. Since the label space $\mathbb{L}$ is discrete, the integral of function $f : \mathbb{X} \times \mathbb{L} \to \mathbb{R}$ is corrected to

$$\int f(\mathbf{x})d\mathbf{x} = \sum_{\ell\in\mathbb{L}} \int_{\mathbb{X}} f(\boldsymbol{x}, \ell)d\boldsymbol{x} \tag{3.3.17}$$

Similarly, the set integral of function $f : \mathcal{F}(\mathbb{X} \times \mathbb{L}) \to \mathbb{R}$ becomes

$$\begin{aligned}\int f(\mathrm{X})\delta\mathrm{X} &= \sum_{n=0}^{\infty} \frac{1}{n!} \int f(\{\mathbf{x}_1, \ldots, \mathbf{x}_n\})d(\mathbf{x}_1, \ldots, \mathbf{x}_n)\\ &= \sum_{n=0}^{\infty} \frac{1}{n!} \sum_{(\ell_1,\ldots,\ell_n)\in\mathbb{L}^n} \int_{\mathbb{X}^n} f(\{(\boldsymbol{x}_1, \ell_1), \ldots, (\boldsymbol{x}_n, \ell_n)\})d(\boldsymbol{x}_1, \ldots, \boldsymbol{x}_n)\end{aligned} \tag{3.3.18}$$

When calculating the set integral involving the labeled RFS, the following Lemma 2 is often employed. Before introducing it, we first introduce Lemma 1, which is commonly used when dealing with the labeled RFS.

Lemma 1 *If $f : \mathbb{L}^n \to \mathbb{R}$ is a symmetric function, that is, its value at any n-tuple parameter is the same as its value at any permutation of that n-tuple, then*

$$\sum_{(\ell_1,\ldots,\ell_n)\in\mathbb{L}^n} \delta_n(|\{\ell_1, \ldots, \ell_n\}|) f(\ell_1, \ldots, \ell_n) = n! \sum_{\{\ell_1,\ldots,\ell_n\}\in\mathcal{F}_n(\mathbb{L})} f(\ell_1, \ldots, \ell_n) \tag{3.3.19}$$

Due to the presence of the term $\delta_n(|\{\ell_1, \ldots, \ell_n\}|)$, the summation in the above equation turns into the summation over indices in $\mathbb{L}^n$ with different components. The symmetry of f means that all $n!$ permutations of $(\ell_1, \ldots, \ell_n)$ have the same function value $f(\ell_1, \ldots, \ell_n)$. Furthermore, all $n!$ permutations of $(\ell_1, \ldots, \ell_n)$ with different components define an equivalent equation $\{\ell_1, \ldots, \ell_n\} \in \mathcal{F}_n(\mathbb{L})$, in which, $\mathcal{F}_n(\mathbb{L})$ represents the set of finite subsets of $\mathbb{L}$ containing exactly n elements.

Lemma 2 *Let $\Delta(\mathrm{X})$ denote the distinct label indicator $\delta_{|\mathrm{X}|}(|\mathcal{L}(\mathrm{X})|)$, for the integrable functions $h : \mathcal{F}(\mathbb{L}) \to \mathbb{R}$ and $g : \mathbb{X} \times \mathbb{L} \to \mathbb{R}$ on $\mathbb{X}$, we have (see Appendix*

D for proof)

$$\int \Delta(\mathrm{X})h(\mathcal{L}(\mathrm{X}))g^{\mathrm{X}}\delta\mathrm{X} = \sum_{L\subseteq\mathbb{L}} h(L)\left[\int g(\boldsymbol{x},\cdot)d\boldsymbol{x}\right]^{L} \tag{3.3.20}$$

More generally, when g has no separable form, the following lemma can be obtained.

Lemma 3 *Let $\mathbb{X}$ be the state space and $\mathbb{L}$ be the discrete label space, then, for function $h : \mathcal{F}(\mathbb{L}) \to \mathbb{R}$ and integrable function $g : \mathcal{F}(\mathbb{L} \times \mathbb{X}) \to \mathbb{R}$, we have*

$$\begin{aligned}
&\int \Delta(\mathrm{X})h(\mathcal{L}(\mathrm{X}))g(\mathrm{X})\delta\mathrm{X} = \int \delta_{|\mathrm{X}|}(|\mathcal{L}(\mathrm{X})|)h(\mathcal{L}(\mathrm{X}))g(\mathrm{X})\delta\mathrm{X} \\
&= \sum_{n=0}^{\infty}\frac{1}{n!}\sum_{(\ell_1,\ldots,\ell_n)\in\mathbb{L}^n} \delta_n(|\{\ell_1,\ldots,\ell_n\}|)h(\{\ell_1,\ldots,\ell_n\})\int g(\{(\boldsymbol{x}_1,\ell_1),\ldots,(\boldsymbol{x}_n,\ell_n)\})d\boldsymbol{x}_{1:n} \\
&= \sum_{n=0}^{\infty}\sum_{\{\ell_1,\ldots,\ell_n\}\in\mathcal{F}_n(\mathbb{L})} h(\{\ell_1,\ldots,\ell_n\})g_{\{\ell_{1:n}\}} = \sum_{L\subseteq\mathbb{L}} h(L)g_L
\end{aligned} \tag{3.3.21}$$

where $\{\ell_{1:n}\} \triangleq \{\ell_1,\ldots,\ell_n\}$, and

$$g_{\{\ell_{1:n}\}} = \int g(\{(\boldsymbol{x}_1,\ell_1),\ldots,(\boldsymbol{x}_n,\ell_n)\})d\boldsymbol{x}_{1:n} \tag{3.3.22}$$

The above lemma generalizes Lemma 2 to the case where g is inseparable.

Similar to the unlabeled RFS, two important statistics for the labeled RFS are the PHD and cardinality distribution.

The PHD of the labeled RFS is [152]

$$\begin{aligned}
v(\mathbf{x}) = v(\boldsymbol{x},\ell) &= \int \pi(\{\boldsymbol{x}\}\cup\mathrm{X})\delta\mathrm{X} \\
&= \int \pi(\{(\boldsymbol{x},\ell)\}\cup\mathrm{X})\delta\mathrm{X}
\end{aligned} \tag{3.3.23}$$

and the PHDs of the labeled RFS and unlabeled RFS have the following relation

$$v(\boldsymbol{x}) = \sum_{\ell\in\mathbb{L}} v(\mathbf{x}) = \sum_{\ell\in\mathbb{L}} v(\boldsymbol{x},\ell) \tag{3.3.24}$$

The cardinality distribution of any RFS X on $\mathbb{X}\times\mathbb{L}$ is [14]

$$\rho(|\mathrm{X}|) = \frac{1}{|\mathrm{X}|!}\int_{(\mathbb{X}\times\mathbb{L})^{|\mathrm{X}|}} \pi\big(\{\mathbf{x}_1,\ldots,\mathbf{x}_{|\mathrm{X}|}\}\big)d\big(\mathbf{x}_1,\ldots,\mathbf{x}_{|\mathrm{X}|}\big) \tag{3.3.25}$$

According to (3.3.16), it is easy to obtain that the cardinality distribution $\rho(|\mathrm{X}|)$ of the labeled RFS is equal to its unlabeled version $\rho(|X|)$ (3.2.15).

Next, the main types of the labeled RFSs are introduced.

3.3.6 Labeled Poisson RFS

A labeled Poisson RFS X with state space $\mathbb{X}$ and label space $\mathbb{L} = \{(\alpha_i : i \in \mathbb{N})\}$ is a Poisson RFS with an appended label $\alpha_i \in \mathbb{L}$ and with an intensity of v on $\mathbb{X}$. A sample of the labeled Poisson RFS can be generated according to the steps listed in Table 3.1.

The process always generates a finite set of augmented states with distinct labels. Intuitively, the set of unlabeled states generated during the above process is a Poisson RFS with intensity v. However, the set of labeled states is not a Poisson RFS on $\mathbb{X}\times\mathbb{L}$. In fact, its density is

$$\begin{aligned}\pi(\{\mathbf{x}_1, \ldots, \mathbf{x}_n\}) &= \pi(\{(\boldsymbol{x}_1, \ell_1), \ldots, (\boldsymbol{x}_n, \ell_n)\}) \\ &= \delta_{\mathbb{L}(n)}(\{\ell_1, \ldots, \ell_n\})\mathcal{PS}(n; \langle v, 1\rangle) \\ &\prod_{i=1}^{n} v(\boldsymbol{x}_i)/\langle v, 1\rangle \end{aligned} \tag{3.3.26}$$

where $\mathcal{PS}(n; \lambda) = \exp(-\lambda)\lambda^n/n!$ represents a Poisson distribution with ratio λ, and $\mathbb{L}(n) = \{\alpha_i \in \mathbb{L}\}_{i=1}^{n}$.

To verify the density (3.3.26) of the labeled Poisson RFS, note that the likelihood (probability density) of points $(\boldsymbol{x}_1, \ell_1), \ldots, (\boldsymbol{x}_n, \ell_n)$ generated in a certain order during the above process is

$$\begin{aligned} p((\boldsymbol{x}_1, \ell_1), \ldots, (\boldsymbol{x}_n, \ell_n)) &= \delta_{(\alpha_1,\ldots,\alpha_n)}((\ell_1, \ldots, \ell_n))\mathcal{PS}(n; \langle v, 1\rangle) \\ &\prod_{i=1}^{n} v(\boldsymbol{x}_i)/\langle v, 1\rangle \end{aligned} \tag{3.3.27}$$

According to [392], $\pi(\{(\boldsymbol{x}_1, \ell_1), \ldots, (\boldsymbol{x}_n, \ell_n)\})$ is symmetric with $p((\boldsymbol{x}_1, \ell_1), \ldots, (\boldsymbol{x}_n, \ell_n))$ over all permutations σ of $\{1, \ldots, n\}$, i.e.,

Table 3.1 Sampling from the labeled Poisson distribution

1	Initialization: $\mathrm{X} = \emptyset$
2	Sampling $n \sim \mathcal{PS}(\cdot; \langle v, 1\rangle)$
3	for $i = 1 : n$
4	Sampling $\boldsymbol{x} \sim v(\cdot)/\langle v, 1\rangle$
5	Set $\mathrm{X} = \mathrm{X} \cup \{(\boldsymbol{x}, \alpha_i)\}$
6	end

$$\pi(\{(\boldsymbol{x}_1, \ell_1), \ldots, (\boldsymbol{x}_n, \ell_n)\}) = \sum_{\sigma} p\big((\boldsymbol{x}_{\sigma(1)}, \ell_{\sigma(1)}), \ldots, (\boldsymbol{x}_{\sigma(n)}, \ell_{\sigma(n)})\big)$$
$$= \mathcal{PS}(n; \langle v, 1\rangle) \sum_{\sigma} \left(\prod_{i=1}^{n} \frac{v(\boldsymbol{x}_{\sigma(i)})}{\langle v, 1\rangle} \right)$$
$$\delta_{(\alpha_i, \ldots, \alpha_n)}\big((\ell_{\sigma(1)}, \ldots, \ell_{\sigma(n)})\big) \tag{3.3.28}$$

where the sums over the permutation σ are all 0. The only exception is $\{\ell_1, \ldots, \ell_n\} = \{\alpha_1, \ldots, \alpha_n\}$, in which case there is a permutation σ such that $(\ell_{\sigma(1)}, \ldots, \ell_{\sigma(n)}) = (\alpha_1, \ldots, \alpha_n)$. Thus, (3.3.26) can be obtained.

To verify that the unlabeled version of the labeled Poisson RFS is indeed the Poisson RFS, (3.3.26) is substituted into (3.3.16), thus (3.3.1) can be obtained by simplifying the sum on the labels.

Remark The labeled Poisson RFS can be generalized to the labeled IIDC RFS by removing the Poisson assumption on the cardinality distribution and specifying an arbitrary cardinality distribution.

3.3.7 Labeled IIDC RFS

The labeled IIDC RFS generalizes the Poisson assumption of the cardinality of the labeled Poisson RFS to a general cardinality distribution $\rho(|\mathrm{X}|)$ and obeys the following distribution

$$\pi(\mathrm{X}) = \pi\left(\{\mathbf{x}_1, \ldots, \mathbf{x}_{|\mathrm{X}|}\}\right) = \pi\left(\{(x_1, \ell_1), \ldots, (\boldsymbol{x}_{|\mathrm{X}|}, \ell_{|\mathrm{X}|})\}\right)$$
$$= \delta_{\mathbb{L}(|\mathrm{X}|)}\left(\{\ell_1, \ldots, \ell_{|\mathrm{X}|}\}\right)\rho(|\mathrm{X}|) \prod_{i=1}^{|\mathrm{X}|} v(\boldsymbol{x}_i)/\langle v, 1\rangle \tag{3.3.29}$$

3.3.8 LMB RFS

Similar to the MB RFS, the LMB RFS can be fully described by a parameter set $\{\varepsilon^{(c)}, p^{(c)} : c \in C\}$ with an index set C. The LMB RFS X with state space $\mathbb{X}$, label space $\mathbb{L}$, and (finite) parameter set $\{\varepsilon^{(c)}, p^{(c)} : c \in C\}$ is a MB RFS corresponding to (non-empty) Bernoulli components on $\mathbb{X}$ and attached with labels. Thus, if the Bernoulli component $\{\varepsilon^{(c)}, p^{(c)}\}$ generates a non-empty set, the label of the corresponding state is given by $\alpha(c)$, where $\alpha : C \to \mathbb{L}$ is a one-to-one map. Table 3.2 illustrates how to generate the LMB RFS samples, wherein $\mathcal{U}$ represents a uniform distribution.

Table 3.2 Sampling of the LMB RFS

1	Initialization: $\mathrm{X} = \emptyset$
2	for $c \in C$
3	Sampling $u \sim \mathcal{U}[0, 1]$
4	if $u \leq \varepsilon^{(c)}$
5	Sampling $\boldsymbol{x} \sim p^{(c)}(\cdot)$ Set $\mathrm{X} = \mathrm{X} \cup \{(\boldsymbol{x}, \alpha(c))\}$
6	end
7	end

Obviously, the above process always generates a finite set of augmented states with distinct labels. Intuitively, the set of unlabeled states is the MB RFS. However, the set of labeled states is not a MB RFS on $\mathbb{X} \times \mathbb{L}$. In fact, its density is

$$\begin{aligned}\pi(\{\mathbf{x}_1, \ldots, \mathbf{x}_n\}) &= \pi(\{(\boldsymbol{x}_1, \ell_1), \ldots, (\boldsymbol{x}_n, \ell_n)\}) \\ &= \delta_n(|\{\ell_1, \ldots, \ell_n\}|) \prod_{c \in C} \left(1 - \varepsilon^{(c)}\right) \prod_{j=1}^{n} \frac{1_{\alpha(C)}(\ell_j) \varepsilon^{(\alpha^{-1}(\ell_j))} p^{(\alpha^{-1}(\ell_j))}(\boldsymbol{x}_j)}{1 - \varepsilon^{(\alpha^{-1}(\ell_j))}}\end{aligned} \tag{3.3.30}$$

To verify the density of the LMB RFS (3.3.30), consider the likelihood of points $(\boldsymbol{x}_1, \ell_1), \ldots, (\boldsymbol{x}_n, \ell_n)$ generated during the above process in a certain order

$$\begin{aligned}p((\boldsymbol{x}_1, \ell_1), \ldots, (\boldsymbol{x}_n, \ell_n)) = \delta_n(|\{\ell_1, \ldots, \ell_n\}|) \operatorname{ord}(\ell_1, \ldots, \ell_n) \prod_{c \in C} \left(1 - \varepsilon^{(c)}\right) \\ \times \prod_{j=1}^{n} \frac{1_{\alpha(C)}(\ell_j) \varepsilon^{(\alpha^{-1}(\ell_j))} p^{(\alpha^{-1}(\ell_j))}(\boldsymbol{x}_j)}{1 - \varepsilon^{(\alpha^{-1}(\ell_j))}}\end{aligned} \tag{3.3.31}$$

where $\operatorname{ord}(\ell_1, \ldots, \ell_n)$ is 1 when $\ell_1 < \cdots < \ell_n$, otherwise it is 0. Because $\mathbb{L}$ is discrete, the order of its elements can always be defined. According to [392], $\pi(\{(\boldsymbol{x}_1, \ell_1), \ldots, (\boldsymbol{x}_n, \ell_n)\})$ is symmetric with $p((\boldsymbol{x}_1, \ell_1), \ldots, (\boldsymbol{x}_n, \ell_n))$ cver all permutations σ of $\{1, \ldots, n\}$, i.e.,

$$\begin{aligned}\pi(\{(\boldsymbol{x}_1, \ell_1), \ldots, (\boldsymbol{x}_n, \ell_n)\}) = \sum_{\sigma} \delta_n\left(\left|\left\{\ell_{\sigma(1)}, \ldots, \ell_{\sigma(n)}\right\}\right|\right) \operatorname{ord}\left(\ell_{\sigma(1)}, \ldots, \ell_{\sigma(n)}\right) \\ \prod_{c \in C} \left(1 - \varepsilon^{(c)}\right) \\ \times \prod_{j=1}^{n} \frac{1_{\alpha(C)}\left(\ell_{\sigma(j)}\right) \varepsilon^{(\alpha^{-1}(\ell_{\sigma(j)}))} p^{(\alpha^{-1}(\ell_{\sigma(j)}))}\left(\boldsymbol{x}_{\sigma(j)}\right)}{1 - \varepsilon^{(\alpha^{-1}(\ell_{\sigma(j)}))}}\end{aligned} \tag{3.3.32}$$

If labels are distinct, there is only one permutation in the correct order such that the sum over σ degenerates to one item. In addition, since $\delta(|\{\ell_1, \ldots, \ell_n\}|)$ and the products on n are symmetric (permutation invariant) with respect to $(\boldsymbol{x}_1, \ell_1), \ldots, (\boldsymbol{x}_n, \ell_n)$, (3.3.30) can be obtained.

Through rearrangement, another compact form of the LMB density can be obtained

$$\pi(\mathrm{X}) = \Delta(\mathrm{X}) 1_{\alpha(C)}(\mathcal{L}(\mathrm{X}))[\Phi(\mathrm{X}; \cdot)]^{C} \tag{3.3.33}$$

where $\Delta(\mathrm{X})$ is the distinct label indicator defined in (3.3.15), $\mathcal{L}(X)$ represents the labels of X defined in Definition 1, $1_Y(X)$ is the generalized indicator function (GIF) defined in (3.2.11), and

$$\begin{aligned}\Phi(\mathrm{X}; c) &= \sum_{(\boldsymbol{x},\ell)\in\mathrm{X}} \delta_{\alpha(c)}(\ell)\varepsilon^{(c)} p^{(c)}(\boldsymbol{x}) + \left[1 - 1_{\mathcal{L}(\mathrm{X})}(\alpha(c))\right]\left(1 - \varepsilon^{(c)}\right) \\ &= \begin{cases} 1 - \varepsilon^{(c)}, & \text{if } \alpha(c) \notin \mathcal{L}(\mathrm{X}) \\ \varepsilon^{(c)} p^{(c)}(\boldsymbol{x}), & \text{if } (\boldsymbol{x}, \alpha(c)) \in \mathrm{X} \end{cases}\end{aligned} \tag{3.3.34}$$

Remark Note that the above summation has only one non-zero term. When $\alpha(c) \notin \mathcal{L}(\mathrm{X})$, $\Phi(\mathrm{X}; \cdot)$ is $1 - \varepsilon^{(c)}$; or when $(\boldsymbol{x}, \alpha(c)) \in \mathrm{X}$, $\Phi(\mathrm{X}; \cdot)$ is $\varepsilon^{(c)} p^{(c)}(x)$. Therefore, the two expressions of $\Phi(\mathrm{X}; \cdot)$ are equivalent. The first expression of $\Phi(\mathrm{X}; \cdot)$ is often used in the multi-target Chapman-Kolmogorov (C-K) equation.

For convenience, the density of the LMB RFS given by (3.3.30) is often expressed in the abbreviated form $\pi = \left\{\varepsilon^{(c)}, p^{(c)}\right\}_{c\in C}$. Although the density of the LMB RFS allows a general map α of labels, in target tracking, it is usually assumed that α is a map of identities. Therefore, the superscript representing the component index corresponds directly to the label ℓ. Thus, the density of the LMB RFS with parameter set $\pi = \left\{\varepsilon^{(\ell)}, p^{(\ell)}\right\}_{\ell\in\mathbb{L}}$ has the following more compact form

$$\pi(\mathrm{X}) = \Delta(\mathrm{X}) w(\mathcal{L}(\mathrm{X})) p^{\mathrm{X}} \tag{3.3.35}$$

where

$$w(L) = \prod_{\ell'\in\mathbb{L}} \left(1 - \varepsilon^{(\ell')}\right) \prod_{\ell\in L} \frac{1_{\mathbb{L}}(\ell)\varepsilon^{(\ell)}}{1 - \varepsilon^{(\ell)}} \tag{3.3.36}$$

$$p(\boldsymbol{x}, \ell) = p^{(\ell)}(\boldsymbol{x}) \tag{3.3.37}$$

To verify that the unlabeled version of the LMB RFS is indeed a MB RFS, substituting (3.3.30) into (3.3.16) and simplifying the sum over the labels will yield (3.3.9).

3.3.9 GLMB RFS

The GLMB is a generalization of the LMB, which is closed under the multi-target C-K equation concerning the multi-target transition kernel and conjugate with respect to the multi-target likelihood function.

Definition 2 The GLMB RFS is a labeled RFS that obeys the following distribution on state-space $\mathbb{X}$ and (discrete) label space $\mathbb{L}$

$$\pi(\mathrm{X}) = \Delta(\mathrm{X}) \sum_{c\in\mathbb{C}} w^{(c)}(\mathcal{L}(\mathrm{X}))\left[p^{(c)}\right]^{\mathrm{X}} \tag{3.3.38}$$

where $\mathbb{C}$ is the set of discrete indices, and non-negative weight $w^{(c)}(L)$ and probability density $p^{(c)}$ satisfy respectively

$$\sum_{L\subseteq\mathbb{L}} \sum_{c\in\mathbb{C}} w^{(c)}(L) = \sum_{(L,c)\in\mathcal{F}(\mathbb{L})\times\mathbb{C}} w^{(c)}(L) = 1 \tag{3.3.39}$$

$$\int p^{(c)}(\boldsymbol{x}, \ell)d\boldsymbol{x} = 1 \tag{3.3.40}$$

The GLMB can be understood as a mixture of the multi-target exponentials (MoE) [69]. Each item in the mixture (3.3.38) consists of weight $w^{(c)}(\mathcal{L}(\mathrm{X}))$ and MoE $\left[p^{(c)}\right]^{\mathrm{X}}$, where the weight depends only on the label of the multi-target state, whereas the MoE depends on the entire multi-target state. The elements of the GLMB RFS are not statistically independent.

The following conclusions give the PHD and cardinality distribution of the GLMB RFS, with proof as shown in Appendix D.

Proposition 1 *The PHD of the GLMB RFS is*

$$v(\boldsymbol{x}, \ell) = \sum_{c\in\mathbb{C}} p^{(c)}(\boldsymbol{x}, \ell) \sum_{L\subseteq\mathbb{L}} 1_L(\ell) w^{(c)}(L) \tag{3.3.41}$$

and the PHD of the corresponding unlabeled version is

$$v(\boldsymbol{x}) = \sum_{c\in\mathbb{C}} \sum_{\ell\in\mathbb{L}} p^{(c)}(\boldsymbol{x}, \ell) \sum_{L\subseteq\mathbb{L}} 1_L(\ell) w^{(c)}(L) \tag{3.3.42}$$

It should be noted that the integral of each $p^{(c)}(\cdot, \ell)$ is 1, and $\sum_{L\subseteq\mathbb{L}}$ is equivalent to $\sum_{|\mathrm{X}|=0}^{\infty} \sum_{L\in\mathcal{F}_{|\mathrm{X}|}(\mathbb{L})}$. Then, the following can be obtained by exchanging the order of summation

$$\int v(\boldsymbol{x})d\boldsymbol{x} = \sum_{|\mathrm{X}|=0}^{\infty} \sum_{c\in\mathbb{C}} \sum_{L\in\mathcal{F}_{|\mathrm{X}|}(\mathbb{L})} \left[\sum_{\ell\in\mathbb{L}} 1_L(\ell)\right] w^{(c)}(L)$$

$$= \sum_{|\mathrm{X}|=0}^{\infty} \sum_{c \in \mathbb{C}} \sum_{L \in \mathcal{F}_{|\mathrm{X}|}(\mathbb{L})} |\mathrm{X}| \cdot w^{(c)}(L) = \sum_{|\mathrm{X}|=0}^{\infty} |\mathrm{X}| \cdot \rho(|\mathrm{X}|) \quad (3.3.43)$$

The above equation shows that the PHD mass is equal to the average cardinality.

Proposition 2 *The cardinality distribution of the GLMB RFS is*

$$\rho(|\mathrm{X}|) = \sum_{L \in \mathcal{F}_{|\mathrm{X}|}(\mathbb{L})} \sum_{c \in \mathbb{C}} w^{(c)}(L) = \sum_{L \in \mathcal{F}(\mathbb{L})} \sum_{c \in \mathbb{C}} \delta_{|\mathrm{X}|}(|L|) w^{(c)}(L) \quad (3.3.44)$$

According to (3.3.39), *it is easy to verify that* Eq. (3.3.44) *is indeed a probability distribution, i.e.,*

$$\sum_{|\mathrm{X}|=0}^{\infty} \rho(|\mathrm{X}|) = \sum_{|\mathrm{X}|=0}^{\infty} \sum_{L \in \mathcal{F}_{|\mathrm{X}|}(\mathbb{L})} \sum_{c \in \mathbb{C}} w^{(c)}(L) = \sum_{L \subseteq \mathbb{L}} \sum_{c \in \mathbb{C}} w^{(c)}(L) = 1 \quad (3.3.45)$$

In addition, the integral of the probability density in (3.3.38) *is* 1, *namely,*

$$\begin{aligned}\int \pi(\mathrm{X}) \delta(\mathrm{X}) &= \sum_{|\mathrm{X}|=0}^{\infty} \frac{1}{|\mathrm{X}|!} \int_{(\mathbb{L} \times \mathbb{X})^{|\mathrm{X}|}} \pi\left(\{\mathbf{x}_1, \ldots, \mathbf{x}_{|\mathrm{X}|}\}\right) d\left(\mathbf{x}_1, \ldots, \mathbf{x}_{|\mathrm{X}|}\right) \\ &= \sum_{|\mathrm{X}|=0}^{\infty} \rho(|\mathrm{X}|) = 1 \end{aligned} \quad (3.3.46)$$

Since $\Delta(\mathrm{X}) = \delta_{|\mathrm{X}|}(|\mathcal{L}(\mathrm{X})|)$ *is the distinct label indicator, the RFS defined in* (3.3.38) *is indeed a labeled RFS.*

The GLMB RFS actually includes the labeled Poisson RFS and the LMB RFS. For the density (3.3.26) of the labeled Poisson RFS, note that $\delta_{\mathbb{L}(|\mathrm{X}|)}(\mathcal{L}(\mathrm{X})) = \Delta(\mathrm{X})\delta_{\mathbb{L}(|\mathcal{L}(\mathrm{X})|)}(\mathcal{L}(\mathrm{X}))$, and thus

$$\pi(\mathrm{X}) = \Delta(\mathrm{X})\delta_{\mathbb{L}(|\mathcal{L}(\mathrm{X})|)}(\mathcal{L}(\mathrm{X}))\mathcal{PS}(|\mathcal{L}(\mathrm{X})|; \langle v, 1\rangle) \prod_{(\boldsymbol{x},\ell) \in \mathrm{X}} v(\boldsymbol{x})/\langle v, 1\rangle \quad (3.3.47)$$

Therefore, the labeled Poisson RFS is a special case of the GLMB RFS with the following weight and density

$$w^{(c)}(L) = \delta_{\mathbb{L}(|L|)}(L)\mathcal{PS}(|L|; \langle v, 1\rangle) \quad (3.3.48)$$

$$p^{(c)}(\boldsymbol{x}, \ell) = v(\boldsymbol{x})/\langle v, 1\rangle \quad (3.3.49)$$

The LMB RFS obeying the distribution of Eq. (3.3.30) is also a special case of the GLMB RFS with the following weight and density

$$w^{(c)}(L) = \prod_{\ell' \in \mathbb{L}} \left(1 - \varepsilon^{(\ell)}\right) \prod_{\ell \in L} \frac{1_{\mathbb{L}}(\ell)\varepsilon^{(\ell)}}{1 - \varepsilon^{(\ell)}} \tag{3.3.50}$$

$$p^{(c)}(\boldsymbol{x}, \ell) = p^{(\ell)}(\boldsymbol{x}) \tag{3.3.51}$$

Note that in (3.3.49) and (3.3.51), there is only one element in the index space $\mathbb{C}$, which means that the labeled Poisson density and the LMB density are special cases of the GLMB density with only one item. In this case, the superscript (c) is unnecessary.

3.3.10 δ-GLMB RFS

Although the GLMB family is closed under the Bayesian recursion, its numerical implementation is not easy. The δ-GLMB RFS is a special case of the GLMB RFS class, which is also closed under the Chapman-Kolmogorov equation and Bayesian rule. With the δ-GLMB RFS, the computation and storage required for the prediction and update steps can be further reduced, and it delivers an expression that is easy to realize numerically. In this sense, it is very suitable for multi-target tracking.

Definition 3 The δ-GLMB RFS with state-space $\mathbb{X}$ and (discrete) label space $\mathbb{L}$ is a special case of the GLMB with the following parameters

$$\mathbb{C} = \mathcal{F}(\mathbb{L}) \times \Xi \tag{3.3.52}$$

$$w^{(c)}(L) = w^{(I,\vartheta)}(L) = w^{(I,\vartheta)}\delta_I(L) \tag{3.3.53}$$

$$p^{(c)} = p^{(I,\vartheta)} = p^{(\vartheta)} \tag{3.3.54}$$

where Ξ is a discrete space, ϑ is a discrete index, and the physical meaning of ϑ can be referred to the following text. Thus, the distribution of the δ-GLMB RFS is

$$\pi(\mathrm{X}) = \Delta(\mathrm{X}) \sum_{(I,\vartheta) \in \mathcal{F}(\mathbb{L}) \times \Xi} w^{(I,\vartheta)} \delta_I(\mathcal{L}(\mathrm{X})) \left[p^{(\vartheta)}\right]^{\mathrm{X}} \tag{3.3.55}$$

Note that for the GLMB RFS with $\mathbb{C} = \mathcal{F}(\mathbb{L}) \times \Xi$, the amounts of $w^{(c)}$ and $p^{(c)}$ that need to be stored/computed are respectively $|\mathcal{F}(\mathbb{L}) \times \Xi|$ and $|\mathcal{F}(\mathbb{L}) \times \Xi|$, while for the δ-GLMB RFS, the amounts of $w^{(c)}$ and $p^{(c)}$ that need to be stored/computed are $|\mathcal{F}(\mathbb{L}) \times \Xi|$ and $|\Xi|$, respectively. In fact, by approximating $\mathcal{F}(\mathbb{L}) \times \Xi$ and Ξ with a smaller subset, which is composed of feasible hypotheses with larger weights, the computational workload and storage volume can be further reduced.

According to Proposition 1, the PHD of the unlabeled version of the δ-GLMB RFS is

$$
\begin{aligned}
v(\boldsymbol{x}) &= \sum_{(I,\vartheta)\in\mathcal{F}(\mathbb{L})\times\Xi} \sum_{\ell\in\mathbb{L}} p^{(\vartheta)}(\boldsymbol{x},\ell) \sum_{L\in\mathbb{L}} 1_L(\ell) w^{(I,\vartheta)} \delta_I(L) \\
&= \sum_{\ell\in\mathbb{L}} \sum_{(I,\vartheta)\in\mathcal{F}(\mathbb{L})\times\Xi} w^{(I,\vartheta)} 1_I(\ell) p^{(\vartheta)}(\boldsymbol{x},\ell)
\end{aligned} \tag{3.3.56}
$$

where, the inner layer summation (i.e., the weighted sum of the densities for track ℓ on all hypotheses containing track ℓ) can be interpreted as the PHD of track ℓ. Therefore, the total PHD is the sum of the PHDs of all the tracks, which can be further simplified to

$$
v(\boldsymbol{x}) = \sum_{(I,\vartheta)\in\mathcal{F}(\mathbb{L})\times\Xi} w^{(I,\vartheta)} \sum_{\ell\in I} p^{(\vartheta)}(\boldsymbol{x},\ell) \tag{3.3.57}
$$

The existence probability of track ℓ can be obtained by summing the weights of the two-tuples composed of the corresponding label set and discrete index, namely,

$$
\varepsilon^{(\ell)} = \sum_{(I,\vartheta)\in\mathcal{F}(\mathbb{L})\times\Xi} w^{(I,\vartheta)} 1_I(\ell) \tag{3.3.58}
$$

According to Proposition 2, the cardinality distribution of the δ-GLMB RFS can be obtained as

$$
\rho(|\mathrm{X}|) = \sum_{(I,\vartheta)\in\mathcal{F}(\mathbb{L})\times\Xi} \sum_{L\in\mathcal{F}_{|\mathrm{X}|}(\mathbb{L})} w^{(I,\vartheta)} \delta_I(L) = \sum_{(I,\vartheta)\in\mathcal{F}_{|\mathrm{X}|}(\mathbb{L})\times\Xi} w^{(I,\vartheta)} \tag{3.3.59}
$$

Therefore, the probability of $n = |\mathrm{X}|$ tracks is the sum of the weights of the hypotheses that have exactly n tracks.

In target tracking, the δ-GLMB can be used to represent the multi-target predicted density and multi-target filtering density at time k, namely,

$$
\pi_{k|k-1}\left(\mathrm{X}_k|Z^{k-1}\right) = \Delta(\mathrm{X}_k) \sum_{(I,\vartheta)\in\mathcal{F}(\mathbb{L}_{0:k})\times\Theta_{0:k-1}} w_{k|k-1}^{(I,\vartheta)} \delta_I(\mathcal{L}(\mathrm{X}_k)) \left[p_{k|k-1}^{(\vartheta)}\left(\cdot|Z^{k-1}\right)\right]^{\mathrm{X}_k} \tag{3.3.60}
$$

$$
\pi_k\left(\mathrm{X}_k|Z^k\right) = \Delta(\mathrm{X}_k) \sum_{(I,\vartheta)\in\mathcal{F}(\mathbb{L}_{0:k})\times\Theta_{0:k}} w_k^{(I,\vartheta)} \delta_I(\mathcal{L}(\mathrm{X}_k)) \left[p_k^{(\vartheta)}\left(\cdot|Z^k\right)\right]^{\mathrm{X}_k} \tag{3.3.61}
$$

where $Z^k = (Z_1, \ldots, Z_k)$, each $I \in \mathcal{F}(\mathbb{I}_{0:k})$ represents the set of track labels at time k, and each $\vartheta \triangleq (\theta_0, \theta_1, \ldots, \theta_{k-1}) \in \Theta_{0:k-1} \triangleq \Theta_0 \times \Theta_1 \times \ldots \times \Theta_{k-1}$ represents the history of the association map (or the association map history) up to time $k-1$, with the association map θ_k being the map function from the track labels to the measurement indices at time k, whose strict definition is detailed in Sect. 3.4.2; $\Theta = \Theta_k$ represents the space of the association map at time k; the subset of all the association maps whose domain of definition is L is denoted as $\Theta(L) = \Theta_k(L)$; and the discrete space Ξ is the space $\Theta_{0:k-1}$ of association map history, namely, $\Xi = \Theta_{0:k-1}$. For the predicted density, each $\vartheta \in \Xi$ represents the history of the association map up to time $k - 1$, namely, $\vartheta = (\theta_0, \theta_1, \ldots, \theta_{k-1})$; while for the filtering density, each ϑ represents the history of the association map up to time k, namely, $\vartheta = (\theta_0, \theta_1, \ldots, \theta_k)$; The two-tuple $(I, \vartheta) \in \mathcal{F}(\mathbb{L}_{0:k}) \times \Theta_{0:k}$ represents the hypothesis that the track set I has the association map history ϑ, and the associated weight $w_k^{(I,\vartheta)}$ can be interpreted as the probability that the hypothesis (I, ϑ) is true, whereas $(I, \vartheta) \in \mathcal{F}(\mathbb{L}_{0:k}) \times \Theta_{0:k}$ is called the predicted hypothesis, with the corresponding probability (weight) being $w_{k|k-1}^{(I,\vartheta)}$. It is worth noting that not all hypotheses are feasible, i.e., not all two-tuples (I, ϑ) are consistent, and for the infeasible two-tuple, its weight is 0. $p^{(\vartheta)}(\cdot, \ell)$ is the probability density of the kinetic state corresponding to track ℓ conditioned on association map history ϑ, and the densities $p_{k|k-1}^{(\vartheta)}(\cdot, \ell|Z^{k-1})$ and $p_k^{(\vartheta)}(\cdot, \ell|Z^k)$ represent the predicted density and filtering density of the kinetic state corresponding to track ℓ under the association history ϑ, respectively.

For example, for the δ-GLMB with initial multi-target prior of $\Xi = \emptyset$, it has the following form

$$\pi_0(\mathrm{X}) = \Delta(\mathrm{X}) \sum_{I \in \mathcal{F}(\mathbb{L}_0)} w_0^{(I)} \delta_I(\mathcal{L}(\mathrm{X}))[p_0]^{\mathrm{X}} \tag{3.3.62}$$

where each $I \in \mathcal{F}(\mathbb{L}_0)$ represents the set of track labels born at time 0, $w_0^{(I)}$ represents the weight of the hypothesis that I is the set of track labels at time 0, and $p_C(\cdot, \ell)$ is the probability density of the kinetic state corresponding to track $\ell \in I$. It is assumed that the following two possibilities may exist:

(1) The probability that there is a target with a label of (0, 2) is 0.3, and the density is $p_0(\cdot, (0, 2)) = \mathcal{N}(\cdot; \boldsymbol{m}, \boldsymbol{P}_2)$;
(2) The probability that there are two targets with labels of (0, 1) and (0, 2) is 0.7, and the corresponding densities are respectively $p_0(\cdot, (0, 1)) = \mathcal{N}(\cdot; \boldsymbol{0}, \boldsymbol{P}_1)$ and $p_0(\cdot, (0, 2)) = \mathcal{N}(\cdot; \boldsymbol{m}, \boldsymbol{P}_2)$.

Then, the δ-GLMB expression is

$$\pi_0(\mathrm{X}) = 0.3\delta_{\{(0,2)\}}(\mathcal{L}(\mathrm{X}))[p_0]^{\mathrm{X}} + 0.7\delta_{\{(0,1),(0,2)\}}(\mathcal{L}(\mathrm{X}))[p_0]^{\mathrm{X}} \tag{3.3.63}$$

3.4 RFS Formulation of Multi-target System

In the single-target system, the state and measurement at time k are modeled as two vectors of possibly different dimensions. These vectors vary with time, but their dimensions are invariable. However, in the multi-target system, the multi-target state and multi-target measurement should be a set of individual targets and measurements. In this case, the multi-target state and multi-target measurement evolve over time, and the number of individual targets and the number of measurements may change, that is, the dimensions of multi-target state and multi-target measurement may also change over time. In addition, there is no order in the elements of multi-target state and multi-target measurement. In many practical systems, although the sensor reports carry a specific measurement order, the result of the target tracking algorithm is independent of this order. Therefore, it is a natural way to model measurements as elements of an order-independent set.

For a multi-target scenario, it is assumed that at time $k-1$, there are N_{k-1} targets with state $\{\boldsymbol{x}_{k-1,1}, \boldsymbol{x}_{k-1,2}, \ldots, \boldsymbol{x}_{k-1,N_{k-1}}\}$, wherein N_{k-1} is the number of targets at time $k-1$. At the next moment, the existing targets may die or continue to survive, among which the surviving targets will evolve into new states or derive new targets. At the same time, independent birth targets may also appear. During this process, N_k new states $\{\boldsymbol{x}_{k,1}, \boldsymbol{x}_{k,2}, \ldots, \boldsymbol{x}_{k,N_k}\}$ will be obtained. Assume that the sensor receives M_k measurements $\{z_{k,1}, z_{k,2}, \ldots, z_{k,M_k}\}$ at time k. Only some of these measurements are actually generated by targets, and the rest of them may come from clutters or false alarms. For the sensor, there is no difference between the target-originated measurements and the clutters. In other words, the measurement source is unknown to the sensor, so the order in which they appear does not matter. Even in an ideal situation where the sensor observes all targets with no clutter received, the single-target filtering method is still not applicable, as there is no information about which target produces which measurement. The purpose of multi-target tracking is to jointly estimate the number of targets and their states based on such measurements with uncertain sources.

It should be noted that the set of target states and the set of measurements at time k are disordered, so they can be naturally expressed as a finite set, namely

$$X_k = \{\boldsymbol{x}_{k,1}, \boldsymbol{x}_{k,2}, \ldots, \boldsymbol{x}_{k,N_k}\} \in \mathcal{F}(\mathbb{X}) \tag{3.4.1}$$

$$Z_k = \{z_{k,1}, z_{k,2}, \ldots, z_{k,M_k}\} \in \mathcal{F}(\mathbb{Z}) \tag{3.4.2}$$

where $\mathcal{F}(\mathbb{X})$ and $\mathcal{F}(\mathbb{Z})$ represent the sets of all finite subsets of $\mathbb{X}$ and $\mathbb{Z}$, respectively. The core of the RFS description is to regard the target set X_k and the measurement set Z_k as multi-target state and multi-target measurement, respectively.

In the context of the labeled RFS, the multi-target state and multi-target measurement are respectively expressed as

$$\mathbf{X}_k = \left\{\mathbf{x}_{k,1}, \mathbf{x}_{k,2}, \ldots, \mathbf{x}_{k,N_k}\right\} \tag{3.4.3}$$

$$Z_k = \{z_{k,1}, z_{k,2}, \ldots, z_{k,M_k}\} \tag{3.4.4}$$

where $\mathbf{x}_{k,1}, \mathbf{x}_{k,2}, \ldots, \mathbf{x}_{k,N_k}$ represents the N_k target states that take values in the state space (including labels) $\mathbb{X} \times \mathbb{L}$ at time k. It should be noted that although the multi-target measurement represented by (3.4.4) and (3.4.2) are the same, the multi-target state represented by (3.4.3) and (3.4.1) are different. The former represents the labeled multi-target state set, while the latter is the unlabeled multi-target state set.

In summary, in the single-target system, the uncertainty is described by modeling the state $\boldsymbol{x}_k$ and measurement $\boldsymbol{z}_k$ as random vectors. Similarly, in the multi-target system, the uncertainty is described by modeling multi-target state X_k (or $\mathbf{X}_k$) and multi-target measurement Z_k respectively as RFSs on state-space $\mathbb{X}$ (or $\mathbb{X} \times \mathbb{L}$) and measurement space $\mathbb{Z}$, respectively.

By modeling the multi-target state and multi-target measurement as RFSs, the multi-target filtering problem can be regarded as a Bayesian filtering problem with (multi-target) state-space and measurement space $\mathcal{F}(\mathbb{X})$ (or $\mathcal{F}(\mathbb{X} \times \mathbb{L})$) and $\mathcal{F}(\mathbb{Z})$, respectively.

3.4.1 Multi-target Motion Model and Multi-target Transition Kernel

The following describes the RFS model corresponding to the temporal evolution of the multi-target state, which incorporates the processes of the target movement, birth, and death. For a given multi-target state X_{k-1} at time $k-1$, each $\boldsymbol{x}_{k-1} \in X_{k-1}$ either dies with probability $1 - p_{S,k}(\boldsymbol{x}_{k-1})$ at time k, or continues to survive with probability $p_{S,k}(\boldsymbol{x}_{k-1})$ and transitions to a new state $\boldsymbol{x}_k$ with probability density $\phi_{k|k-1}(\boldsymbol{x}_k \ \boldsymbol{x}_{k-1})$. Therefore, for a given state $\boldsymbol{x}_{k-1} \in X_{k-1}$ at time $k-1$, its behavior at the next time can be modeled as RFS $S_{k|k-1}(\boldsymbol{x}_{k-1})$, which is taken as $\emptyset$ when the target dies, or $\{\boldsymbol{x}_k\}$ when the target survives. The new target born at time k may be a birth target (which is independent of any survival target) or spawned from a target at time $k-1$. Therefore, given the multi-target state X_{k-1} at time $k-1$, the multi-target state X_k at time k is given by the union of the surviving targets, the spawning targets, and the birth targets, namely,

$$
\begin{aligned}
X_k &= \left[\bigcup_{\boldsymbol{\xi} \in X_{k-1}} S_{k|k-1}(\boldsymbol{\xi})\right] \cup \left[\bigcup_{\boldsymbol{\xi} \in X_{k-1}} Q_{k|k-1}(\boldsymbol{\xi})\right] \cup B_k \\
&= S_{k|k-1}(X_{k-1}) \cup Q_{k|k-1}(X_{k-1}) \cup B_k \\
&= S_{k|k-1}(X_{k-1}) \cup \Gamma_{k|k-1}(X_{k-1})
\end{aligned}
\tag{3.4.5}
$$

where $S_{k|k-1}(X_{k-1})$ represents the RFS of the surviving targets at time k; $Q_{k|k-1}(X_{k-1})$ and B_k respectively represent the RFS of targets spawned from X_{k-1} and the RFS of birth targets at time k; the specific forms of $Q_{k|k-1}(\cdot)$ and B_k depend on the specific problem; $\Gamma_{k|k-1}(X_{k-1})$ represents the RFS of newly emerging targets at time k, including the birth targets and spawning targets, namely,

$$\Gamma_{k|k-1}(X_{k-1}) = Q_{k|k-1}(X_{k-1}) \bigcup B_k \tag{3.4.6}$$

It is assumed that the RFSs that make up the union (3.4.5) are independent of each other. According to various target dynamics, target birth and death models under different assumptions, $S_{k|k-1}(X_{k-1})$, $Q_{k|k-1}(X_{k-1})$ and B_k can be explicitly determined [14, 392]. RFS X_k encapsulates all aspects of information about multi-target motion, such as time-varying number of targets, motion, birth and spawning of targets and their interaction.

The statistical behavior of RFS X_k modeled by (3.4.5) can also be described by the following Markov multi-target transition density[2]

$$\phi_{k|k-1}(X_k|X_{k-1}) \tag{3.4.7}$$

that is the probability density of evolution from a given multi-target state X_{k-1} to X_k, which encapsulates the underlying models of motion, birth, and death. Under different assumptions, the multi-target transition density $\phi_{k|k-1}(\cdot|\cdot)$ can be derived from the multi-target transition equation using the FISST. For the detailed derivation process, refer to [14, 392]. Generally, according to the multi-target dynamic model (3.4.5), the following multi-target Markov transition probability can be obtained as

$$\phi_{k|k-1}(X_k|X_{k-1}) = \sum_{W \subseteq X_k} \pi_S(W|X_{k-1})\pi_\Gamma(X_k - W|X_{k-1}) \tag{3.4.8}$$

where $\pi_S(\cdot|X_{k-1})$ is the FISST density of RFS $S_{k|k-1}(X_{k-1})$ of survival targets, and $\pi_\Gamma(\cdot|X_{k-1})$ is the FISST density of RFS $\Gamma_{k|k-1}(X_{k-1})$ of birth targets. Note that the minus sign in (3.4.8) represents the set difference.

Taking the labeled RFS background as an example, the specific equation of the multi-target Markov transition probability (3.4.8) is given below. For the sake of conciseness, the subscript k representing the timestamp is omitted when no confusion is caused. Let X denote the labeled RFS of the targets at time $k-1$ and the label space be $\mathbb{L}$. Given the multi-target state X at the previous moment, each state $(\boldsymbol{x}, \ell) \in \mathrm{X}$ either dies with probability $q_s(\boldsymbol{x}, \ell) = 1 - p_s(\boldsymbol{x}, \ell)$, or survives at the next moment with probability $p_s(\boldsymbol{x}, \ell)$ and evolves to a new state $(\boldsymbol{x}_+, \ell_+)$ with probability $\phi_+(\boldsymbol{x}_+, \ell_+|\boldsymbol{x}, \ell) = \phi(\boldsymbol{x}_+|\boldsymbol{x}, \ell)\delta_\ell(\ell_+)$, where $\phi(\boldsymbol{x}_+|\boldsymbol{x}, \ell)$ is the single-target transition kernel. Therefore, the distribution of the set S of surviving targets at the next moment is

$$\pi_S(\mathrm{S}|\mathrm{X}) = \Delta(\mathrm{S})\Delta(\mathrm{X})1_{\mathcal{L}(\mathrm{X})}(\mathcal{L}(\mathrm{S}))[\Phi(\mathrm{S};\cdot)]^{\mathrm{X}} \tag{3.4.9}$$

---

[2] For simplicity, the same symbol $\phi_{k|k-1}$ is used for the multi-target transition function (3.4.7) and the single-target transition function (2.2.2) (other symbols are similar), which will not cause confusion. It is easy to distinguish them according to the parameter in the bracket of the function, because in the single-target case, the parameter is a vector, while in the multi-target case, the parameter is a finite set.

where

$$\Phi(\mathrm{S};\boldsymbol{x},\ell) = \sum_{(\boldsymbol{x}_+,\ell_+)\in \mathrm{S}} \delta_\ell(\ell_+) p_S(\boldsymbol{x},\ell)\phi(\boldsymbol{x}_+|\boldsymbol{x},\ell) + (1 - 1_{\mathcal{L}(\mathrm{S})}(\ell)) q_S(\boldsymbol{x},\ell)$$
$$= \begin{cases} p_S(\boldsymbol{x},\ell)\phi(\boldsymbol{x}_+|\boldsymbol{x},\ell), & if\ (\boldsymbol{x}_+,\ell) \in \mathrm{S} \\ q_S(\boldsymbol{x},\ell), & if\ \ell \notin \mathcal{L}(\mathrm{S}) \end{cases} \tag{3.4.10}$$

Let B denote the labeled RFS of birth targets and the label space be $\mathbb{B}$. The following labeling scheme is adopted: each birth target is labeled with a two-tuple (k, n), wherein k is the current time, and n is the unique index that distinguishes the birth targets born at the same time. Since the birth label space $\mathbb{B}$ and the surviving label space $\mathbb{L}$ change every time, they are always mutually exclusive, namely, $\mathbb{L} \cap \mathbb{B} = \emptyset$. Because the birth targets has different labels and their states are assumed to be independent, B can be modeled as a LMB distribution, i.e.,

$$\pi_\gamma(\mathrm{B}) = \Delta(\mathrm{B})\omega_\gamma(\mathcal{L}(\mathrm{B}))p_\gamma^{\mathrm{B}} \tag{3.4.11}$$

where w_γ and p_γ are given parameters of the multi-target birth density π_γ defined on $\mathbb{X} \times \mathbb{B}$. Specifically, $w_\gamma(\cdot)$ is the birth weight and $p_\gamma(\cdot, \ell)$ is the single-target birth density. Note that $\pi_\gamma(\mathrm{B}) = 0$ if B contains any element $\mathbf{b}$ that satisfies $\mathcal{L}(\mathbf{b}) \notin \mathbb{B}$. The birth model (3.4.11) may include the labeled Poisson, labeled IIDC, LMB and GLMB models [69]. The LMB birth model is adopted here, however, it is easy to generalize to the case of the GLMB birth model.

If the target spawning is ignored, the multi-target state X_+ at the next moment is the union of the surviving targets S and the birth targets B, that is, $\mathrm{X}_+ = \mathrm{S} \cup \mathrm{B}$. Since the label spaces $\mathbb{L}$ and $\mathbb{B}$ are mutually exclusive, the states of birth targets are independent of those of surviving targets, and thus, S nd B are independent of each other. According to the FISST, the multi-target transition kernel $\phi(\cdot|\cdot)$: $\mathcal{F}(\mathbb{X} \times \mathbb{L}) \times \mathcal{F}(\mathbb{X} \times \mathbb{L}) \to [0, \infty)$ is

$$\phi(\mathrm{X}_+|\mathrm{X}) = \sum_{\mathrm{S}\subseteq \mathrm{X}_+} \pi_s(\mathrm{S}|\mathrm{X})\pi_\gamma(\mathrm{X}_+ - \mathrm{S}) \tag{3.4.12}$$

Consider a subset $\mathrm{X}_+ \cap (\mathbb{X} \times \mathbb{L}) = \{\mathbf{x}_+ \in \mathrm{X}_+ : \mathcal{L}(\mathbf{x}_+) \in \mathbb{L}\}$ of X_+ composed of surviving targets, for any $\mathrm{S} \subseteq \mathrm{X}_+$, if S is not a subset of surviving targets, i.e., $\mathrm{S} \not\subseteq \mathrm{X}_+ \cap (\mathbb{X} \times \mathbb{L})$, then the labels of S will not be a subset of the current label space, namely, $\mathcal{L}(\mathrm{S}) \not\subseteq \mathbb{L}$, then $1_{\mathcal{L}(\mathrm{X})}(\mathcal{L}(\mathrm{S})) = 0$, so $\pi_S(\mathrm{S}|\mathrm{X}) = 0$ according to (3.4.9). Therefore, only the case of $\mathrm{S} \subseteq \mathrm{X}_+ \cap (\mathbb{X} \times \mathbb{L})$ needs to be considered. Further, for any non-empty S, if $\mathrm{S} \subseteq \mathrm{X}_+ \cap (\mathbb{X} \times \mathbb{L})$, then there exists $\mathbf{x}_+ \in \mathrm{X}_+ - \mathrm{S}$ such that $\mathcal{L}(\mathbf{x}_+) \in \mathbb{L}$. Because $\mathbb{B}$ and $\mathbb{L}$ are mutually exclusive, $\mathcal{L}(\mathbf{x}_+) \notin \mathbb{B}$, and thus $\pi_\gamma(\mathrm{X}_+ - \mathrm{S}) = 0$. According to the above analysis, the multi-target transition density (transition kernel) eventually degenerates into the product of the surviving target transition density and the birth target density, namely,

$$\phi(\mathrm{X}_+|\mathrm{X}) = \pi_S\Big(\mathrm{X}_+ \bigcap (\mathbb{X} \times \mathbb{L})|\mathrm{X}\Big)\pi_\gamma(\mathrm{X}_+ - (\mathbb{X} \times \mathbb{L})) \tag{3.4.13}$$

3.4.2 Multi-target Measurement Model and Multi-target Measurement Likelihood

The following describes the RFS measurement model, which takes into account the detection uncertainty and clutter. Due to missed detection and clutter interference, the multi-target measurement Z_k at time k is a finite subset of the measurement space $\mathbb{Z}$. In the multi-target measurement model, a given target with state $\boldsymbol{x}_k \in X_k$ is either missed with probability $1 - p_{D,k}(\boldsymbol{x}_k)$ or detected with probability $p_{D,k}(\boldsymbol{x}_k)$ and generates a measurement $\boldsymbol{z}_k$ with likelihood $g_k(\boldsymbol{z}_k|\boldsymbol{x}_k)$. Therefore, at time k, each state $\boldsymbol{x}_k \in X_k$ generates RFS $D_k(\boldsymbol{x}_k)$, which is taken as $\emptyset$ when the target is missed, or as $\{\boldsymbol{z}_k\}$ when the target is detected. In addition to the target-originated measurements, the sensor also receives a set K_k of false alarm measurements or clutters. Therefore, given the multi-target state X_k at time k, the multi-target measurement Z_k received by the sensor is given by the union of the target-originated measurements and the clutters, i.e.,

$$Z_k = \left[\bigcup_{\boldsymbol{x}_k \in X_k} D_k(\boldsymbol{x}_k)\right] \bigcup K_k = D_k(X_k) \bigcup K_k \tag{3.4.14}$$

where the RFSs that make up the union (3.4.14) are assumed to be independent of each other. The actual form of K_k depends on the specific problem, and $D_k(X_k)$ and K_k can be determined according to the underlying sensor physical model [14, 392]. RFS Z_k encapsulates all sensor features, such as target detection, measurement noise, sensor field of view (i.e., state-dependent detection probability), and false alarms.

The statistical behavior of RFS Z_k modeled in (3.4.14) can also be described by the following multi-target likelihood

$$g_k(Z_k|X_k) \tag{3.4.15}$$

which is the likelihood that the measurement Z_k is generated by the multi-target state X_k. Under different assumptions, the multi-target likelihood $g_k(Z_k|X_k)$ can be derived from the underlying sensor physical model. For the detailed derivation process, please refer to [14, 392]. Generally, according to the multi-target measurement model (3.4.14), the following multi-target likelihood can be obtained

$$g_k(Z_k|X_k) = \sum_{D \subseteq Z_k} \pi_D(D|X_k)\pi_C(Z_k - D) \tag{3.4.16}$$

where $\pi_D(\cdot|X_k)$ is the FISST density of measurement RFS $D_k(X_k)$ generated by targets, and $\pi_C(\cdot)$ is the FISST density of false alarm RFS K_k.

The following also takes the labeled RFS background as an example to give the specific formula of the multi-target likelihood (3.4.16). Let X denote the labeled RFS of surviving targets at the measurement time. A particular target $\mathbf{x} = (\boldsymbol{x}, \ell) \in \mathrm{X}$ is either missed with probability $q_D(\mathbf{x}) = 1 - p_D(\mathbf{x}) = 1 - p_D(\boldsymbol{x}, \ell)$ or is detected with probability $p_D(\mathbf{x}) = p_D(\boldsymbol{x}, \ell)$ and then generates a measurement z with likelihood $g(z|\mathbf{x}) = g(z|\boldsymbol{x}, \ell)$. Thus, $\mathbf{x}$ generates a Bernoulli RFS with parameter $\{(p_D(\mathbf{x}), g(\cdot|\mathbf{x}))\}$. Let D denote a set of target detections (non-clutter measurements). If the elements of D are assumed to be conditionally independent, D is a MB RFS with a parameter set $\{(p_D(\mathbf{x}), g(\mathbf{z}|\mathbf{x})) : \mathbf{x} \in \mathrm{X}\}$, and its probability density is

$$\pi_D(D|\mathrm{X}) = \{(p_D(\mathbf{x}), g(z|\mathbf{x})) : \mathbf{x} \in \mathrm{X}\}_{z \in D} \tag{3.4.17}$$

Let K denote the set of clutter measurements, which is independent of the target measurements. K is modeled as a Poisson RFS with intensity $\kappa(\cdot)$, and thus K obeys the following distribution

$$\pi_C(K) = \exp(-\langle\kappa, 1\rangle)\kappa^K \tag{3.4.18}$$

For a given multi-target state X at time k, the multi-target measurement $Z = \{z_1, z_2, \ldots, z_{|Z|}\}$ is the union of the target measurement D and the clutter measurement K, that is, $Z = D \bigcup K$. Since D is independent of K, the multi-target likelihood is the convolution of π_D and π_C, namely,

$$g(\mathrm{Z}|\mathrm{X}) = \sum_{D \subseteq Z} \pi_D(D|\mathrm{X})\pi_C(\mathrm{Z} - D) \tag{3.4.19}$$

where $Z - D$ represents the set difference between set Z and set D. The above equation can be equivalently expressed as [14]

$$g(\mathrm{Z}|\mathrm{X}) = \exp(-\langle\kappa, 1\rangle)\kappa^Z \sum_{\theta \in \Theta(\mathcal{L}(\mathrm{X}))} [\varphi_Z(\cdot; \theta)]^{\mathrm{X}} \tag{3.4.20}$$

where Θ is the space of map $\theta : \mathbb{L} \to \{0 : |\mathrm{Z}|\} \triangleq \{0, 1, \ldots, |Z|\}$, $\Theta(I)$ is a subset of association maps with a defined domain I, $\Theta(\mathcal{L}(\mathrm{X}))$ is the set of all one-to-one maps $\theta : \mathcal{L}(\mathrm{X}) \to \{0, 1, \ldots, |Z|\}$ from labels in X to measurement indices in Z, and θ satisfies $\theta(i) = \theta(j) > 0 \Rightarrow i = j$, that is, when the range is restricted to positive integers, θ has a one-to-one correspondence property. An association map θ describes which track produces which measurement, namely, track ℓ generates measurement $z_{\theta(\ell)} \in Z$, and the integer 0 is assigned to the missed tracks. The condition "$\theta(i) = \theta(j) > 0 \Rightarrow i = j$" means that a track produces at most one measurement at any time, and a measurement is assigned to at most one track. Besides,

$$\begin{aligned}\varphi_Z(\boldsymbol{x},\ell;\theta) &= \delta_0(\theta(\ell))(1-p_D(\boldsymbol{x},\ell)) \\ &\quad + (1-\delta_0(\theta(\ell)))p_D(\boldsymbol{x},\ell)g\big(z_{\theta(\ell)}|\boldsymbol{x},\ell\big)/\kappa\big(z_{\theta(\ell)}\big) \\ &= \begin{cases} p_D(\boldsymbol{x},\ell)g\big(z_{\theta(\ell)}|\boldsymbol{x},\ell\big)/\kappa\big(z_{\theta(\ell)}\big), & if\,\theta(\ell)>0 \\ 1-p_D(\boldsymbol{x},\ell), & if\,\theta(\ell)=0 \end{cases}\end{aligned} \tag{3.4.21}$$

When considering the association gate, $\varphi_z(\boldsymbol{x},\ell;\theta)$ needs to be incorporated into the gate probability p_G, which becomes

$$\varphi_z(\boldsymbol{x},\ell;\theta) = \begin{cases} p_G p_D(\boldsymbol{x},\ell)g\big(z_{\theta(\ell)}|\boldsymbol{x},\ell\big)/\kappa\big(z_{\theta(\ell)}\big), & if\,\theta(\ell)>0 \\ 1-p_G p_D(\boldsymbol{x},\ell), & if\,\theta(\ell)=0 \end{cases} \tag{3.4.22}$$

3.5 Multi-target Bayesian Recursion

In the Bayesian estimation, what is of interest is the posterior probability density. The space of finite subsets of $\mathbb{X}$ does not inherit the commonly used concepts of Euclidean integral and Euclidean density. Therefore, standard tools for random vectors are no longer suitable for the RFS, and the application of Bayesian inference to multi-target estimation requires a suitable concept of probability density for the RFS. The FISST proposed by Mahler provides a practical and powerful mathematical tool to deal with the RFS. The FISST introduces the non-measure-theoretic concept of the density through the set integral and set derivative [392].

The multi-target posterior density conditioned on the multi-target measurement history $Z^k = Z_{1:k} = \{Z_1, \ldots, Z_k\}$ captures all the information about the target state set, which can be calculated recursively by the following equation

$$\pi_{0:k}(X_{0:k}|Z_{1:k}) \propto g_k(Z_k|X_k)\phi_{k|k-1}(X_k|X_{k-1})\pi_{0:k-1}(X_{0:k-1}|Z_{1:k-1}) \tag{3.5.1}$$

where $X_{0:k} = \{X_0, X_1, \ldots, X_k\}$, $\phi_{k|k-1}(\cdot|\cdot)$ is the multi-target transition density from time $k-1$ to time k, and $g_k(\cdot|\cdot)$ is the multi-target likelihood function at time k. The multi-target posterior density captures all the information about the number of targets and their states. The multi-target transition density encapsulates potential models such as target motion, birth and death, whereas the multi-target likelihood function encapsulates potential models such as target detection and false alarms.

There is a certain connection between the FISST theory and the conventional probability theory. Specifically, the set derivative of the belief mass function of the RFS is closely related to its probability density [62]. For any closed subsets $S \subseteq \mathbb{X}$ and $T \subseteq \mathbb{Z}$, let $\beta_{k|k}(S|Z_{1:k}) \triangleq P(X_k \subseteq S|Z_{1:k})$ denote the (posterior) belief mass function of RFS X_k given a set of all measurements $Z_{1:k} = \{Z_1, Z_2, \ldots, Z_k\}$ up to time k, and $\beta_{k|k-1}(S|X_{k-1}) \triangleq P(X_k \subseteq S|X_{k-1})$ denote the (predicted) belief mass function of RFS X_k modeled by (3.4.5); $\beta_k(T|X_k) \triangleq P(Z_k \subseteq T|X_k)$ represents the belief mass function of RFS Z_k modeled by (3.4.14), then the multi-target posterior density $\pi_k(\cdot|Z_{1:k})$, multi-target transition density $\phi_{k|k-1}(\cdot|X_{k-1})$ and multi-target likelihood $g_k(\cdot|X_k)$ are the set derivatives of $\beta_{k|k}(\cdot|Z_{1:k})$, $\beta_{k|k-1}(\cdot|X_{k-1})$ and $\beta_k(\cdot|X_k)$, respectively.

Multi-target filtering focuses on the marginal density of the multi-target posterior density at the current moment. Let $\pi_{k|k-1}(X_k|Z_{1:k-1})$ denote the multi-target predicted density up to time k, and $\pi_k(X_k|Z_{1:k})$ denote the multi-target filtering density at time k. The multi-target filtering density captures all the information of the multi-target state (including the number of targets and their states) at the current moment. The multi-target Bayesian filter propagates the filtering density π_k over time according to the following prediction and update equations [6, 392]

$$\pi_{k|k-1}(X_k|Z_{1:k-1}) = \int \phi_{k|k-1}(X_k|X)\pi_{k-1}(X|Z_{1:k-1})\delta X \tag{3.5.2}$$

$$\pi_k(X_k|Z_{1:k}) = \frac{g_k(Z_k|X_k)\pi_{k|k-1}(X_k|Z_{1:k-1})}{\int g_k(Z_k|X)\pi_{k|k-1}(X|Z_{1:k-1})\delta X} \tag{3.5.3}$$

The multi-target Bayesian filter is the core of the RFS method [14], which recursively propagates the filtering density of the multi-target state forward in time. The main difference between recursion (3.5.1–3.5.3) and standard (clutter free) single-target filtering (2.2.6–2.2.8) lies in that the dimensions of X_k and Z_k are variable as k changes. In addition, it should be noted that the integral in (3.5.2–3.5.3) is the set integral defined in (3.2.3), and this multidimensional integral makes the multi-target Bayesian filter generally incapable of obtaining to a closed analytic form. Besides, the relevant functions of Eqs. (3.5.2–3.5.3) are dimensional. Specifically, $\pi_k(X|Z_{1:k})$, $\pi_{k-1}(X|Z_{1:k-1})$,$\pi_{k|k-1}(X|Z_{1:k-1})$, and $\phi_{k|k-1}(X|X_{k-1})$ have dimension $K_x^{-|X|}$, and $g_k(Z_k|X_k)$ has dimension $K_z^{-|Z_k|}$, where K_x and K_z respectively represent the dimensions of the volumes in spaces $\mathbb{X}$ and $\mathbb{Z}$. However, the corresponding functions in Eqs. (2.2.7–2.2.8), namely, $p_k(\boldsymbol{x}|\boldsymbol{z}_{1:k})$, $p_{k-1}(\boldsymbol{x}|\boldsymbol{z}_{1:k-1})$, $p_{k|k-1}(\boldsymbol{x}|\boldsymbol{z}_{1:k-1})$, $\phi_{k|k-1}(\boldsymbol{x}|\boldsymbol{x}_{k-1})$, and $g_k(\boldsymbol{z}_k|\boldsymbol{x}_k)$ are all dimensionless.

For the labeled RFS, the multi-target Bayesian filter in Eqs. (3.5.1–3.5.3) is corrected to

$$\pi_{kk-1}(\mathrm{X}_k|Z_{1:k-1}) = \int \phi_{k|k-1}(\mathrm{X}_k|\mathrm{X}_{k-1})\pi_{k-1}(\mathrm{X}_{k-1}|Z_{1:k-1})\delta\mathrm{X}_{k-1} \tag{3.5.4}$$

$$\pi_k(\mathrm{X}_k|Z_{1:k}) = \frac{g_k(Z_k|\mathrm{X}_k)\pi_{k|k-1}(\mathrm{X}_k|Z_{1:k-1})}{\int g_k(Z_k|\mathrm{X})\pi_{k|k-1}(\mathrm{X}|Z_{1:k-1})\delta\mathrm{X}} \tag{3.5.5}$$

It should be noted that integral in the above equations is the "set integral" for the labeled RFS defined in (3.3.18).

3.6 Multi-target Formal Modeling Paradigm

The formal Bayes modeling paradigm applicable to the single-target case provides a set of systematic, general procedures for the development of single-target filtering algorithms: First, the measurement model describing the sensor behavior and unrelated to the specific implementation and the motion model describing the target dynamics are constructed respectively; Then, according to the constructed models, the corresponding probability mass functions (PMF) are derived; After that, the true likelihood function and the true Markov transition density that can faithfully reflect the probability density functions of the original models are derived by using the conventional calculus; Thus, the recursion of the Bayesian filter that propagates the posterior density over time is obtained. Please refer to Fig. 3.1a.

Similar to the single-target formal Bayes modeling, the modeling procedure can be extended to multi-source multi-target systems [14, 15]: first, the multi-target measurement model describing the sensor behavior and independent of the specific implementation and the multi-target motion model describing the multi-target dynamics are constructed respectively. According to the constructed models, the corresponding belief mass functions are derived. Then, using the FISST tool dedicated to the RFS, the specific expressions of the true multi-target likelihood function that can truthfully reflect the multi-target measurement model and the true multi-target Markov transition density that can truthfully reflect the multi-target motion model are derived, respectively; Thus, the recursion of the multi-target Bayesian filter that propagates the posterior multi-target density over time is obtained. Please refer to Fig. 3.1b. The above-mentioned multi-source multi-target formal Bayes modeling paradigm has properties independent of specific implementations, so it is convenient to establish a general mathematical representation.

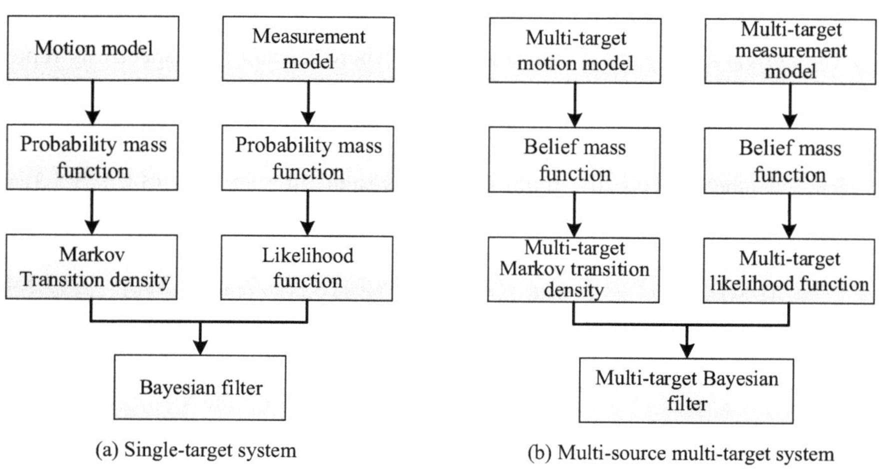

Fig. 3.1 Formal modeling process **a** single-target system, **b** multi-source multi-target system

It should be noted that in the single-target formal Bayesian modeling, state and measurement variables are modeled using random vectors, while in the multi-source multi-target formal Bayes modeling, state and measurement variables are modeled using random finite sets. Therefore, a set of FISST tools dedicated to the RFS is needed.

It can be seen from the above that there are many direct mathematical correspondences between single-sensor single-target statistics and multi-sensor multi-target statistics. In this sense, general statistical methods can be directly "transplanted" from the single-sensor single-target case to the multi-sensor multi-target case after proper processing. In other words, this correspondence can be regarded as a dictionary. The dictionary establishes a direct mapping relationship between word and grammars in the context of random vectors and synonyms and grammars in the context of random finite sets. Thereby, in principle, any "sentence" (any concept or algorithm) expressed in the context of random vectors can be directly "translated" into the corresponding "sentence" in the context of random finite sets. This process can be regarded as a general strategy to solve the problem of multi-source multi-target information fusion. For example, both the true multi-target likelihood function and the true multi-target Markov density can be directly analogized to the likelihood function and Markov density in the single-target case. The analogy with the fixed gain Kalman filter in the single target case directly stimulates the PHD and CPHD approximation techniques, and so on.

Although some statistical concepts and techniques in the filtering theory and information theory can be easily transplanted under the RFS paradigm, the correspondence between dictionaries is not exactly one-to-one correspondence in the transplantation process. For example, vectors can be added and subtracted, yet the RFS can not. For another example, conventional single-target estimators (such as maximum a posterior and expected a posterior estimators) are undefined for RFS-based multi-target estimators [14].

3.7 Particle Multi-target Filter

This section introduces a sequential Monte Carlo (SMC) implementation of the multi-target Bayesian filter. The SMC filtering technique allows the recursive propagation of weighted particle sets that approximate the posterior density, and the core idea of the technique is to approximate the integral of interest with random samples. Since the dimensionless FISST multi-target density is actually a probability density [62], random samples can be used to construct a Monte Carlo approximation of the integral of interest. Therefore, the single-target particle filter can be directly generalized to the multi-target case. However, under the multi-target condition, each particle is a finite set, and therefore, the particles themselves have variable dimensions. The recursive propagation of multi-target posterior density over time involves the calculation of the multiple set integrals, which is much more computationally intensive than that of single-target filtering.

It is assumed that the weighted particle set $\{(w_{k-1}^{(i)}, X_{k-1}^{(i)})\}_{i=1}^{\nu}$ representing the multi-target posterior π_{k-1} is available at time $k-1$, that is,

$$\pi_{k-1}(X_{k-1}|Z_{1:k-1}) \approx \sum_{i=1}^{\nu} w_{k-1}^{(i)} \delta_{X_{k-1}^{(i)}}(X_{k-1}) \tag{3.7.1}$$

where ν is the total number of particles. The particle multi-target filter approximates the multi-target posterior π_k at time k through a new weighted particle set $\{(w_k^{(i)}, X_k^{(i)})\}_{i=1}^{\nu}$, and the specific steps are as follows.

- Prediction
  - Sample $\tilde{X}_k^{(i)} \sim q_k\left(\cdot|X_{k-1}^{(i)}, Z_k\right)$, $i = 1, \ldots, \nu$ and calculate predicted weights

$$w_{k|k-1}^{(i)} = \frac{\phi_{k|k-1}(\tilde{X}_k^{(i)}|X_{k-1}^{(i)})}{q_k(\tilde{X}_k^{(i)}|X_{k-1}^{(i)}, Z_k)} w_{k-1}^{(i)} \tag{3.7.2}$$

- Update
  - Update the weights

$$w_k^{(i)} = w_{k|k-1}^{(i)} g_k(Z_k|\tilde{X}_k^{(i)}) \tag{3.7.3}$$

  - Normalize the weights

$$\tilde{w}_k^{(i)} = w_k^{(i)} / \sum_{j=1}^{N} w_k^{(j)}, i = 1, \ldots, N \tag{3.7.4}$$

- Re-sampling
  - Resample $\{(\tilde{w}_k^{(i)}, \tilde{X}_k^{(i)})\}_{i=1}^{\nu}$ to obtain $\{(w_k^{(i)}, X_k^{(i)})\}_{i=1}^{\nu}$.

In the above steps, $\sup_{X,X'} \left|\phi_{k|k-1}(X|X') / q_k(X|X', Z_k)\right|$ is limited by default, thus the weights are well-defined. $\tilde{X}_k^{(i)}$ is the sample from the RFS or point process, and the specific content about sampling from point process can refer to the field of spatial statistics [392, 394]. The importance sampling density $q_k(\cdot|X_{k-1}^{(i)}, Z_k)$ can be simply set as the multi-target transition density. However, it should be noted that compared with the single-target case, choosing the dynamic prior of multi-target transition density as the proposed density has more serious problems. If the number of targets is large, importance sampling needs to be implemented in a high-dimensional space, which is usually difficult to find an effective importance density. For a fixed number of particles, conventional choices of importance densities, such as $q_k(\cdot|X_{k-1}^{(i)}, Z_k) = \phi_{k|k-1}(\cdot|X_{k-1}^{(i)})$, may typically leads to an exponential decrease in the effectiveness of the algorithm with the number of targets [62]. In addition, owing to the combination characteristics of multi-target Markov transition (3.4.8) and multi-target likelihood (3.4.16), the amount of calculation for weight update is also large. Therefore, the simple bootstrapping method works well only in case of a small number of targets.

The above algorithm seems to be similar to the particle filter described in Sect. 2.9 in form, however, the complexity of the multi-target particle system is actually much higher than that of the single-target one. It is assumed that the number of particles used to represent the n targets is $\tilde{\nu}_n$, and the maximum number of particles corresponding to the n targets is denoted as ν_n. $X_k^{(0)}, X_k^{(1)}, \ldots, X_k^{(\nu)}$ can be sorted according to the incremental number of targets. The first multi-target particle $X_k^{(0)} = \emptyset$ represents the sample when there is no target (0 target), hence, $\tilde{\nu}_0 = 1$. The subsequent $\tilde{\nu}_1$ multi-target particles $X_k^{(1)} = \{\boldsymbol{x}_1^{(1)}\}, \ldots, X_k^{(\nu_1)} = \{\boldsymbol{x}_1^{(\nu_1)}\}$ represent the sample when there is 1 target (thus, $\tilde{\nu}_1 = \nu_1$), and the next $\tilde{\nu}_2$ multi-target particles $X_k^{(\nu_1+1)} = \{\boldsymbol{x}_1^{(\nu_1+1)}, \boldsymbol{x}_2^{(\nu_1+1)}\}, \ldots, X_k^{(\nu_2)} = \{\boldsymbol{x}_1^{(\nu_2)}, \boldsymbol{x}_2^{(\nu_2)}\}$ represent the sample when there are 2 targets (thus, $\tilde{\nu}_2 = \nu_2 - \nu_1$). Generally, $\tilde{\nu}_n$ multi-target particles $X_k^{(\nu_{n-1}+1)} = \{\boldsymbol{x}_1^{(\nu_{n-1}+1)}, \ldots, \boldsymbol{x}_n^{(\nu_{n-1}+1)}\}, \ldots, X_k^{(\nu_n)} = \{\boldsymbol{x}_1^{(\nu_n)}, \ldots, \boldsymbol{x}_n^{(\nu_n)}\}$ represent the sample when there are n targets (thus, $\tilde{\nu}_n = \nu_n - \nu_{n-1}$). This still applies to the case that $\tilde{\nu}_{\widehat{n}}$ multi-target particles $X_k^{(\nu_{\widehat{n}-1}+1)} = \{\boldsymbol{x}_1^{(\nu_{\widehat{n}-1}+1)}, \ldots, \boldsymbol{x}_{\widehat{n}}^{(\nu_{\widehat{n}-1}+1)}\}, \ldots, X_{k|k}^{(\nu_{\widehat{n}})} = \{\boldsymbol{x}_1^{(\nu_{\widehat{n}})}, \ldots, \boldsymbol{x}_n^{(\nu_{\widehat{n}})}\}$ represent the sample when there are $\widehat{n}$ targets (thus, $\tilde{\nu}_{\widehat{n}} = \nu_{\widehat{n}} - \nu_{\widehat{n}-1}$), where $\widehat{n}$ represents the maximum possible number of targets. Thus, the total number of samples is

$$\nu = \tilde{\nu}_0 + \tilde{\nu}_1 + \cdots + \tilde{\nu}_{\widehat{n}} = 1 + \nu_1 + (\nu_2 - \nu_1) + \cdots + (\nu_{\widehat{n}} - \nu_{\widehat{n}-1}) = 1 + \nu_{\widehat{n}}$$

3.7.1 Prediction of Particle Multi-target Filter

It is assumed that the prior multi-target particle system is

$$\begin{aligned}\pi_{k-1}(X_{k-1}|Z_{1:k-1}) &\approx \omega_{k-1}^{(0)}\delta_{\emptyset}(X_{k-1}) + \frac{1-\omega_{k-1}^{(0)}}{\nu}\sum_{i=1}^{\nu}\delta_{X_{k-1}^{(i)}}(X_{k-1}) \\ &= \sum_{i=0}^{\nu} w_{k-1}^{(i)}\delta_{X_{k-1}^{(i)}}(X_{k-1})\end{aligned} \tag{3.7.5}$$

where $X_{k-1}^{(0)} = \emptyset$, $w_{k-1}^{(0)} = \omega_{k-1}^{(0)}$, $w_{k-1}^{(i)} = (1-\omega_{k-1}^{(0)})/\nu$, $i = 1, \ldots, \nu$ and $\omega_{k-1}^{(0)}$ represent the weight when there is no target.

Remark It is generally assumed that all particles have equal importance weights, i.e., $w_{k-1}^{(i)} = 1/\nu, \quad i = 1, \ldots, \nu$. The equal weight assumption is equivalent to treating the particles as a random sampling of uniform distribution, in other words, more particles will be sampled in a larger region of the state space, and less particles will be sampled otherwise. Therefore, the equal weight sampling process is actually equivalent to probability weighting. In the particle multi-target filter, however, the weight $w_{k-1}^{(0)}$ of empty particle $X_{k-1}^{(0)} = \emptyset$ needs to be handled carefully [2]. Since $\emptyset$ is a discrete state, sampling from multi-target density $\pi_{k-1}(X_{k-1}|Z_{1:k-1})$ will generate multiple copies of $\emptyset$. If these copies are merged into a single particle, $w_{k-1}^{(0)}$ should be different from the weights of other particles. In fact, the concept of equal weight in this sense is adopted here.

Similar to the single-target case, a single multi-target random sample can be drawn from the multi-target dynamic prior, i.e., $X_{k|k-1}^{(i)} \sim \phi_{k|k-1}\left(\cdot|X_{k-1}^{(i)}\right), i = 0, 1, \ldots, \nu_{\widehat{n}}$. However, since the multi-target Markov density integrates the target survival, birth, spawning, and death models, the complexity implied by the expression $X_{k|k-1}^{(i)} \sim \phi_{k|k-1}\left(\cdot|X_{k-1}^{(i)}\right)$ is far beyond intuitive understanding. According to the multi-target motion model, the predicted multi-target particles have the following form

$$X_{k|k-1}^{(i)} = X_{s,+}^{(i)} \bigcup X_{\beta,+}^{(i)} \bigcup X_{\gamma,+}^{(i)} \tag{3.7.6}$$

where each item $X_{s,+}^{(i)}$, $X_{\beta,+}^{(i)}$ and $X_{\gamma,+}^{(i)}$ in the union respectively represents the sets of survived targets, spawning targets and birth targets, which are introduced separately as follows.

3.7.1.1 Target Survival and Death

According to the standard multi-target motion model, the probability that the target whose state is $\boldsymbol{x}'$ at time $k-1$ continues to survive at time k is $p_s(\boldsymbol{x}')$, which means that the multi-target probability distribution of $X_{s,+}^{(i)}$ is a multi-target multi-Bernoulli distribution of the form (3.3.9). Let $X_{k-1}^{(i)} = \{\boldsymbol{x}_1'^{(i)}, \ldots, \boldsymbol{x}_{n_i}'^{(i)}\}$ be the multi-target particles with the number n_i of targets at time $k-1$, and let $p_{s,j}^{(i)} = p_s(\boldsymbol{x}_j'^{(i)})$, then the probability that all the targets with state $\{\boldsymbol{x}_1'^{(i)}, \ldots, \boldsymbol{x}_{n_i}'^{(i)}\}$ will survive is $\prod_{j=1}^{n_i} p_{s,j}^{(i)}$, and the probability that all the targets will die is $\prod_{j=1}^{n_i}(1 - p_{s,j}^{(i)})$. More generally, the survival probability of subset $\{\boldsymbol{x}_{j_1}'^{(i)}, \ldots, \boldsymbol{x}_{j_m}'^{(i)}\}$ $(1 \le j_1 < \ldots j_m \le m)$ containing $m \le n_i$ elements is

$$p_{n_i}(m, j_1, \ldots, j_m) = \frac{p_{s,j_1}^{(i)} \cdots p_{s,j_m}^{(i)}}{(1 - p_{s,j_1}^{(i)}) \ldots (1 - p_{s,j_m}^{(i)})} \prod_{j=1}^{n_i} (1 - p_{s,j}^{(i)}) \tag{3.7.7}$$

Then, according to the above distribution, a sample is extracted as follows

$$\left(\tilde{m}, \tilde{j}_1, \ldots, \tilde{j}_{\tilde{m}}\right) \sim p_{n_i}(\cdot, \cdot, \ldots, \cdot) \tag{3.7.8}$$

and for each $\tilde{j}_1, \ldots, \tilde{j}_{\tilde{m}}$, the following particles are drawn

$$\begin{aligned} \boldsymbol{x}_{\tilde{j}_1}^{(i)} &\sim \phi_{k|k-1}(\cdot|\boldsymbol{x}_{\tilde{j}_1}^{\prime(i)}) \\ &\vdots \\ \boldsymbol{x}_{\tilde{j}_{\tilde{m}}}^{(i)} &\sim \phi_{k|k-1}(\cdot|\boldsymbol{x}_{\tilde{j}_{\tilde{m}}}^{\prime(i)}) \end{aligned} \tag{3.7.9}$$

Finally, the set of surviving multi-target particles is obtained as

$$X_{s,+}^{(i)} = \{\boldsymbol{x}_{\tilde{j}_1}^{(i)}, \ldots, \boldsymbol{x}_{\tilde{j}_{\tilde{m}}}^{(i)}\} \tag{3.7.10}$$

The above process is repeated for $i = 1, \ldots, \nu_{\frown n}$.

3.7.1.2 Target Birth

According to the standard multi-target motion model, the multi-target probability density $q_{\gamma,k}(X)$ models birth targets that are not spawned from existing targets. The simplest method to extract $X_{\gamma,+}^{(i)}$ from a known $q_{\gamma,k}^{(i)}(X)$ is as follows. Let $\rho_{\gamma,k}^{(i)}(n)$ be the cardinality distribution of $q_{\gamma,k}^{(i)}(X)$, which is defined by (3.2.15). According to the cardinality distribution, a sample of the number of targets is drawn

$$\tilde{n} \sim \rho_{\gamma,k}^{(i)}(\cdot) \tag{3.7.11}$$

and then a multi-target state sample is drawn from the following distribution

$$q_{\gamma,k}^{(i)}(\boldsymbol{x}_1^{(i)}, \ldots, \boldsymbol{x}_{\tilde{n}}^{(i)}) = q_{\gamma,k}^{(i)}(\{\boldsymbol{x}_1^{(i)}, \ldots, \boldsymbol{x}_{\tilde{n}}^{(i)}\}) \big/ \tilde{n}! \tag{3.7.12}$$

that is,

$$\left(\tilde{\boldsymbol{x}}_1^{(i)}, \ldots, \tilde{\boldsymbol{x}}_{\tilde{n}}^{(i)}\right) \sim q_{\gamma,k}^{(i)}(\cdot, \ldots, \cdot) \tag{3.7.13}$$

Finally, the set of the birth multi-target particles is obtained as

$$X_{\gamma,+}^{(i)} = \{\tilde{\boldsymbol{x}}_1^{(i)}, \ldots, \tilde{\boldsymbol{x}}_{\tilde{n}}^{(i)}\} \tag{3.7.14}$$

The above process is repeated for $i = 1, \ldots, \nu_{\frown n}$.

3.7.1.3 Target Spawning

The probability density $q_{\beta,+}^{(i)}(X|\boldsymbol{x}')$ models the set of targets spawned from the existing target in state $\boldsymbol{x}'$ at time $k-1$. In the same way that $X_{\gamma,+}^{(i)}$ is drawn from $q_{\gamma,k}^{(i)}(X)$, sample $X_{\beta,+}^{(i)}$ can be drawn from $q_{\beta,+}^{(i)}(X|\boldsymbol{x}')$.

Let $\rho_{\beta,+}^{(i)}(n|\boldsymbol{x}')$ be the cardinality distribution of $q_{\beta,+}^{(i)}(X|\boldsymbol{x}')$, and $X_{k-1}^{(i)} = \{\boldsymbol{x}_1'^{(i)}, \ldots, \boldsymbol{x}_{n'}'^{(i)}\}$ denote the multi-target particle at time $k-1$. From the cardinality distribution, the samples of the numbers of targets corresponding to each state are drawn

$$\begin{aligned} \tilde{n}_1 &\sim \rho_{\beta,+}^{(i)}\left(\cdot|\boldsymbol{x}_1'^{(i)}\right) \\ &\vdots \\ \tilde{n}_{n'} &\sim \rho_{\beta,+}^{(i)}\left(\cdot|\boldsymbol{x}_{n'}'^{(i)}\right) \end{aligned} \tag{3.7.15}$$

Then, for $j = 1, \ldots, n'$, a multi-target state sample is drawn from the following distribution

$$q_{\beta,+}^{(i,j)}(\boldsymbol{x}_1^{(i)}, \ldots, \boldsymbol{x}_{\tilde{n}_j}^{(i)}) = q_{\beta,+}^{(i)}(\{\boldsymbol{x}_1^{(i)}, \ldots, \boldsymbol{x}_{\tilde{n}_j}^{(i)}\}|\boldsymbol{x}_j'^{(i)}) \Big/ \tilde{n}_j! \tag{3.7.16}$$

that is,

$$\begin{aligned} \left(\tilde{\boldsymbol{x}}_1^{(i,1)}, \ldots, \tilde{\boldsymbol{x}}_{\tilde{n}_1}^{(i,1)}\right) &\sim q_{\beta,+}^{(i,1)}(\cdot, \ldots, \cdot) \\ &\vdots \\ \left(\tilde{\boldsymbol{x}}_1^{(i,n')}, \ldots, \tilde{\boldsymbol{x}}_{\tilde{n}_{n'}}^{(i,n')}\right) &\sim q_{\beta,+}^{(i,n')}(\cdot, \ldots, \cdot) \end{aligned} \tag{3.7.17}$$

Finally, the spawning multi-target particle set is obtained as

$$X_{\beta,+}^{(i)} = \{\tilde{\boldsymbol{x}}_1^{(i,1)}, \ldots, \tilde{\boldsymbol{x}}_{\tilde{n}_1}^{(i,1)}, \ldots, \tilde{\boldsymbol{x}}_1^{(i,n')}, \ldots, \tilde{\boldsymbol{x}}_{\tilde{n}_{n'}}^{(i,n')}\} \tag{3.7.18}$$

The above process is repeated for $i = 1, \ldots, \nu_{\widehat{n}}$.

3.7.2 *Update of Particle Multi-target Filter*

Assume that the predicted multi-target particle system is

$$\pi_{k|k-1}(X_k|Z_{1:k-1}) \approx w_+^{(0)}\delta_\emptyset(X_k) + \frac{1-w_+^{(0)}}{\nu}\sum_{i=1}^{\nu}\delta_{X_{k|k-1}^{(i)}}(X_k)$$

$$= \sum_{i=0}^{\nu} w_{k|k-1}^{(i)} \delta_{X_{k|k-1}^{(i)}}(X_k) \tag{3.7.19}$$

where $X_{k|k-1}^{(0)} = \emptyset$, $w_{k|k-1}^{(0)} = w_{+}^{(0)}$ and $w_{k|k-1}^{(i)} = (1 - w_{+}^{(0)})/\nu$. According to (3.7.2), when $q_k(\tilde{X}_k^{(i)}|X_{k-1}^{(i)}, Z_k)$ adopts dynamic prior $\phi_{k|k-1}(\tilde{X}_k^{(i)}|X_{k-1}^{(i)})$, $w_{k|k-1}^{(i)} = w_{k-1}^{(i)}$, $i = 0, \ldots, \nu$.

Substituting (3.7.19) into the normalization factor of the multi-target Bayesian recursion can obtain

$$\int g_k(Z_k|X)\pi_{k|k-1}(X|Z_{1:k-1})\delta X = \int g_k(Z_k|X)\left[\sum_{i=0}^{\nu} w_{k|k-1}^{(i)} \delta_{X_{k|k-1}^{(i)}}(X_k)\right]\delta X$$

$$= \sum_{i=0}^{\nu} w_{k|k-1}^{(i)} g_k(Z_k|X_{k|k-1}^{(i)}) \tag{3.7.20}$$

Thus, the multi-target distribution after the multi-target Bayesian update is

$$\pi_k(X_k|Z_{1:k}) = \frac{\sum_{i=0}^{\nu} g_k(Z_k|X_{k|k-1}^{(i)}) w_{k|k-1}^{(i)} \delta_{X_{k|k-1}^{(i)}}(X_k)}{\sum_{i=0}^{\nu} w_{k|k-1}^{(i)} g_k(Z_k|X_{k|k-1}^{(i)})}$$

$$= \sum_{i=0}^{\nu} w_k^{(i)} \delta_{X_k^{(i)}}(X_k) \tag{3.7.21}$$

where $X_k^{(i)} = X_{k|k-1}^{(i)}$, and the corresponding weight is

$$w_k^{(i)} = \frac{g_k(Z_k|X_{k|k-1}^{(i)}) w_{k|k-1}^{(i)}}{\sum_{i=0}^{\nu} w_{k|k-1}^{(i)} g_k(Z_k|X_{k|k-1}^{(i)})}, \quad i = 0, 1, \ldots, \nu \tag{3.7.22}$$

The particles now have unequal weights, so they must be replaced with new equally weighted particle systems. The replaced particles, however, should reflect the influence of $w_k^{(i)}$, which can be achieved by resampling techniques, such as removing small-weighted particles while copying large-weighted particles, so that a new posterior particle system $\tilde{X}_k^{(0)}, \tilde{X}_k^{(1)}, \ldots, \tilde{X}_k^{(\nu)}$ and the following new approximation can be obtained

$$\pi_k(X_k|Z_{1:k}) = \omega_k^{(0)} \delta_{\emptyset}(X_k) + \frac{1 - \omega_k^{(0)}}{\nu} \sum_{i=1}^{\nu} \delta_{X_k^{(i)}}(X_k) \tag{3.7.23}$$

After the resampling step, because resampling duplicates identical copies of particles, it results in "particle depletion" such that all but a few particles have almost negligible weights. To increase the statistical diversity of resampling particles[397], similar to the single-target case, a Markov chain Monte Carlo (MCMC) step can be chosen to increase particle diversity. Since the particles belong to different dimensional spaces, the reversible jump MCMC step needs to be used [398]. Under standard assumptions, the mean square error of the SMC approximation is inversely proportional to the number of particles [62].

After obtaining the posterior density, the multi-target state estimation also needs to be considered. Given the particle approximation $\{(w_k^{(i)}, X_k^{(i)})\}_{i=0}^{\nu}$ of π_k, the particle approximation of v_k can be obtained by

$$v_k(\boldsymbol{x}) \approx \sum_{i=0}^{\nu} w_k^{(i)} \left[\sum_{\boldsymbol{x}' \in X_k^{(i)}} \delta_{\boldsymbol{x}'}(\boldsymbol{x}) \right] = \sum_{i=0}^{\nu} \sum_{\boldsymbol{x}' \in X_k^{(i)}} w_k^{(i)} \delta_{\boldsymbol{x}'}(\boldsymbol{x}) \tag{3.7.24}$$

Thus, the state estimation of each target can be obtained based on the peak value of v_k. Please refer to Sect. 4.4.4 for details.

3.8 Performance Metrics of Multi-target Tracking

In the filtering/control problem, the miss-distance between the reference quantity and its estimated/control value is an important concept, sometimes simply referred to as the error. The multi-target miss-distance is detailed as follows. From a mathematical point of view, the basic necessary condition for allowing the consensus distance to be measurable is that the miss-distance is a metric on the space of a finite set of targets. Let $\mathcal{X}$ be any non-empty set, if a function $d : \mathcal{X} \times \mathcal{X} \to \mathbb{R}_+ = [0,\infty)$ satisfies the following 3 conditions, then the function is called a metric:

(1) Identity: If and only if $\boldsymbol{x} = \boldsymbol{y}$, $d(\boldsymbol{x},\boldsymbol{y}) = 0$;
(2) Symmetry: For all $\boldsymbol{x},\boldsymbol{y} \in \mathcal{X}$, $d(\boldsymbol{x},\boldsymbol{y}) = d(\boldsymbol{y},\boldsymbol{x})$;
(3) Triangle inequality: For all $\boldsymbol{x},\boldsymbol{y},\boldsymbol{z} \in \mathcal{X}$, $d(\boldsymbol{x},\boldsymbol{z}) \leq d(\boldsymbol{x},\boldsymbol{y}) + d(\boldsymbol{y},\boldsymbol{z})$.

In the context of multi-target miss-distance, a closed and bounded observation window $W \subset \mathbb{R}^N$ is fixed, and $\mathcal{X}$ is chosen to be a set of finite subsets of W. In this section, d always represents a metric on W (such as the typical Euclidean metric $d(\boldsymbol{x},\boldsymbol{y}) = ||\boldsymbol{x} - \boldsymbol{y}||$), and for different metrics considered on $\mathcal{X}$, the corresponding subscripts and superscripts are attached to distinguish them, such as d_H, d_p or $\overline{d}_p^c$.

3.8.1 Hausdorff Metric

The Hausdorff metric has a long history in random geometry due to its theoretical advantages. It yields a standard topology [399] over a set of closed subsets of W, often referred to as the Matheron topology [14] in the FISST context, which can be used to define random sets. For finite non-empty subsets X and Y of W, the Hausdorff metric d_H is defined as

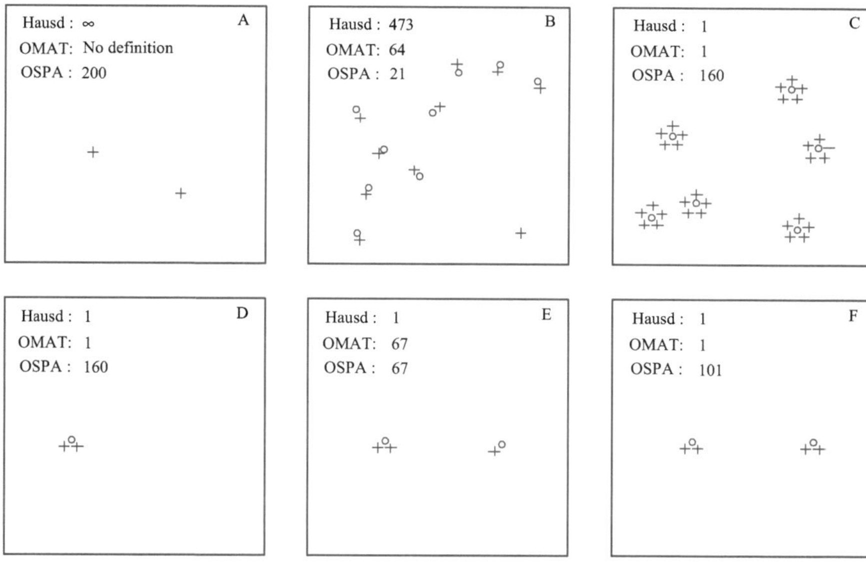

Fig. 3.2 Advantages and disadvantages of multi-target tracking performance metrics in different scenarios [359] o stands for a target and + stands for an estimation. **a** Two false estimates; **b** Several accurate estimates accompanied by an outlier; **c** Each target corresponds to multiple estimates; **d–f** Comparison on balance and imbalance of the assignment between estimates and targets

$$d_H(X,Y) = \max\left\{\max_{x\in X}\min_{y\in Y} d(\boldsymbol{x},\boldsymbol{y}),\ \max_{y\in Y}\min_{x\in X} d(\boldsymbol{x},\boldsymbol{y})\right\} \tag{3.8.1}$$

The proof that d_H is a metric is given in [358], which also discusses some of its advantages and difficulties in multi-target filtering. Only the most important points are highlighted here. First, the Hausdorff metric is traditionally used as a measure of the dissimilarity between binary images, for example, it is very suitable for describing the difference between two images in optical suppression. However, it is very insensitive to the difference in the cardinalities of a finite set (as shown in scenarios C-F in Fig. 3.2), which is obviously undesirable for the performance evaluation of multi-target filters. In addition, it penalizes outliers too heavily (see scenario B), and it can not give a reasonable result if a set is empty (see scenario A). At this time, some studies simply set this case as ∞.

3.8.2 Optimal Mass Transfer (OMAT) Metric

To overcome the problems of the Hausdorff metric in the performance evaluation of multi-target filtering, Hoffman and Mahler introduced a new metric in 2004: optimal mass transfer (OMAT) [358].

For $1 \leq p < \infty$ and finite non-empty subsets $X = \{\boldsymbol{x}_1, \ldots, \boldsymbol{x}_m\}$ and $Y = \{\boldsymbol{y}_1, \ldots, \boldsymbol{y}_n\}$ of W, define

$$d_p(X,Y) \triangleq \min_{\boldsymbol{C}} \left\{ \sum_{i=1}^{m} \sum_{j=1}^{n} \boldsymbol{C}_{i,j} [d(\boldsymbol{x}_i, \boldsymbol{y}_j)]^p \right\}^{1/p} \tag{3.8.2}$$

$$d_\infty(X,Y) \triangleq \min_{\boldsymbol{C}} \max_{1 \leq i \leq m, 1 \leq j \leq n} \tilde{\boldsymbol{C}}_{i,j} d(\boldsymbol{x}_i, \boldsymbol{y}_j) \tag{3.8.3}$$

where the minimum symbol $\min(\cdot)$ traverses all transfer matrices $\boldsymbol{C} = (\boldsymbol{C}_{i,j})$ of size $m \times n$. If $\boldsymbol{C}_{i,j} \neq 0$, $\tilde{\boldsymbol{C}}_{i,j} = 1$; otherwise, $\tilde{\boldsymbol{C}}_{i,j} = 0$. If all elements of the matrix $\boldsymbol{C}$ with dimension $m \times n$ are non-negative, and if $\sum_{j=1}^{n} \boldsymbol{C}_{i,j} = 1/m$ for $1 \leq i \leq m$ and $\sum_{i=1}^{m} \boldsymbol{C}_{i,j} = 1/n$ for $1 \leq j \leq n$, then $\boldsymbol{C}$ is called a transfer matrix.

Equation (3.8.2) actually defines the p-order Wasserstein metric between the empirical distributions of point patterns X and Y, thus Hoffman and Mahler also call it the Wasserstein metric. Since the optimal subpattern assignment (OSPA) metric introduced later is also based on the Wasserstein metric, to avoid potential confusion, the function d_p is referred to as the OMAT metric of order p.

The advantage of the OMAT metric is that it partially addresses the undesired cardinality behavior of the Hausdorff metric (see scenario E in Fig. 3.2), and can handle outliers by introducing the parameter p (see scenario B in Fig. 3.2). However, the OMAT metric still suffers from many problems, such as consistency, intuitive explanation, geometry dependence, zero cardinality, compatibility with mathematical theory, etc. The description of these problems is detailed in [359].

3.8.3 Optimal Subpattern Assignment (OSPA) Metric

The optimal subpattern assignment (OSPA) metric is also constructed based on the Wasserstein metric, but completely overcomes the aforementioned problems of the OMAT metric. The OSPA metric is increasingly being used as a performance measure of multi-target estimation, and has played an important role in the development of novel filters and estimators (such as the set JPDA filter [400], minimum average OSPA estimator [401] and other filters based on the RFS framework).

Denote $d^{(c)}(\boldsymbol{x}, \boldsymbol{y}) \triangleq \min(c, d(\boldsymbol{x}, \boldsymbol{y}))$ as the distance between $\boldsymbol{x}, \boldsymbol{y} \in W$, which is capped at $c > 0$, and use Ω_k to represent the set of permutations on $\{1, 2, \ldots, k\}$, where $k \in \mathbb{N} = \{1, 2, \ldots\}$. For $1 \leq p < \infty$, $c > 0$, and any finite subsets $X = \{\boldsymbol{x}_1, \ldots, \boldsymbol{x}_m\}$ and $Y = \{\boldsymbol{y}_1, \ldots, \boldsymbol{y}_n\}$ of W with $m, n \in \mathbb{N}_0 = \{0, 1, 2, \ldots\}$, if $m \leq n$, define

$$\overline{d}_p^{(c)}(X,Y) \triangleq \{n^{-1}[\min_{\sigma \in \Omega_n} \sum_{i=1}^{m} (d^{(c)}(\boldsymbol{x}_i, \boldsymbol{y}_{\sigma(i)}))^p + c^p(n-m)]\}^{1/p} \tag{3.8.4}$$

Conversely, if $m > n$, $\overline{d}_p^{(c)}(X,Y) \triangleq \overline{d}_p^{(c)}(Y,X)$. In addition,

$$\overline{d}_\infty^{(c)}(X,Y) \triangleq \begin{cases} \min_{\sigma\in\Omega_n} \max_{1\leq i\leq n} d^{(c)}(\boldsymbol{x}_i,\boldsymbol{y}_{\sigma(i)}), & \text{if } m = n \\ c, & \text{if } m \neq n \end{cases} \tag{3.8.5}$$

and if $m = n = 0$, $\overline{d}_\infty^{(c)}(X,Y) = 0$ by convention. The function $\overline{d}_p^{(c)}$ is called the p-order OSPA metric with a cut-off value c, and [359] has proved that the OSPA satisfies the metric axiom.

For $p < \infty$ and assuming $m \leq n$, the OSPA distance between two point patterns X and Y is solved by the following 3 steps.

(1) From the perspective of the p-order Wasserstein metric (i.e., the p-order OMAT metric), the sub-pattern (a subset composed of m elements) of m points of Y that is closest to X is found out to obtain the optimal point assignment.
(2) For each point $\boldsymbol{y}_j$ of Y, if no point is assigned to it, α_j is set as a cut-off value c, otherwise, α_j is set as the minimum of both the distance between it and the assigned point in X and c.
(3) The p-order mean $(n^{-1}\sum_{j=1}^{n}\alpha_j^p)^{1/p}$ of $\alpha_1, \ldots, \alpha_n$ is calculated.

Another slightly different calculation method of the OSPA metric is as follows: first, fill in the point pattern X with a smaller cardinality (whose cardinality is m), and let $n-m$ dummy points locate at a distance $\geq c$ from any point in Y (hence, it is usually necessary to expand the window W). Then, calculate the p-order Wasserstein metric between the obtained point patterns. The first method is more suitable for practical use, while the second method is convenient for theoretical derivation.

Calculate the OSPA distance between the two point patterns on the left in Fig. 3.3. The window is assumed as 1000×1000 m^2, $p = 1$, and $c = 200$. Examining each point $+$ of Y (point pattern with larger cardinality), it can be concluded that three α_j s are equal to the cut-off value of 200 m, and seven α_j s are equal to the length 90 m of the dotted line. Thus, $\overline{d}_1^{(c)}(X,Y) = 10^{-1}(3 \cdot 200 + 7 \cdot 90)\text{m} = 123\,\text{m}$. For the left panel, the optimal point assignment is easily obtained by observation because, for any other assignments, the average length of the dotted line will be greater. However, the right panel is more complicated, and it is difficult to directly obtain the optimal point assignment by the observation method, and the Hungarian method needs to be used.

The OSPA metric $\overline{d}_p^{(c)}$ can be efficiently computed by calculating the optimal point assignment using the Hungarian method. For $p < \infty$ and two point patterns $X = \{\boldsymbol{x}_1, \ldots, \boldsymbol{x}_m\}$ and $Y = \left\{\boldsymbol{y}_1, \ldots; \boldsymbol{y}_n\right\}$ with $m \leq n$, use the distance matrix $\boldsymbol{D} = \{\boldsymbol{D}_{i,j}\}_{1\leq i,j\leq n}$, where $\boldsymbol{D}_{i,j} = (d^{(c)}(\boldsymbol{x}_i,\boldsymbol{y}_j))^p$ if $1 \leq i \leq m$ and $1 \leq j \leq n$, and $\boldsymbol{D}_{i,j} = c^p$ otherwise. This corresponds to introducing $n - m$ dummy points at distance $\geq c$ from any point in Y. The complexity of the Hungarian method is cubic with the dimension of the distance matrix, so the complexity of calculating the OSPA distance is $\mathcal{O}(\max(m, n)^3)$.

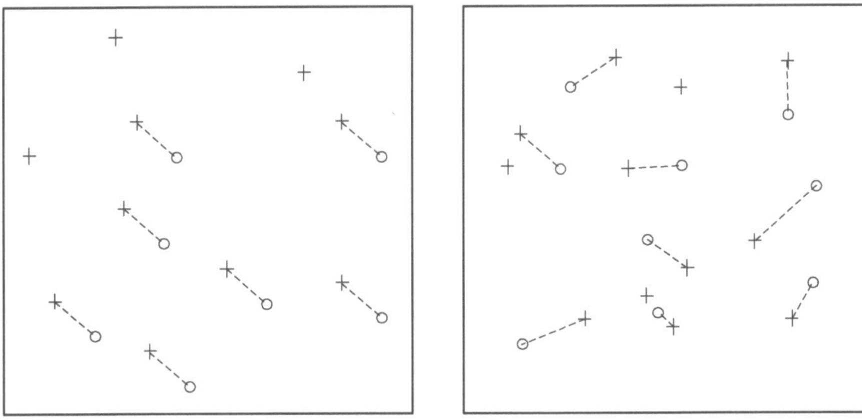

Fig. 3.3 Optimal subpattern assignment [359]

3.8.3.1 Application of OSPA Metric in Multi-target Estimation

In the context of multi-target performance evaluation, the OSPA distance can be interpreted as the p-order error of "per-target".[3] This error consists of two parts, namely, "localization" error and "cardinality" error. Precisely, for $p < \infty$, if $m \leq n$, the two components are respectively

$$\bar{e}^{(c)}_{p,\mathrm{loc}}(X,Y) \triangleq \{n^{-1} \min_{\sigma \in \Omega_n} \sum_{i=1}^{m} [d^{(c)}(\boldsymbol{x}_i, \boldsymbol{y}_{\sigma(i)})]^p\}^{1/p} \tag{3.8.6}$$

$$\bar{e}^{(c)}_{p,\mathrm{card}}(X,Y) \triangleq [n^{-1}(n-m)c^p]^{1/p} \tag{3.8.7}$$

Otherwise, if $m > n$, $\bar{e}^{(c)}_{p,\mathrm{loc}}(X,Y) \triangleq \bar{e}^{(c)}_{p,\mathrm{loc}}(Y,X)$, $\bar{e}^{(c)}_{p,\mathrm{card}}(X,Y) \triangleq \bar{e}^{(c)}_{p,\mathrm{card}}(Y,X)$. Therefore, the two components can be respectively interpreted as the miss-distance caused purely by localization (within the OSPA) and purely by cardinality (penalty at the maximum distance). Note that the functions $\bar{e}^{(c)}_{p,\mathrm{loc}}$ and $\bar{e}^{(c)}_{p,\mathrm{card}}$ are not metrics over the space of finite subsets. However, $\bar{e}^{(c)}_{p,\mathrm{loc}}$ can be regarded as a metric over the space of finite subsets with a fixed cardinality (that is, once the optimal assignment is determined), while $\bar{e}^{(c)}_{p,\mathrm{card}}(X,Y)$ can be considered as a metric over the space of non-negative integers (i.e., only the cardinality of the set). Also, it should be noted that it is usually not necessary to decompose the OSPA metric into individual components when evaluating performance, but the decomposition may provide additional valuable information.

[3] Strictly speaking, the "per-target" error is the "per-estimated-target" error when the cardinality is overestimated, and the "per-true-target" error when the cardinality is underestimated.

3.8.3.2 Meaning of Parameters p and c

The role played by the order parameter p is similar to its role in the OMAT metric. Keeping c fixed, as p increases, it becomes more difficult for metric $\overline{d}_p^{(c)}$ to accept an outlier estimates not close to any true target. Because the larger the p, the p-order mean assigns the greater weight to abnormally large values in α_j. Note, however, that in the OSPA metric, this behavior is curbed to some extent, because the distance between two points is cut off at c, which is also the penalty deserved for a point when it is deemed "unassignable". According to Hölder's inequality, it is easy to know that the OSPA distance increases as p value grows, that is, for $1 \leq p_1 < p_2 < \infty$ and $c > 0$,

$$\overline{d}_{p_1}^{(c)}(X, Y) \leq \overline{d}_{p_2}^{(c)}(X, Y)$$

Thus, order parameter p controls the penalty assigned to outlier estimates (estimates that are not close to any true track). The value of p determines the sensitivity of $\overline{d}_p^{(c)}$ to outlier estimates, with a higher p value increasing the sensitivity to outliers. Two important choices of p shall be noted here. For $p = 1$, the OSPA measures the first-order error of "per-target". At this time, the sum of the localization component and the cardinality component is equal to the total metric, which facilitates the direct interpretation of the total OSPA metric and its components. However, $p = 2$ is usually a more practical choice, as a smooth distance curve can be obtained in this case, which is a typical problem faced by other metrics when constructing the p-order mean. In the subsequent parts, no attention will be paid to the impact of choosing different p, and $p = 2$ is assumed by default.

The cut-off parameter c determines the relative weight of the cardinality error component as a part of the total error relative to the localization error component. The smaller c value tends to highlight the localization error, thus making the metric less sensitive to the cardinality error, while a larger c value mainly strengthens the cardinality error and weakens the localization error. The following principle can be followed when selecting the cut-off value c: the c value corresponding to the typical localization error can be deemed as a smaller c value, which emphasizes the influence of the localization error; the larger c value corresponding to the maximum distance between targets can be considered as a larger c value, which emphasizes the impact of the cardinality error; while any c value that is significantly larger than the typical localization error yet significantly smaller than the maximum distance between targets can be considered moderate, which trades off between the two components. Note that the cardinality error comes as a result of an unequal number of elements in X and Y, so c can also be interpreted as a penalty assigned to missed detections or false alarms.

According to the definition of the OSPA metric, it is obvious that for any $c > 0$,

$$\overline{d}_p^{(c)}(X, Y) \in [0, c] \tag{3.8.8}$$

and for $0 < c_1 < c_2 < \infty$,

$$\overline{d}_p^{(c_1)}(X, Y) \leq \overline{d}_p^{(c_2)}(X, Y) \tag{3.8.9}$$

Equation (3.8.8) can be used to adjust $\overline{d}_p^{(c)}$.

3.8.4 OSPA Metric Incorporating Labeling Errors

The OSPA metric does not consider the track label error. To solve this problem, a new metric needs to be defined on the space of finite sets containing the track information. The track is defined on the discrete-time support points $\mathcal{T} = (t_1, t_2, \ldots, t_K)$, and a track $\mathbf{x}$ on $\mathcal{T}$ is a labeled sequence of length K, i.e.,

$$\mathbf{x} = (\mathbf{x}_1, \mathbf{x}_2, \ldots, \mathbf{x}_K) \tag{3.8.10}$$

where $\mathbf{x}_k$ is an empty set or a singleton whose element is $(\ell, \boldsymbol{x}_k)$. Here, $\ell \in \mathbb{L}$ is the track label, which usually does not change with time, and $\boldsymbol{x}_k$ is a time-varying state vector that corresponds to a point in N-dimensional state space $W : \boldsymbol{x}_k \in W \subseteq \mathbb{R}^N$.

For convenience, an indicator e_k is introduced and defined as follows: if a target exists at time t_k, e_k takes the value of 1, otherwise, takes the value of 0, namely,

$$\mathbf{x}_k = \begin{cases} \{(\ell, \boldsymbol{x}_k)\}, & \text{if } e_k = 1 \\ \emptyset, & \text{if } e_k = 0 \end{cases} \tag{3.8.11}$$

To define a metric at specific time t_k, $k = (1, 2, \ldots, K)$, let the set of all tracks on $\mathcal{T}$ at a specified time t_k be denoted as $\mathbb{X}_k$ (namely, $\mathbf{x}_k \in \mathbb{X}_k$), and the set of finite subsets of $\mathbb{X}_k$ be $\mathcal{X}_k$; Define a metric space $(\mathcal{X}_k, d)$, where function $d : \mathcal{X}_k \times \mathcal{X}_k \to \mathbb{R}_+ = [0, \infty)$ satisfies the 3 metric axioms of identity, symmetry, and triangle inequality.

If $\mathrm{X}_k \in \mathcal{X}_k$ represents the set of true tracks at time t_k, and $\mathrm{Y}_k \in \mathcal{X}_k$ is the set of estimated tracks obtained by an algorithm at time t_k, the metric $d(\mathrm{X}_k, \mathrm{Y}_k)$ should quantify the total estimation error of the tracking algorithm at time t_k, and should integrate different aspects (such as, real-time, tracking accuracy, continuity, data association, false tracks, etc.) of tracking performance into the metric in a mathematically rigorous manner.

The multi-target tracking metric (denoted as the OSPA-L metric) introduced herein is based on the OSPA metric. The OSPA-L metric on $\mathcal{X}_k$ is the distance between any two target sets. At this time, the target is the track at time t_k, and the two sets are

$$\mathrm{X}_k = \left\{\left(\ell_1, \boldsymbol{x}_{k,1}\right), \ldots, \left(\ell_m, \boldsymbol{x}_{k,m}\right)\right\} \tag{3.8.12}$$

$$\mathrm{Y}_k = \left\{\left(\ell'_1, y_{k,1}\right), \ldots, \left(\ell'_n, y_{k,n}\right)\right\} \tag{3.8.13}$$

where $\mathrm{X}_k \in \mathcal{X}_k$ and $\mathrm{Y}_k \in \mathcal{X}_k$ respectively represent the true track and the track generated by the tracker at time t_k. According to (3.8.11), the cardinalities of set X_k and set Y_k are m and n, respectively, both related to k. When $m \leq n$, the OSPA-L distance between X_k and Y_k is defined as

$$\overline{d}_p^{(c)}(\mathrm{X}_k, \mathrm{Y}_k) \triangleq \left\{n^{-1}\left[\min_{\sigma\in\Omega_n}\sum_{i=1}^{m}\left(d^{(c)}\left(\mathbf{x}_{k,i}, \mathbf{y}_{k,\sigma(i)}\right)\right)^p + c^p(n-m)\right]\right\}^{1/p} \tag{3.8.14}$$

where $\mathbf{x}_{k,i} = \left(\ell_i, \boldsymbol{x}_{k,i}\right)$, $\mathbf{y}_{k,\sigma(i)} = \left(\ell'_{\sigma(i)}, y_{k,\sigma(i)}\right)$, and $d^{(c)}(\mathbf{x}, \mathbf{y}) = \min(c, d(\mathbf{x}, \mathbf{y}))$ is the cutoff distance between two tracks at time t_k; $c > 0$ is the cutoff parameter; $d(\mathbf{x},\mathbf{y})$ represents the base distance between two tracks at time t_k; Ω_n denotes the set of permutations on $\{1, 2, \ldots, n\}$; p is the order parameter of the OSPA metric, satisfying the condition $1 \leq p < \infty$. The selection principle of parameters c and p is consistent with the original OSPA metric.

For the case of $m > n$, define $\overline{d}_p^{(c)}(\mathrm{X}_k, \mathrm{Y}_k) \triangleq \overline{d}_p^{(c)}(\mathrm{Y}_k, \mathrm{X}_k)$. If both X_k and Y_k are empty sets (namely, $m = n = 0$), the OSPA-L distance is 0.

To determine the OSPA-L metric for tracks, it is necessary to define the base distance $d(\mathbf{x},\mathbf{y})$ between tracks $\mathbf{x} = (\ell, \boldsymbol{x})$ and $\mathbf{y} = \left(\ell', y\right)$.

3.8.4.1 Base Distance Between Two Labeled Vectors

The base distance $d(\mathbf{x},\mathbf{y})$ is a metric on space $\mathbb{R}^N \times \mathbb{L}$, defined as

$$d(\mathbf{x},\mathbf{y}) = \left(d^{p'}(\mathbf{x}, \mathbf{y}) + d^{p'}\left(\ell, \ell'\right)\right)^{1/p'} \tag{3.8.15}$$

where $1 \leq p' < \infty$ is the order parameter of the base distance; $d(\boldsymbol{x},\boldsymbol{y})$ is the positional base distance, which is a metric defined on $\mathbb{R}^N$, and is often taken as the p'-norm: $d(\boldsymbol{x}, \boldsymbol{y}) = \|\boldsymbol{x} - \boldsymbol{y}\|_{p'}$; $d\left(\ell, \ell'\right)$ is the label error, which is a metric defined on $\mathbb{L}$ and can be taken as $d(\ell,\ell') = \alpha\ \bar{\delta}[\ell,\ell']$, where $\bar{\delta}\left[\ell, \ell'\right]$ is the complement of the Kronecker delta, i.e., if $\ell = \ell'$, $\bar{\delta}\left[\ell, \ell'\right] = 0$, otherwise, $\bar{\delta}\left[\ell, \ell'\right] = 1$. Parameter $\alpha \in [0,c]$ controls the degree of penalty assigned to the label error $d\left(\ell, \ell'\right)$ relative to positional miss-distance $d(\boldsymbol{x},\boldsymbol{y})$. The case $\alpha = 0$ means no penalty is assigned, while $\alpha = c$ assigns the heaviest penalty. For the specific selection of α, refer to [360]. It is easy to prove that the base distance defined in (3.8.15) satisfies the metric axioms [360].

After defining the base distance (3.8.15), the OSPA-L metric for tracks still cannot be determined, and the label of the estimated track also needs to be determined.

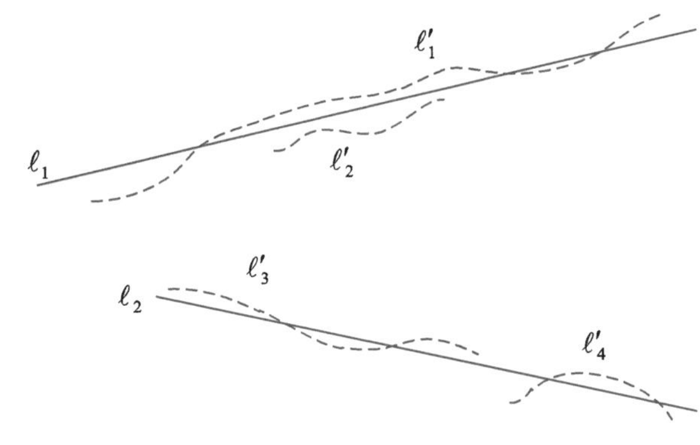

Fig. 3.4 Two true tracks ℓ_1, ℓ_2 (solid lines) and four estimated tracks $\ell'_1, \ell'_2, \ell'_3, \ell'_4$ (dotted lines) [360]

3.8.4.2 Label of Estimated Track

When evaluating the tracking performance, it is necessary to assign the tracker output to the true tracks. To explain the need to label the estimated tracks, consider an explanatory example as shown in Fig. 3.4. In the figure, there are 2 true tracks labeled with ℓ_1 and ℓ_2 (as shown by the solid lines) and 4 estimated tracks labeled with $\ell'_1, \ldots, \ell'_4$ (as shown by dotted lines). To determine the OSPA-L metric for tracks, it is necessary to assign the labels of true tracks (ℓ_1 and ℓ_2 in the figure) to some estimated tracks in a globally optimal way. For the scenario in the figure, it is expected that ℓ_1 is assigned to track ℓ'_1, as estimated track ℓ'_1 is the optimal approximation of ℓ_1. Similarly, ℓ_2 should be assigned to ℓ'_3, and it is necessary to ensure that labels different from ℓ_1 and ℓ_2 are assigned to other two estimated tracks.

A simple assignment method is to use the existing two-dimensional assignment algorithm, such as the Munkres method, Jonker-Volgenant-Castanon (JVC) method, auction method, etc. Assume that the sets of true tracks and estimated tracks are respectively $\{\mathbf{x}_1, \ldots, \mathbf{x}_m\}$ and $\{\mathbf{y}_1, \ldots, \mathbf{y}_n\}$. Denote the track state $\mathbf{x}_\ell$ at time t_k as $\mathbf{x}_{k,\ell}$. Based on the existence indicator for tracks defined in (3.8.11), when $m \leq n$, the optimal global assignment σ^* can be obtained from the following equation

$$\sigma^* = \arg\min_{\sigma \in \Omega_n} \sum_{\ell=1}^{m} \sum_{k=1}^{K} \Big[e_k^\ell e_k^{\sigma(\ell)} \min(\Delta, ||\boldsymbol{x}_{k,\ell} - \boldsymbol{y}_{k,\sigma(\ell)}||_2) \\ + (1 - e_k^\ell) e_k^{\sigma(\ell)} \Delta + e_k^\ell (1 - e_k^{\sigma(\ell)}) \Delta \Big] \tag{3.8.16}$$

where Δ corresponds to the penalty assigned to missed or false tracks, which is relatively insensitive to the results of the OSPA-L [360]; e_k^ℓ and $e_k^{\sigma(l)}$ are the presence indicators of true tracks and estimated tracks from σ-assignment, respectively. For the case of $m > n$, (3.8.16) needs to be modified properly.

Through careful examination of (3.8.16), it can be seen that the item in the square bracket is actually the OSPA distance between $\mathbf{x}_{k,\ell}$ and $\mathbf{y}_{k,\sigma(\ell)}$ when $p = 2$ and $c = \Delta$. According to (3.8.11), these two sets are either empty sets or singletons, thus they

Table 3.3 Calculation steps of the OSPA-L metric

1	Function OSPA-L $(\{\mathbf{x}_1, \ldots, \mathbf{x}_m\}, \{\mathbf{y}_1, \ldots, \mathbf{y}_n\})$
2	% label the estimated tracks
3	for $j = 1, \ldots, n$
4	label $[\mathbf{y}^{(i)}] = I$(initial value, different from labels of all true tracks)
5	end
6	Find the global optimal assignment σ^* for tracks $\{\mathbf{x}_1, \ldots, \mathbf{x}_m\}$ to $\{\mathbf{y}_1, \ldots, \mathbf{y}_n\}$
7	for $i = 1, \ldots, m$
8	label $[\mathbf{y}^{(\sigma^*(i))}] =$ label $[\mathbf{x}^{(i)}]$
9	end
10	% calculate the distance
11	for $k = 1, \ldots, K$
12	According to (3.8.12) and (3.8.13), label sets X_k and Y_k at time t_k are formed
13	Based on the base distance (3.8.15), calculate the OSPA distance at time t_k according to (3.8.14)
14	end

are in the form of (3.8.16). Hence, by minimizing the overall OSPA distance between the paired tracks accumulated at discrete time points $\mathcal{T}$, the assignment of estimated tracks to true tracks is obtained. A larger Δ value will help to assign estimated tracks with a longer duration to true tracks. If an estimated track corresponds to a true track labeled ℓ through σ^* assignment, then its label is also set to ℓ. According to the σ^*-assignment, the labels different from those of all true tracks are assigned to the remaining unassigned estimated tracks.

Table 3.3 summarizes the basic steps for computing the OSPA-L metric.

3.9 Summary

The chapter introduces the basic knowledge involved in the RFS-based target tracking algorithms in detail, including the statistical descriptors of RFSs and the important concepts such as first-order moment (PHD) and second-order moment (cardinality), and gives the common RFS types related to the algorithms to be introduced later, mainly including unlabeled RFSs such as Poisson RFS, IIDC RFS, Bernoulli RFS, and MB RFS, and labeled RFS such as labeled Poisson RFS, LMB RFS, δ-GLMB RFS, and -GLMB RFS. Then, based on the multi-target system model (including multi-target motion model and multi-target measurement model), the multi-target Bayesian recursion is introduced, the multi-target formal modeling paradigm is sorted out, and the implementation method of the general particle multi-target filter is provided, which is helpful to better understand the RFS-based target tracking algorithms in a more comprehensive way. Finally, the performance metrics of multi-target tracking to measure the performance of RFS-based target tracking algorithms are described.

Chapter 4
Probability Hypothesis Density Filter

4.1 Introduction

Although the particle multi-target filter introduced in the previous chapter provides a general solution for the multi-target Bayesian recursion, due to the combinatorial complexity of multi-target Bayesian recursion, the computational load is too heavy. Hence, this filter is typically only suitable for relatively ideal scenarios where the number of targets is small or the signal to noise ratio is high for example. To alleviate the problem that the calculation of the multi-target Bayesian filter cannot be tractable, Mahler used the FISST to derive and develop the first-order moment approximation of the complete multi-target Bayesian filter, i.e., the PHD filter [6]. More specifically, similar to the constant gain Kalman filter that propagates the first-order moment (mean) of a single target state density, the PHD filter does not propagate the multi-target posterior density over time, but propagates first-order statistical moment (i.e., posterior intensity) of the posterior multi-target state. As it does not require track-measurement association, naturally incorporates the track initiation and track termination, and can estimate the time-varying number of targets online, it immediately attracted widespread attention among the academic community once it is realized, starting an upsurge in the research of RFS-based tracking algorithms.

This chapter first introduces the PHD recursion and then its SMC implementation and GM implementation, namely the SMC-PHD filter and the GM-PHD filter, which are respectively suitable for nonlinear non-Gaussian and linear Gaussian system models. At last, the chapter introduces the extension of the GM-PHD filter, which can be applied respectively to the survival probability (and detection probability) in exponential mixture form and mildly nonlinear Gaussian models.

W. Wu et al., *Target Tracking with Random Finite Sets*,
https://doi.org/10.1007/978-981-19-9815-7_4

4.2 PHD Recursion

As mentioned earlier, the PHD filter propagates the first-order moment (that is, posterior intensity PHD $v_k(\cdot|Z^k)$ of target RFS), rather than the multi-target filtering density $\pi_k(\cdot|Z^k)$. Besides, the PHD filter does not require data association. In multi-target state space model, the PHD recursion assumes that false alarms are Poisson distributed, and that the updated and predicted multi-target RFS are also Poisson distributed. The Poisson hypothesis is a mathematic simplification that allows the update step of the PHD filter to have a closed form expression. More specifically, the PHD filter is based on the following assumptions:

A.1: Each target evolves independently of each other and generates independent measurements;
A.2: The clutter is of Poisson type and independent of the target-originated measurements;
A.3: The predicted multi-target RFS dominated by the multi-target predicted density $\pi_{k|k-1}$ is also of Poisson type.

Assumptions A.1 and A.2 are widely used in the bulk of target tracking algorithms [9]. Besides, assumption A.3 is also a reasonable approximation under the condition where the interaction between targets is negligible [6]. In fact, assumption A.3 can be fully satisfied, when there is no spawning target and RFS X_{k-1} and B_k are of Poisson type, where X_{k-1} and B_k represent the survival target RFS and the birth target RFS defined in (3.4.5), respectively.

$v_{k|k-1}$ and v_k respectively represent the intensities corresponding to multi-target predicted density $\pi_{k|k-1}$ and multi-target posterior density π_k in recursion (3.5.2)–(3.5.3). The PHD filter recursively propagates intensity function v_k over time according to the prediction and update steps. Based on assumptions A.1–A.3 and by using the FISST or classical probability tools, the posterior intensity is propagated over time as the following PHD recursion

$$v_{k|k-1}(\boldsymbol{x}) = \int p_{S,k}(\boldsymbol{\xi})\phi_{k|k-1}(\boldsymbol{x}|\boldsymbol{\xi})v_{k-1}(\boldsymbol{\xi})d\boldsymbol{\xi} + \int v_{\beta,k}(\boldsymbol{x}|\boldsymbol{\xi})v_{k-1}(\boldsymbol{\xi})d\boldsymbol{\xi} + v_{\gamma,k}(\boldsymbol{x}) \tag{4.2.1}$$

$$v_k(\boldsymbol{x}) = [1 - p_{D,k}(\boldsymbol{x})]v_{k|k-1}(\boldsymbol{x}) + \sum_{\boldsymbol{z}\in Z_k} \frac{p_{D,k}(\boldsymbol{x})g_k(\boldsymbol{z}|\boldsymbol{x})v_{k|k-1}(\boldsymbol{x})}{\kappa_k(\boldsymbol{z}) + \int p_{D,k}(\boldsymbol{\xi})g_k(\boldsymbol{z}|\boldsymbol{\xi})v_{k|k-1}(\boldsymbol{\xi})d\boldsymbol{\xi}} \tag{4.2.2}$$

Another expression for $v_k(\cdot)$ can be written as

$$v_k(\boldsymbol{x}) = \left[(1 - p_{D,k}(\boldsymbol{x})) + \sum_{\boldsymbol{z}\in Z_k} \frac{\psi_{k,\boldsymbol{z}}(\boldsymbol{x})}{\kappa_k(\boldsymbol{z}) + \langle \psi_{k,\boldsymbol{z}}, v_{k|k-1}\rangle}\right] v_{k|k-1}(\boldsymbol{x}) \tag{4.2.3}$$

where

$$\psi_{k,z}(\boldsymbol{x}) = p_{D,k}(\boldsymbol{x})g_k(z|\boldsymbol{x}) \tag{4.2.4}$$

$$\langle f, h\rangle = \int f(\boldsymbol{x})h(\boldsymbol{x})d\boldsymbol{x} \tag{4.2.5}$$

In (4.2.1)–(4.2.2), $v_{\gamma,k}(\cdot)$ is the intensity of birth RFS B_k at time k, $v_{\beta,k}(\cdot|\boldsymbol{\xi})$ is the intensity of RFS $Q_{k|k-1}(\boldsymbol{\xi})$ at time k spawned from the target whose previous state is $\boldsymbol{\xi}$ [see (3.4.5)], $p_{S,k}(\boldsymbol{\xi})$ is the probability that the target still exists at time k given the previous state $\boldsymbol{\xi}$, and $\phi_{k|k-1}(\cdot|\cdot)$ represents the target transition probability density. $p_{D,k}(\boldsymbol{x})$ is the detection probability of state $\boldsymbol{x}$ at time k; $g_k(\cdot|\boldsymbol{x})$ is the likelihood of single-target measurement at time k when the current state $\boldsymbol{x}$ is given; $\kappa_k(\cdot)$ is the intensity function of (Poisson) clutter RFS C_k at time k [see (3.4.14)]. $\kappa_k(\cdot)$ can also be written as $\kappa_k(\cdot) = \lambda_{C,k}p_{C,k}(\cdot)$, wherein, $\lambda_{C,k} = \int \kappa_k(z)dz$ is the expected number of clutter measurement (also known as the clutter ratio); $p_{C,k}(\cdot) = \kappa_k(\cdot)/\lambda_{C,k}$ is the spatial distribution of clutters on the surveillance area.

According to Eqs. (4.2.1) and (4.2.2), the PHD filter avoids combinatorial computations resulting from the unknown association between measurements and targets. In addition, since the posterior intensity is a function on single-target state space $\mathbb{X}$, the computational workload of the PHD recursion is much smaller than that of multi-target Bayesian recursion (3.5.2)–(3.5.3) operating on $\mathcal{F}(\mathbb{X})$. However, the PHD recursion (4.2.1)–(4.2.2) still involves multiple integration. Generally speaking, there is no closed-form expression, and numerical integration faces the problem of the "curse of dimensionality". The SMC implementation (SMC-PHD) and the GM implementation (GM-PHD) of the PHD recursion are described below, respectively.

4.3 SMC-PHD Filter

For highly nonlinear non-Gaussian problems, v_k can be approximated by a set of weighted particles $\{w_k^{(i)}, \boldsymbol{x}_k^{(i)}\}_{i=1}^{L_k}$. The basic idea of the SMC-PHD filter is to propagate multi-target posterior intensity function (PHD) through the weighted particle set.

4.3.1 Prediction Step

The PHD prediction (4.2.1) can be re-written into the following equation

$$v_{k|k-1}(\boldsymbol{x}_k) = \int \Phi_{k|k-1}(\boldsymbol{x}_k, \boldsymbol{x}_{k-1})v_{k-1}(\boldsymbol{x}_{k-1})d\boldsymbol{x}_{k-1} + v_{\gamma,k}(\boldsymbol{x}_k) \tag{4.3.1}$$

where

$$\Phi_{k|k-1}(\boldsymbol{x}, \boldsymbol{\xi}) = p_{S,k}(\boldsymbol{\xi})\phi_{k|k-1}(\boldsymbol{x}|\boldsymbol{\xi}) + v_{\beta,k}(\boldsymbol{x}|\boldsymbol{\xi}) \tag{4.3.2}$$

Given the particle representation of v_{k-1}, namely,

$$v_{k-1}(\boldsymbol{x}_{k-1}) \approx \sum_{i=1}^{L_{k-1}} w_{k-1}^{(i)} \delta_{\boldsymbol{x}_{k-1}^{(i)}}(\boldsymbol{x}_{k-1}) \tag{4.3.3}$$

then we have

$$v_{k|k-1}(\boldsymbol{x}_k) = \sum_{i=1}^{L_{k-1}} w_{k-1}^{(i)} \Phi_{k|k-1}(\boldsymbol{x}_k, \boldsymbol{x}_{k-1}^{(i)}) + v_{\gamma,k}(\boldsymbol{x}_k) \tag{4.3.4}$$

where a particle approximation of $v_{k|k-1}$ can be obtained by applying the importance sampling to each term. When the importance (or proposed) density $q_k(\cdot|\boldsymbol{x}_{k-1}, Z_k)$ and $q_{\gamma,k}(\cdot|Z_k)$ are given and respectively satisfy $q_k(\boldsymbol{x}_k|\boldsymbol{x}_{k-1}, Z_k) > 0$ when $\Phi_{k|k-1}(\boldsymbol{x}_k, \boldsymbol{x}_{k-1}) > 0$ and $q_{\gamma,k}(\cdot|Z_k) > 0$ when $v_{\gamma,k}(\boldsymbol{x}_k) > 0$, Eq. (4.3.4) can be re-written as

$$v_{k|k-1}(\boldsymbol{x}_k) = \sum_{i=1}^{L_{k-1}} w_{k-1}^{(i)} \frac{\Phi_{k|k-1}(\boldsymbol{x}_k, \boldsymbol{x}_{k-1}^{(i)})}{q_k(\boldsymbol{x}_k|\boldsymbol{x}_{k-1}^{(i)}, Z_k)} q_k(\boldsymbol{x}_k|\boldsymbol{x}_{k-1}^{(i)}, Z_k) + \frac{v_{\gamma,k}(\boldsymbol{x}_k)}{q_{\gamma,k}(\boldsymbol{x}_k|Z_k)} q_{\gamma,k}(\boldsymbol{x}_k|Z_k) \tag{4.3.5}$$

Thus, the following particle approximation can be obtained

$$v_{k|k-1}(\boldsymbol{x}_k) \approx \sum_{i=1}^{L_{k-1}+L_{\gamma,k}} w_{k|k-1}^{(i)} \delta_{\boldsymbol{x}_k^{(i)}}(\boldsymbol{x}_k) \tag{4.3.6}$$

where

$$\boldsymbol{x}_k^{(i)} \sim \begin{cases} q_k\left(\cdot\middle|\boldsymbol{x}_{k-1}^{(i)}, Z_k\right), & i = 1, \ldots, L_{k-1} \\ q_{\gamma,k}(\cdot|Z_k), & i = L_{k-1}+1, \ldots, L_{k-1}+L_{\gamma,k} \end{cases} \tag{4.3.7}$$

$$w_{k|k-1}^{(i)} = \begin{cases} \dfrac{\Phi_{k|k-1}\left(\boldsymbol{x}_k^{(i)}, \boldsymbol{x}_{k-1}^{(i)}\right) w_{k-1}^{(i)}}{q_k\left(\boldsymbol{x}_k^{(i)}\middle|\boldsymbol{x}_{k-1}^{(i)}, Z_k\right)}, & i = 1, \ldots, L_{k-1} \\ \dfrac{v_{\gamma,k}\left(\boldsymbol{x}_k^{(i)}\right)}{L_{\gamma,k} q_{\gamma,k}\left(\boldsymbol{x}_k^{(i)}|Z_k\right)}, & i = L_{k-1}+1, \ldots, L_{k-1}+L_{\gamma,k} \end{cases} \tag{4.3.8}$$

It should be noted that starting from v_{k-1} (4.3.3) with L_{k-1} particles, another set composed of L_{k-1} particles can be predicted forward through kernel $\Phi_{k|k-1}$; In addition, the birth process generates $L_{\gamma,k}$ new particles. The number of new particles

$L_{\gamma,k}$ can be a function of k to accommodate birth targets whose number changes at each moment. Assume that the total mass of $v_{\gamma,k}$ has a closed form, then we can typically make $L_{\gamma,k}$ directly proportional to this mass, namely, $L_{\gamma,k} = \alpha \int v_{\gamma,k}(\boldsymbol{x})d\boldsymbol{x}$, so that each birth target corresponds to an average of α particles.

4.3.2 Update Step

For the update step, based on $v_{k|k-1}$ characterized by $\{w_{k|k-1}^{(i)}, \boldsymbol{x}_k^{(i)}\}_{i=1}^{L_{k-1}+L_{\gamma,k}}$ obtained from the prediction step, applying the update Eq. (4.2.3) yields

$$v_k(\boldsymbol{x}) \approx \sum_{i=1}^{L_{k-1}+L_{\gamma,k}} w_k^{(i)} \delta_{\boldsymbol{x}_k^{(i)}}(\boldsymbol{x}) \tag{4.3.9}$$

where

$$w_k^{(i)} = \left[(1 - p_{D,k}(\boldsymbol{x}_k^{(i)})) + \sum_{z \in Z_k} \frac{\psi_{k,z}(\boldsymbol{x}_k^{(i)})}{\kappa_k(z) + \sum_{j=1}^{L_{k-1}+L_{\gamma,k}} \psi_{k,z}(\boldsymbol{x}_k^{(j)}) w_{k|k-1}^{(j)}} \right] w_{k|k-1}^{(i)} \tag{4.3.10}$$

According to (4.3.10), the update step maps the intensity function represented by particles $\{w_{k|k-1}^{(i)}, \boldsymbol{x}_k^{(i)}\}_{i=1}^{L_{k-1}+L_{\gamma,k}}$ to the intensity function represented by particles $\{w_k^{(i)}, \boldsymbol{x}_k^{(i)}\}_{i=1}^{L_{k-1}+L_{\gamma,k}}$, by modifying the weights of these particles. According to the particle concentration in a given area S in the single-target state-space, the expected number of targets in this space can be obtained, that is, $\mathrm{E}[|\mathbb{X}_k \cap S||Z_{1:k}] \approx \sum_{i=1}^{L_{k-1}+L_{\gamma,k}} 1_S\left(\boldsymbol{x}_k^{(i)}\right) w_k^{(i)}$.

4.3.3 Resampling and Multi-target State Extraction

Through the SMC-PHD recursion, the particle approximation of the intensity function at time $k > 0$ can be obtained from the particle approximation at the previous moment. It should be noted that since $v_{k|k}$ is characterized by $L_k = L_{k-1} + L_{\gamma,k}$ particles, even if the number L_k of targets does not increase, the number of particles may still continue to grow over time. This leads to an extremely low efficiency, which is attributed to the waste of computing resources in areas where the targets do not present. On the one hand, if L_k is fixed, the ratio of the number of particles to the number of targets will fluctuate as the number of targets changes. Therefore, sometimes the number of particles may be insufficient when dealing with the targets; on the other hand, when the number of targets is small or there is no target at all, there may still be too many particles. In this sense, a more effective computing strategy is to adaptively allocate α particles to per-object at each time.

Since the expected number $N_{k|k} = \int v_{k|k}(\boldsymbol{x})d\boldsymbol{x}$ of targets can be obtained by estimation of $\hat{N}_{k|k} = \sum_{i=1}^{L_k} w_k^{(i)} = \sum_{i=1}^{L_{k-1}+L_{\gamma,k}} w_k^{(i)}$, the intuitive choice is to make the number of particles satisfy $L_k \approx \alpha \hat{N}_{k|k}$. Another alternative way is to eliminate low-weighted particles and copy high-weighted particles, so that the particles are concentrated in important areas. This can be achieved by re-sampling $L_k \approx \alpha \hat{N}_{k|k}$ particles from $\{w_k^{(i)}, \boldsymbol{x}_k^{(i)}\}_{i=1}^{L_k}$ and redistributing the total mass $\hat{N}_{k|k}$ among the L_k re-sampled particles.

After obtaining the posterior intensity v_k, the next task is to extract the multi-target state estimation. Generally, this task is not easy. For the SMC-PHD (or particle PHD) filter [62], multi-target state extraction requires additional particle clustering. In the SMC-PHD filter, the estimation $\hat{N}_k$ of number of targets is obtained from the total mass of the particles representing v_k, and then the standard clustering algorithm is used to group these particles into $\hat{N}_k$ clusters to obtain the estimated states. The clustering technique can be implemented through the k-means or expectation maximization (EM) algorithm. The SMC-PHD filter performs better when posterior intensity v_k is naturally formed as $\hat{N}_k$ clusters. Otherwise, when $\hat{N}_k$ is not the same as the number of clusters formed, the state estimation will become unreliable. Partial solutions are given by measurement-driven PHD [163] and auxiliary particle PHD [402].

4.3.4 Algorithm Steps

An SMC implementation algorithm of the PHD recursion, called SMC-PHD (or particle PHD) filter, can be obtained based on the above content. The specific steps are as follows:

- Prediction
  - For $i = 1, \ldots, L_{k-1}$ perform sampling $\tilde{\boldsymbol{x}}_k^{(i)} \sim q_k\left(\cdot \middle| \boldsymbol{x}_{k-1}^{(i)}, Z_k\right)$ and calculate the predicted weights

$$w_{k|k-1}^{(i)} = \frac{\Phi_{k|k-1}(\tilde{\boldsymbol{x}}_k^{(i)}, \boldsymbol{x}_{k-1}^{(i)})}{q_k(\tilde{\boldsymbol{x}}_k^{(i)}|\boldsymbol{x}_{k-1}^{(i)}, Z_k)} w_{k-1}^{(i)} \tag{4.3.11}$$

  - For $i = L_{k-1}+1, \ldots, L_{k-1}+L_{\gamma,k}$, perform sampling $\tilde{\boldsymbol{x}}_k^{(i)} \sim q_{\gamma,k}(\cdot|Z_k)$ and calculate the weights of birth particles

$$w_{k|k-1}^{(i)} = \frac{v_{\gamma,k}(\tilde{\boldsymbol{x}}_k^{(i)})}{L_{\gamma,k} q_{\gamma,k}(\tilde{\boldsymbol{x}}_k^{(i)}|Z_k)} \tag{4.3.12}$$

- Update
  - For $i = 1, \ldots, L_{k-1} + L_{\gamma,k}$, the weights are updated as follows

$$\tilde{w}_k^{(i)} = \left[(1 - p_{D,k}(\tilde{\boldsymbol{x}}_k^{(i)})) + \sum_{z \in Z_k} \frac{\psi_{k,z}(\tilde{\boldsymbol{x}}_k^{(i)})}{\kappa_k(z) + \sum_{j=1}^{L_{k-1}+L_{\gamma,k}} \psi_{k,z}(\tilde{\boldsymbol{x}}_k^{(j)}) w_{k|k-1}^{(j)}} \right] w_{k|k-1}^{(i)} \tag{4.3.13}$$

- Re-sampling
  - Calculating the total mass $\hat{N}_{k|k} = \sum_{i=1}^{L_{k-1}+L_{\gamma,k}} \tilde{w}_k^{(i)}$;
  - Resampling $\{\tilde{w}_k^{(i)}/\hat{N}_{k|k}, \tilde{\boldsymbol{x}}_k^{(i)}\}_{i=1}^{L_{k-1}+L_{\gamma,k}}$ to get $\{w_k^{(i)}/\hat{N}_{k|k}, \boldsymbol{x}_k^{(i)}\}_{i=1}^{L_k}$;
  - Multiplying the weight by $\hat{N}_{k|k}$ to get $\{w_k^{(i)}, \boldsymbol{x}_k^{(i)}\}_{i=1}^{L_k}$.

In the prediction step of the above algorithm, it is assumed that

$$\sup_{\boldsymbol{x},\boldsymbol{\xi}} \left| \frac{\Phi_{k|k-1}(\boldsymbol{x}, \boldsymbol{\xi})}{q_k(\boldsymbol{x}|\boldsymbol{\xi}, Z_k)} \right| \leq \zeta_k \tag{4.3.14}$$

$$\sup_{\boldsymbol{x}} \left| \frac{v_{\gamma,k}(\boldsymbol{x})}{q_{\gamma,k}(\boldsymbol{x}|Z_k)} \right| \leq \sigma_k \tag{4.3.15}$$

where sup represents the supremum, and ζ_k and σ_k are finite. Thus, weights (4.3.11)–(4.3.12) are strictly defined.

Attention needs to be paid when performing the resampling step of the SMC-PHD filter. At this time, new weights $\{w_k^{(i)}\}_{i=1}^{L_k}$ are not normalized to 1. On the contrary, the sum of the weights is $\hat{N}_{k|k} = \sum_{i=1}^{L_{k-1}+L_{\gamma,k}} w_k^{(i)}$. Similar to the standard particle filter, under the constraint $\sum_{i=1}^{L_{k-1}+L_{\gamma,k}} n_k^{(i)} = L_k$, each particle $\tilde{\boldsymbol{x}}_k^{(i)}$ is duplicated for $n_k^{(i)}$ times, thus finally $\{\boldsymbol{x}_k^{(i)}\}_{i=1}^{L_k}$ is obtained. A random sampling method that satisfies the condition $\mathrm{E}\left[n_k^{(i)}\right] = L_k \omega_k^{(i)}$ can be selected, where $\omega_k^{(i)} > 0$, $\sum_{i=1}^{L_{k-1}+L_{\gamma,k}} \omega_k^{(i)} = 1$ is a series of weights set by the user. Then, the new weight is set as $w_k^{(i)} \propto \tilde{w}_k^{(i)}/\omega_k^{(i)}$, where $\sum_{i=1}^{L_k} w_k^{(i)} = \hat{N}_{k|k}$, rather than $\sum_{i=1}^{L_k} w_k^{(i)} = 1$. Typically, $\omega_k^{(i)} = \tilde{w}_k^{(i)}/\hat{N}_{k|k}$ is set, or $\omega_k^{(i)} \propto (\tilde{w}_k^{(i)})^\tau$ can be selected where $\tau \in (0, 1)$.

For the initialization, the importance sampling can be employed to obtain a particle approximation of initial intensity function. If there is no prior information, the initial intensity function can also be set to 0, thus no particle is needed. In this case, the algorithm starts sampling from the birth process during the next iteration. A better strategy is to estimate the number $\hat{N}_0$ of targets based on the measurements, and set the initial intensity function to be the uniform intensity with total mass $\hat{N}_0$.

When there is only one target and no birth, death, clutter, and the detection probability is 1, the SMC-PHD filter will degenerate into the standard particle filter. In the context of the standard particle filter, it is important to select the importance distribution that minimizes the (conditional) variance of weights, whereas in the context of the PHD filter, this becomes very difficult, thus requiring further research.

4.4 GM-PHD Filter

Under a specific type of linear Gaussian multi-target (LGM) model, a PHD recursion with a closed form solution can be obtained, and the resulting multi-target filter is called a Gaussian mixture PHD (GM-PHD) filter.

4.4.1 Model Assumptions on GM-PHD Recursion

The closed form solution of the PHD recursion (4.2.1)–(4.2.2) requires the LGM model in addition to the assumptions of A.1–A.3. The model includes not only the standard linear Gaussian model for each target but also some assumptions on the target birth, death and detection, which are detailed as follows.

A.4: Each target follows a linear Gaussian dynamic model, and the sensor has a linear Gaussian measurement model, namely,

$$\phi_{k|k-1}(\boldsymbol{x}_k|\boldsymbol{x}_{k-1}) = \mathcal{N}(\boldsymbol{x}_k; \boldsymbol{F}_{k-1}\boldsymbol{x}_{k-1}, \boldsymbol{Q}_{k-1}) \tag{4.4.1}$$

$$g_k(z_k|\boldsymbol{x}_k) = \mathcal{N}(z_k; \boldsymbol{H}_k\boldsymbol{x}_k, \boldsymbol{R}_k) \tag{4.4.2}$$

where $\mathcal{N}(\cdot; \boldsymbol{m}, \boldsymbol{P})$ represents the Gaussian density with mean $\boldsymbol{m}$ and covariance $\boldsymbol{P}$, $\boldsymbol{F}_{k-1}$ is the state transition matrix, $\boldsymbol{Q}_{k-1}$ is the process noise covariance, $\boldsymbol{H}_k$ is the measurement matrix, and $\boldsymbol{R}_k$ is the measurement noise covariance.

A.5: Both the survival probability and the detection probability are independent of the target state, namely,

$$p_{S,k}(\boldsymbol{x}) = p_{S,k} \tag{4.4.3}$$

$$p_{D,k}(\boldsymbol{x}) = p_{D,k} \tag{4.4.4}$$

A.6: Both the intensities of the birth and spawning target RFSs are in Gaussian mixture form, namely,

$$v_{\gamma,k}(\boldsymbol{x}) = \sum_{i=1}^{J_{\gamma,k}} w_{\gamma,k}^{(i)} \mathcal{N}(\boldsymbol{x}; \boldsymbol{m}_{\gamma,k}^{(i)}, \boldsymbol{P}_{\gamma,k}^{(i)}) \tag{4.4.5}$$

$$v_{\beta,k}(\boldsymbol{x}|\boldsymbol{\xi}) = \sum_{i=1}^{J_{\beta,k}} w_{\beta,k}^{(i)} \mathcal{N}(\boldsymbol{x}; \boldsymbol{F}_{\beta,k-1}^{(i)}\boldsymbol{\xi} + \boldsymbol{d}_{\beta,k-1}^{(i)}, \boldsymbol{P}_{\beta,k-1}^{(i)}) \tag{4.4.6}$$

where $J_{\gamma,k}, w_{\gamma,k}^{(i)}, \boldsymbol{m}_{\gamma,k}^{(i)}, \boldsymbol{P}_{\gamma,k}^{(i)}, i = 1, \ldots, J_{\gamma,k}$ are the given model parameters, which determine the shape of the birth intensity; $w_{\gamma,k}^{(i)}$, $\boldsymbol{m}_{\gamma,k}^{(i)}$ and $\boldsymbol{P}_{\gamma,k}^{(i)}$ respectively represent the weight, mean and covariance of the ith mixed component of the birth intensity; $J_{\gamma,k}$ is the total number of components; Similarly, $J_{\beta,k}, w_{\beta,k}^{(i)}, \boldsymbol{F}_{\beta,k-1}^{(i)}, \boldsymbol{d}_{\beta,k-1}^{(i)}, \boldsymbol{P}_{\beta,k-1}^{(i)}, i = 1, \ldots, J_{\beta,k}$ determine the shape of target intensity derived from the previous state $\boldsymbol{\xi}$.

Remark 1 The assumptions A.4 and A.5 are often adopted by most tracking algorithms [9]. For illustration purposes, this section only focuses on state-independent $p_{S,k}$ and $p_{D,k}$ for the time being. The closed form PHD recursion under general scenarios will be introduced in Sect. 4.5.1.

Remark 2 In Assumption A.6, $\boldsymbol{m}_{\gamma,k}^{(i)}, i = 1, \ldots, J_{\gamma,k}$ are the peaks of the birth target intensity in (4.4.5). These points correspond to positions where birth targets are most likely to appear, for example, the air bases or airports. Covariance matrix $\boldsymbol{P}_{\gamma,k}^{(i)}$ determines the degree of dispersion of birth intensity adjacent to peak $\boldsymbol{m}_{\gamma,k}^{(i)}$. Weight $w_{\gamma,k}^{(i)}$ gives the expected number of birth targets originated from $\boldsymbol{m}_{\gamma,k}^{(i)}$. For the intensity of spawning target from the previous state $\boldsymbol{\xi}$ represented by (4.4.6), an explanation similar to (4.4.5) can be applied, except that the ith peak $\boldsymbol{F}_{\beta,k-1}^{(i)}\boldsymbol{\xi} + \boldsymbol{d}_{\beta,k-1}^{(i)}$ is an affine function of $\boldsymbol{\xi}$. Generally speaking, the spawning target is modeled in the vicinity of its parent state, for example, $\boldsymbol{\xi}$ may correspond to the state of an air carrier at time $k-1$, then $\boldsymbol{F}_{\beta,k-1}^{(i)}\boldsymbol{\xi} + \boldsymbol{d}_{\beta,k-1}^{(i)}$ is the expected state of the spawned air carrier at time k.

4.4.2 GM-PHD Recursion

For the linear Gaussian multi-target model, the following two propositions give the closed form solution of PHD recursion (4.2.1)–(4.2.2). More precisely, these propositions show how the Gaussian components of posterior intensity are analytically propagated to the next moment.

Proposition 3 *Assume that assumptions* A.4–A.6 *are established, and posterior intensity v_{k-1} at time $k-1$ is the following Gaussian mixture form*

$$v_{k-1}(\boldsymbol{x}) = \sum_{i=1}^{J_{k-1}} w_{k-1}^{(i)} \mathcal{N}(\boldsymbol{x}; \boldsymbol{m}_{k-1}^{(i)}, \boldsymbol{P}_{k-1}^{(i)}) \tag{4.4.7}$$

then predicted intensity $v_{k|k-1}$ at time k is also in the form of Gaussian mixture, which is

$$v_{k|k-1}(\boldsymbol{x}) = v_{S,k|k-1}(\boldsymbol{x}) + v_{\beta,k|k-1}(\boldsymbol{x}) + v_{\gamma,k}(\boldsymbol{x}) \tag{4.4.8}$$

where $v_{\gamma,k}(\cdot)$ is the birth target intensity at time k, given by (4.4.5), *and*

$$v_{S,k|k-1}(\boldsymbol{x}) = p_{S,k} \sum_{i=1}^{J_{k-1}} w_{k-1}^{(i)} \mathcal{N}(\boldsymbol{x}; \boldsymbol{m}_{S,k|k-1}^{(i)}, \boldsymbol{P}_{S,k|k-1}^{(i)}) \tag{4.4.9}$$

$$\boldsymbol{m}_{S,k|k-1}^{(i)} = \boldsymbol{F}_{k-1} \boldsymbol{m}_{k-1}^{(i)} \tag{4.4.10}$$

$$\boldsymbol{P}_{S,k|k-1}^{(i)} = \boldsymbol{F}_{k-1} \boldsymbol{P}_{k-1}^{(i)} \boldsymbol{F}_{k-1}^{\mathrm{T}} + \boldsymbol{Q}_{k-1} \tag{4.4.11}$$

$$v_{\beta,k|k-1}(\boldsymbol{x}) = \sum_{i=1}^{J_{k-1}} \sum_{j=1}^{J_{\beta,k}} w_{k-1}^{(i)} w_{\beta,k}^{(j)} \mathcal{N}(\boldsymbol{x}; \boldsymbol{m}_{\beta,k|k-1}^{(i,j)}, \boldsymbol{P}_{\beta,k|k-1}^{(i,j)}) \tag{4.4.12}$$

$$\boldsymbol{m}_{\beta,k|k-1}^{(i,j)} = \boldsymbol{F}_{\beta,k-1}^{(j)} \boldsymbol{m}_{k-1}^{(i)} + \boldsymbol{d}_{\beta,k-1}^{(j)} \tag{4.4.13}$$

$$\boldsymbol{P}_{\beta,k|k-1}^{(i,j)} = \boldsymbol{F}_{\beta,k-1}^{(j)} \boldsymbol{P}_{k-1}^{(i)} (\boldsymbol{F}_{\beta,k-1}^{(j)})^{\mathrm{T}} + \boldsymbol{P}_{\beta,k-1}^{(j)} \tag{4.4.14}$$

It should be noted that $v_{\beta,k}(\boldsymbol{x}|\boldsymbol{\xi})$ in (4.4.6) *should be differentiated from $v_{\beta,k|k-1}(\boldsymbol{x})$ in* (4.4.12).

Proposition 4 *Assume that assumptions* A.4–A.6 *are established, and the predicted intensity $v_{k|k-1}$ at time k is in the following Gaussian mixture*

$$v_{k|k-1}(\boldsymbol{x}) = \sum_{i=1}^{J_{k|k-1}} w_{k|k-1}^{(i)} \mathcal{N}(\boldsymbol{x}; \boldsymbol{m}_{k|k-1}^{(i)}, \boldsymbol{P}_{k|k-1}^{(i)}) \tag{4.4.15}$$

then, posterior (or updated) intensity v_k at time k is also in the form of Gaussian mixture, which is

$$v_k(\boldsymbol{x}) = (1 - p_{D,k}) v_{k|k-1}(\boldsymbol{x}) + \sum_{z \in Z_k} v_{D,k}(\boldsymbol{x}; z) \tag{4.4.16}$$

where

$$v_{D,k}(\boldsymbol{x}; z) = \sum_{i=1}^{J_{k|k-1}} w_k^{(i)}(z) \mathcal{N}(\boldsymbol{x}; \boldsymbol{m}_{k|k}^{(i)}(z), \boldsymbol{P}_{k|k}^{(i)}) \tag{4.4.17}$$

$$w_k^{(i)}(z) = \frac{p_{D,k} w_{k|k-1}^{(i)} q_k^{(i)}(z)}{\kappa_k(z) + p_{D,k} \sum_{j=1}^{J_{k|k-1}} w_{k|k-1}^{(j)} q_k^{(j)}(z)} \tag{4.4.18}$$

$$\boldsymbol{m}_{k|k}^{(i)}(z) = \boldsymbol{m}_{k|k-1}^{(i)} + \boldsymbol{G}_k^{(i)}(z - \boldsymbol{\eta}_{k|k-1}^{(i)}) \tag{4.4.19}$$

$$\boldsymbol{P}_{k|k}^{(i)} = [\boldsymbol{I} - \boldsymbol{G}_k^{(i)} \boldsymbol{H}_k] \boldsymbol{P}_{k|k-1}^{(i)} \tag{4.4.20}$$

$$\boldsymbol{G}_k^{(i)} = \boldsymbol{P}_{k|k-1}^{(i)} \boldsymbol{H}_k^{\mathrm{T}} [\boldsymbol{S}_{k|k-1}^{(i)}]^{-1} \tag{4.4.21}$$

$$q_k^{(i)}(z) = \mathcal{N}(z; \boldsymbol{\eta}_{k|k-1}^{(i)}, \boldsymbol{S}_{k|k-1}^{(i)}) \tag{4.4.22}$$

$$\boldsymbol{\eta}_{k|k-1}^{(i)} = \boldsymbol{H}_k \boldsymbol{m}_{k|k-1}^{(i)} \tag{4.4.23}$$

$$\boldsymbol{S}_{k|k-1}^{(i)} = \boldsymbol{H}_k \boldsymbol{P}_{k|k-1}^{(i)} \boldsymbol{H}_k^{\mathrm{T}} + \boldsymbol{R}_k \tag{4.4.24}$$

Proposition 3 and Proposition 4 can be derived by applying the Gaussian function product formula in Appendix A. Proposition 3 can be obtained by substituting Eqs. (4.4.1), (4.4.3), and (4.4.5)–(4.4.7) into the PHD prediction (4.2.1), and replacing integral of the form (A.1) with the proper Gaussian function provided by Lemma 4 in Appendix A. Similarly, Proposition 4 can be obtained by substituting Eqs. (4.4.2), (4.4.4) and (4.4.15) into the PHD update (4.2.3), and replacing integral of the form (A.1) and the product of Gaussian functions of the form (A.2) with the proper Gaussian functions provided by Lemma 4 and Lemma 5 respectively.

Proposition 3 and Proposition 4 are the prediction and update steps of the PHD recursion under the linear Gaussian multi-target model, so they are called the Gaussian mixture PHD (GM-PHD) recursion. According to Proposition 3 and Proposition 4, if initial intensity v_0 is a mixture of Gaussians (including case $v_0 = 0$), then all subsequent predicted intensities $v_{k|k-1}$ and posterior intensities v_k are also mixtures of Gaussians. Proposition 3 provides a closed expression for computing the means, covariances and weights of $v_{k|k-1}$ from the means, covariances and weights of v_{k-1}. Once a new measurement set is obtained, Proposition 4 provides a closed expression, in which the means, covariances and weights of v_k are calculated by means, covariances and weights of $v_{k|k-1}$. For the sake of completeness, Table 4.1 summarizes the key steps of the GM-PHD filter.

Remark 3 In Proposition 3, predicted intensity $v_{k|k-1}$ is composed of three items $v_{S,k|k-1}$, $v_{\beta,k|k-1}$, and $v_{\gamma,k}$ corresponding to survival targets, spawning targets and birth targets respectively; Similarly, in Proposition 4, the updated posterior intensity v_k is composed of the missed item $(1 - p_{D,k})v_{k|k-1}$ and $|Z_k|$ detected items $v_{D,k}(\cdot; z)$ (one item for each measurement $z \in Z_k$). The recursion of means and covariances of $v_{S,k|k-1}$ and $v_{\beta,k|k-1}$ corresponds to the Kalman prediction, while the recursion of means and covariances of $v_{D,k}(\cdot; z)$ corresponds to the Kalman update.

Table 4.1 Pseudocode of the GM-PHD Filter

% given $\{w_{k-1}^{(i)}, \boldsymbol{m}_{k-1}^{(i)}, \boldsymbol{P}_{k-1}^{(i)}\}_{i=1}^{J_{k-1}}$ and measurement set Z_k
% Step 1. Prediction for birth targets
$i = 0$,
for $j = 1, \ldots, J_{\gamma,k}$
$i \leftarrow i + 1$,
$w_{k\|k-1}^{(i)} = w_{\gamma,k}^{(j)}, \boldsymbol{m}_{k\|k-1}^{(i)} = \boldsymbol{m}_{\gamma,k}^{(j)}, \boldsymbol{P}_{k\|k-1}^{(i)} = \boldsymbol{P}_{\gamma,k}^{(j)}$
end
for $j = 1, \ldots, J_{\beta,k}$
for $n = 1, \ldots, J_{k-1}$
$i \leftarrow i + 1$,
$w_{k\|k-1}^{(i)} = w_{\beta,k}^{(j)} w_{k-1}^{(n)}$,
$\boldsymbol{m}_{k\|k-1}^{(i)} = \boldsymbol{F}_{\beta,k-1}^{(j)} \boldsymbol{m}_{k-1}^{(n)} + \boldsymbol{d}_{\beta,k-1}^{(j)}$,
$\boldsymbol{P}_{k\|k-1}^{(i)} = \boldsymbol{F}_{\beta,k-1}^{(j)} \boldsymbol{P}_{k-1}^{(n)} (\boldsymbol{F}_{\beta,k-1}^{(j)})^{\mathrm{T}} + \boldsymbol{P}_{\beta,k-1}^{(j)}$
end
end
% Step 2. Prediction for survival targets
for $j = 1, \ldots, J_{k-1}$
$i \leftarrow i + 1$,
$w_{k\|k-1}^{(i)} = p_{S,k} w_{k-1}^{(j)}$,
$\boldsymbol{m}_{k\|k-1}^{(i)} = \boldsymbol{F}_{k-1} \boldsymbol{m}_{k-1}^{(j)}, \boldsymbol{P}_{k\|k-1}^{(i)} = \boldsymbol{F}_{k-1} \boldsymbol{P}_{k-1}^{(j)} \boldsymbol{F}_{k-1}^{\mathrm{T}} + \boldsymbol{Q}_{k-1}$
end
$J_{k\|k-1} = i$,
% Step 3. Preparation for the update step
for $j = 1, \ldots, J_{k\|k-1}$
$\boldsymbol{\eta}_{k\|k-1}^{(j)} = \boldsymbol{H}_k \boldsymbol{m}_{k\|k-1}^{(j)}, \boldsymbol{S}_{k\|k-1}^{(j)} = \boldsymbol{H}_k \boldsymbol{P}_{k\|k-1}^{(j)} \boldsymbol{H}_k^{\mathrm{T}} + \boldsymbol{R}_k$,
$\boldsymbol{G}_k^{(j)} = \boldsymbol{P}_{k\|k-1}^{(j)} \boldsymbol{H}_k^{\mathrm{T}} [\boldsymbol{S}_{k\|k-1}^{(j)}]^{-1}, \boldsymbol{P}_{k\|k}^{(j)} = [\boldsymbol{I} - \boldsymbol{K}_k^{(j)} \boldsymbol{H}_k] \boldsymbol{P}_{k\|k-1}^{(j)}$
end
% Step 4. Update

(continued)

Table 4.1 (continued)

for $j = 1, \ldots, J_{k\|k-1}$
$\quad w_k^{(j)} = (1 - p_{D,k}) w_{k\|k-1}^{(j)}$,
$\quad \boldsymbol{m}_k^{(j)} = \boldsymbol{m}_{k\|k-1}^{(j)}, \boldsymbol{P}_k^{(j)} = \boldsymbol{P}_{k\|k-1}^{(j)}$
end
$n = 0$
for $z \in Z_k$
$\quad n \leftarrow n + 1$,
$\quad$ for $j = 1, \ldots, J_{k\|k-1}$
$\quad\quad w_k^{(nJ_{k\|k-1}+j)} = p_{D,k} w_{k\|k-1}^{(j)} \mathcal{N}(z; \boldsymbol{\eta}_{k\|k-1}^{(j)}, \boldsymbol{S}_{k\|k-1}^{(j)})$,
$\quad\quad \boldsymbol{m}_k^{(nJ_{k\|k-1}+j)} = \boldsymbol{m}_{k\|k-1}^{(j)} + \boldsymbol{G}_k^{(j)}(z - \boldsymbol{\eta}_{k\|k-1}^{(j)})$,
$\quad\quad \boldsymbol{P}_k^{(nJ_{k\|k-1}+j)} = \boldsymbol{P}_{k\|k}^{(j)}$
end
for $j = 1, \ldots, J_{k\|k-1}$
$\quad\quad w_k^{(nJ_{k\|k-1}+j)} = \frac{w_k^{(nJ_{k\|k-1}+j)}}{\kappa_k(z) + \sum_{i=1}^{J_{k\|k-1}} w_k^{(nJ_{k\|k-1}+i)}}$
$\quad$ end
end
$J_k = (n+1) J_{k\|k-1} = n J_{k\|k-1} + J_{k\|k-1}$
% Output: $\{w_k^{(i)}, \boldsymbol{m}_k^{(i)}, \boldsymbol{P}_k^{(i)}\}_{i=1}^{J_k}$

When Gaussian mixture intensity $v_{k|k-1}$ and v_k are given, the corresponding expected number $\hat{N}_{k|k-1}$ and $\hat{N}_k$ of target can be obtained by summation of corresponding weights. According to Proposition 3 and Proposition 4, the closed form recursion of $\hat{N}_{k|k-1}$ and $\hat{N}_k$ can be obtained.

Corollary 1 *According to* Proposition 3, *the predicted mean of the number of targets is*

$$\hat{N}_{k|k-1} = \hat{N}_{k-1}\left(p_{S,k} + \sum_{i=1}^{J_{\beta,k}} w_{\beta,k}^{(i)}\right) + \sum_{i=1}^{J_{\gamma,k}} w_{\gamma,k}^{(i)} \tag{4.4.25}$$

Corollary 2 *According to* Proposition 4, *the updated mean of the number of targets is*

$$\hat{N}_k = \hat{N}_{k|k-1}(1 - p_{D,k}) + \sum_{z \in Z_k} \sum_{j=1}^{J_{k|k-1}} w_k^{(j)}(z) \tag{4.4.26}$$

In Corollary 1, the predicted mean of number of targets is obtained by adding the mean of the number of survived targets, the mean of the number of spawning targets, and the mean of the number of birth targets. The similar explanation can be obtained for Corollary 2. When there is no clutter, the updated mean of the number of targets is the number of measurements plus the mean of the number of missed targets.

4.4.3 Pruning for GM-PHD Filter

From the point of view that both the GM-PHD filter and the GSF filter described in Sect. 2.8 propagate Gaussian mixtures over time, the GM-PHD filter is similar to the GSF filter. Like the GSF, the GM-PHD filter also faces the computational problem that the number of Gaussian components continues to increase over time. In fact, at time k, the number of Gaussian components required to represent v_k in the GM-PHD filter is

$$[J_{k-1}(1 + J_{\beta,k}) + J_{\gamma,k}](1 + |Z_k|) = \mathcal{O}(J_{k-1}|Z_k|) \tag{4.4.27}$$

where J_{k-1} is the number of components representing v_{k-1}. This indicates that the number of components in the posterior intensity will grow infinitely.

To manage the ever-growing number of Gaussian mixture components, Gaussian mixture pruning techniques, such as pruning secondary components and merging similar components, are required. By pruning the component with smaller weight $w_k^{(i)}$, a desirable approximation of the following Gaussian mixture posterior intensity can be obtained

$$v_k(\boldsymbol{x}) = \sum_{i=1}^{J_k} w_k^{(i)} \mathcal{N}(\boldsymbol{x}; \boldsymbol{m}_k^{(i)}, \boldsymbol{P}_k^{(i)}) \tag{4.4.28}$$

This can be achieved by discarding components with weights below a certain preset threshold or by retaining only a certain number of components with largest weights. Besides, some Gaussian components that are close enough can be accurately approximated by a single Gaussian component, so these components can be merged. These ideas lead to a simple heuristic pruning algorithm, as shown in Table 4.2.

4.4.4 Multi-target State Extraction

Unlike the SMC-PHD filter, which requires an additional particle clustering step in multi-target state extraction and has the disadvantage of unstable performance, the GM-PHD filter is easy to extract the multi-target state estimation according to the Gaussian mixture expression of the posterior intensity v_k. It should be noted that

Table 4.2 Pruning steps of the GM-PHD filter

Given $\{w_k^{(i)}, \boldsymbol{m}_k^{(i)}, \boldsymbol{P}_k^{(i)}\}_{i=1}^{J_k}$, pruning threshold T, merging threshold U, and maximizing allowable number of Gaussian terms $J_{\max}$
Set $n = 0$, and $I = \{i = 1, \ldots, J_k \mid w_k^{(i)} > T\}$
repeat
$\quad n \leftarrow n + 1$
$\quad j = \arg_{i \in I} \max w_k^{(i)}$
$\quad L = \{i \in I \mid (\boldsymbol{m}_k^{(i)} - \boldsymbol{m}_k^{(j)})^{\mathrm{T}} (\boldsymbol{P}_k^{(i)})^{-1} (\boldsymbol{m}_k^{(i)} - \boldsymbol{m}_k^{(j)}) \leq U\}$
$\quad \tilde{w}_k^{(n)} = \sum_{i \in L} w_k^{(i)}$
$\quad \tilde{\boldsymbol{m}}_k^{(n)} = \frac{1}{\tilde{w}_k^{(n)}} \sum_{i \in L} w_k^{(i)} \boldsymbol{m}_k^{(i)}$
$\quad \tilde{\boldsymbol{P}}_k^{(n)} = \frac{1}{\tilde{w}_k^{(n)}} \sum_{i \in L} w_k^{(i)} (\boldsymbol{P}_k^{(i)} + (\tilde{\boldsymbol{m}}_k^{(n)} - \boldsymbol{m}_k^{(i)})(\tilde{\boldsymbol{m}}_k^{(n)} - \boldsymbol{m}_k^{(i)})^{\mathrm{T}})$
$\quad I \leftarrow I \backslash L$
until $I = \emptyset$
If $n > J_{\max}$, only retain $J_{\max}$ Gaussian components with the largest weights in $\{\tilde{w}_k^{(i)}, \tilde{\boldsymbol{m}}_k^{(i)}, \tilde{\boldsymbol{P}}_k^{(i)}\}_{i=1}^{n}$
Output pruned Gaussian components: $\{\tilde{w}_k^{(i)}, \tilde{\boldsymbol{m}}_k^{(i)}, \tilde{\boldsymbol{P}}_k^{(i)}\}_{i=1}^{n}$

after the pruning step (see Table 4.2), Gaussian components that are close together will be merged, and when the means of the Gaussian components are far apart, these means are actually local extrema of v_k. Because the height of each peak depends on the weight and the covariance, selecting $\hat{N}_k$ highest peaks of v_k may result in state estimates corresponding to small weight components. This is not an expected result, because even if the peaks are high, the expected number of targets corresponding to these peaks may be small. A better approach is to choose the means of the Gaussian components with weights greater than some threshold (such as 0.5). In summary, in the GM-PHD filter, multi-target state estimation first estimates the number of targets according to the sum of the weights, and then extracts the corresponding number of components with the largest weights from the PHD as the state estimation. Table 4.3 concludes the state estimation step of the GM-PHD filter, in which round$(\cdot)$ stands for the rounding operation.

4.5 Extension of GM-PHD Filter

In the derivation of the GM-PHD filter, it is assumed that the survival probability and the detection probability are independent of the target state (see Assumption A.7). In fact, by using a similar derivation, a general form of the GM-PHD filter can

Table 4.3 Multi-target state extraction of the GM-PHD filter

Given $\{w_k^{(i)}, \boldsymbol{m}_k^{(i)}, \boldsymbol{P}_k^{(i)}\}_{i=1}^{J_k}$
Set $\hat{X}_k = \emptyset$
for $i = 1, \ldots, J_k$
if $w_k^{(i)} > 0.5$
for $j = 1, \ldots,$ round$\left(w_k^{(i)}\right)$
Update $\hat{X}_k \leftarrow \{\hat{X}_k, \boldsymbol{m}_k^{(i)}\}$
end
end
end
Output multi-target state estimation $\hat{X}_k$

be obtained for the more general exponential mixture form of survival and detection probabilities. In addition, the GM-PHD filter can also be extended to make it compatible with a mildly nonlinear Gaussian model using the methods similar to the EKF and the UKF.

4.5.1 Extension to Exponential Mixture Survival and Detection Probabilities

For certain types of state-dependent survival and detection probabilities, closed-form solutions for the PHD recursion can still be obtained. Actually, Proposition 3 and Proposition 4 can be easily extended to deal with $p_{S,k}(\boldsymbol{x})$ and $p_{D,k}(\boldsymbol{x})$ of the following exponential mixture form

$$p_{S,k}(\boldsymbol{\xi}) = w_{S,k}^{(0)} + \sum_{i=1}^{J_{S,k}} w_{S,k}^{(i)} \mathcal{N}(\boldsymbol{\xi}; \boldsymbol{m}_{S,k}^{(i)}, \boldsymbol{P}_{S,k}^{(i)}) \tag{4.5.1}$$

$$p_{D,k}(\boldsymbol{\xi}) = w_{D,k}^{(0)} + \sum_{i=1}^{J_{D,k}} w_{D,k}^{(i)} \mathcal{N}(\boldsymbol{\xi}; \boldsymbol{m}_{D,k}^{(i)}, \boldsymbol{P}_{D,k}^{(i)}) \tag{4.5.2}$$

where $J_{S,k}, w_{S,k}^{(0)}, w_{S,k}^{(i)}, \boldsymbol{m}_{S,k}^{(i)}, \boldsymbol{P}_{S,k}^{(i)}, i = 1, \ldots, J_{S,k}$ and $J_{D,k}, w_{D,k}^{(0)}, w_{D,k}^{(i)}, \boldsymbol{m}_{D,k}^{(i)}, \boldsymbol{P}_{D,k}^{(i)}, i = 1, \ldots, J_{D,k}$ are given model parameters. $p_{S,k}(\boldsymbol{x})$ and $p_{D,k}(\boldsymbol{x})$ take values between 0 and 1 for all $\boldsymbol{x}$.

By applying Lemma 5 in Appendix A, $p_{S,k}(\boldsymbol{\xi})v_{k-1}(\boldsymbol{\xi})$ can be converted into the Gaussian mixture form, and then the product of it and transition density $\phi_{k|k-1}(\boldsymbol{x}|\boldsymbol{\xi})$ is integrated according to Lemma 4 in Appendix A. In this way, a closed-form predicted intensity $v_{k|k-1}$ can be obtained. By applying Lemma 5 once and twice for $p_{D,k}(\boldsymbol{x})v_{k|k-1}(\boldsymbol{x})$ and $p_{D,k}(\boldsymbol{x})g_k(z|\boldsymbol{x})v_{k|k-1}(\boldsymbol{x})$ respectively, these products are converted into Gaussian mixture forms, thus the closed-form updated intensity v_k can be obtained. Specifically, the following propositions give the Gaussian mixture expressions of $v_{k|k-1}$ and v_k. For the sake of brevity, the implementation steps are not described.

Proposition 5 *Based on* Proposition 3, $p_{S,k}(\boldsymbol{x})$ *of* Eq. (4.4.3) *is replaced with* $p_{S,k}(\boldsymbol{x})$ *of* Eq. (4.5.1), *and the predicted intensity* $v_{k|k-1}$ *is given by* Eq. (4.4.8). *Nevertheless,* $v_{S,k|k-1}$ *is replaced as*

$$v_{S,k|k-1}(\boldsymbol{x}) = \sum_{i=1}^{J_{k-1}} \sum_{j=0}^{J_{S,k}} w_{k-1}^{(i)} w_{S,k}^{(j)} q_{k-1}^{(i,j)} \mathcal{N}(\boldsymbol{x}; \boldsymbol{m}_{S,k|k-1}^{(i,j)}, \boldsymbol{P}_{S,k|k-1}^{(i,j)}) \tag{4.5.3}$$

where

$$\boldsymbol{m}_{S,k|k-1}^{(i,j)} = \boldsymbol{F}_{k-1} \boldsymbol{m}_{k-1}^{(i,j)} \tag{4.5.4}$$

$$\boldsymbol{P}_{S,k|k-1}^{(i,j)} = \boldsymbol{F}_{k-1} \boldsymbol{P}_{k-1}^{(i,j)} \boldsymbol{F}_{k-1}^{\mathrm{T}} + \boldsymbol{Q}_{k-1} \tag{4.5.5}$$

$$q_{k-1}^{(i,j)} = \mathcal{N}(\boldsymbol{m}_{S,k}^{(j)}; \boldsymbol{m}_{k-1}^{(i)}, \boldsymbol{P}_{S,k}^{(j)} + \boldsymbol{P}_{k-1}^{(i)}), \quad q_{k-1}^{(i,0)} = 1 \tag{4.5.6}$$

$$\boldsymbol{m}_{k-1}^{(i,j)} = \boldsymbol{m}_{k-1}^{(i)} + \boldsymbol{G}_{k-1}^{(i,j)} (\boldsymbol{m}_{S,k}^{(j)} - \boldsymbol{m}_{k-1}^{(i)}), \quad \boldsymbol{m}_{k-1}^{(i,0)} = \boldsymbol{m}_{k-1}^{(i)} \tag{4.5.7}$$

$$\boldsymbol{P}_{k-1}^{(i,j)} = (\boldsymbol{I} - \boldsymbol{G}_{k-1}^{(i,j)}) \boldsymbol{P}_{k-1}^{(i)}, \quad \boldsymbol{P}_{k-1}^{(i,0)} = \boldsymbol{P}_{k-1}^{(i)} \tag{4.5.8}$$

$$\boldsymbol{G}_{k-1}^{(i,j)} = \boldsymbol{P}_{k-1}^{(i)} (\boldsymbol{P}_{k-1}^{(i)} + \boldsymbol{P}_{S,k}^{(j)})^{-1} \tag{4.5.9}$$

Proposition 6 *Based on* Proposition 4, $p_{D,k}(\boldsymbol{x})$ *of* Eq. (4.4.4) *is replaced with* $p_{D,k}(\boldsymbol{x})$ *of* (4.5.2), *thus updated intensity* v_k *of* Eq. (4.4.16) *becomes*

$$v_k(\boldsymbol{x}) = v_{k|k-1}(\boldsymbol{x}) - v_{D,k}(\boldsymbol{x}) + \sum_{z \in Z_k} v_{D,k}(\boldsymbol{x}; z) \tag{4.5.10}$$

where

$$v_{D,k}(\boldsymbol{x}) = \sum_{i=1}^{J_{k|k-1}} \sum_{j=0}^{J_{D,k}} w_{k|k-1}^{(i,j)} \mathcal{N}(\boldsymbol{x}; \boldsymbol{m}_{k|k-1}^{(i,j)}, \boldsymbol{P}_{k|k-1}^{(i,j)}) \tag{4.5.11}$$

$$w_{k|k-1}^{(i,j)} = w_{D,k}^{(j)} w_{k|k-1}^{(i)} q_{k|k-1}^{(i,j)} \tag{4.5.12}$$

$$q_{k|k-1}^{(i,j)} = \mathcal{N}(\boldsymbol{m}_{D,k}^{(j)}; \boldsymbol{m}_{k|k-1}^{(i)}, \boldsymbol{P}_{D,k}^{(j)} + \boldsymbol{P}_{k|k-1}^{(i)}), \quad q_{k|k-1}^{(i,0)} = 1 \tag{4.5.13}$$

$$\boldsymbol{m}_{k|k-1}^{(i,j)} = \boldsymbol{m}_{k|k-1}^{(i)} + \boldsymbol{G}_{k|k-1}^{(i,j)}(\boldsymbol{m}_{D,k}^{(j)} - \boldsymbol{m}_{k|k-1}^{(i)}), \quad \boldsymbol{m}_{k|k-1}^{(i,0)} = \boldsymbol{m}_{k|k-1}^{(i)} \tag{4.5.14}$$

$$\boldsymbol{P}_{k|k-1}^{(i,j)} = [\boldsymbol{I} - \boldsymbol{G}_{k|k-1}^{(i,j)}]\boldsymbol{P}_{k|k-1}^{(i)}, \quad \boldsymbol{P}_{k|k-1}^{(i,0)} = \boldsymbol{P}_{k|k-1}^{(i)} \tag{4.5.15}$$

$$\boldsymbol{G}_{k|k-1}^{(i,j)} = \boldsymbol{P}_{k|k-1}^{(i)}(\boldsymbol{P}_{k|k-1}^{(i)} + \boldsymbol{P}_{D,k}^{(j)})^{-1} \tag{4.5.16}$$

$$v_{D,k}(\boldsymbol{x}; \boldsymbol{z}) = \sum_{i=1}^{J_{k|k-1}} \sum_{j=0}^{J_{D,k}} w_k^{(i,j)}(\boldsymbol{z})\mathcal{N}(\boldsymbol{x}; \boldsymbol{m}_{k|k}^{(i,j)}(\boldsymbol{z}), \boldsymbol{P}_{k|k}^{(i,j)}) \tag{4.5.17}$$

$$w_k^{(i,j)}(\boldsymbol{z}) = \frac{w_{k|k-1}^{(i,j)} q_k^{(i,j)}(\boldsymbol{z})}{\kappa_k(\boldsymbol{z}) + \sum_{r=1}^{J_{k|k-1}} \sum_{s=0}^{J_{D,k}} w_{k|k-1}^{(r,s)} q_k^{(r,s)}(\boldsymbol{z})} \tag{4.5.18}$$

$$q_k^{(i,j)}(\boldsymbol{z}) = \mathcal{N}(\boldsymbol{z}; \boldsymbol{\eta}_{k|k-1}^{(i,j)}, \boldsymbol{S}_{k|k-1}^{(i,j)}) \tag{4.5.19}$$

$$\boldsymbol{m}_{k|k}^{(i,j)}(\boldsymbol{z}) = \boldsymbol{m}_{k|k-1}^{(i,j)} + \boldsymbol{G}_k^{(i,j)}(\boldsymbol{z} - \boldsymbol{\eta}_{k|k-1}^{(i,j)}) \tag{4.5.20}$$

$$\boldsymbol{P}_{k|k}^{(i,j)} = (\boldsymbol{I} - \boldsymbol{G}_k^{(i,j)}\boldsymbol{H}_k)\boldsymbol{P}_{k|k-1}^{(i,j)} \tag{4.5.21}$$

$$\boldsymbol{G}_k^{(i,j)} = \boldsymbol{P}_{k|k-1}^{(i,j)}\boldsymbol{H}_k^{\mathrm{T}}[\boldsymbol{S}_{k|k-1}^{(i,j)}]^{-1} \tag{4.5.22}$$

$$\boldsymbol{\eta}_{k|k-1}^{(i,j)} = \boldsymbol{H}_k\boldsymbol{m}_{k|k-1}^{(i,j)} \tag{4.5.23}$$

$$\boldsymbol{S}_{k|k-1}^{(i,j)} = \boldsymbol{H}_k\boldsymbol{P}_{k|k-1}^{(i,j)}\boldsymbol{H}_k^{\mathrm{T}} + \boldsymbol{R}_k \tag{4.5.24}$$

It can be seen from the above that the GM-PHD filter is easy to be extended to the survival probability in exponential mixture form. However, for the detection probability in exponential mixture form, although the updated intensity itself is non-negative (therefore, the sum of weights is non-negative), the updated intensity contains Gaussian components with both positive and negative weights. In this case, care must be taken to ensure the non-negativity of the intensity function after pruning and merging in the specific implementation.

4.5.2 Generalization to Nonlinear Gaussian Model

In addition to being generalizable to exponential mixture forms of survival probability and detection probability, the GM-PHD filter can also be extended to the nonlinear Gaussian motion and measurement models. Specifically, model assumptions A.5 and

A.6 are still necessary, but the state process and measurement process can be relaxed to the nonlinear model described in (2.2.1) and (2.2.3), namely,

$$\boldsymbol{x}_k = f_k(\boldsymbol{x}_{k-1}, \boldsymbol{v}_{k-1}) \tag{4.5.25}$$

$$z_k = h_k(\boldsymbol{x}_k, \boldsymbol{n}_k) \tag{4.5.26}$$

where $f_{k|k-1}$ and h_k are non-linear state and measurement functions, Gaussian noises $\boldsymbol{v}_{k-1}$ and $\boldsymbol{n}_k$ respectively represent the zero-mean process noise and measurement noise, and their corresponding covariance matrices are $\boldsymbol{Q}_{k-1}$ and $\boldsymbol{R}_k$, respectively. Since state function $f_{k|k-1}$ and measurement function h_k are non-linear functions, the posterior intensity is no longer in the form of Gaussian mixture. However, the GM-PHD filter can be modified to make it suitable for nonlinear Gaussian models.

In the single-target filtering, the analytical approximations of nonlinear Bayes filter mainly include the EKF and the UKF [22]. The EKF approximates the posterior density with a single Gaussian whose propagation over time is accomplished by applying the Kalman recursion to the local linearization of the nonlinear maps $f_{k|k-1}$ and h_k. The UKF also approximates the posterior density through a single Gaussian, but instead of using a linearized model, it uses the unscented transformation to calculate the Gaussian approximation of the posterior density at the next moment.

In the non-linear model, the posterior intensity of multi-target state propagated by the PHD recursion (4.2.1)–(4.2.2) is a weighted sum of different non-Gaussian functions. By using the methods similar to the EKF and the UKF, each one of these non-Gaussian constituent function can be approximated with a single Gaussian. If the EKF method is adopted, the approximation of the posterior intensity at the next moment can be achieved by applying the locally linearized models to the GM-PHD recursion, whereas in a way similar to the UKF, the components of Gaussian mixture approximation of the posterior intensity at the next moment can be calculated by the unscented transformation. In both cases, the weights of these components are approximate.

Based on the above analysis, two nonlinear Gaussian mixture PHD filters can be obtained, namely, the extended Kalman PHD (EK-PHD) filter and the unscented Kalman PHD (UK-PHD) filter. The key steps of these two filters are summarized respectively in Tables 4.4 and 4.5.

Remark 4 Similar to the single-target case, the EK-PHD filter can only be applied to differentiable nonlinear models. In addition, the calculation of Jacobian matrices can be tedious and error-prone. On the other hand, the UK-PHD filter does not suffer from these limitations and can even be applied to discontinuous models.

Remark 5 Unlike the SMC-PHD filter, whose particle approximation (in a certain sense) converges to the posterior intensity as the number of particles tends to infinity [62, 67], the EK-PHD and UK-PHD filters cannot guarantee that their particle approximations can converge to the posterior intensity. However, for mildly nonlinear problems, the EK-PHD and UK-PHD filters provide the desirable approximations, with

Table 4.4 Key steps of the EK-PHD filter

Given: $\{w_{k-1}^{(i)}, \boldsymbol{m}_{k-1}^{(i)}, \boldsymbol{P}_{k-1}^{(i)}\}_{i=1}^{J_{k-1}}$ and measurement set Z_k
Step 1. Birth target prediction
As same as Step 1 in Table 4.1
Step 2. Survival target prediction
for $j = 1, \ldots, J_{k-1}$
$i \leftarrow i + 1$
$w_{k\|k-1}^{(i)} = p_{S,k} w_{k-1}^{(j)}, \boldsymbol{m}_{k\|k-1}^{(i)} = f_{k\|k-1}(\boldsymbol{m}_{k-1}^{(j)}, \boldsymbol{0})$
$\boldsymbol{F}_{k-1}^{(j)} = \frac{\partial f_{k\|k-1}(\boldsymbol{x}_{k-1},\boldsymbol{0})}{\partial \boldsymbol{x}_{k-1}}\Big\|_{\boldsymbol{x}_{k-1}=\boldsymbol{m}_{k-1}^{(j)}}, \boldsymbol{V}_{k-1}^{(j)} = \frac{\partial f_{k\|k-1}(\boldsymbol{m}_{k-1}^{(j)},\boldsymbol{v}_{k-1})}{\partial \boldsymbol{v}_{k-1}}\Big\|_{\boldsymbol{v}_{k-1}=\boldsymbol{0}}$
$\boldsymbol{P}_{k\|k-1}^{(i)} = \boldsymbol{F}_{k-1}^{(j)} \boldsymbol{P}_{k-1}^{(j)} (\boldsymbol{F}_{k-1}^{(j)})^{\mathrm{T}} + \boldsymbol{V}_{k-1}^{(j)} \boldsymbol{Q}_{k-1} (\boldsymbol{V}_{k-1}^{(j)})^{\mathrm{T}}$
end
$J_{k\|k-1} = i$
Step 3 Construction of PHD updated components
for $j = 1, \ldots, J_{k-1}$
$\eta_{k\|k-1}^{(j)} = h_k(\boldsymbol{m}_{k\|k-1}^{(j)}, \boldsymbol{0})$
$\boldsymbol{H}_k^{(j)} = \frac{\partial h_k(\boldsymbol{x}_k,\boldsymbol{0})}{\partial \boldsymbol{x}_k}\Big\|_{\boldsymbol{x}_k=\boldsymbol{m}_{k\|k-1}^{(j)}}, \boldsymbol{N}_k^{(j)} = \frac{\partial h_k(\boldsymbol{m}_{k\|k-1}^{(j)},\boldsymbol{n}_k)}{\partial \boldsymbol{n}_k}\Big\|_{\boldsymbol{n}_k=\boldsymbol{0}}$
$\boldsymbol{S}_{k\|k-1}^{(j)} = \boldsymbol{H}_k^{(j)} \boldsymbol{P}_{k\|k-1}^{(j)} (\boldsymbol{H}_k^{(j)})^{\mathrm{T}} + \boldsymbol{N}_k^{(j)} \boldsymbol{R}_k (\boldsymbol{N}_k^{(j)})^{\mathrm{T}}$
$\boldsymbol{G}_k^{(j)} = \boldsymbol{P}_{k\|k-1}^{(j)} (\boldsymbol{H}_k^{(j)})^{\mathrm{T}} (\boldsymbol{S}_{k\|k-1}^{(j)})^{-1}$
$\boldsymbol{P}_{k\|k}^{(j)} = (\boldsymbol{I} - \boldsymbol{K}_k^{(j)} \boldsymbol{H}_k^{(j)}) \boldsymbol{P}_{k\|k-1}^{(j)}$
end
Step 4. Update
Same as Step 4 in Table 4.1
Output:$\{w_k^{(i)}, \boldsymbol{m}_k^{(i)}, \boldsymbol{P}_k^{(i)}\}_{i=1}^{J_k}$

lower computational workloads than that of the SMC-PHD filter. In addition, the SMC-PHD filter require a large number of particles and an additional clustering operation to extract multi-target state estimation.

4.6 Summary

The chapter, on the basis of the PHD recursion, respectively describes the SMC-PHD and GM-PHD filters for the nonlinear non-Gaussian model and the linear Gaussian model, and introduces the extension of the GM-PHD filter in detail, including a more complex GM-PHD filter applicable to survival probability and detection probability

Table 4.5 Key steps of the UK-PHD Filter

Given: $\{w_{k-1}^{(i)}, \boldsymbol{m}_{k-1}^{(i)}, \boldsymbol{P}_{k-1}^{(i)}\}_{i=1}^{J_{k-1}}$ and measurement set Z_k
Step 1. Construction of birth target components
Same as Step 1 in Table 4.1
for $j = 1, \ldots, i$
Set $\boldsymbol{u} = [(\boldsymbol{m}_{k\|k-1}^{(j)})^{\mathrm{T}} \quad \boldsymbol{0}]^{\mathrm{T}}$, $\boldsymbol{\Sigma} = \mathrm{blkdiag}(\boldsymbol{P}_{k\|k-1}^{(j)}, \boldsymbol{R}_k)$
By using the UT, a set of σ points and their weights are generated based on mean $\boldsymbol{u}$ and covariance $\boldsymbol{\Sigma}$, denoted as $\{\boldsymbol{y}_k^{(n)}, \omega^{(n)}\}_{n=0}^{N}$
split $\boldsymbol{y}_k^{(n)} = \left[\left(\boldsymbol{x}_{k\|\mathrm{k}-1}^{(n)}\right)^{\mathrm{T}}, \left(\boldsymbol{n}_k^{(n)}\right)^{\mathrm{T}}\right]^{\mathrm{T}}, n = 0, 1, \ldots, N$
$z_{k\|k-1}^{(n)} = h_k\left(\boldsymbol{x}_{k\|k-1}^{(n)}, \boldsymbol{n}_k^{(n)}\right), \quad n = 0, 1, \ldots, N$
$\boldsymbol{\eta}_{k\|k-1}^{(j)} = \sum_{n=0}^{N} \omega^{(n)} \boldsymbol{z}_{k\|k-1}^{(n)}$
$\boldsymbol{S}_{k\|k-1}^{(j)} = \sum_{n=0}^{N} \omega^{(n)} (\boldsymbol{z}_{k\|k-1}^{(n)} - \boldsymbol{\eta}_{k\|k-1}^{(j)})(\boldsymbol{z}_{k\|k-1}^{(n)} - \boldsymbol{\eta}_{k\|k-1}^{(j)})^{\mathrm{T}}$
$\boldsymbol{C}_k^{(j)} = \sum_{n=0}^{L} \omega^{(n)} (\boldsymbol{x}_{k\|k-1}^{(n)} - \boldsymbol{m}_{k\|k-1}^{(j)})(\boldsymbol{z}_{k\|k-1}^{(n)} - \boldsymbol{\eta}_{k\|k-1}^{(j)})^{\mathrm{T}}$
$\boldsymbol{G}_k^{(j)} = \boldsymbol{C}_k^{(j)} (\boldsymbol{S}_{k\|k-1}^{(j)})^{-1}$
$\boldsymbol{P}_{k\|k}^{(j)} = \boldsymbol{P}_{k\|k-1}^{(j)} - \boldsymbol{G}_k^{(j)} (\boldsymbol{S}_{k\|k-1}^{(j)})^{-1} (\boldsymbol{G}_k^{(j)})^{\mathrm{T}}$
end
Step 2 Construction of survived target components
for $j = 1, \ldots, J_{k-1}$
$i \leftarrow i + 1$,
$w_{k\|k-1}^{(i)} = p_{S,k} w_{k-1}^{(j)}$,
Set $\boldsymbol{u} = [(\boldsymbol{m}_{k-1}^{(i)})^{\mathrm{T}} \quad \boldsymbol{0} \quad \boldsymbol{0}]^{\mathrm{T}}$, $\boldsymbol{\Sigma} = \mathrm{blkdiag}(\boldsymbol{P}_{k-1}^{(j)}, \boldsymbol{Q}_{k-1}, \boldsymbol{R}_k)$
By using the UT, a set of σ points and their weights are generated based on mean $\boldsymbol{u}$ and covariance $\boldsymbol{\Sigma}$, denoted as $\{\boldsymbol{y}_k^{(n)}, \omega^{(n)}\}_{n=0}^{N}$
split $\boldsymbol{y}_k^{(n)} = [(\boldsymbol{x}_{k-1}^{(n)})^{\mathrm{T}}, (\boldsymbol{v}_{k-1}^{(n)})^{\mathrm{T}}, (\boldsymbol{n}_k^{(n)})^{\mathrm{T}}]^{\mathrm{T}}, n = 0, 1, \ldots, N$
$\boldsymbol{x}_{k\|k-1}^{(n)} = f_{k\|k-1}\left(\boldsymbol{x}_{k-1}^{(n)}, v_{k-1}^{(n)}\right), \quad n = 0, 1, \ldots, N$
$\boldsymbol{z}_{k\|k-1}^{(n)} = h_k\left(\boldsymbol{x}_{k\|k-1}^{(n)}, \boldsymbol{n}_k^{(n)}\right), \quad n = 0, 1, \ldots, N$
$\boldsymbol{m}_{k\|k-1}^{(i)} = \sum_{n=0}^{N} \omega^{(n)} \boldsymbol{x}_{k\|k-1}^{(n)}$
$\boldsymbol{P}_{k\|k-1}^{(i)} = \sum_{n=0}^{N} \omega^{(n)} (\boldsymbol{x}_{k\|k-1}^{(n)} - \boldsymbol{m}_{k\|k-1}^{(j)})(\boldsymbol{x}_{k\|k-1}^{(n)} - \boldsymbol{m}_{k\|k-1}^{(j)})^{\mathrm{T}}$
$\boldsymbol{\eta}_{k\|k-1}^{(i)} = \sum_{n=0}^{N} \omega^{(n)} \boldsymbol{z}_{k\|k-1}^{(n)}$
$\boldsymbol{S}_{k\|k-1}^{(i)} = \sum_{n=0}^{N} \omega^{(n)} (\boldsymbol{z}_{k\|k-1}^{(n)} - \boldsymbol{\eta}_{k\|k-1}^{(j)})(\boldsymbol{z}_{k\|k-1}^{(n)} - \boldsymbol{\eta}_{k\|k-1}^{(j)})^{\mathrm{T}}$
$\boldsymbol{C}_k^{(i)} = \sum_{n=0}^{N} \omega^{(n)} (\boldsymbol{x}_{k\|k-1}^{(n)} - \boldsymbol{m}_{k\|k-1}^{(j)})(\boldsymbol{z}_{k\|k-1}^{(n)} - \boldsymbol{\eta}_{k\|k-1}^{(j)})^{\mathrm{T}}$

(continued)

Table 4.5 (continued)

$\boldsymbol{G}_k^{(i)} = \boldsymbol{C}_k^{(i)}(\boldsymbol{S}_{k\|k-1}^{(i)})^{-1}$
$\boldsymbol{P}_{k\|k}^{(i)} = \boldsymbol{P}_{k\|k-1}^{(i)} - \boldsymbol{G}_k^{(i)}(\boldsymbol{S}_{k\|k-1}^{(i)})^{-1}(\boldsymbol{G}_k^{(i)})^{\mathrm{T}}$
end
$J_{k\|k-1} = i$
Step 3. Update
Same as Step 4 in Table 4.1
Output: $\{w_k^{(i)}, \boldsymbol{m}_k^{(i)}, \boldsymbol{P}_k^{(i)}\}_{i=1}^{J_k}$

in exponential mixture form, and extended Kalman implementation and unscented Kalman implementation of the PHD filter. It should be noted that when extracting target states, the SMC-PHD filter requires an additional particle swarm clustering technology, whereas the GM-PHD filter is simpler and more reliable.

The PHD filter is mainly disadvantaged by the lack of high-order cardinality information, because the PHD recursion is a first-order moment approximation, which propagates the cardinality information through a single parameter, and effectively approximates the cardinality distribution through a mean-matched Poisson distribution. Since the mean and covariance of the Poisson distribution are equal, when the number of targets is large, the cardinality variance estimated by the PHD filter is correspondingly large. Furthermore, in the absence of the higher-order cardinality distribution, only the mean of the number of targets can be used as a valid expected a posterior (EAP) estimation. At this time, under the condition of low signal-to-noise ratio (SNR), the estimation tends to become abnormal due to the minor modes caused by clutters.

Chapter 5
Cardinalized Probability Hypothesis Density Filter

5.1 Introduction

Although the PHD filter enjoys the significant advantage of low computational workloads, it potentially assumes that the posterior multi-target distribution approximately conforms with the Poisson distribution. Such an approximation inevitably damages a large amount of information of the complete multi-target distribution, leaving the estimation extremely unstable (namely, large variance) in the presence of false alarms, especially missed detections of targets. The order of the approximate multi-target moments can be increased to overcome the disadvantage of the PHD recursion. The complete second-order multi-target moment filter refers to that the filter propagates not only the first-order moment (i.e., the PHD), but also the second-order multi-target moment (i.e., the multi-target covariance density). However, in the case of a large number of targets, the calculation of the complete second-order multi-target moment filter is very difficult. Different from the complete second-order multi-target moment filter, which propagates the covariance density, the CPHD filter [3] propagates both the intensity function (PHD) and the cardinality distribution (the probability distribution of the number of targets). Therefore, strictly speaking, the CPHD filter is a partial second-order multi-target moment filter, which is a generalization of the PHD filter [3]. In the CPHD filter, the propagation of the cardinality distribution (compared with the covariance density) greatly relieves the computational workloads. On the other hand, when compared with the PHD filter, as the CPHD filter can directly obtain the cardinality distribution, it delivers a better performance at the cost of higher computational complexity. For this reason, the CPHD filter is a compromise between the information loss of the first-order multi-target moment approximation and the computational complexity of the complete second-order moment approximation. The PHD/CPHD filters can be applied to joint estimation of clutter parameters, detection probability and multi-target state [403, 173], and have also been extended to multiple model, extended target, unresolved measurement, and distributed multi-sensor multi-target filtering. Please refer to [15] for detailed progress about the PHD/CPHD filters.

W. Wu et al., *Target Tracking with Random Finite Sets*,
https://doi.org/10.1007/978-981-19-9815-7_5

Similar to the previous chapter, this chapter first gives the CPHD recursion, and on this basis, the SMC implementation and GM implementation of the CPHD recursion are introduced, namely the SMC-CPHD and GM-CPHD filters. The two are respectively applicable to the nonlinear non-Gaussian and linear Gaussian system models. At last, two extended versions of the GM-CPHD filter are given, namely, the extended Kalman CPHD (EK-CPHD) and unscented Kalman CPHD (UK-CPHD) filters, which can be applied to a mildly nonlinear Gaussian model.

5.2 CPHD Recursion

Instead of propagating the multi-target filtering density $\pi_k(\cdot|Z^k)$, the CPHD filter propagates the (partial) second-order moment. Specifically, what it propagates is the posterior intensity (PHD) v_k and cardinality distribution ρ_k of an RFS density. The CPHD recursion relies on the following assumptions regarding target dynamics and sensor measurement:

- Each target evolves and generates measurements independently;
- The birth RFS and survived RFS are independent of each other;
- The clutter RFS is an IIDC process and is independent of target-originated measurement RFS;
- The prior multi-target RFS and predicted multi-target RFS are the IIDC processes.

It should be noted that the above assumptions are similar to those of the PHD recursion described in Sect. 5.4.2, except that the Poisson process is relaxed to the IIDC process.

$v_{k|k-1}$ and $\rho_{k|k-1}$ respectively represent the intensity and the cardinality distribution related to the predicted multi-target state. v_k and ρ_k respectively represent the intensity and the cardinality distribution related to the posterior multi-target state. The following two propositions (see Appendix E for the proof) clearly show how posterior intensity and posterior cardinality distribution are jointly propagated over time. The CPHD recursion form provided in [64] is adopted, as it is easier to implement.

Proposition 7 *When posterior intensity v_{k-1} and posterior cardinality distribution ρ_{k-1} at time $k-1$ are given, the predicted intensity $v_{k|k-1}$ and predicted cardinality distribution $\rho_{k|k-1}$ are respectively given by*[1]

$$v_{k|k-1}(\boldsymbol{x}) = \int p_{S,k}(\boldsymbol{\xi})\phi_{k|k-1}(\boldsymbol{x}|\boldsymbol{\xi})v_{k-1}(\boldsymbol{\xi})d\boldsymbol{\xi} + v_{\gamma,k}(\boldsymbol{x}) \tag{5.2.1}$$

[1] It should be noted that the target spawning is not taken into consideration herein. In general, the RFS framework of multi-target filtering includes the target spawning case. Please refer to [63] for specific details. For the CPHD filter that considers the target spawning, refer to [123].

$$\rho_{k|k-1}(n) = \sum_{j=0}^{n} \rho_{\gamma,k}(n-j)\Psi_{k|k-1}[v_{k-1}, \rho_{k-1}](j) \tag{5.2.2}$$

where $p_{S,k}(\boldsymbol{\xi})$ is the probability that the target survives at time k when the previous state $\boldsymbol{\xi}$ is given; $\phi_{k|k-1}(\cdot|\boldsymbol{\xi})$ represents the single-target transition density at time k under the condition of the previous state $\boldsymbol{\xi}$; $v_{\gamma,k}(\cdot)$ is the birth target intensity at time k; $\rho_{\gamma,k}(\cdot)$ is the cardinality distribution of birth targets at time k; $\Psi_{k|k-1}[v_{k-1}, \rho_{k-1}](j)$ is calculated through the following equation

$$\Psi_{k|k-1}[v, \rho](j) = \sum_{i=j}^{n} C_j^i \frac{\langle p_{S,k}, v\rangle^j \langle 1 - p_{S,k}, v\rangle^{i-j}}{\langle 1, v\rangle^i} \rho(i) \tag{5.2.3}$$

with $C_j^n = n!/[j!(n-j)!]$ being the combination coefficient (or binomial coefficient).

Proposition 8 *When predicted density $v_{k|k-1}$ and predicted cardinality distribution $\rho_{k|k-1}$ at time k are given, updated intensity v_k and updated cardinality distribution ρ_k are respectively*

$$\begin{aligned} v_k(\boldsymbol{x}) = & \frac{\langle \Upsilon_k^{(1)}[v_{k|k-1}, Z_k], \rho_{k|k-1}\rangle}{\langle \Upsilon_k^{(0)}[v_{k|k-1}, Z_k], \rho_{k|k-1}\rangle}[1 - p_{D,k}(\boldsymbol{x})]v_{k|k-1}(\boldsymbol{x}) \\ & + \sum_{z \in Z_k} \frac{\langle \Upsilon_k^{(1)}[v_{k|k-1}, Z_k - \{z\}], \rho_{k|k-1}\rangle}{\langle \Upsilon_k^{(0)}[v_{k|k-1}, Z_k], \rho_{k|k-1}\rangle} \varphi_{k,z}(\boldsymbol{x}) v_{k|k-1}(\boldsymbol{x}) \end{aligned} \tag{5.2.4}$$

$$\rho_k(n) = \frac{\Upsilon_k^{(0)}[v_{k|k-1}, Z_k](n)\rho_{k|k-1}(n)}{\langle \Upsilon_k^{(0)}[v_{k|k-1}, Z_k], \rho_{k|k-1}\rangle} \tag{5.2.5}$$

where

$$\begin{aligned} & \Upsilon_k^{(u)}[v, Z](n) \\ & = \sum_{j=0}^{\min(|Z|,n)} (|Z|-j)!\rho_{C,k}(|Z|-j)P_{j+u}^n \frac{\langle 1 - p_{D,k}, v\rangle^{n-(j+u)}}{\langle 1, v\rangle^n} e_j(\alpha_k(v, Z)) \end{aligned} \tag{5.2.6}$$

$$\varphi_{k,z}(\boldsymbol{x}) = \frac{\langle 1, \kappa_k\rangle}{\kappa_k(z)} p_{D,k}(\boldsymbol{x}) g_k(z|\boldsymbol{x}) \tag{5.2.7}$$

$$\alpha_k(v, Z) = \{\langle v, \varphi_{k,z}\rangle : z \in Z\} \tag{5.2.8}$$

In the above equations, Z_k is a measurement set at time k, $g_k(\cdot|\boldsymbol{x})$ is the single-target measurement likelihood function at time k when the current state $\boldsymbol{x}$ is given; $p_{D,k}(\boldsymbol{x})$ is the target detection probability at time k given current state $\boldsymbol{x}$; $\kappa_k(\cdot)$ is the intensity of clutter measurements at time k; $\rho_{C,k}(\cdot)$ is the clutter cardinality distribution at time k; $P_j^n = n!/(n-j)!$ is the permutation coefficient; $e_j(\cdot)$ represents the j-order elementary symmetric function (ESF); for a finite set Z of real numbers, $e_j(\cdot)$ is defined as

$$e_j(Z) = \sum_{S \subseteq Z, |S|=j} \left(\prod_{\zeta \in S} \zeta \right) \tag{5.2.9}$$

and $e_0(Z) = 1$ by default.

Propositions 7 and 8 are respectively the prediction and update steps of the CPHD recursion. Without considering the target spawning scenario, the CPHD intensity prediction (5.2.1) is exactly the same as the PHD prediction (4.2.1). The CPHD cardinality prediction (5.2.2) is the convolution of cardinality distributions of birth and survived targets, because the predicted cardinality is the sum of cardinalities of birth and survived targets. It should be noted that the CPHD intensity prediction and cardinality prediction (5.2.1)–(5.2.2) are decoupled, whereas the CPHD intensity update and cardinality update (5.2.4)–(5.2.5) are coupled. However, the CPHD intensity update (5.2.4) is similar to the PHD update (4.2.2), both of which consist of 1 missed detection item and $|Z_k|$ detection items. The cardinality update (5.2.5) incorporates the clutter cardinality, measurement set, predicted intensity, and predicted cardinality distribution. In fact, Eq. (5.2.5) is the Bayesian update, in which, $\Upsilon_k^{(0)}[v_{k|k-1}, Z_k](n)$ is the likelihood of multi-target measurement Z_k when n targets are given, and $\left\langle \Upsilon_k^{(0)}[v_{k|k-1}, Z_k], \rho_{k|k-1} \right\rangle$ is a normalization constant.

5.3 SMC-CPHD Filter

The SMC implementation of the PHD filter is easy to be generalized to the CPHD filter. For the general nonlinear non-Gaussian multi-target model, the SMC implementation of the CPHD recursion (namely, SMC-CPHD filter) can be obtained by following the SMC implementation of the PHD recursion. The basic idea is to recursively propagate over time a set of weighted samples representing posterior intensity and posterior cardinality distribution.

5.3.1 SMC-CPHD Recursion

Assume that posterior intensity v_{k-1} and posterior cardinality distribution ρ_{k-1} at time $k-1$ are given, and

$$v_{k-1}(\boldsymbol{x}) = \sum_{j=1}^{L_{k-1}} w_{k-1}^{(j)} \delta_{\boldsymbol{x}_{k-1}^{(j)}}(\boldsymbol{x}) \tag{5.3.1}$$

In addition, the importance (or proposed) densities $q_k(\cdot|\boldsymbol{x}_{k-1}, Z_k)$ and $q_{\gamma,k}(\cdot|Z_k)$ that satisfy support$(v_k) \subseteq$ support(q_k) and support$(v_{\gamma,k}) \subseteq$ support$(q_{\gamma,k})$ are also given. Then, according to Eqs. (5.2.1)–(5.2.2), it can be concluded that predicted intensity $v_{k|k-1}$ and predicted cardinality distribution $\rho_{k|k-1}$ are respectively

$$v_{k|k-1}(\boldsymbol{x}) \approx \sum_{j=1}^{L_{k-1}+L_{\gamma,k}} w_{k|k-1}^{(j)} \delta_{\boldsymbol{x}_k^{(j)}}(\boldsymbol{x}) \tag{5.3.2}$$

$$\rho_{k|k-1}(n) \approx \sum_{j=0}^{n} \rho_{\gamma,k}(n-j) \sum_{i=j}^{\infty} C_j^i \frac{\langle p_{S,k}^{(1:L_{k-1})}, w_{k-1}\rangle^j \langle 1 - p_{S,k}^{(1:L_{k-1})}, w_{k-1}\rangle^{i-j}}{\langle 1, w_{k-1}\rangle^i} \rho_{k-1}(i) \tag{5.3.3}$$

where

$$\boldsymbol{x}_k^{(j)} \sim \begin{cases} q_k(\cdot|\boldsymbol{x}_{k-1}^{(j)}, Z_k), & j = 1, \ldots, L_{k-1} \\ q_{\gamma,k}(\cdot|Z_k), & j = L_{k-1}+1, \ldots, L_{k-1}+L_{\gamma,k} \end{cases} \tag{5.3.4}$$

$$w_{k|k-1}^{(j)} = \begin{cases} \dfrac{p_{S,k}(\boldsymbol{x}_{k-1}^{(j)})\phi_{k|k-1}(\boldsymbol{x}_k^{(j)}|\boldsymbol{x}_{k-1}^{(j)})}{q_k(\boldsymbol{x}_k^{(j)}|\boldsymbol{x}_{k-1}^{(j)}, Z_k)} w_{k-1}^{(j)}, & j = 1, \ldots, L_{k-1} \\ \dfrac{v_{\gamma,k}(\boldsymbol{x}_k^{(j)})}{L_{\gamma,k} q_{\gamma,k}(\boldsymbol{x}_k^{(j)}|Z_k)}, & j = L_{k-1}+1, \ldots, L_{k-1}+L_{\gamma,k} \end{cases} \tag{5.3.5}$$

$$w_{k-1} = [w_{k-1}^{(1)}, \ldots, w_{k-1}^{(L_{k-1})}]^{\mathrm{T}} \tag{5.3.6}$$

$$p_{S,k}^{(1:L_{k-1})} = [p_{S,k}(\boldsymbol{x}_{k-1}^{(1)}), \ldots, p_{S,k}(\boldsymbol{x}_{k-1}^{(L_{k-1})})]^{\mathrm{T}} \tag{5.3.7}$$

and $v_{\gamma,k}(\cdot)$ is the birth target intensity at time k.

Assume that predicted intensity $v_{k|k-1}$ and predicted cardinality distribution $\rho_{k|k-1}$ at time k are given, and

$$v_{k|k-1}(\boldsymbol{x}) = \sum_{j=1}^{L_{k|k-1}} w_{k|k-1}^{(j)} \delta_{\boldsymbol{x}_k^{(j)}}(\boldsymbol{x}) \tag{5.3.8}$$

Then, according to Eqs. (5.2.4)–(5.2.5), updated intensity v_k and updated cardinality distribution ρ_k are respectively

$$v_k(\boldsymbol{x}) = \sum_{j=1}^{L_{k|k-1}} w_k^{(j)} \delta_{\boldsymbol{x}_k^{(j)}}(\boldsymbol{x}) \tag{5.3.9}$$

$$\rho_k(n) = \frac{\Upsilon_k^{(0)}[w_{k|k-1}, Z_k](n)\rho_{k|k-1}(n)}{\langle \Upsilon_k^{(0)}[w_{k|k-1}, Z_k], \rho_{k|k-1}\rangle} \tag{5.3.10}$$

where

$$\begin{aligned} w_k^{(j)} = w_{k|k-1}^{(j)} \Bigg[& (1 - p_{D,k}(\boldsymbol{x}_k^{(j)})) \frac{\langle \Upsilon_k^{(1)}[w_{k|k-1}, Z_k], \rho_{k|k-1}\rangle}{\langle \Upsilon_k^{(0)}[w_{k|k-1}, Z_k], \rho_{k|k-1}\rangle} + \\ & \sum_{z \in Z_k} \varphi_{k,z}(\boldsymbol{x}_k^{(j)}) \frac{\langle \Upsilon_k^{(1)}[w_{k|k-1}, Z_k - \{z\}], \rho_{k|k-1}\rangle}{\langle \Upsilon_k^{(0)}[w_{k|k-1}, Z_k], \rho_{k|k-1}\rangle} \Bigg] \end{aligned} \tag{5.3.11}$$

$$\begin{aligned} & \Upsilon_k^{(u)}[w, Z](n) \\ & = \sum_{j=0}^{\min(|Z|,n)} (|Z| - j)!\rho_{C,k}(|Z| - j) P_{j+u}^n \frac{\langle 1 - p_{D,k}^{(1:L_{k|k-1})}, w\rangle^{n-(j+u)}}{\langle 1, w\rangle^n} e_j(\alpha_k(w, Z)) \end{aligned} \tag{5.3.12}$$

$$\alpha_k(w, Z) = \{\langle w, \varphi_{k,z}^{(1:L_{k|k-1})}\rangle : z \in Z\} \tag{5.3.13}$$

$$w_{k|k-1} = [w_{k|k-1}^{(1)}, \ldots, w_{k|k-1}^{(L_{k|k-1})}]^{\mathrm{T}} \tag{5.3.14}$$

$$p_{D,k}^{(1:L_{k|k-1})} = [p_{D,k}(\boldsymbol{x}_k^{(1)}), \ldots, p_{D,k}(\boldsymbol{x}_k^{(L_{k|k-1})})]^{\mathrm{T}} \tag{5.3.15}$$

$$\varphi_{k,z}^{(1:L_{k|k-1})} = [\varphi_{k,z}(\boldsymbol{x}_k^{(1)}), \ldots, \varphi_{k,z}(\boldsymbol{x}_k^{(L_{k|k-1})})]^{\mathrm{T}} \tag{5.3.16}$$

$$\varphi_{k,z}(\boldsymbol{x}_k^{(j)}) = \frac{\langle 1, \kappa_k\rangle}{\kappa_k(z)} p_{D,k}(\boldsymbol{x}_k^{(j)}) g_k(z|\boldsymbol{x}_k^{(j)}) \tag{5.3.17}$$

5.3.2 Multi-target State Estimation

Like the SMC-PHD filter, in the SMC-CPHD filter, the number $L_k = L_{k-1} + L_{\gamma,k}$ of particles required to represent the posterior intensity at time k will keep increasing over time. Therefore, the adaptive allocation scheme of the number of particles for the SMC-PHD filter can be employed. In addition, the SMC-CPHD filter also requires re-sampling and clustering techniques for state extraction. When re-sampling based on the posterior intensity, the new weight should also be scaled to $\overline{N}_k = \sum_{j=1}^{L_{k|k-1}} w_k^{(j)}$, rather than 1. In the state extraction, the first-order moment visualization technique can be adopted to extract the multi-target state. Due to the extra information about the cardinality distribution, the state extraction of the SMC-CPHD filter is slightly different from that of the SMC-PHD filter.

First, the number of targets is estimated through the EAP estimator $\hat{N}_k = \sum_{n=0}^{\infty} n\rho_k(n)$ or the MAP estimator $\hat{N}_k = \arg\max_n \rho_k(n)$. It should be noted that in the case of low signal-to-noise ratio, the EAP estimator is prone to fluctuate and unreliable. This is because false alarms and missed detections of targets can easily induce minor modes in the posterior cardinality. Therefore, the mean may randomly deviate from the primary modes corresponding to the targets. On the other hand, the MAP estimator ignores the minor modes and directly locks to the primary modes corresponding to the targets, which is more reliable. In this sense, the MAP estimator is generally more popular than the EAP estimators [3]. Then, the clustering technology is used to group the particle swarm into a given number (the estimated number $\hat{N}_k$ of targets) of clusters, and then $\hat{N}_k$ means $(\boldsymbol{m}_k^{(1)}, \ldots, \boldsymbol{m}_k^{(\hat{N}_k)})$ of clusters are selected as the multi-target state estimation, i.e., $\hat{X}_k = \{\boldsymbol{m}_k^{(1)}, \ldots, \boldsymbol{m}_k^{(\hat{N}_k)}\}$.

As mentioned above, the state extraction through the clustering technique may result in an unreliable estimation. Thereby, for the linear Gaussian multi-target (LGM) model, the Gaussian mixture CPHD (GM-CPHD) filter described in the next section can be adopted. For the mildly nonlinear model, the approximate EK-CPHD and UK-CPHD filters described in Sect. 5.5 require lower computational workloads, with a better performance than the SMC implementation. This is because their potential Gaussian mixture representation dispenses with the demands for clustering.

5.4 GM-CPHD Filter

Under the assumption of the LGM model (see the assumptions A.4–A.6 in Sect. 4.4.1 for details, but as the target spawning is not considered here, there is no assumption on the target spawning in A.6), a closed form solution of the CPHD recursion can be obtained according to the CPHD recursion [see (5.2.1), (5.2.2), (5.2.4) and (5.2.5)].

5.4.1 GM-CPHD Recursion

For the LGM model, the following two propositions give the closed form solution of the CPHD recursion. Specifically, these propositions show how posterior intensity and posterior cardinality distribution (in the form of Gaussian mixture) are propagated analytically over time.

Proposition 9 *Assume that posterior intensity v_{k-1} and posterior cardinality distribution ρ_{k-1} at time $k-1$ are given, and v_{k-1} is in the form of Gaussian mixture*

$$v_{k-1}(\boldsymbol{x}) = \sum_{i=1}^{J_{k-1}} w_{k-1}^{(i)} \mathcal{N}(\boldsymbol{x}; \boldsymbol{m}_{k-1}^{(i)}, \boldsymbol{P}_{k-1}^{(i)}) \tag{5.4.1}$$

Then, predicted intensity $v_{k|k-1}$ is also in the form of Gaussian mixture, and $v_{k|k-1}$ and predicted cardinality distribution $\rho_{k|k-1}$ are respectively

$$v_{k|k-1}(\boldsymbol{x}) = v_{S,k|k-1}(\boldsymbol{x}) + v_{\gamma,k}(\boldsymbol{x}) \tag{5.4.2}$$

$$\rho_{k|k-1}(n) = \sum_{j=0}^{n} \left\{ \rho_{\gamma,k}(n-j) \sum_{i=j}^{\infty} C_j^i \rho_{k-1}(i) p_{S,k}^j (1 - p_{S,k})^{i-j} \right\} \tag{5.4.3}$$

where $v_{\gamma,k}(\cdot)$ is the birth target intensity at time k, which is given by (4.4.5), and $v_{S,k|k-1}(\boldsymbol{x})$ is given by (4.4.9).

Proposition 10 *Assume that predicted intensity $v_{k|k-1}$ and predicted cardinality distribution $\rho_{k|k-1}$ at time k are given, and $v_{k|k-1}$ is in Gaussian mixture form*

$$v_{k|k-1}(\boldsymbol{x}) = \sum_{i=1}^{J_{k|k-1}} w_{k|k-1}^{(i)} \mathcal{N}\left(\boldsymbol{x}; \boldsymbol{m}_{k|k-1}^{(i)}, \boldsymbol{P}_{k|k-1}^{(i)}\right) \tag{5.4.4}$$

Then, updated intensity v_k is also in the form of Gaussian mixture, and v_k and updated cardinality distribution ρ_k are respectively

$$\begin{aligned} v_k(\boldsymbol{x}) = {} & \frac{\left\langle \Upsilon_k^{(1)}[w_{k|k-1}, Z_k], \rho_{k|k-1} \right\rangle}{\left\langle \Upsilon_k^{(0)}[w_{k|k-1}, Z_k], \rho_{k|k-1} \right\rangle} (1 - p_{D,k}) v_{k|k-1}(\boldsymbol{x}) \\ & + \sum_{z \in Z_k} \sum_{i=1}^{J_{k|k-1}} w_k^{(i)}(z) \mathcal{N}\left(\boldsymbol{x}; \boldsymbol{m}_{k|k}^{(i)}(z), \boldsymbol{P}_{k|k}^{(i)}\right) \end{aligned} \tag{5.4.5}$$

$$\rho_k(n) = \frac{\Upsilon_k^{(0)}[w_{k|k-1}, Z_k](n) \rho_{k|k-1}(n)}{\langle \Upsilon_k^{(0)}[w_{k|k-1}, Z_k], \rho_{k|k-1} \rangle} \tag{5.4.6}$$

where $\boldsymbol{m}_{k|k}^{(i)}(z)$, $\boldsymbol{P}_{k|k}^{(i)}$, and $q_k^{(i)}(z)$ are respectively described in (4.4.19), (4.4.20), and (4.4.22), and

$$\begin{aligned} & \Upsilon_k^{(u)}[w, Z](n) \\ & = \sum_{j=0}^{\min(|Z|,n)} (|Z| - j)! \rho_{C,k}(|Z| - j) P_{j+u}^n \frac{(1 - p_{D,k})^{n-(j+u)}}{\langle 1, w \rangle^{j+u}} e_j(\Lambda_k(w, Z)) \end{aligned} \tag{5.4.7}$$

$$\Lambda_k(w, Z) = \left\{ \frac{\langle 1, \kappa_k \rangle}{\kappa_k(z)} p_{D,k} w^{\mathrm{T}} q_k(z) : z \in Z \right\} \tag{5.4.8}$$

$$w_{k|k-1} = [w_{k|k-1}^{(1)}, \ldots, w_{k|k-1}^{(J_{k|k-1})}]^{\mathrm{T}} \tag{5.4.9}$$

$$q_k(z) = [q_k^{(1)}(z), \ldots, q_k^{(J_{k|k-1})}(z)]^{\mathrm{T}} \tag{5.4.10}$$

$$w_k^{(i)}(z) = p_{D,k} w_{k|k-1}^{(i)} q_k^{(i)}(z) \frac{\langle 1, \kappa_k \rangle}{\kappa_k(z)} \frac{\langle \Upsilon_k^{(1)}[w_{k|k-1}, Z_k - \{z\}], \rho_{k|k-1} \rangle}{\langle \Upsilon_k^{(0)}[w_{k|k-1}, Z_k], \rho_{k|k-1} \rangle} \tag{5.4.11}$$

Propositions 9 and 10 are respectively the prediction and update steps of the CPHD recursion under the LGM model. The PHD recursion is a special case of the CPHD recursion [3]. Similarly, the GM-PHD recursion is a special case of the GM-CPHD recursion given in Propositions 9 and 10.

Similar to the derivation of the GM-PHD filter, Propositions 9 and 10 can be obtained by applying the product equation for Gaussian distributions. Specifically, Proposition 9 can be derived as follows: the intensity prediction can be obtained by substituting Eqs. (4.4.1), (4.4.3) and (5.4.1) into the CPHD intensity prediction (5.2.1), and replacing integrals involving Gaussian products with the appropriate Gaussians given in Lemma 4 of Appendix A; the cardinality prediction can be obtained by simplifying the expression of the CPHD cardinality prediction (5.2.2) with the assumptions in (4.4.3). Proposition 10 is derived by the following steps: First, Eq. (5.4.7) can be obtained by substituting Eqs. (4.4.2), (4.4.4) and (5.4.4) into Eq. (5.2.6), and using Lemma 5 in Appendix A to simplify the resulting expression; Then, the intensity update can be obtained by substituting Eqs. (4.4.2), (4.4.4), (5.4.4) and the result of (5.4.7) into the CPHD intensity update (5.2.4), and replacing Gaussian products with the appropriate Gaussians provided in Lemma 5; the cardinality update can be obtained by substituting the result of Eq. (5.4.7) into the CPHD cardinality update (5.2.5).

According to Propositions 9 and 10, if the initial intensity v_0 is a Gaussian mixture (including the case of $v_0 = 0$), all subsequent predicted intensity $v_{k|k-1}$ and posterior intensity v_k are also Gaussian mixture. Proposition 9 provides a closed expression, in which the mean, covariance and weight of $v_{k|k-1}$ is calculated by the mean, covariance and weight of v_{k-1}, and $\rho_{k|k-1}$ is calculated by cardinality distribution ρ_{k-1}. When a new measurement set is collected, Proposition 10 provides a closed expression for calculating the mean, covariance and weight of v_k from the mean, covariance and weight of $v_{k|k-1}$ and for calculating ρ_k from cardinality distribution $\rho_{k|k-1}$.

5.4.2 *Multi-target State Extraction*

Similar to the GM-PHD filter, the state extraction from the GM-CPHD filter involves first estimating the number of targets, and then extracting the corresponding number of mixed components with the largest weight from the posterior intensity as the state estimation. The number of targets can be estimated by the EAP estimator $\hat{N}_k = \mathrm{E}[|X_k|]$ or the MAP estimator $\hat{N}_k = \arg\max \rho_k(\cdot)$. As mentioned earlier, the latter is usually more popular. In fact, as the CPHD filter directly estimates the cardinality distribution, the estimation of the number of targets can be obtained through the MAP estimator, namely,

$$\hat{N}_k = \arg_n \max \rho_k(n) \tag{5.4.12}$$

When the MAP estimation $\hat{N}_k$ of the number of targets is given, the state estimation can be obtained through the estimation extraction steps as same as the GM-PHD filter.

5.4.3 Implementation Issues of GM-CPHD Filter

Problems arising from the specific implementation of the GM-CPHD filter are as follows.

(1) Calculation of the cardinality distribution: The propagation of the cardinality distribution essentially involves the use of Eqs. (5.4.3) and (5.4.6) to recursively predict and update the mass of the distribution. However, if the cardinality distribution is infinitely tailed, the calculation of the entire posterior cardinality is not feasible, as this involves the calculation of an infinite number of terms. In fact, if the cardinality distribution is shortly or moderately tailed, it can be pruned by restriction $n = N_{\max}$ and approximated by $\{\rho_k(n)\}_{n=0}^{N_{\max}}$, which has a finite number of terms. When $n = N_{\max}$ is remarkably greater than the number of targets at any time in a scene, the approximation is reasonable.

(2) Calculation of the ESF: Calculating the ESF directly based on Definition (5.2.9) is absolutely not feasible. According to basic conclusions of combinatorial theory (such as the well-known Newton-Girard equation or the equivalent Vieta theory), ESF $e_j(\cdot)$ can be calculated through the steps described in [64]. When $r_1, r_2, \ldots, r_M$ represent different roots of polynomial $\alpha_M x^M + \alpha_{M-1}x^{M-1} + \ldots + \alpha_1 x + \alpha_0$, j-order $e_j(\cdot)$ $(j = 0, 1, \ldots, M)$ is given by $e_j(r_1, r_2, \ldots, r_M) = (-1)^j \alpha_{M-j} / \alpha_M$. Therefore, the value of $e_j(Z)$ can be obtained by expanding the polynomial determined by roots given by elements of Z. This can be implemented through proper recursion or convolution. For finite set Z, the computational complexity of $e_j(Z)$ is $\mathcal{O}(|Z|^2)$. Through the proper decomposition and recursion, the complexity can be further reduced to $\mathcal{O}(|Z| \log^2 |Z|)$ [64].

In the CPHD recursion, each measurement update step needs to calculate $|Z|+1$ ESFs, that is, each EFS for Z and one EFS for the set $\{Z - \{z\}\}$, wherein $z \in Z$. Therefore, the complexity of the CPHD recursion is $\mathcal{O}(|Z|^3)$. Through the proper decomposition and recursion, the complexity of the CPHD filter can be reduced to $\mathcal{O}(|Z|^2 \log^2 |Z|)$. When $|Z|$ is relatively large, the computational workloads can be saved moderately, and the reduction in the complexity delivers a certain advantage. In addition, the gate technique [9] used in traditional tracking algorithms can be employed to reduce the number of measurements, thereby further reducing the computational workloads.

Compared with the PHD filter whose complexity is proportional to the number of targets and the number of measurements, the complexity of the CPHD filter is proportional to the number of targets, yet its relation with the number of measurements is $\mathcal{O}(|Z|^2 \log |Z|)$ [64].

(3) Management of mixed components: similar to the GM-PHD filter, Gaussian components required to represent the posterior will increase infinitely. To alleviate this problem, the pruning and merging steps adopted by the GM-PHD filter can be directly applied to the GM-CPHD filter. The basic idea is to discard components with minor weights and merge the close components.

5.5 Extension of GM-CPHD Filter

Similar to the extension of the GM-PHD filter, the GM-CPHD recursion can be extended accordingly through linearization and unscented transform techniques. The linear single-target dynamic model and linear measurement model can be relaxed to non-linear functions with respect to state and noise variables [see (4.5.25)–(4.5.26)].

5.5.1 Extended Kalman CPHD Recursion

Similar to the EKF, non-linear approximation of the GM-CPHD recursion can be obtained by applying the local linearization technique to f_k and h_k (that is, the EK-CPHD recursion).

In Proposition 9, by using the first-order approximation to the nonlinear equation to predict the mixture components of the survival targets, the nonlinear target motion model (4.5.25) is approximated as follows to perform the prediction step, that is, the original formulas (4.4.10) and (4.4.11) are replaced by approximations (5.5.1) and (5.5.2)

$$\boldsymbol{m}_{S,k|k-1}^{(i)} = f_k(\boldsymbol{m}_{k-1}^{(i)}, \boldsymbol{0}) \tag{5.5.1}$$

$$\boldsymbol{P}_{S,k|k-1}^{(i)} = \boldsymbol{F}_{k-1}^{(i)} \boldsymbol{P}_{k-1}^{(i)} (\boldsymbol{F}_{k-1}^{(i)})^{\mathrm{T}} + \boldsymbol{V}_{k-1}^{(i)} \boldsymbol{Q}_{k-1} (\boldsymbol{V}_{k-1}^{(i)})^{\mathrm{T}} \tag{5.5.2}$$

where

$$\boldsymbol{F}_{k-1}^{(i)} = \frac{\partial f_k(\boldsymbol{x}, \boldsymbol{0})}{\partial \boldsymbol{x}}\bigg|_{\boldsymbol{x}=\boldsymbol{m}_{k-1}^{(i)}} \tag{5.5.3}$$

$$\boldsymbol{V}_{k-1}^{(i)} = \frac{\partial f_k(\boldsymbol{m}_{k-1}^{(i)}, \boldsymbol{v})}{\partial \boldsymbol{v}}|_{\boldsymbol{v}=\boldsymbol{0}} \tag{5.5.4}$$

In Proposition 10, by applying the first-order approximation of the nonlinear equation to update each predicted mixture component, the nonlinear measurement model (4.5.26) can be approximated as follows to obtain the update step, namely, replacing the original formulas (4.4.23) and (4.4.24) with approximations (5.5.5) and (5.5.6), and calculating (4.4.20) and (4.4.21) based on the linearization results of Eqs. (5.5.7) and (5.5.8)

$$\boldsymbol{\eta}_{k|k-1}^{(i)} = h_k(\boldsymbol{m}_{k|k-1}^{(i)}, \boldsymbol{0}) \tag{5.5.5}$$

$$\boldsymbol{S}_{k|k-1}^{(i)} = \boldsymbol{H}_k^{(i)} \boldsymbol{P}_{k|k-1}^{(i)} (\boldsymbol{H}_k^{(i)})^{\mathrm{T}} + \boldsymbol{N}_k^{(i)} \boldsymbol{R}_k (\boldsymbol{N}_k^{(i)})^{\mathrm{T}} \tag{5.5.6}$$

where

$$\boldsymbol{H}_k^{(i)} = \frac{\partial h_k(\boldsymbol{x}, \boldsymbol{0})}{\partial \boldsymbol{x}}\bigg|_{\boldsymbol{x}=\boldsymbol{m}_{k|k-1}^{(i)}} \tag{5.5.7}$$

$$\boldsymbol{N}_k^{(i)} = \frac{\partial h_k(\boldsymbol{m}_{k|k-1}^{(i)}, \boldsymbol{n})}{\partial \boldsymbol{n}}|_{\boldsymbol{n}=\boldsymbol{0}} \tag{5.5.8}$$

5.5.2 Unscented Kalman CPHD Recursion

Similar to the UKF, based on the UT, a non-linear approximation of the GM-CPHD recursion, that is, unscented Kalman CPHD (UK-CPHD) recursion, can be obtained. The strategy adopted is to use the UT to propagate the first- and second-order moments of each mixture component after nonlinear transform f_k and h_k, wherein the specific steps are as follows.

First, for each mixture component of the posterior intensity, its mean and covariance are respectively set as $\boldsymbol{u}_k^{(i)}$ and $\sum_k^{(i)}$, and a series of σ points (sigma points) $\{\boldsymbol{y}_k^{(n)}\}_{n=0}^N$ and corresponding weights $\{\omega^{(n)}\}_{n=0}^N$ are generated through the UT technique, where,

$$\boldsymbol{u}_k^{(i)} = [\boldsymbol{m}_{k-1}^{(i)} \ \boldsymbol{0} \ \boldsymbol{0}]^{\mathrm{T}} \tag{5.5.9}$$

$$\sum\nolimits_k^{(i)} = \mathrm{blkdiag}(\boldsymbol{P}_{k-1}^{(j)}, \boldsymbol{Q}_{k-1}, \boldsymbol{R}_k) \tag{5.5.10}$$

And then, the σ point is split into

$$\boldsymbol{y}_k^{(n)} = [(\boldsymbol{x}_{k-1}^{(n)})^{\mathrm{T}}, (\boldsymbol{v}_{k-1}^{(n)})^{\mathrm{T}}, (\boldsymbol{n}_k^{(n)})^{\mathrm{T}}]^{\mathrm{T}}, \quad n = 0, \ldots, N \tag{5.5.11}$$

For the prediction step, σ point is propagated via non-linear state transition function $\boldsymbol{x}_{k|k-1}^{(n)} = f_k(\boldsymbol{x}_{k-1}^{(n)}, \boldsymbol{v}_{k-1}^{(n)}), \quad n = 0, \ldots, N$. Thus, in Proposition 9, the prediction step can be implemented by approximating the nonlinear motion model, that is, replacing original Eqs. (4.4.10) and (4.4.11) with approximate Eqs. (5.5.12) and (5.5.13)

$$\boldsymbol{m}_{S,k|k-1}^{(i)} = \sum_{n=0}^{N} \omega^{(n)} \boldsymbol{x}_{k|k-1}^{(n)} \tag{5.5.12}$$

$$\boldsymbol{P}_{S,k|k-1}^{(i)} = \sum_{n=0}^{N} \omega^{(n)} (\boldsymbol{x}_{k|k-1}^{(n)} - \boldsymbol{m}_{S,k|k-1}^{(i)})(\boldsymbol{x}_{k|k-1}^{(n)} - \boldsymbol{m}_{S,k|k-1}^{(i)})^{\mathrm{T}} \tag{5.5.13}$$

For the update step, σ point is propagated through non-linear measurement function $\boldsymbol{z}_{k|k-1}^{(n)} = h_k(\boldsymbol{x}_{k|k-1}^{(n)}, \boldsymbol{n}_k^{(n)}), \quad n = 0, \ldots, N$. Thus, in Proposition 10, the update step can be implemented by approximating the nonlinear measurement model, that is, replacing the original Eqs. (4.4.23) and (4.4.24) with approximations (5.5.14) and (5.5.15), and replacing original Eqs. (4.4.20) and (4.4.21) with Eqs. (5.5.16) and (5.5.17)

$$\boldsymbol{\eta}_{k|k-1}^{(i)} = \sum_{n=0}^{N} \omega^{(n)} \boldsymbol{z}_{k|k-1}^{(n)} \tag{5.5.14}$$

$$\boldsymbol{S}_{k|k-1}^{(i)} = \sum_{n=0}^{N} \omega^{(n)} (\boldsymbol{z}_{k|k-1}^{(n)} - \boldsymbol{\eta}_{k|k-1}^{(i)})(\boldsymbol{z}_{k|k-1}^{(n)} - \boldsymbol{\eta}_{k|k-1}^{(i)})^{\mathrm{T}} \tag{5.5.15}$$

$$\boldsymbol{P}_k^{(i)} = \boldsymbol{P}_{k|k-1}^{(i)} - \boldsymbol{G}_k^{(i)} (\boldsymbol{S}_{k|k-1}^{(i)})^{-1} (\boldsymbol{G}_k^{(i)})^{\mathrm{T}} \tag{5.5.16}$$

$$\boldsymbol{G}_k^{(i)} = \boldsymbol{C}_k^{(i)} (\boldsymbol{S}_{k|k-1}^{(i)})^{-1} \tag{5.5.17}$$

$$\boldsymbol{C}_k^{(i)} = \sum_{n=0}^{N} \omega^{(n)} (\boldsymbol{x}_{k|k-1}^{(n)} - \boldsymbol{m}_{k|k-1}^{(i)})(\boldsymbol{z}_{k|k-1}^{(n)} - \boldsymbol{\eta}_{k|k-1}^{(i)})^{\mathrm{T}} \tag{5.5.18}$$

Note: the EK-CPHD and UK-CPHD recursions have advantages and disadvantages similar to the EKF and the UKF. The EK-CPHD recursion needs to calculate the Jacobian matrix, so it can only be applied to the differentiable state model and measurement model; The UK-CPHD completely avoids the derivative operation, and can even be applied to discontinuous models. When dealing with non-linear problems, the EK-CPHD and UK-CPHD approximations require much less computational workloads than the SMC version, and they can extract state estimation more easily due to the potential Gaussian mixture implementation.

5.6 Summary

The chapter, on the basis of the PHD recursion, respectively describes the SMC-CPHD and GM-CPHD filters for the nonlinear non-Gaussian model and the linear Gaussian model, and introduces the extension of the GM-CPHD filter in detail, including the extended Kalman implementation and unscented Kalman implementation of the CPHD filter. Besides, the CPHD recursion can also be extended to survival probability and detection probability in the form of exponential mixture as described in Sect. 4.5.1.

Similar to the SMC-PHD and GM-PHD filters, the SMC-CPHD filter also requires an additional particle swarm clustering technique when extracting target states, while the GM-CPHD filter is simpler and more reliable. In addition, it should be noted that the CPHD filter is not a simple extension of the PHD filter. Although the propagation of its intensity function (PHD) and cardinality distribution is decoupled in the prediction step, the two are mutually coupled in the update step.

Chapter 6
Multi-Bernoulli Filter

6.1 Introduction

Unlike the PHD and CPHD filters, which are respectively the first- and second-order moment approximations of the multi-target Bayesian recursion, the multi-Bernoulli (MB) filter is the probability density approximation of the complete multi-target Bayesian recursion. It propagates a finite yet time-varying number of hypothetical tracks during the recursion, wherein each hypothetical track is characterized by the existence probability and probability density of the current hypothetical state. The MB filter, which is better-known as the MeMBer filter [14], was first proposed by Mahler. The filter has a "cardinality bias" problem: overestimating the number of targets. To solve the problem, Vo et al. proposed a cardinality balance MeMBer (CBMeMBer) filter. Thanks to its linear complexity and its SMC implementation that requires no clustering operation during multi-target state extraction, it is highly suitable for the nonlinear non-Gaussian model.

This chapter first introduces the MeMBer filter and gives the reason of the cardinality bias. On this basis, the improved CBMeMBer filter is provided. Then, the SMC and GM implementations of this filter are introduced respectively.

6.2 Multi-target Multi-Bernoulli (MeMBer) Filter

The following briefly introduces the MeMBer recursion proposed by Mahler. First, the model assumptions of the MeMBer recursion is given, and then the cardinality bias problem during the update step is analyzed.

W. Wu et al., *Target Tracking with Random Finite Sets*,
https://doi.org/10.1007/978-981-19-9815-7_6

6.2.1 Model Assumptions on MeMBer Approximation

The core of the MeMBer recursion is to approximate the multi-target RFS at each moment by using the MB RFS. It is based on the following model assumptions:

- Each target evolves and generates the measurement independently;
- Birth targets are the MB RFS and are independent of survived targets;
- The clutters are the sparse Poisson RFS and are independent of the measurements generated by the targets.

6.2.2 MeMBer Recursion

Propositions 11 and 12 provide the explicit expressions for the MeMBer recursion.

Proposition 11 *(MeMBer prediction): if the posterior multi-target density at time $k-1$ is in the following MB form*

$$\pi_{k-1} = \{(\varepsilon_{k-1}^{(i)}, p_{k-1}^{(i)})\}_{i=1}^{M_{k-1}} \tag{6.2.1}$$

the predicted multi-target density also has the MB form, i.e.,[1]

$$\pi_{k|k-1} = \{(\varepsilon_{S,k|k-1}^{(i)}, p_{S,k|k-1}^{(i)})\}_{i=1}^{M_{k-1}} \cup \{(\varepsilon_{\gamma,k}^{(i)}, p_{\gamma,k}^{(i)})\}_{i=1}^{M_{\gamma,k}} \tag{6.2.2}$$

with

$$\varepsilon_{S,k|k-1}^{(i)} = \varepsilon_{k-1}^{(i)} \langle p_{S,k}, p_{k-1}^{(i)} \rangle \tag{6.2.3}$$

$$p_{S,k|k-1}^{(i)}(\boldsymbol{x}) = \langle p_{S,k}\phi_{k|k-1}(\boldsymbol{x}|\cdot), p_{k-1}^{(i)} \rangle \Big/ \langle p_{S,k}, p_{k-1}^{(i)} \rangle \tag{6.2.4}$$

where $\phi_{k|k-1}(\cdot|\boldsymbol{\xi})$ is the single-target transition density at time k given previous state $\boldsymbol{\xi}$, $p_{S,k}(\boldsymbol{\xi})$ is the target survival probability at time k given state $\boldsymbol{\xi}$, and $\{(\varepsilon_{\gamma,k}^{(i)}, p_{\gamma,k}^{(i)})\}_{i=1}^{M_{\gamma,k}}$ are the parameters of the birth MB RFS at time k.

Equation (6.2.2) indicates that, the MB parameter set of multi-target predicted density $\pi_{k|k-1}$ is composed of the union of the MB parameter sets of both survival targets [the first term in (6.2.2)] and birth targets [the second term in (6.2.2)]. Therefore, the total number of predicted hypothetical tracks is $M_{k|k-1} = M_{k-1} + M_{\gamma,k}$.

Proposition 12 *(MeMBer update): if the predicted multi-target density at time k has the following MB form*

[1] Please note that the target spawning is not taken into consideration herein.

$$\pi_{k|k-1} = \{(\varepsilon^{(i)}_{k|k-1}, p^{(i)}_{k|k-1})\}_{i=1}^{M_{k|k-1}} \tag{6.2.5}$$

the posterior multi-target density can be approximated in the following MB form

$$\pi_k \approx \{(\varepsilon^{(i)}_{L,k}, p^{(i)}_{L,k})\}_{i=1}^{M_{k|k-1}} \cup \{(\varepsilon_{U,k}(z), p_{U,k}(\cdot; z))\}_{z\in Z_k} \tag{6.2.6}$$

in which

$$\varepsilon^{(i)}_{L,k} = \varepsilon^{(i)}_{k|k-1} \frac{1 - \langle p^{(i)}_{k|k-1}, p_{D,k}\rangle}{1 - \varepsilon^{(i)}_{k|k-1}\langle p^{(i)}_{k|k-1}, p_{D,k}\rangle} \tag{6.2.7}$$

$$p^{(i)}_{L,k}(\boldsymbol{x}) = p^{(i)}_{k|k-1}(\boldsymbol{x}) \frac{1 - p_{D,k}(\boldsymbol{x})}{1 - \langle p^{(i)}_{k|k-1}, p_{D,k}\rangle} \tag{6.2.8}$$

$$\varepsilon_{U,k}(z) = \frac{\sum_{i=1}^{M_{k|k-1}} \frac{\varepsilon^{(i)}_{k|k-1}\langle p^{(i)}_{k|k-1}, \psi_{k,z}\rangle}{1-\varepsilon^{(i)}_{k|k-1}\langle p^{(i)}_{k|k-1}, p_{D,k}\rangle}}{\kappa_k(z) + \sum_{i=1}^{M_{k|k-1}} \frac{\varepsilon^{(i)}_{k|k-1}\langle p^{(i)}_{k|k-1}, \psi_{k,z}\rangle}{1-\varepsilon^{(i)}_{k|k-1}\langle p^{(i)}_{k|k-1}, p_{D,k}\rangle}} \tag{6.2.9}$$

$$p_{U,k}(\boldsymbol{x}; z) = \frac{\sum_{i=1}^{M_{k|k-1}} \frac{\varepsilon^{(i)}_{k|k-1} p^{(i)}_{k|k-1}(\boldsymbol{x})\psi_{k,z}(\boldsymbol{x})}{1-\varepsilon^{(i)}_{k|k-1}\langle p^{(i)}_{k|k-1}, p_{D,k}\rangle}}{\sum_{i=1}^{M_{k|k-1}} \frac{\varepsilon^{(i)}_{k|k-1}\langle p^{(i)}_{k|k-1}, \psi_{k,z}\rangle}{1-\varepsilon^{(i)}_{k|k-1}\langle p^{(i)}_{k|k-1}, p_{D,k}\rangle}} \tag{6.2.10}$$

where Z_k is the set of measurements at time k, $p_{D,k}(\boldsymbol{x})$ is the target detection probability at time k given current state $\boldsymbol{x}$, $\psi_{k,z}(\boldsymbol{x})$ is given by (4.2.4), and $\kappa_k(\cdot)$ is the Poisson clutter intensity at time k. In the above equations, it is assumed that both $p_{D,k}(\boldsymbol{x})$ and $\varepsilon^{(i)}_{k|k-1}, i = 1, \cdots, M_{k|k-1}$ cannot be equal to 1 by default.

Equation (6.2.6) indicates that, the MB parameter set of the updated multi-target density π_k is composed of the union of MB parameter sets of both legacy tracks [the first term in (6.2.6)] and measurement update tracks [the second term in (6.2.6)]. Therefore, the total number of the posterior hypothetical tracks is $M_k = M_{k|k-1} + |Z_k|$.

Although the time prediction step of the MeMBer recursion is accurate, the measurement update step is based on the following approximation of the PGFl of the posterior multi-target state at time k [14]

$$G_k[h] \approx \prod_{i=1}^{M_{k|k-1}} G^{(i)}_{L,k}[h] \prod_{z\in Z_k} G_{U,k}[h; z] \tag{6.2.11}$$

where

$$G^{(i)}_{L,k}[h] = \frac{1 - \varepsilon^{(i)}_{k|k-1} + \varepsilon^{(i)}_{k|k-1}\langle p^{(i)}_{k|k-1}, hq_{D,k}\rangle}{1 - \varepsilon^{(i)}_{k|k-1} + \varepsilon^{(i)}_{k|k-1}\langle p^{(i)}_{k|k-1}, q_{D,k}\rangle} \tag{6.2.12}$$

$$G_{U,k}[h;z] = \frac{\kappa_k(z) + \sum_{i=1}^{M_{k|k-1}} G_{U,k}^{(i)}[h;z]}{\kappa_k(z) + \sum_{i=1}^{M_{k|k-1}} G_{U,k}^{(i)}[1;z]} \tag{6.2.13}$$

$$G_{U,k}^{(i)}[h;z] = \frac{\varepsilon_{k|k-1}^{(i)}\langle p_{k|k-1}^{(i)}, h\psi_{k,z}\rangle}{1 - \varepsilon_{k|k-1}^{(i)} + \varepsilon_{k|k-1}^{(i)}\langle p_{k|k-1}^{(i)}, hq_{D,k}\rangle} \tag{6.2.14}$$

$$q_{D,k} = 1 - p_{D,k} \tag{6.2.15}$$

When the clutter is sparse (not too dense), the above approximation is reasonable. However, it shall be noted that, the first product in (6.2.11) is an MB but the second product is not. In fact, each factor $G_{U,k}[h;z]$ in the second product is no longer a PGFl of an RFS. However, we can find the Bernoulli approximation of $G_{U,k}[\cdot;z]$ and allow the second product in (6.2.11) to be approximated as an MB, so that $G_k[\cdot]$ in (6.2.11) is also an MB approximation. In the approximation of the MeMBer update, Mahler simply sets h in the denominator of (6.2.14) to $h = 1$ [14], i.e.,

$$G_{U,k}^{(i)}[h;z] = \frac{\varepsilon_{k|k-1}^{(i)}\langle p_{k|k-1}^{(i)}, h\psi_{k,z}\rangle}{1 - \varepsilon_{k|k-1}^{(i)}\langle p_{k|k-1}^{(i)}, p_{D,k}\rangle} \tag{6.2.16}$$

Then, by substituting the resulting (6.2.16) into (6.2.13), the following MB approximation is obtained

$$G_{U,k}[h;z] \approx 1 - \varepsilon_{U,k}(z) + \varepsilon_{U,k}(z)\langle p_{U,k}(\cdot;z), h\rangle \tag{6.2.17}$$

where $\varepsilon_{U,k}(z)$ and $p_{U,k}(\cdot;z)$ are given by (6.2.9) and (6.2.10), respectively. However, the above approximation will lead to a bias in the cardinality of measurement update tracks, thus resulting in a bias of the posterior cardinality.

6.2.3 Cardinality Bias Problem of MeMBer Filter

According to Proposition 11, the mean of the cardinality of the predicted multi-target state is

$$\overline{N}_{k|k-1} = \sum_{i=1}^{M_{k-1}} \varepsilon_{S,k|k-1}^{(i)} + \sum_{i=1}^{M_{\gamma,k}} \varepsilon_{\gamma,k}^{(i)} \tag{6.2.18}$$

According to Proposition 12, the posterior multi-target density can be approximated to an MB with the following cardinality mean

$$\widetilde{N}_k = \sum_{i=1}^{M_{k|k-1}} \varepsilon_{L,k}^{(i)} + \sum_{z \in Z_k} \varepsilon_{U,k}(z) \tag{6.2.19}$$

However, even if Eq. (6.2.11) holds, the above mean is not the cardinality mean of the posterior multi-target state (posterior cardinality mean). The following proposition provides the posterior cardinality mean when Eq. (6.2.11) is assumed to hold.

Proposition 13 *(see Appendix F for proof): if the PGFl (6.2.11) of the posterior multi-target density holds, the cardinality mean of the posterior multi-target state at time k is*

$$\overline{N}_k = \sum_{i=1}^{M_{k|k-1}} \varepsilon_{L,k}^{(i)} + \sum_{z \in Z_k} \varepsilon_{U,k}^{*}(z) \tag{6.2.20}$$

where $\varepsilon_{L,k}^{(i)}$ is given by (6.2.7), and

$$\varepsilon_{U,k}^{*}(z) = \frac{\sum_{i=1}^{M_{k|k-1}} \frac{\varepsilon_{k|k-1}^{(i)}(1-\varepsilon_{k|k-1}^{(i)})\langle p_{k|k-1}^{(i)}, \psi_{k,z}\rangle}{(1-\varepsilon_{k|k-1}^{(i)}\langle p_{k|k-1}^{(i)}, p_{D,k}\rangle)^2}}{\kappa_k(z) + \sum_{i=1}^{M_{k|k-1}} \frac{\varepsilon_{k|k-1}^{(i)}\langle p_{k|k-1}^{(i)}, \psi_{k,z}\rangle}{1-\varepsilon_{k|k-1}^{(i)}\langle p_{k|k-1}^{(i)}, p_{D,k}\rangle}} \tag{6.2.21}$$

Corollary of Proposition 13 *According to Proposition* 13, *the posterior cardinality bias at time k can be obtained as*

$$\widetilde{N}_k - \overline{N}_k = \sum_{z \in Z_k} \frac{\sum_{i=1}^{M_{k|k-1}} \frac{(\varepsilon_{k|k-1}^{(i)})^2(1-\langle p_{k|k-1}^{(i)}, p_{D,k}\rangle)\langle p_{k|k-1}^{(i)}, \psi_{k,z}\rangle}{(1-\varepsilon_{k|k-1}^{(i)}\langle p_{k|k-1}^{(i)}, p_{D,k}\rangle)^2}}{\kappa_k(z) + \sum_{i=1}^{M_{k|k-1}} \frac{\varepsilon_{k|k-1}^{(i)}\langle p_{k|k-1}^{(i)}, \psi_{k,z}\rangle}{1-\varepsilon_{k|k-1}^{(i)}\langle p_{k|k-1}^{(i)}, p_{D,k}\rangle}} \tag{6.2.22}$$

where each term summed over z is not negative and it equals to 0 only when $p_{D,k} = 1$. Therefore, the bias is always not negative and equals to 0 only when $p_{D,k} = 1$.

6.3 Cardinality Balanced MeMBer (CBMeMBer) Filter

In view of the cardinality bias problem of the MeMBer filter, the following introduces the cardinality unbiased MeMBer update procedure, which is called the cardinality balance MeMBer update, and the corresponding multi-target filter is called the CBMeMBer filter.

6.3.1 Cardinality Balancing of MeMBer Filter

Similar to the method proposed by Mahler in [14], [5] suggests to approximate PGFl $G_{U,k}[h; z]$ via Bernoulli $1 - \varepsilon_{U,k}(z) + \varepsilon_{U,k}(z)\langle p_{U,k}(\cdot; z), h\rangle$, and selects parameters ($\varepsilon_{U,k}(z)$ and $p_{U,k}(\cdot; z)$) to enable the proposed Bernoulli approximation to share the same intensity function and the same cardinality mean with the original PGFl, thus eliminating cardinality bias. Specifically, let $v_{U,k}(\cdot; z)$ represent the intensity function of $G_{U,k}[\cdot; z]$, and we can notice that the intensity function for this Bernoulli is $\varepsilon_{U,k}(z)p_{U,k}(\cdot; z)$, i.e.,

$$\varepsilon_{U,k}(z)p_{U,k}(\cdot; z) = v_{U,k}(\cdot; z) \tag{6.3.1}$$

Integration and normalization of the above equation leads to the following Bernoulli parameters

$$\varepsilon_{U,k}(z) = \int v_{U,k}(\boldsymbol{x}; z)d\boldsymbol{x} \tag{6.3.2}$$

$$p_{U,k}(\cdot; z) = v_{U,k}(\cdot; z)/\, \varepsilon_{U,k}(z) \tag{6.3.3}$$

If $G_{U,k}[\cdot; z]$ is indeed a PGFl of the RFS, this method will obtain the best Bernoulli approximation of the first-order moment of $G_{U,k}[\cdot; z]$.

By taking Frechet derivative[2] at $h = 1$ in the direction of $\zeta=\delta_x$ (i.e., functional derivative at $\boldsymbol{x}$), the intensity function $v_{U,k}(\cdot; z)$ of PGFl $G_{U,k}[\cdot; z]$ can be obtained as follows

$$v_{U,k}(\boldsymbol{x}; z) = \frac{\sum_{i=1}^{M_{k|k-1}} v_{U,k}^{(i)}(\boldsymbol{x}; z)}{\kappa_k(z) + \sum_{i=1}^{M_{k|k-1}} G_{U,k}^{(i)}(1; z)} \tag{6.3.4}$$

where

$$\begin{aligned} v_{U,k}^{(i)}(\boldsymbol{x}; z) = {} & p_{k|k-1}^{(i)}(\boldsymbol{x})(1 - \varepsilon_{k|k-1}^{(i)}\langle p_{k|k-1}^{(i)}, p_{D,k}\rangle)^{-2} \\ & \times [(1 - \varepsilon_{k|k-1}^{(i)}\langle p_{k|k-1}^{(i)}, p_{D,k}\rangle)\varepsilon_{k|k-1}^{(i)}\psi_{k,z}(\boldsymbol{x}) \\ & - (\varepsilon_{k|k-1}^{(i)})^2\langle p_{k|k-1}^{(i)}, \psi_{k,z}\rangle(1 - p_{D,k}(\boldsymbol{x}))] \end{aligned} \tag{6.3.5}$$

If $p_{D,k}(\boldsymbol{x}) = 0$, $v_{U,k}(\boldsymbol{x}; z)$ is generally negative. Therefore, in (6.3.3), $p_{U,k}(\cdot; z)$ is not a valid probability density. However, the cardinality mean (6.2.21) of $G_{U,k}[\cdot; z]$ and $\varepsilon_{U,k}(z)$ given by Formula (6.3.2) are consistent. In order to make $p_{U,k}(\cdot; z)$ valid, we can approximate $p_{D,k}(\boldsymbol{x}) \approx 1$ to eliminate negative terms in (6.3.5), i.e.,

---

[2] The Frechet derivative is defined as $\lim_{\lambda\to 0^+} (G_{U,k}[h + \lambda\zeta; z] - G_{U,k}[h; z])/\,\lambda$ [14].

$$p_{U,k}(\cdot; z) = \frac{\sum_{i=1}^{M_{k|k-1}} \frac{\varepsilon_{k|k-1}^{(i)} p_{k|k-1}^{(i)}(\boldsymbol{x})\psi_{k,z}(\boldsymbol{x})}{1-\varepsilon_{k|k-1}^{(i)}\langle p_{k|k-1}^{(i)}, p_{D,k}\rangle}}{\sum_{i=1}^{M_{k|k-1}} \frac{\varepsilon_{k|k-1}^{(i)}(1-\varepsilon_{k|k-1}^{(i)})\langle p_{k|k-1}^{(i)}, \psi_{k,z}\rangle}{(1-\varepsilon_{k|k-1}^{(i)}\langle p_{k|k-1}^{(i)}, p_{D,k}\rangle)^2}} \tag{6.3.6}$$

Additionally, $\langle p_{k|k-1}^{(i)}, p_{D,k}\rangle \approx 1$ can be obtained. Therefore, the above equation is a valid probability density.

6.3.2 CBMeMBer Recursion

The obtained MB approximation of the updated multi-target density is given by the following proposition.

Proposition 14 *(cardinality balanced MeMBer update): under the premise of Proposition 13, if the predicted multi-target density at time k is in the following MB form*

$$\pi_{k|k-1} = \{(\varepsilon_{k|k-1}^{(i)}, p_{k|k-1}^{(i)})\}_{i=1}^{M_{k|k-1}} \tag{6.3.7}$$

then the posterior multi-target density can be approximated to the following MB with unbiased cardinality

$$\pi_k \approx \{(\varepsilon_{L,k}^{(i)}, p_{L,k}^{(i)})\}_{i=1}^{M_{k|k-1}} \cup \{(\varepsilon_{U,k}^{*}(z), p_{U,k}^{*}(\cdot; z))\}_{z\in Z_k} \tag{6.3.8}$$

where $\varepsilon_{L,k}^{(i)}$, $p_{L,k}^{(i)}(\boldsymbol{x})$ and $\varepsilon_{U,k}^{*}(z)$ are respectively given by (6.2.7), (6.2.8) and (6.2.21), and

$$p_{U,k}^{*}(\boldsymbol{x}; z) = \frac{\sum_{i=1}^{M_{k|k-1}} \frac{\varepsilon_{k|k-1}^{(i)}}{1-\varepsilon_{k|k-1}^{(i)}} p_{k|k-1}^{(i)}(\boldsymbol{x})\psi_{k,z}(\boldsymbol{x})}{\sum_{i=1}^{M_{k|k-1}} \frac{\varepsilon_{k|k-1}^{(i)}}{1-\varepsilon_{k|k-1}^{(i)}} \langle p_{k|k-1}^{(i)}, \psi_{k,z}\rangle} \tag{6.3.9}$$

Propositions 11 and 14 constitute the prediction and update steps of the cardinality balanced MeMBer (CBMeMBer) filter, respectively. The filter propagates the MB parameters of the posterior multi-target density forward over time.

Although the time prediction step of the CBMeMBer filter is accurate, the posterior is approximated by using the probability generating functional (PGFl) twice in the process of the measurement update. In the first approximation step, the posterior PGFl is expressed as the product of the legacy term and the update term (it is not necessarily in the MB form), as shown in (6.2.11). In the second approximation step, the MB PGFl matching the posterior cardinality mean of the first approximate PGFl is selected. Through these two steps, the multi-target tracking filter with low computational burden is obtained. However, it is based on the assumption of high signal-to-noise ratio (SNR) of sensor detections (relatively sparse clutter and higher detection probability). Though the performance is acceptable at high SNR, the cardinality bias is significant in the case of low SNR, which is mainly due to the approximation of the PGFl in the first step. In brief, although the posterior PGFl approximation may not be an appropriate PGFl, it is close to the true posterior PGFl. Therefore, at high SNR, the approximation given by Proposition 14 is reasonable.

The complexity of the CBMeMBer recursion has a linear relationship with the number of targets and the number of measurements, and is similar to that of the PHD filter. Compared with the CPHD filter whose complexity has a linear relation with the number of targets and a cubic relation with the number of measurements, the CBMeMBer filter has a lower complexity.

The MB representation $\pi_k = \{(\varepsilon_k^{(i)}, p_k^{(i)})\}_{i=1}^{M_k}$ of the posterior multi-target density has an intuitive interpretation and it is easy to extract the multi-target state estimation. The existence probability $\varepsilon_k^{(i)}$ indicates the possibility that the ith hypothetical track is a true track, and posterior density $p_k^{(i)}$ describes the current track state to be estimated. Therefore, according to the posterior density of the hypothetical track whose existence probability exceeds a given threshold (e.g., 0.5), by selecting its mean or mode (the mode is more popular as it is more stable than the mean), we can realize the multi-target state estimation. In addition, the following two steps can also be used: first, based on the posterior cardinality distribution, the number of targets can be estimated by taking the mean or mode; then, the corresponding number of hypothetical tracks with the highest existence probabilities are selected, and their own means or modes are calculated based on each posterior density.

6.3.3 Track Labeling for CBMeMBer Filter

By incorporating the label information, the CBMeMBer recursion can be extended to propagate the track continuity. Specifically, given the MB density $\pi = \{(\varepsilon^{(i)}, p^{(i)})\}_{i=1}^{M}$, each Bernoulli component $(\varepsilon^{(i)}, p^{(i)})$ is assigned with an unique track label $\ell^{(i)}$ to distinguish hypothetical tracks, and the triple set $\mathcal{T} = \left\{(\ell^{(i)}, \varepsilon^{(i)}, p^{(i)})\right\}_{i=1}^{M}$ is referred to as a track table. A simple track label propagation strategy is described as follows.

Prediction: if at time $k-1$ the posterior track table is $\mathcal{T}_{k-1} = \left\{\left(\ell_{k-1}^{(i)}, \varepsilon_{k-1}^{(i)}, p_{k-1}^{(i)}\right)\right\}_{i=1}^{M_{k-1}}$, the predicted track table at time k is $\mathcal{T}_{k|k-1} = \left\{\left(\ell_{S,k|k-1}^{(i)}, \varepsilon_{S,k|k-1}^{(i)}, p_{S,k|k-1}^{(i)}\right)\right\}_{i=1}^{M_{k-1}} \cup \left\{\left(\ell_{\gamma,k}^{(i)}, \varepsilon_{\gamma,k}^{(i)}, p_{\gamma,k}^{(i)}\right)\right\}_{i=1}^{M_{\gamma,k}}$, where, $\ell_{S,k|k-1}^{(i)} = \ell_{k-1}^{(i)}$, $\ell_{\gamma,k}^{(i)}$ is the birth track label, and $\varepsilon_{S,k|k-1}^{(i)}, p_{S,k|k-1}^{(i)}, \varepsilon_{\gamma,k}^{(i)}$ and $p_{\gamma,k}^{(i)}$ are given by Proposition 11. In other words, the survival components keep their original labels while birth components are assigned with new labels.

Update: if at time k the predicted track table is $\mathcal{T}_{k|k-1} = \left\{\left(\ell_{k|k-1}^{(i)}, \varepsilon_{k|k-1}^{(i)}, p_{k|k-1}^{(i)}\right)\right\}_{i-1}^{M_{k|k-1}}$, the updated track table at time k is $\mathcal{T}_k = \left\{\left(\ell_{L,k}^{(i)}, \varepsilon_{L,k}^{(i)}, p_{L,k}^{(i)}\right)\right\}_{i=1}^{M_{k|k-1}} \cup \left\{\left(\ell_{U,k}(z), \varepsilon_{U,k}(z), p_{U,k}(\cdot, z)\right)\right\}_{z \in Z_k}$, where, $\ell_{L,k}^{(i)} = \ell_{k|k-1}^{(i)}$, $\ell_{U,k}(z) = \ell_{k|k-1}^{(n)}$, $n = \arg_i \max \varepsilon_{k|k-1}^{(i)}(1 - \varepsilon_{k|k-1}^{(i)})\langle p_{k|k-1}^{(i)}, \psi_{k,z}\rangle \Big/ (1 - \varepsilon_{k|k-1}^{(i)}\langle p_{k|k-1}^{(i)}, p_{D,k}\rangle)^2$, while $\varepsilon_{L,k}^{(i)}$, $p_{L,k}^{(i)}$, $\varepsilon_{U,k}(z)$ and $p_{U,k}(\cdot, z)$ are given by Proposition 14. In other words, the legacy tracks keep their original labels, and the predicted track label with the greatest contribution to current updated existence probability (6.2.21) is assigned to the measurement update component.

Although the above method is easy to implement, the performance of this method is poor when the targets are close to each other, which can be improved by the track association strategy in [114]. In addition, we shall pay attention to the difference between label-included CBMeMBer filter described here and the labeled multi-Bernoulli filter that will be introduced in Chap. 7. The former is merely a simple heuristic scheme while the latter is based on a strict labeled RFS theory and has a sound theoretical foundation.

The following introduces the sequential Monte Carlo implementation of the CMBeMBer filter (SMC-CBMeMBer filter) in a general condition and the analytical Gaussian mixture implementation of the CMBeMBer filter (GM-CBMeMBer filter) in the linear Gaussian condition.

6.4 SMC-CBMeMBer Filter

The SMC-CBMeMBer filter can be applicable to the non-linear dynamic and measurement models as well as the state-dependent survival probability and detection probability.

6.4.1 SMC-CBMeMBer Recursion

The SMC-CBMeMBer recursion involves two steps, i.e., prediction and update.

(1) Prediction. It is assumed that, the (MB) posterior multi-target density at time $k-1$ is $\pi_{k-1} = \{(\varepsilon_{k-1}^{(i)}, p_{k-1}^{(i)})\}_{i=1}^{M_{k-1}}$ and each $p_{k-1}^{(i)}$, $i = 1, \cdots, M_{k-1}$ is composed of a set of weighted samples $\{(w_{k-1}^{(i,j)}, \boldsymbol{x}_{k-1}^{(i,j)})\}_{j=1}^{L_{k-1}^{(i)}}$, i.e.,

$$p_{k-1}^{(i)}(\boldsymbol{x}) = \sum_{j=1}^{L_{k-1}^{(i)}} w_{k-1}^{(i,j)} \delta_{\boldsymbol{x}_{k-1}^{(i,j)}}(\boldsymbol{x}) \tag{6.4.1}$$

Given importance density (or proposed density) ($q_k^{(i)}(\cdot|\boldsymbol{x}_{k-1}, Z_k)$ and $q_{\gamma,k}^{(i)}(\cdot|Z_k)$), the predicted (MB) multi-target density $\pi_{k|k-1} = \{(\varepsilon_{S,k|k-1}^{(i)}, p_{S,k|k-1}^{(i)})\}_{i=1}^{M_{k-1}} \cup \{(\varepsilon_{\gamma,k}^{(i)}, p_{\gamma,k}^{(i)})\}_{i=1}^{M_{\gamma,k}}$ can be calculated as follows

$$\varepsilon_{S,k|k-1}^{(i)} = \varepsilon_{k-1}^{(i)} \sum_{j=1}^{L_{k-1}^{(i)}} w_{k-1}^{(i,j)} p_{S,k}(\boldsymbol{x}_{k-1}^{(i,j)}) \tag{6.4.2}$$

$$p_{S,k|k-1}^{(i)}(\boldsymbol{x}) = \sum_{j=1}^{L_{k-1}^{(i)}} \tilde{w}_{S,k|k-1}^{(i,j)} \delta_{\boldsymbol{x}_{S,k|k-1}^{(i,j)}}(\boldsymbol{x}) \tag{6.4.3}$$

$$p_{\gamma,k}^{(i)}(\boldsymbol{x}) = \sum_{j=1}^{L_{\gamma,k}^{(i)}} \tilde{w}_{\gamma,k}^{(i,j)} \delta_{\boldsymbol{x}_{\gamma,k}^{(i,j)}}(\boldsymbol{x}) \tag{6.4.4}$$

where $\varepsilon_{\gamma,k}^{(i)}$ is the birth model parameter, and

$$\boldsymbol{x}_{S,k|k-1}^{(i,j)} \sim q_k^{(i)}\left(\cdot|\boldsymbol{x}_{k-1}^{(i,j)}, Z_k\right), j = 1, \ldots, L_{k-1}^{(i)} \tag{6.4.5}$$

$$\tilde{w}_{S,k|k-1}^{(i,j)} = w_{S,k|k-1}^{(i,j)} \bigg/ \sum_{j=1}^{L_{k-1}^{(i)}} w_{S,k|k-1}^{(i,j)} \tag{6.4.6}$$

$$w_{S,k|k-1}^{(i,j)} = w_{k-1}^{(i,j)} \phi_{k|k-1}(\boldsymbol{x}_{S,k|k-1}^{(i,j)}|\boldsymbol{x}_{k-1}^{(i,j)}) p_{S,k}(\boldsymbol{x}_{k-1}^{(i,j)}) \Big/ q_k^{(i)}(\boldsymbol{x}_{S,k|k-1}^{(i,j)}|\boldsymbol{x}_{k-1}^{(i,j)}, Z_k) \tag{6.4.7}$$

$$\boldsymbol{x}_{\gamma,k}^{(i,j)} \sim q_{\gamma,k}^{(i)}(\cdot|Z_k), j = 1, \ldots, L_{\gamma,k}^{(i)} \tag{6.4.8}$$

$$\tilde{w}_{\gamma,k}^{(i,j)} = w_{\gamma,k}^{(i,j)} \bigg/ \sum_{j=1}^{L_{\gamma,k}^{(i)}} w_{\gamma,k}^{(i,j)} \tag{6.4.9}$$

$$w_{\gamma,k}^{(i,j)} = \frac{p_{\gamma,k}(\boldsymbol{x}_{\gamma,k}^{(i,j)})}{q_{\gamma,k}^{(i)}(\boldsymbol{x}_{\gamma,k}^{(i,j)}|Z_k)} \tag{6.4.10}$$

(2) Update. It is assumed that the (MB) multi-target predicted density at time k is $\pi_{k|k-1} = \{(\varepsilon_{k|k-1}^{(i)}, p_{k|k-1}^{(i)})\}_{i=1}^{M_{k|k-1}}$ and each $p_{k|k-1}^{(i)}, i = 1, \cdots, M_{k|k-1}$ is composed of a set of weighted samples $\{(w_{k|k-1}^{(i,j)}, \boldsymbol{x}_{k|k-1}^{(i,j)})\}_{j=1}^{L_{k|k-1}^{(i)}}$, i.e.,

$$p_{k|k-1}^{(i)}(\boldsymbol{x}) = \sum_{j=1}^{L_{k|k-1}^{(i)}} w_{k|k-1}^{(i,j)} \delta_{\boldsymbol{x}_{k|k-1}^{(i,j)}}(\boldsymbol{x}) \tag{6.4.11}$$

The (MB approximation of) updated multi-target density $\pi_k = \{(\varepsilon_{L,k}^{(i)}, p_{L,k}^{(i)}(\boldsymbol{x}))\}_{i=1}^{M_{k|k-1}} \cup \{(\varepsilon_{U,k}^{*}(\boldsymbol{z}), p_{U,k}^{*}(\cdot, \boldsymbol{z}))\}_{\boldsymbol{z}\in Z_k}$ can be calculated as follows

$$\varepsilon_{L,k}^{(i)} = \varepsilon_{k|k-1}^{(i)} \frac{1 - \eta_{L,k}^{(i)}}{1 - \varepsilon_{k|k-1}^{(i)} \eta_{L,k}^{(i)}} \tag{6.4.12}$$

$$p_{L,k}^{(i)}(\boldsymbol{x}) = \sum_{j=1}^{L_{k|k-1}^{(i)}} \tilde{w}_{L,k}^{(i,j)} \delta_{\boldsymbol{x}_{k|k-1}^{(i,j)}}(\boldsymbol{x}) \tag{6.4.13}$$

$$\varepsilon_{U,k}^{*}(\boldsymbol{z}) = \frac{\sum_{i=1}^{M_{k|k-1}} \frac{\varepsilon_{k|k-1}^{(i)}(1-\varepsilon_{k|k-1}^{(i)})\eta_{U,k}^{(i)}(\boldsymbol{z})}{(1-\varepsilon_{k|k-1}^{(i)}\eta_{L,k}^{(i)})^2}}{\kappa_k(\boldsymbol{z}) + \sum_{i=1}^{M_{k|k-1}} \frac{\varepsilon_{k|k-1}^{(i)}\eta_{U,k}^{(i)}(\boldsymbol{z})}{1-\varepsilon_{k|k-1}^{(i)}\eta_{L,k}^{(i)}}} \tag{6.4.14}$$

$$p_{U,k}^{*}(\boldsymbol{x}; \boldsymbol{z}) = \sum_{i=1}^{M_{k|k-1}} \sum_{j=1}^{L_{k|k-1}^{(i)}} \tilde{w}_{U,k}^{*(i,j)}(\boldsymbol{z}) \delta_{\boldsymbol{x}_{k|k-1}^{(i,j)}}(\boldsymbol{x}) \tag{6.4.15}$$

where

$$\eta_{L,k}^{(i)} = \sum_{j=1}^{L_{k|k-1}^{(i)}} w_{k|k-1}^{(i,j)} p_{D,k}(\boldsymbol{x}_{k|k-1}^{(i,j)}) \tag{6.4.16}$$

$$\tilde{w}_{L,k}^{(i,j)} = w_{L,k}^{(i,j)} \Bigg/ \sum_{j=1}^{L_{k|k-1}^{(i)}} w_{L,k}^{(i,j)} \tag{6.4.17}$$

$$w_{L,k}^{(i,j)} = w_{k|k-1}^{(i,j)}(1 - p_{D,k}(\boldsymbol{x}_{k|k-1}^{(i,j)})) \tag{6.4.18}$$

$$\eta_{U,k}^{(i)}(\boldsymbol{z}) = \sum_{j=1}^{L_{k|k-1}^{(i)}} w_{k|k-1}^{(i,j)} \psi_{k,\boldsymbol{z}}(\boldsymbol{x}_{k|k-1}^{(i,j)}) \tag{6.4.19}$$

$$\tilde{w}_{U,k}^{*(i,j)}(\boldsymbol{z}) = w_{U,k}^{*(i,j)}(\boldsymbol{z}) \Bigg/ \sum_{i=1}^{M_{k|k-1}} \sum_{j=1}^{L_{k|k-1}^{(i)}} w_{U,k}^{*(i,j)}(\boldsymbol{z}) \tag{6.4.20}$$

$$w_{U,k}^{*(i,j)}(\boldsymbol{z}) = w_{k|k-1}^{(i,j)} \frac{\varepsilon_{k|k-1}^{(i)}}{1 - \varepsilon_{k|k-1}^{(i)}} \psi_{k,\boldsymbol{z}}(\boldsymbol{x}_{k|k-1}^{(i,j)}) \tag{6.4.21}$$

6.4.2 Resampling and Implementation Issues

Similar to the standard particle filter, the sample degradation problem is inevitable [384]. To mitigate the degradation effect, the resample step is implemented for each hypothetical track after the update step. This will effectively eliminate the low-weight particles, duplicate the high-weight particles, and cause the particles to concentrate in an important zone of (single-target) state-space. There are a variety of re-sampling schemes and the choice of different schemes will influence both the computational burden and the Monte Carlo approximation error [384].

It shall be noted that, due to the target birth in the prediction step and averaging of the number of hypothetical tracks in the update step, the number of particles required to represent the posterior multi-target density will be increasing. In order to reduce the number of particles, the hypothetical tracks are pruned by discarding tracks whose existence probability is below a certain threshold ε_{Th} (e.g., 10^{-3}). For the remaining hypothetical tracks, similar to the SMC-PHD/CPHD filter, it is expected that the number of particles allocated in the track density is proportional to the desired number of surviving targets. Therefore, at each time, the number of particles in each hypothetical track is re-assigned in proportion to its existence probability. Namely, $L_{\gamma,k}^{(i)} = \varepsilon_{\gamma,k}^{(i)} L_{\max}$ particles are sampled for each birth term in the process of prediction, while $L_k^{(i)} = \varepsilon_k^{(i)} L_{\max}$ particles are re-sampled for each updated track in the process of re-sampling. Furthermore, for each hypothetical track, it is often necessary to limit both the maximum $L_{\max}$ and the minimum $L_{\min}$ of the number of particles.

In the SMC-PHD/CPHD filter, the multi-target state estimation is realized through the following steps: firstly, the number of targets is estimated according to the cardinality mean or mode; then, the particles are clustered according to the intensity function to form the corresponding number of clusters; finally, the center of each cluster constitutes the multi-target state estimation. It is obvious that, when the estimated number of targets does not match the number of clusters naturally formed in the particle swarm, the output estimation of clustering is prone to errors. In addition, the clustering operation has a large amount of calculation, and its complexity has no specific relationship with the number of targets. In contrast, in the SMC-CBMeMBer filter, based on the MB form of the posterior density, existence probability $\varepsilon_k^{(i)}$ indicates the probability that the ith hypothetical track is a true track, while posterior density $p_k^{(i)}$ describes the statistics of the current state estimation of this track. Thus, an intuitive multi-target state estimation can be realized in the following way: firstly, the number of targets is estimated by the EAP or MAP cardinality estimation, and then the corresponding number of means or modes are selected from the state or track density with the highest existence probability to obtain multi-target estimation. However, since it is difficult to obtain each state estimate through the calculation of mode according to the particles, the mean of the corresponding posterior density can be used to obtain each state estimate. It is quite easy to calculate the mean of each posterior density, and the computational complexity has a linear relation with the number of hypothetical tracks. Therefore, this is a significant advantage of the SMC-CBMeMBer filter over the SMC-PHD/CPHD filter.

The SMC-CBMeMBer filter is equivalent to running many particle filters in parallel, which is linear with the number of targets. However, since the track merging and removal require calculating the distance between two estimations pairwise, its computational complexity is quadratic with the number of targets. But, by using a certain computational strategy, the complexity of the merging and removal can be reduced to $\mathcal{O}(n\log(n))$ [249].

6.5 GM-CBMeMBer Filter

The GM-CBMeMBer filter, as a closed form solution to the CBMeMBer recursion, is suitable for the linear Gaussian multi-target (LGM) model. Note that in the context of different RFS algorithms, the specific contents contained in the LGM model are not exactly the same.

6.5.1 *Model Assumptions on GM-CBMeMBer Recursion*

The linear Gaussian multi-target model assumes that the state transition model and the measurement model of each target are the standard linear Gaussian model (see Assumption A.4). Moreover, the corresponding assumptions have been made for the target birth, death and detection (see A.5 and A.6 for details). However, since the target spawning is not taken into consideration, there is no assumption about the target spawning in A.6, and the birth model (4.4.5) in A.6 shall be modified as follows.

- The birth model is an MB with the parameters $\{(\varepsilon_{\gamma,k}^{(i)}, p_{\gamma,k}^{(i)})\}_{i=1}^{M_{\gamma,k}}$, where, $p_{\gamma,k}^{(i)}$, $i = 1, \cdots, M_{\gamma,k}$ is in the Gaussian mixture form

$$p_{\gamma,k}^{(i)}(\boldsymbol{x}) = \sum_{j=1}^{J_{\gamma,k}^{(i)}} w_{\gamma,k}^{(i,j)} \mathcal{N}(\boldsymbol{x}; \boldsymbol{m}_{\gamma,k}^{(i,j)}, \boldsymbol{P}_{\gamma,k}^{(i,j)}) \tag{6.5.1}$$

where, $w_{\gamma,k}^{(i,j)}$, $\boldsymbol{m}_{\gamma,k}^{(i,j)}$ and $\boldsymbol{P}_{\gamma,k}^{(i,j)}$ represent the weight, mean and covariance of the jth component.

6.5.2 *GM-CBMeMBer Recursion*

Based on the LGM model, a closed-form solution of the CBMeMBer recursion is available, which shows how the posterior density is analytically propagated over time.

(1) Prediction. Given (MB) posterior multi-target density $\pi_{k-1} = \{(\varepsilon_{k-1}^{(i)}, p_{k-1}^{(i)})\}_{i=1}^{M_{k-1}}$ at time $k-1$, assume that each probability density $p_{k-1}^{(i)}, i = 1, \cdots, M_{k-1}$ is in the form of a Gaussian mixture

$$p_{k-1}^{(i)}(\boldsymbol{x}) = \sum_{j=1}^{J_{k-1}^{(i)}} w_{k-1}^{(i,j)} \mathcal{N}(\boldsymbol{x}; \boldsymbol{m}_{k-1}^{(i,j)}, \boldsymbol{P}_{k-1}^{(i,j)}) \tag{6.5.2}$$

then, the (MB) multi-target predicted density $\pi_{k|k-1} = \{(\varepsilon_{S,k|k-1}^{(i)}, p_{S,k|k-1}^{(i)})\}_{i=1}^{M_{k-1}} \cup \{(\varepsilon_{\gamma,k}^{(i)}, p_{\gamma,k}^{(i)})\}_{i=1}^{M_{\gamma,k}}$ can be calculated as follows

$$\varepsilon_{S,k|k-1}^{(i)} = p_{S,k}\varepsilon_{k-1}^{(i)} \tag{6.5.3}$$

$$p_{S,k|k-1}^{(i)}(\boldsymbol{x}) = \sum_{j=1}^{J_{k-1}^{(i)}} w_{k-1}^{(i,j)} \mathcal{N}(\boldsymbol{x}; \boldsymbol{m}_{S,k|k-1}^{(i,j)}, \boldsymbol{P}_{S,k|k-1}^{(i,j)}) \tag{6.5.4}$$

where $\{(\varepsilon_{\gamma,k}^{(i)}, p_{\gamma,k}^{(i)})\}_{i=1}^{M_{\gamma,k}}$ is given by the birth model (6.5.1), and

$$\boldsymbol{m}_{S,k|k-1}^{(i,j)} = \boldsymbol{F}_{k-1}\boldsymbol{m}_{k-1}^{(i,j)} \tag{6.5.5}$$

$$\boldsymbol{P}_{S,k|k-1}^{(i,j)} = \boldsymbol{F}_{k-1}\boldsymbol{P}_{k-1}^{(i,j)}\boldsymbol{F}_{k-1}^{\mathrm{T}} + \boldsymbol{Q}_{k-1} \tag{6.5.6}$$

(2) Update. Given the (MB) multi-target predicted density $\pi_{k|k-1} = \{(\varepsilon_{k|k-1}^{(i)}, p_{k|k-1}^{(i)})\}_{i=1}^{M_{k|k-1}}$ at time k, assume that each $p_{k|k-1}^{(i)}, i = 1, \cdots, M_{k|k-1}$ is in the following Gaussian mixture form

$$p_{k|k-1}^{(i)}(\boldsymbol{x}) = \sum_{j=1}^{J_{k|k-1}^{(i)}} w_{k|k-1}^{(i,j)} \mathcal{N}(\boldsymbol{x}; \boldsymbol{m}_{k|k-1}^{(i,j)}, \boldsymbol{P}_{k|k-1}^{(i,j)}) \tag{6.5.7}$$

then, the updated (MB approximation of) multi-target density $\pi_k = \{(\varepsilon_{L,k}^{(i)}, p_{L,k}^{(i)}(\boldsymbol{x}))\}_{i=1}^{M_{k|k-1}} \cup \{(\varepsilon_{U,k}^{*}(z), p_{U,k}^{*}(\cdot, z))\}_{z \in Z_k}$ can be calculated as follows

$$\varepsilon_{L,k}^{(i)} = \varepsilon_{k|k-1}^{(i)} \frac{1 - p_{D,k}}{1 - \varepsilon_{k|k-1}^{(i)} p_{D,k}} \tag{6.5.8}$$

$$p_{L,k}^{(i)}(\boldsymbol{x}) = p_{k|k-1}^{(i)}(\boldsymbol{x}) \tag{6.5.9}$$

$$\varepsilon_{U,k}^{*}(z)=\frac{\sum_{i=1}^{M_{k|k-1}}\frac{\varepsilon_{k|k-1}^{(i)}(1-\varepsilon_{k|k-1}^{(i)})\eta_{U,k}^{(i)}(z)}{(1-\varepsilon_{k|k-1}^{(i)}p_{D,k})^{2}}}{\kappa_{k}(z)+\sum_{i=1}^{M_{k|k-1}}\frac{\varepsilon_{k|k-1}^{(i)}\eta_{U,k}^{(i)}(z)}{1-\varepsilon_{k|k-1}^{(i)}p_{D,k}}} \tag{6.5.10}$$

$$p_{U,k}^{*}(\boldsymbol{x};z)=\frac{\sum_{i=1}^{M_{k|k-1}}\sum_{j=1}^{J_{k|k-1}^{(i)}}w_{U,k}^{(i,j)}(z)\mathcal{N}(\boldsymbol{x};\boldsymbol{m}_{U,k}^{(i,j)},\boldsymbol{P}_{U,k}^{(i,j)})}{\sum_{i=1}^{M_{k|k-1}}\sum_{j=1}^{J_{k|k-1}^{(i)}}w_{U,k}^{(i,j)}(z)} \tag{6.5.11}$$

where

$$\eta_{U,k}^{(i)}(z)=p_{D,k}\sum_{j=1}^{J_{k|k-1}^{(i)}}w_{k|k-1}^{(i,j)}q_{k}^{(i,j)}(z) \tag{6.5.12}$$

$$q_{k}^{(i,j)}(z)=\mathcal{N}(z;\boldsymbol{H}_{k}\boldsymbol{m}_{k|k-1}^{(i,j)},\boldsymbol{S}_{k}^{(i,j)}) \tag{6.5.13}$$

$$w_{U,k}^{(i,j)}(z)=\frac{\varepsilon_{k|k-1}^{(i)}}{1-\varepsilon_{k|k-1}^{(i)}}p_{D,k}w_{k|k-1}^{(i,j)}q_{k}^{(i,j)}(z) \tag{6.5.14}$$

$$\boldsymbol{m}_{U,k}^{(i,j)}(z)=\boldsymbol{m}_{k|k-1}^{(i,j)}+\boldsymbol{G}_{U,k}^{(i,j)}(z-\boldsymbol{H}_{k}\boldsymbol{m}_{k|k-1}^{(i,j)}) \tag{6.5.15}$$

$$\boldsymbol{P}_{U,k}^{(i,j)}=(\boldsymbol{I}-\boldsymbol{G}_{U,k}^{(i,j)}\boldsymbol{H}_{k})\boldsymbol{P}_{k|k-1}^{(i,j)} \tag{6.5.16}$$

$$\boldsymbol{G}_{U,k}^{(i,j)}=\boldsymbol{P}_{k|k-1}^{(i,j)}\boldsymbol{H}_{k}^{\mathrm{T}}(\boldsymbol{S}_{k}^{(i,j)})^{-1} \tag{6.5.17}$$

$$\boldsymbol{S}_{k}^{(i,j)}=\boldsymbol{H}_{k}\boldsymbol{P}_{k|k-1}^{(i,j)}\boldsymbol{H}_{k}^{\mathrm{T}}+\boldsymbol{R}_{k} \tag{6.5.18}$$

The derivation of closed-form prediction and update formulas involves analytic computation of Gaussian product and Gaussian integral and it is conducted in the same way as the GM-PHD and GM-CPHD filters. Refer to Sects. 4.4 and 5.4 for the specific derivation process.

With the help of standard approximations, the closed form solution to the CBMeMBer recursion for the linear Gaussian model can be extended to the non-linear Gaussian dynamic and measurement models. Specifically, by using strategies similar to linearization and unscented transformation adopted by the EKF and UKF filters, the extended Kalman CBMeMBer filter and unscented Kalman CBMeMBer filter can be obtained, respectively. Since these extensions are conceptually intuitive, no specific equations are given here, and only the basic method of approximating this recursion is briefly described. For the extended Kalman CBMeMBer (EK-CBMeMBer) filter, by substituting the local linear approximations of the nonlinear dynamic model and the measurement model respectively, the closed-form expressions for the prediction and update of each Gaussian component can be obtained. For the unscented

Kalman CBMeMBer (UK-CBMeMBer) filter, by using the unscented transformation to analytically propagate means and covariances through non-linear dynamic model and measurement model, the prediction and update of each Gaussian component can be obtained. In addition, using a similar approach in Sect. 4.5.1, it is easy to extend the GM-CBMeMBer closed-form recursion to the case where $p_{S,k}(\boldsymbol{x})$ and $p_{D,k}(\boldsymbol{x})$ are in the exponential mixture form.

6.5.3 Implementation Issues of GM-CBMeMBer Filter

(1) Pruning of hypothetical track components. Due to the target birth in the prediction process and averaging of the number of hypothetical tracks in the update process, the number of Gaussian components required to express the MB posterior density will be increasing without limit. In order to reduce the number of components, it is important to prune hypothetical tracks at each time and discard hypothetical tracks with existence probabilities below a certain threshold ε_{Th} (e.g., 10^{-3}). For the rest of tracks, similar to the GM-PHD/CPHD filter, components with weights below a certain threshold T are discarded, and components within a certain distance U are merged, so as to reduce the number of components. Moreover, the maximum number of components of each hypothetical track shall be capped to $J_{\max}$.

(2) Multi-target state estimation. Similar to the SMC implementation of the CBMeMBer recursion, the estimated number of targets may be either the cardinality mean or mode. The latter is usually preferred. Unlike the SMC implementation, the GM implementation can obtain the state mean or state mode. For example, since each posterior density is in Gaussian mixture form, if the Gaussian components are far apart from each other, the mode can be calculated. For simplicity, the mean of the Gaussian component with the highest weight can be used.

6.6 Summary

Based on the MeMBer filter, this Chapter describes the improved CBMeMBer filter and then introduces the SMC and GM implementations of this filter. The CBMeMBer filter has the same linear complexity as the PHD filter, which is significantly lower than the cubic complexity of the CPHD filter. Furthermore, unlike the SMC-PHD/CPHD filter, the SMC-CBMeMBer filter does not require the clustering operation during the target state extraction, making it especially suitable for the nonlinear non-Gaussian case. It should be noted that in the derivation of the (CB)MeMBer filter, in order to obtain the reasonable form of the PGFl of a posterior distribution, the assumption of high signal-to-noise ratio is potentially used, which limits the application of the filter.

Chapter 7
Labeled RFS Filters

7.1 Introduction

The core of random finite sets (RFS) methods is the multi-target Bayesian filter [14, 15], which recursively propagates the filtering density of the multi-target state over time. Strictly speaking, the aforementioned PHD, CPHD and CBMeMBer filters are all multi-target filters rather than multi-target trackers, because these filters themselves cannot provide the target identity (or track) information, which once became the main reason why RFS-based target tracking algorithms were widely criticized. To solve this problem, Vo et al. incorporated target attributes or labels into each target state and developed a series of labeled RFS filters, including the GLMB filter, δ-GLMB filter, LMB filter and Mδ-GLMB filter. These filters are multi-target Bayesian filtering solutions that can provide the estimation of target trajectories.

In the multi-target Bayesian recursion, the multi-target density captures the uncertainty in the number of targets and their states, as well as the statistical correlation between targets. In general, the accurate calculation of multi-target density is not tractable, whereas a tractable implementation usually requires the assumption of statistical independence between targets. For example, the PHD, CPHD and CBMeMBer filters are derived from the statistical independence of multi-target densities. On the other hand, although the typical multi-target tracking methods such as the MHT and the JPDA are able to model the statistical correlation between targets, the MHT does not have the concept of multi-target density, while the JPDA only has the concept of multi-target density with a known number of targets. Currently, only the GLMB family is a tractable multi-target density family capable of capturing the statistical correlation between targets. Unlike the PHD, CPHD and CBMeMBer filters, which are all approximations of full multi-target Bayesian recursion, the GLMB filter is the first exact solution, which exploits the conjugate property of the GLMB distribution family. Therefore, a closed form solution of the full multi-target Bayesian recursion can be obtained. As one of the special forms of the GLMB filter, the δ-GLMB filter has a form more suitable for multi-target tracking. However, due to the combinatorial nature of the multi-target Bayesian recursion, the accurate GLMB

W. Wu et al., *Target Tracking with Random Finite Sets*,
https://doi.org/10.1007/978-981-19-9815-7_7

and δ-GLMB filters are computationally prohibitive. In order to reduce the amount of computation, the LMB filter has been developed based on the δ-GLMB filter, which has been successfully applied to multi-target tracking under severe conditions. However, compared with the δ-GLMB filter, it is at the sacrifice of some tracking performance. Drawing on the idea on the evolution from the PHD filter to the CPHD filter, the Mδ-GLMB filter is a compromise between the tracking performance (relative to the LMB filter) and the computational burden (relative to the δ-GLMB filter).

This chapter follows the development history of the labeled RFS filters, first introduces the GLMB filter with theoretical significance, then gives the δ-GLMB filter and elaborates the specific implementation process. Based on this, the LMB filter and the Mδ-GLMB filter are introduced respectively. Finally, various RFS filters (including the unlabeled RFS filters described in the previous chapters) are compared comprehensively in terms of the tracking performance and computational efficiency through simulations.

7.2 Generalized Labeled Multi-Bernoulli (GLMB) Filter

The generalized labeled multi-Bernoulli (GLMB) filter is a generalization of the MB filter and it can output the target track by incorporating the track label information into the state.

With the transition kernel defined by (3.4.13), the GLMB density is closed under the Chapman-Kolmogorov (C-K) prediction equation.

Proposition 15 *(see Appendix G for proof): let $p_S(\boldsymbol{x}, \ell)$ denote the survival probability of the target with state $(\boldsymbol{x}, \ell)$ at the next time, $q_S(\boldsymbol{x}, \ell) = 1 - p_S(\boldsymbol{x}, \ell)$ the non-survival probability of the target, and the birth density is the LMB with label space $\mathbb{B}$, weight $w_\gamma(\cdot)$, and single-target density $p_\gamma(\cdot, \ell)$.* If the current multi-target prior is a GLMB with label space $\mathbb{L}$ of the form (3.3.38), then the multi-target predicted density is also a GLMB with label space $\mathbb{L}_+ = \mathbb{L} \cup \mathbb{B}$, i.e.,

$$\pi_+(\mathrm{X}_+) = \Delta(\mathrm{X}_+) \sum_{c \in \mathbb{C}} w_+^{(c)}(\mathcal{L}(\mathrm{X}_+))[p_+^{(c)}]^{\mathrm{X}_+} \tag{7.2.1}$$

where

$$w_+^{(c)}(L) = w_S^{(c)}(L \cap \mathbb{L}) w_\gamma(L - \mathbb{L}) \tag{7.2.2}$$

$$\begin{aligned} w_S^{(c)}(L) &= [\eta_S^{(c)}]^L \sum_{I \subseteq \mathbb{L}} 1_I(L)[q_S^{(c)}]^{I-L} w^{(c)}(I) \\ &= [\eta_S^{(c)}]^L \sum_{I \supseteq L} [1 - \eta_S^{(c)}]^{I-L} w^{(c)}(I) \end{aligned} \tag{7.2.3}$$

$$p_{+}^{(c)}(\boldsymbol{x},\ell)=1_{\mathbb{L}}(\ell)p_{+,S}^{(c)}(\boldsymbol{x},\ell)+(1-1_{\mathbb{L}}(\ell))p_{\gamma}(\boldsymbol{x},\ell) \tag{7.2.4}$$

$$p_{+,S}^{(c)}(\boldsymbol{x},\ell)=\langle p_S(\cdot,\ell)\phi(\boldsymbol{x}|\cdot,\ell),p^{(c)}(\cdot,\ell)\rangle\Big/\eta_S^{(c)}(\ell) \tag{7.2.5}$$

$$\begin{aligned}\eta_S^{(c)}(\ell)&=\int\langle p_S(\cdot,\ell)\phi(\boldsymbol{x}|\cdot,\ell),p^{(c)}(\cdot,\ell)\rangle d\boldsymbol{x}\\&=\langle p_S(\cdot,\ell),p^{(c)}(\cdot,\ell)\rangle\end{aligned} \tag{7.2.6}$$

$$q_S^{(c)}(\ell)=\langle q_S(\cdot,\ell),p^{(c)}(\cdot,\ell)\rangle=1-\eta_S^{(c)}(\ell) \tag{7.2.7}$$

Proposition 15 explicitly describes how the parameters $w_{+}^{(c)}(\cdot)$ and $p_{+}^{(c)}(\cdot,\cdot)$ of the predicted multi-target density are computed from the parameters $w^{(c)}(\cdot)$ and $p^{(c)}(\cdot,\cdot)$ of the prior multi-target density. For a given label set L, the weight $w_{+}^{(c)}(L)$ is the product of weight $w_{\gamma}(L-\mathbb{L})$ of the birth labels $L-\mathbb{L}=L\cap\mathbb{B}$ and weight $w_S^{(c)}(L\cap\mathbb{L})$ of the survival labels $L\cap\mathbb{L}$. The weight $w_S^{(c)}(L)$ involves a weighted summation of the prior weights over all label sets containing the survival label set L. For a given label ℓ, the predicted single-target density $p_{+}^{(c)}(\cdot,\ell)$ is either the birth target density $p_{\gamma}(\cdot,\ell)$ or the survival target density $p_{+,S}^{(c)}(\cdot,\ell)$. The latter is calculated by the single-target prediction based on the prior density $p^{(c)}(\cdot,\ell)$ and the transition density $\phi(\cdot|\cdot,\ell)$ weighted by survival probability $p_S(\cdot,\ell)$.

Using the likelihood function defined in (3.4.20), the GLMB density is closed under the Bayesian update.

Proposition 16 *(see Appendix G for proof): if the prior distribution is a GLMB of the form* (3.3.38)*, after the multi-target likelihood function* (3.4.20) *is applied, the posterior distribution is also a GLMB, which is*

$$\pi(\mathrm{X}|\mathrm{Z})=\Delta(\mathrm{X})\sum_{c\in\mathbb{C}}\sum_{\theta\in\Theta}w_Z^{(c,\theta)}(\mathcal{L}(\mathrm{X}))[p^{(c,\theta)}(\cdot|\mathrm{Z})]^{\mathrm{X}} \tag{7.2.8}$$

where Θ is the space of maps $\theta:\mathbb{L}\rightarrow\{0:|\mathrm{Z}|\}\triangleq\{0,1,\ldots,|\mathrm{Z}|\}$, satisfying $i=i'$ when $\theta(i)=\theta(i')>0$, and

$$w_Z^{(c,\theta)}(L)=\frac{\delta_{\theta^{-1}(\{0:|\mathrm{Z}|\})}(L)[\eta_Z^{(c,\theta)}]^L w^{(c)}(L)}{\sum_{c\in\mathbb{C}}\sum_{\theta\in\Theta}\sum_{J\subseteq\mathbb{L}}\delta_{\theta^{-1}(\{0:|\mathrm{Z}|\})}(J)[\eta_Z^{(c,\theta)}]^J w^{(c)}(J)} \tag{7.2.9}$$

$$p^{(c,\theta)}(\boldsymbol{x},\ell|\mathrm{Z})=p^{(c)}(\boldsymbol{x},\ell)\varphi_Z(\boldsymbol{x},\ell;\theta)\Big/\eta_Z^{(c,\theta)}(\ell) \tag{7.2.10}$$

$$\eta_Z^{(c,\theta)}(\ell)=\langle p^{(c)}(\cdot,\ell),\varphi_Z(\cdot,\ell;\theta)\rangle \tag{7.2.11}$$

with $\varphi_Z(\boldsymbol{x},\ell;\theta)$ being defined by (3.4.21).

Proposition 16 explicitly describes how to calculate the parameters $w_Z^{(c,\theta)}(\cdot)$ and $p^{(c,\theta)}(\cdot, \cdot|Z)$ of the posterior multi-target density from the parameters $w^{(c)}(\cdot)$ and $p^{(c)}(\cdot, \cdot)$ of the prior multi-target density. The domain of definition for map θ is $\theta^{-1}(\{0 : |Z|\})$, i.e., the θ inverse image of $\{0 : |Z|\}$, and term $\delta_{\theta^{-1}(\{0:|Z|\})}(\mathcal{L}(\mathrm{X}))$ in (7.2.9) means that only those maps with the domain of definition $\mathcal{L}(\mathrm{X}) \subseteq \mathbb{L}$ are considered. For a valid label set L, i.e., $\delta_{\theta^{-1}(\{0:|Z|\})}(L) = 1$, the updated weight $w_Z^{(c,\theta)}(L)$ is proportional to the prior weight $w^{(c)}(L)$, and the scaling factor is the product $[\eta_Z^{(c,\theta)}]^L$ of the single-target normalization constants. For a given label ℓ, according to the Bayes' rule, the updated single-target density $p^{(c,\theta)}(\cdot, \ell|Z)$ is calculated based on the prior single-target density $p^{(c)}(\cdot, \ell)$ and the likelihood function $\varphi_Z(\cdot, \ell; \theta)$.

In the context of the dynamic multi-target estimation, Propositions 15 and 16 show that starting with a GLMB initial prior, all subsequent predicted and posterior densities are also GLMBs. This implies that the GLMB is the conjugate prior with respect to the standard multi-target transition kernel (3.4.13) and the multi-target measurement likelihood (3.4.20), i.e., the form of the density remains unchanged during both the prediction and the update. The GLMB recursion is the first exact closed form solution of the multi-target Bayesian filter.

Similar to the PHD/CPHD filter, the GLMB filter also requires the pruning of the GLMB density in order to manage the growing number of components. Reference [70] details the implementation step for the GLMB filter based on discarding of secondary components, where it is shown that the pruning operation minimizes the L_1 error of the multi-target density. Recently, the GLMB filter has been extended to the more practical and extremely challenging problem of multi-target tracking with merged measurement [94]. Furthermore, it has been developed in real-time multi-target tracker for autonomous safety system [153].

7.3 δ-GLMB Filter

The δ-generalized label multi-Bernoulli (δ-GLMB) filter [69] is based on the family of GLMB distributions and is also an analytic solution to the multi-target Bayesian recursion. This section details an effective implementation of the δ-GLMB multi-target tracking filter. Each iteration of this filter involves the prediction and update steps, both of which generate a weighted sum of multi-object exponentials (MoE) with a large number of terms. In order to reduce the number of terms in the weighted sum, the K shortest path [404] and ranked assignment algorithms are respectively invoked in the prediction and update steps to determine the most important number of terms, so that it is not necessary to exhaustively calculate all terms. The more effective method of combining the prediction and update steps into a single step and pruning the GLMB filtering density using a random Gibbs sampler can be found in [146]. In addition, the computational complexity can be significantly reduced by using other filters (such as the PHD filter) under the same framework as a relatively low-computational look-ahead strategy.

7.3.1 δ-GLMB Recursion

The δ-GLMB filter propagates the δ-GLMB filtering density forward recursively over time through the Bayesian prediction Eq. (3.5.4) and update Eq. (3.5.5). The closed form solution to the prediction and update of the δ-GLMB filter is given by the following two propositions.

Proposition 17 *(see Appendix H for proof): if the multi-target filtering density at the current moment is a δ-GLMB of the form* (3.3.55)*, then the multi-target predicted density at the next moment is also a δ-GLMB, which is*

$$\pi_+(\mathrm{X}_+) = \Delta(\mathrm{X}_+) \sum_{(I_+,\vartheta)\in\mathcal{F}(\mathbb{L}_+)\times\Xi} w_+^{(I_+,\vartheta)}\delta_{I_+}(\mathcal{L}(\mathrm{X}_+))[p_+^{(\vartheta)}]^{\mathrm{X}_+} \tag{7.3.1}$$

where

$$w_+^{(I_+,\vartheta)} = w_S^{(\vartheta)}(I_+\cap\mathbb{L})w_\gamma(I_+\cap\mathbb{B}) \tag{7.3.2}$$

$$\begin{aligned} w_S^{(\vartheta)}(L) &= [\eta_S^{(\vartheta)}]^L \sum_{I\subseteq\mathbb{L}} 1_I(L)[q_S^{(\vartheta)}]^{I-L} w^{(I,\vartheta)} \\ &= [\eta_S^{(\vartheta)}]^L \sum_{I\supseteq L} [1-\eta_S^{(\vartheta)}]^{I-L} w^{(I,\vartheta)} \end{aligned} \tag{7.3.3}$$

$$p_+^{(\vartheta)}(\boldsymbol{x},\ell) = 1_{\mathbb{L}}(\ell)p_{+,S}^{(\vartheta)}(\boldsymbol{x},\ell) + 1_{\mathbb{B}}(\ell)p_\gamma(\boldsymbol{x},\ell) \tag{7.3.4}$$

$$p_{+,S}^{(\vartheta)}(\boldsymbol{x},\ell) = \langle p_S(\cdot,\ell)\phi(\boldsymbol{x}|\cdot,\ell), p^{(\vartheta)}(\cdot,\ell)\rangle \Big/ \eta_S^{(\vartheta)}(\ell) \tag{7.3.5}$$

$$\begin{aligned} \eta_S^{(\vartheta)}(\ell) &= \int \langle p_S(\cdot,\ell)\phi(\boldsymbol{x}|\cdot,\ell), p^{(\vartheta)}(\cdot,\ell)\rangle d\boldsymbol{x} \\ &= \langle p_S(\cdot,\ell), p^{(\vartheta)}(\cdot,\ell)\rangle \end{aligned} \tag{7.3.6}$$

$$q_S^{(\vartheta)}(\ell) = \langle q_S(\cdot,\ell), p^{(\vartheta)}(\cdot,\ell)\rangle \tag{7.3.7}$$

Proposition 15 states that the predicted GLMB is a sum over $c\in\mathbb{C}$, while Proposition 17 provides a more specific conclusion, which is a sum over the label space $(I_+,\vartheta)\in\mathcal{F}(\mathbb{L}_+)\times\Xi$ at the next time, where $\mathbb{L}_+=\mathbb{L}\cup\mathbb{B}$. The predicted multi-target density involves a new double sum over index $(I_+,\vartheta)\in\mathcal{F}(\mathbb{L}_+)\times\Xi$, which includes the sum over the set of survival labels $I \subset \mathbb{L}$. From the perspective of multi-target tracking, this is more intuitive because it shows how the prediction introduces new target labels.

Proposition 18 *(see Appendix H for proof): if the multi-target predicted density at the current moment is a δ-GLMB of the form* (3.3.55)*, then the multi-target filtering density is also a δ-GLMB, which is*

$$\pi(\mathbf{X}|Z) = \Delta(\mathbf{X}) \sum_{(I,\vartheta)\in\mathcal{F}(\mathbb{L})\times\Xi} \sum_{\theta\in\Theta(I)} w^{(I,\vartheta,\theta)}(Z)\delta_I(\mathcal{L}(\mathbf{X}))[p^{(\vartheta,\theta)}(\cdot|Z)]^{\mathbf{X}} \tag{7.3.8}$$

where

$$w^{(I,\vartheta,\theta)}(Z) = \frac{\delta_{\theta^{-1}(\{0:|Z|\})}(I)[\eta_Z^{(\vartheta,\theta)}]^I w^{(I,\vartheta)}}{\sum_{(I,\vartheta)\in\mathcal{F}(\mathbb{L})\times\Xi}\sum_{\theta\in\Theta(I)}\delta_{\theta^{-1}(\{0:|Z|\})}(I)[\eta_Z^{(\vartheta,\theta)}]^I w^{(I,\vartheta)}} \propto [\eta_Z^{(\vartheta,\theta)}]^I w^{(I,\vartheta)} \tag{7.3.9}$$

$$\eta_Z^{(\vartheta,\theta)}(\ell) = \langle p^{(\vartheta)}(\cdot,\ell), \varphi_Z(\cdot,\ell;\theta)\rangle \tag{7.3.10}$$

$$p^{(\vartheta,\theta)}(\boldsymbol{x},\ell|Z) = p^{(\vartheta)}(\boldsymbol{x},\ell)\varphi_Z(\boldsymbol{x},\ell;\theta)\Big/\eta_Z^{(\vartheta,\theta)}(\ell) \tag{7.3.11}$$

In the above equations, $\Theta(I)$ is a subset of the current association map with the domain of definition I, and $\varphi_Z(\boldsymbol{x},\ell;\theta)$ is defined by (3.4.21).

Note that, according to Propositions 17 and 18, the actual value of the association history ϑ is not used in the calculation, it is only used as an index variable. Instead, the value of label set I need to be used in the calculation process.

7.3.2 *Implementation of δ-GLMB Recursion*

The δ-GLMB can be fully characterized by the parameter set $\{(w^{(I,\vartheta)}, p^{(\vartheta)}) : (I,\vartheta) \in \mathcal{F}(\mathbb{L}) \times \Xi\}$. From an implementation perspective, the δ-GLMB parameter set can be conveniently regarded as an enumeration of all hypotheses and their associated (positive) weights and track densities $\{(I^{(h)}, \vartheta^{(h)}, w^{(h)}, p^{(h)})\}_{h=1}^{H}$, as shown in Table 7.1, where $w^{(h)} \triangleq w^{(I^{(h)},\vartheta^{(h)})}$ and $p^{(h)} \triangleq p^{(\vartheta^{(h)})}$, the hypothesis of component h is $(I^{(h)}, \vartheta^{(h)})$, and the associated weight and track density are $w^{(h)}$ and $p^{(h)}(\cdot,\ell)$, $\ell \in I^{(h)}$, respectively. Thus, implementing the δ-GLMB filter is equivalent to recursively propagating the δ-GLMB parameter set forward in time.

Since the number of hypotheses grows super-exponentially with time, the number of components in the δ-GLMB parameter set must be reduced at each instant. A simple solution is to prune the δ-GLMB density by discarding secondary hypotheses. However, in the δ-GLMB recursion, the strategy of computing all components

Table 7.1 Enumeration of δ-GLMB parameter set (integer index variable h distinguishes different components)

$I^{(1)} = \{\ell_1^{(1)}, \ldots, \ell_{\lvert I^{(1)}\rvert}^{(1)}\}$	$I^{(2)} = \{\ell_1^{(2)}, \ldots, \ell_{\lvert I^{(2)}\rvert}^{(2)}\}$	...	$I^{(h)} = \{\ell_1^{(h)}, \ldots, \ell_{\lvert I^{(h)}\rvert}^{(h)}\}$	...
$\vartheta^{(1)}$	$\vartheta^{(2)}$	...	$\vartheta^{(h)}$	...
$w^{(1)}$	$w^{(2)}$	...	$w^{(h)}$	...
$p^{(1)}(\cdot, \ell_1^{(1)})$ $\vdots$ $p^{(1)}(\cdot, \ell_{\lvert I^{(1)}\rvert}^{(1)})$	$p^{(2)}(\cdot, \ell_1^{(2)})$ $\vdots$ $p^{(2)}(\cdot, \ell_{\lvert I^{(2)}\rvert}^{(2)})$	...	$p^{(h)}(\cdot, \ell_1^{(h)})$ $\vdots$ $p^{(h)}(\cdot, \ell_{\lvert I^{(h)}\rvert}^{(h)})$	...

exhaustively first and then discarding components with small weights is not feasible. An efficient pruning method without computing all components is given below.

7.3.2.1 δ-GLMB Prediction

A concrete implementation of the δ-GLMB prediction is described here, which uses the K shortest path algorithm to prune the predicted δ-GLMB without calculating all predicted hypotheses and their weights.

The predicted density given in Proposition 17 has a compact form, but it is difficult to implement because all supersets of L need to be summed in (7.3.3). The equivalent form given by (H.6) is used here [69]

$$\pi_+(\mathbf{X}_+) = \Delta(\mathbf{X}_+) \sum_{(I,\vartheta)\in\mathcal{F}(\mathbb{L})\times\Xi} w^{(I,\vartheta)} \sum_{J\in\mathcal{F}(I)} [\eta_S^{(\vartheta)}]^J [1-\eta_S^{(\vartheta)}]^{I-J} \sum_{L\in\mathcal{F}(\mathbb{B})} w_\gamma(L)\delta_{J\cup L}(\mathcal{L}(\mathbf{X}_+))[p_+^{(\vartheta)}]^{\mathbf{X}_+} \tag{7.3.12}$$

Note that each current hypothesis (I, ϑ) with weight $w^{(I,\vartheta)}$ yields a set of predicted hypotheses $(J \cup L, \vartheta)$, $J \subseteq I$, $L \subseteq \mathbb{B}$ with weight $w_S^{(I,\vartheta)}(J)w_\gamma(L)$, where

$$w_S^{(I,\vartheta)}(J) = w^{(I,\vartheta)}[\eta_S^{(\vartheta)}]^J[1-\eta_S^{(\vartheta)}]^{I-J} \tag{7.3.13}$$

Intuitively, each predicted label set $J \cup L$ is composed of the survival label set J with weight $w_S^{(I,\vartheta)}(J)$ and a birth label set L with weight $w_\gamma(L)$. The weight $w_S^{(I,\vartheta)}(J)$ can be interpreted as the probability that the current label set is I, and the labels in J will survive at the next moment while the rest of labels $I - J$ will die. Since no survival target label is contained in the birth label space $\mathbb{B}$, the birth label set and the survival label set are mutually exclusive. Since the weight of $J \cup L$ is the product $w_S^{(I,\vartheta)}(J)w_\gamma(L)$, pruning of the double sums over J and L is equivalent to pruning of the sum over J and that over L separately.

The following introduces the K shortest path problem in the context of pruning the δ-GLMB predicted densities, then describes the calculation of the δ-GLMB predicted parameters in detail, and finally gives the δ-GLMB prediction algorithm.

A. K shortest path problem

Consider a given hypothesis (I, ϑ), noting that the weight of the survival label set $J \subseteq I$ can be rewritten as

$$w_S^{(I,\vartheta)}(J) = w^{(I,\vartheta)}[1 - \eta_S^{(\vartheta)}]^I [\eta_S^{(\vartheta)} \Big/ (1 - \eta_S^{(\vartheta)})]^J \tag{7.3.14}$$

If the survival label set $J \subseteq I$ is generated in a non-increasing order of $[\eta_S^{(\vartheta)}/(1 - \eta_S^{(\vartheta)})]^J$, the survival set with the highest weight generated by hypothesis (I, ϑ) can be selected, so that there is no need to exhaustively calculate the weights of all survival hypotheses. This can be done by solving the K shortest path problem in the directed graph shown in Fig. 7.1, where S and E stand for the starting and end nodes, respectively.

Define the cost vector $\boldsymbol{C}^{(I,\vartheta)} = [c^{(I,\vartheta)}(\ell_1), \ldots, c^{(I,\vartheta)}(\ell_{|I|})]$, where $c^{(I,\vartheta)}(\ell_j)$ is the cost of node $\ell_j \in I$, which is

$$c^{(I,\vartheta)}(\ell_j) = -\ln[\eta_S^{(\vartheta)}(\ell_j)/(1 - \eta_S^{(\vartheta)}(\ell_j))] \tag{7.3.15}$$

Nodes are ordered with a non-decreasing cost and the distance from node ℓ_i to node ℓ_j is defined as

$$d(\ell_i, \ell_j) = \begin{cases} c^{(I,\vartheta)}(\ell_j), & \text{If } j > i \\ \infty, & \text{others} \end{cases} \tag{7.3.16}$$

Therefore, the total accumulative distance of a path from S to E traversing the node set $J \subseteq I$ is

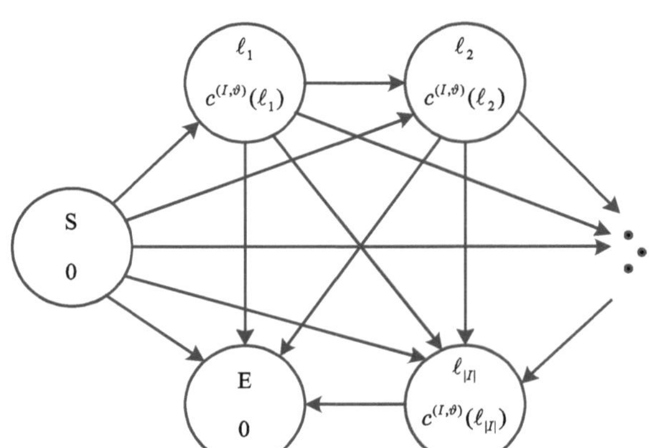

Fig. 7.1 A directed graph with node $\ell_1, \ldots, \ell_{|I|} \in I$ and corresponding cost $c^{(I,\vartheta)}(\ell_1), \ldots, c^{(I,\vartheta)}(\ell_{|I|})$

$$\sum_{\ell \in J} c^{(I,\vartheta)}(\ell) = -\sum_{\ell \in J} \ln\{\eta_S^{(\vartheta)}(\ell) \Big/ [1 - \eta_S^{(\vartheta)}(\ell)]\}$$
$$= -\ln\{[\eta_S^{(\vartheta)} \Big/ (1 - \eta_S^{(\vartheta)})]^J\} \tag{7.3.17}$$

The shortest path from S to E traverses the set of nodes $J^* \subseteq I$ with the shortest distance $\sum_{\ell \in J^*} c^{(I,\vartheta)}(\ell)$, and the maximum value is $[\eta_S^{(\vartheta)} \Big/ (1 - \eta_S^{(\vartheta)})]^{J^*}$. The K shortest path problem seeks to find out a subset of I with the K shortest distances in non-descending order. Therefore, solving the K shortest path problem yields such an enumeration, which is a subset J of I starting at J^* and arranged in non-increasing order of $[\eta_S^{(\vartheta)} \Big/ (1 - \eta_S^{(\vartheta)})]^J$.

For the birth targets, the LMB birth model is adopted, i.e.,

$$w_\gamma(L) = \prod_{\ell \in \mathbb{B}} (1 - \varepsilon_\gamma^{(\ell)}) \prod_{\ell \in L} \frac{1_{\mathbb{B}}(\ell)\varepsilon_\gamma^{(\ell)}}{1 - \varepsilon_\gamma^{(\ell)}} \tag{7.3.18}$$

$$p_\gamma(\boldsymbol{x}, \ell) = p_\gamma^{(\ell)}(\boldsymbol{x}) \tag{7.3.19}$$

Therefore, solving the K shortest path problem with cost vector $\boldsymbol{C}_\gamma = [c_\gamma(\ell_1), \ldots, c_\gamma(\ell_{|\mathbb{B}|})]$ yields the subset of $\mathbb{B}$ with the optimal birth weight, where $c_\gamma(\ell_j)$ is the cost of node ℓ_j, which is

$$c_\gamma(\ell_j) = -\ln[\varepsilon_\gamma^{(\ell_j)} \Big/ (1 - \varepsilon_\gamma^{(\ell_j)})] \tag{7.3.20}$$

By extending the directed digraph to cover both survival nodes and birth nodes with suitable costs, it is possible to obtain the overall K best components. However, compared to survival weight $w_S^{(I,\vartheta)}(J)$, birth weight $w_\gamma(L)$ is quite small, and many birth components will be discarded, making the birth targets difficult to be detected by the filter. To avoid discarding new tracks, a very large value of K is required to preserve the birth hypotheses. Conversely, the separation and pruning strategy introduced above ensures that there are certain birth hypothese to deal with new tracks, and it is highly parallel.

The K shortest path algorithm is a well-known solution to the following combinatorial problem of finding out K paths with the minimum total cost from a given source point to a given destination point in a weighted network. The computational complexity of this algorithm is $\mathcal{O}(|I| \log(|I|)+K)$. Here, the node may have negative weights, so the Bellman-Ford algorithm [405] needs to be called.

B. Calculation of predicted parameters

Next, the parameters ($\eta_S^{(\vartheta)}(\ell)$ and $p_+^{(\vartheta)}(\cdot, \ell)$) of the δ-GLMB predicted components are calculated.

(1) Gaussian mixture implementation. For the linear Gaussian multi-target model, $p_S(\boldsymbol{x}, \ell) = p_S$, $\phi(\boldsymbol{x}_+|\boldsymbol{x}, \ell) = \mathcal{N}(\boldsymbol{x}_+; \boldsymbol{F}\boldsymbol{x}, \boldsymbol{Q})$, where $\mathcal{N}(\cdot; \boldsymbol{m}, \boldsymbol{P})$ is the Gaussian density with mean $\boldsymbol{m}$ and covariance $\boldsymbol{P}$, $\boldsymbol{F}$ is the state transition matrix, $\boldsymbol{Q}$ denotes the process noise covariance and the birth density $p_\gamma^{(\ell)}(\boldsymbol{x})$ is in the Gaussian mixture form. If the single-target density $p^{(\vartheta)}(\cdot, \ell)$ is in the Gaussian mixture form, i.e.,

$$p^{(\vartheta)}(\cdot, \ell) = \sum_{i=1}^{J^{(\vartheta)}(\ell)} \omega_i^{(\vartheta)}(\ell)\mathcal{N}(\boldsymbol{x}; \boldsymbol{m}_i^{(\vartheta)}(\ell), \boldsymbol{P}_i^{(\vartheta)}(\ell)) \tag{7.3.21}$$

then

$$\eta_S^{(\vartheta)}(\ell) = p_S \tag{7.3.22}$$

$$p_+^{(\vartheta)}(\boldsymbol{x}, \ell) = 1_{\mathbb{L}}(\ell) \sum_{i=1}^{J^{(\vartheta)}(\ell)} \omega_i^{(\vartheta)}(\ell)\mathcal{N}(\boldsymbol{x}; \boldsymbol{m}_{S,i}^{(\vartheta)}(\ell), \boldsymbol{P}_{S,i}^{(\vartheta)}(\ell)) + 1_{\mathbb{B}}(\ell)p_\gamma^{(\ell)}(\boldsymbol{x}) \tag{7.3.23}$$

where

$$\boldsymbol{m}_{S,i}^{(\vartheta)}(\ell) = \boldsymbol{F}\boldsymbol{m}_i^{(\vartheta)}(\ell) \tag{7.3.24}$$

$$\boldsymbol{P}_{S,i}^{(\vartheta)}(\ell) = \boldsymbol{F}\boldsymbol{P}_i^{(\vartheta)}(\ell)\boldsymbol{F}^{\mathrm{T}} + \boldsymbol{Q} \tag{7.3.25}$$

When the motion model parameters are related to labels, it is only necessary to substitute $p_S = p_S(\ell)$, $\boldsymbol{F} = \boldsymbol{F}(\ell)$, and $\boldsymbol{Q} = \boldsymbol{Q}(\ell)$ into the above equations.

(2) Sequential Monte Carlo implementation. For the sequential Monte Carlo (SMC) approximation, assume that each single-target density $p^{(\vartheta)}(\cdot, \ell)$ is represented by a set of weighted samples $\{\omega_i^{(\vartheta)}(\ell), \boldsymbol{x}_i^{(\vartheta)}(\ell)\}_{i=1}^{J^{(\vartheta)}(\ell)}$, and the birth density $p_\gamma^{(\ell)}(\cdot)$ can be expressed as a set of weighted samples $\{\omega_{\gamma,i}^{(\vartheta)}(\ell), \boldsymbol{x}_{\gamma,i}^{(\vartheta)}(\ell)\}_{i=1}^{J_\gamma^{(\vartheta)}(\ell)}$, then

$$\eta_S^{(\vartheta)}(\ell) = \sum_{i=1}^{J^{(\vartheta)}(\ell)} \omega_i^{(\vartheta)}(\ell)p_S(\boldsymbol{x}_i^{(\vartheta)}(\ell), \ell) \tag{7.3.26}$$

and $p_+^{(\vartheta)}(\boldsymbol{x}, \ell)$ can be represented by

$$\left\{1_{\mathbb{L}}(\ell)\tilde{\omega}_{S,i}^{(\vartheta)}(\ell), \boldsymbol{x}_{S,i}^{(\vartheta)}(\ell)\right\}_{i=1}^{J^{(\vartheta)}(\ell)} \cup \left\{1_{\mathbb{B}}(\ell)\omega_{\gamma,i}^{(\vartheta)}(\ell), \boldsymbol{x}_{\gamma,i}^{(\vartheta)}(\ell)\right\}_{i=1}^{J_\gamma^{(\vartheta)}(\ell)} \tag{7.3.27}$$

where

$$\boldsymbol{x}_{S,i}^{(\vartheta)}(\ell) \sim q^{(\vartheta)}(\cdot|\boldsymbol{x}_i^{(\vartheta)}(\ell), \ell, Z), i = 1, \ldots, J^{(\vartheta)}(\ell) \tag{7.3.28}$$

$$\omega_{S,i}^{(\vartheta)}(\ell) = \frac{\omega_i^{(\vartheta)}(\ell)\phi(\boldsymbol{x}_{S,i}^{(\vartheta)}(\ell)|\boldsymbol{x}_i^{(\vartheta)}(\ell))p_S(\boldsymbol{x}_i^{(\vartheta)}(\ell), \ell)}{q^{(\vartheta)}(\boldsymbol{x}_{S,i}^{(\vartheta)}(\ell)|\boldsymbol{x}_i^{(\vartheta)}(\ell), \ell, Z)} \tag{7.3.29}$$

$$\tilde{\omega}_{S,i}^{(\vartheta)}(\ell) = \omega_{S,i}^{(\vartheta)}(\ell) \Big/ \sum_{i=1}^{J^{(\vartheta)}(\ell)} \omega_{S,i}^{(\vartheta)}(\ell) \tag{7.3.30}$$

with $q^{(\vartheta)}(\cdot|\boldsymbol{x}_i^{(\vartheta)}(\ell), \ell, Z)$ being the proposed density.

Using $\eta_S^{(\vartheta)}(\ell)$ obtained from (7.3.22) or (7.3.26), the node cost $c^{(I,\vartheta)}(\ell)$ of the K shortest path problem can be calculated according to (7.3.15).

C. Pruning of predicted densities

Given the δ-GLMB filtering density with enumerated parameter set $\{(I^{(h)}, \vartheta^{(h)}, w^{(h)}, p^{(h)})\}_{h=1}^{H}$, the δ-GLMB predicted density (7.3.12) can be written as

$$\pi(\mathrm{X}_+) = \sum_{h=1}^{H} \pi_+^{(h)}(\mathrm{X}_+) \tag{7.3.31}$$

where

$$\pi_+^{(h)}(\mathrm{X}_+) = \Delta(\mathrm{X}_+) \sum_{J \subseteq I^{(h)}} \sum_{L \subseteq \mathbb{B}} w_S^{(I^{(h)},\vartheta^{(h)})}(J) w_\gamma(L) \delta_{J \cup L}(\mathcal{L}(\mathrm{X}_+))[p_+^{(\vartheta^{(h)})}]^{\mathrm{X}_+} \tag{7.3.32}$$

For the δ-GLMB prediction, the δ-GLMB filtering component h produces $2^{|I^{(h)}|+|\mathbb{B}|}$ components.

For the purpose of pruning the predicted δ-GLMB π_+, a simple and highly parallel strategy is to prune each $\pi_+^{(h)}$ as follows. For each $h = 1, \ldots, H$, the K shortest path problem with cost vector $\boldsymbol{C}^{(I^{(h)},\vartheta^{(h)})}$ is solved to obtain $J^{(h,j)}$, $j = 1, \ldots, K^{(h)}$, which is $K^{(h)}$ subsets of $I^{(h)}$ with the highest survival weights, as shown in Fig. 7.2. In the figure, the prior component h generates all subsets of $I^{(h)}$, i.e., $J^{(h,j)}$, $j = 1, \ldots, 2^{|I^{(h)}|}$ with weight $w_S^{(h,j)} \triangleq w_S^{(I^{(h)},\vartheta^{(h)})}(J^{(h,j)})$. The K shortest path algorithm determines $K^{(h)}$ subsets with the maximum weights $w_S^{(h,1)} \geq w_S^{(h,2)} \geq \ldots \geq w_S^{(h,K^{(h)})}$. Similarly, the K shortest path problem with cost vector $\boldsymbol{C}_\gamma$ can also be solved to obtain $L^{(b)}$, $b = 1, \ldots, K_\gamma$, which is K_γ birth subsets with the highest birth weights. Therefore, for each h, the pruned version of $\pi_+^{(h)}$ is

$$\hat{\pi}_+^{(h)}(\mathrm{X}_+) = \Delta(\mathrm{X}_+) \sum_{j=1}^{K^{(h)}} \sum_{b=1}^{K_\gamma} w_+^{(h,j,b)} \delta_{J^{(h,j)} \cup L^{(b)}}(\mathcal{L}(\mathrm{X}_+))[p_+^{(h)}]^{\mathrm{X}_+} \tag{7.3.33}$$

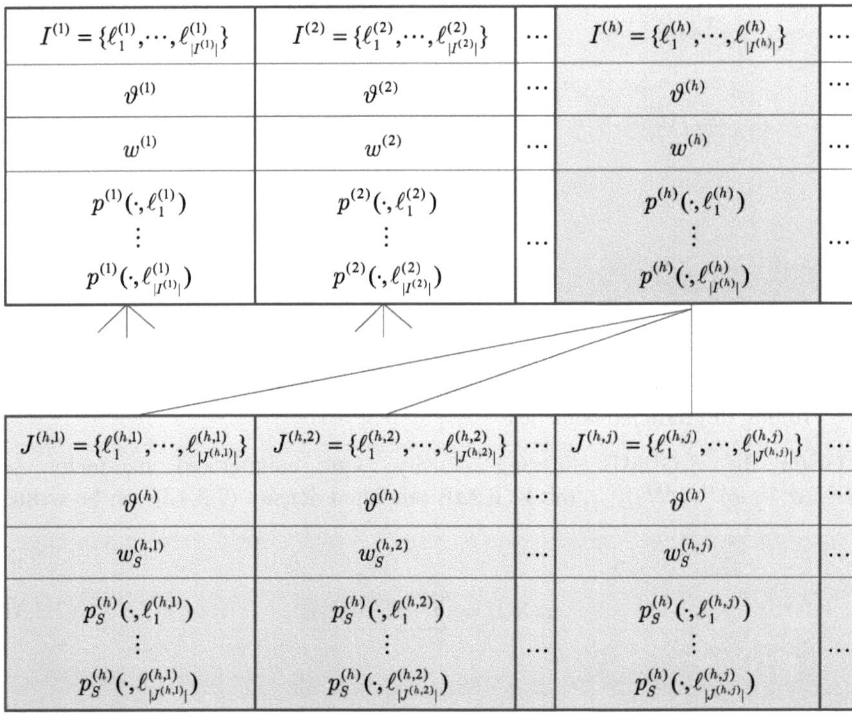

Fig. 7.2 Prediction of survival components

where

$$w_+^{(h,j,b)} \triangleq w_S^{(I^{(h)},\vartheta^{(h)})}(J^{(h,j)}) w_\gamma(L^{(b)}) \tag{7.3.34}$$

$$p_+^{(h)} \triangleq p_+^{(\vartheta^{(h)})} \tag{7.3.35}$$

Since the weights of the (un-pruned) predicted densities sum to 1, the resulting pruned density $\hat{\pi}_+ = \sum_{h=1}^{H} \hat{\pi}_+^{(h)}$ has $T = K_\gamma \sum_{h=1}^{H} K^{(h)}$ components, resulting in a pruning error $1 - \sum_{h=1}^{H} \sum_{j=1}^{K^{(h)}} \sum_{b=1}^{K_\gamma} w_+^{(h,j,b)}$. The final approximate expression is obtained by normalizing the pruned density. Table 7.2 gives the pseudo codes for the prediction step. It is noted that all 3 for loops run in parallel.

The concrete values of the required number of components $K^{(h)}$ and K_γ are generally user-specified or application-specific. A general strategy is to choose $K^{(h)} = \lceil w^{(h)} J_{\max} \rceil$, where $J_{\max}$ is the expected total number of hypotheses. For K_γ, K_γ can be selected so that the resulting pruning density can capture the probability mass of a desired proportion (e.g., 99%) of the birth density. Another strategy is to always keep $T = J_{\max}$ components of π_+ with the maximum weights, which produces a smaller pruning error than the aforementioned strategy [70]. However,

Table 7.2 Pseudo code of the δ-GLMB prediction

• Input: $\{(I^{(h)}, \vartheta^{(h)}, w^{(h)}, p^{(h)}, K^{(h)})\}_{h=1}^{H}$, $K_\gamma, \{(\varepsilon_\gamma^{(\ell)}, p_\gamma^{(\ell)})\}_{\ell\in\mathbb{B}}$
• Output: $\{(I_+^{(h,j,b)}, w_+^{(h,j,b)}, p_+^{(h)})\}_{(h,j,b)=(1,1,1)}^{(H,K^{(h)},K_\gamma)}$
Calculate $\boldsymbol{C}_\gamma$ according to (7.3.20)
$\{L^{(b)}\}_{b=1}^{K_\gamma}$:= k_shortest_path($\mathbb{B}, \boldsymbol{C}_\gamma, K_\gamma$)
for $b = 1 : K_\gamma$
$\quad w_\gamma^{(b)} := \prod_{\ell\in L^{(b)}} \varepsilon_\gamma^{(\ell)} \prod_{\ell\in\mathbb{B}-L^{(b)}} (1-\varepsilon_\gamma^{(\ell)})$
end
for $h = 1 : H$
$\quad$ Calculate $\eta_S^{(h)} := \eta_S^{(\vartheta^{(h)})}$ according to (7.3.22) or (7.3.26)
$\quad$ Calculate $\boldsymbol{C}^{(h)} := \boldsymbol{C}^{(I^{(h)},\vartheta^{(h)})}$ according to (7.3.15)
$\quad \{J^{(h,j)}\}_{j=1}^{K^{(h)}}$:= k_shortest_path($I^{(h)}, \boldsymbol{C}^{(h)}, K^{(h)}$),
$\quad$ for $(j, b) = (1, 1) : (K^{(h)}, K_\gamma)$
$\qquad w_+^{(h,j,b)} := w^{(h)}[\eta_S^{(h)}]^{J^{(h,j)}}[1-\eta_S^{(h)}]^{I^{(h)}-J^{(h,j)}} w_\gamma^{(b)}$,
$\qquad I_+^{(h,j,b)} := J^{(h,j)} \cup L^{(b)}$
$\quad$ end
$\quad$ Calculate $p_+^{(h)} \triangleq p_+^{(\vartheta^{(h)})}$ according to (7.3.23) or (7.3.27)
end
Normalize the weights $\{w_+^{(h,j,b)}\}_{(h,j,b)=(1,1,1)}^{(H,K^{(h)},K_\gamma)}$

this strategy not only increases the dimension of the problem by $(H + K_\gamma)$ times, but also loses the parallelism.

7.3.2.2 δ-GLMB Update

The following introduces a feasible implementation of the δ-GLMB update, which prunes the multi-target filtering density through a ranked assignment algorithm without calculating all hypotheses and their weights. First, the ranked assignment problem in the context of pruning the δ-GLMB filtering density is introduced, then the calculation of the updated δ-GLMB parameters is given in detail, and finally the δ-GLMB update algorithm is summarized.

A. Ranked assignment problem

Note that each hypothesis (I, ϑ) with weight $w^{(I,\vartheta)}$ generates a new hypothesis set $(I, (\vartheta, \theta)), \theta \in \Theta(I)$ with weight $w^{(I,\vartheta,\theta)}(Z) \propto w^{(I,\vartheta)}[\eta_Z^{(\vartheta,\theta)}]^I$ from the δ-GLMB weight update (7.3.9). For a given hypothesis (I, ϑ), if the association map $\theta \in \Theta(I)$

in descending order of $[\eta_Z^{(\vartheta,\theta)}]^I$ can be generated, the components with the maximum weights can be selected without exhaustively calculating all new hypotheses and their weights. This can be accomplished by solving the following ranked assignment problem.

By enumerating $I = \{\ell_1, \ldots, \ell_{|I|}\}$ and $Z = \{z_1, \ldots, z_{|Z|}\}$, each association map $\theta \in \Theta(I)$ can be described by an assignment matrix $\boldsymbol{S}$ of size $|I| \times |Z|$, which consists of 0s and 1s such that each row (or column) sums to 0 or 1. For $i \in \{1, 2, \ldots, |I|\}$, $j \in \{1, 2, \ldots, |Z|\}$, if and only if the jth measurement is assigned to track ℓ_i, i.e., $\theta(\ell_i) = j$, then $s_{i,j} = 1$. The all-zero row i means that the track ℓ_i is missed, while the all-zero column j means that the measurement z_j is a false alarm. The conversion from $\boldsymbol{S}$ to θ can be obtained from $\theta(\ell_i) = \sum_{j=1}^{|Z|} j\delta_1(s_{i,j})$.

The cost matrix of the optimal assignment problem is a matrix of size $|I| \times |Z|$, i.e.,

$$\boldsymbol{C}_Z^{(I,\xi)} = \begin{bmatrix} c_{1,1} & \cdots & c_{1,|Z|} \\ \vdots & \ddots & \vdots \\ c_{|I|,1} & \cdots & c_{|I|,|Z|} \end{bmatrix} \tag{7.3.36}$$

where $c_{i,j}$ means the cost of the $j \in \{1, \ldots, |Z|\}$ th measurement assigned to track ℓ_i, $i \in \{1, \ldots, |I|\}$, which is

$$c_{i,j} = -\ln\left[\frac{\langle p^{(\vartheta)}(\cdot, \ell_i), p_D(\cdot, \ell_i) g(z_j|\cdot, \ell_i)\rangle}{\langle p^{(\vartheta)}(\cdot, \ell_i), 1 - p_D(\cdot, \ell_i)\rangle \kappa(z_j)}\right] \tag{7.3.37}$$

The numerical calculation of $c_{i,j}$ is detailed in (7.3.41) and (7.3.49) below.

The cost of the assignment matrix $\boldsymbol{S}$ is the total cost of the assignment of each measurement to each target, which can be written succinctly as the following Frobenius inner product

$$\mathrm{tr}(\boldsymbol{S}^{\mathrm{T}} \boldsymbol{C}_Z^{(I,\xi)}) = \sum_{i=1}^{|I|} \sum_{j=1}^{|Z|} c_{i,j} s_{i,j} \tag{7.3.38}$$

where $\mathrm{tr}(\cdot)$ represents the trace of a matrix (the sum of diagonal elements of the matrix).

Substituting (3.4.21) into (7.3.10), it can be obtained that the cost of $\boldsymbol{S}$ (and the corresponding association map θ) is related to the filtering hypothesis weight $w^{(I,\vartheta,\theta)}(Z) \propto w^{(I,\vartheta)}[\eta_Z^{(\vartheta,\theta)}]^I$, where

$$[\eta_Z^{(\vartheta,\theta)}]^I = \exp\left(-\mathrm{tr}(\boldsymbol{S}^{\mathrm{T}} \boldsymbol{C}_Z^{(I,\vartheta)})\right) \prod_{\ell \in I} \langle p^{(\vartheta)}(\cdot, \ell), 1 - p_D(\cdot, \ell)\rangle \tag{7.3.39}$$

The optimal assignment problem seeks the assignment matrix $\boldsymbol{S}^*$ (and its corresponding association map θ^*) that minimizes the cost $\mathrm{tr}((\boldsymbol{S}^*)^{\mathrm{T}} \boldsymbol{C}_Z^{(I,\vartheta)})$. The ranked assignment problem seeks an enumeration of least-cost assignment matrices in non-decreasing order. Therefore, solving a ranked optimal assignment problem with the cost matrix $\boldsymbol{C}_Z^{(I,\vartheta)}$ will yield such an enumeration of the association maps θ starting from θ^* and arranged in the order of non-increasing $[\eta_Z^{(\vartheta,\theta)}]^I$ (or weight $w^{(I,\vartheta,\theta)}(Z) \propto w^{(I,\vartheta)}[\eta_Z^{(\vartheta,\theta)}]^I$).

The standard ranked assignment involves the square (equal numbers of rows and columns) cost and assignment matrices, and the sum of rows or columns of the assignment matrix is 1. For the ranked assignment problem of a non-square matrix (i.e., matrix with unequal numbers of rows and columns), it can be restated as a square matrix problem by introducing pseudo variables.

The optimal assignment problem is a well-known combinatorial problem that can be solved by the Hungarian algorithm with polynomial computational complexity [406]. The ranked assignment problem generalizes this problem and enumerates T minimum cost assignments, which are first solved by Murty. Murty's algorithm requires an efficient bipartite assignment algorithm, such as the Munkres [406] or Jonker-Volgenant algorithms [407]. In the context of multi-target tracking, the ranked assignment algorithm with complexity $\mathcal{O}(T|Z|^4)$ has been applied to the MHT [10, 408], and more efficient algorithms with complexity $\mathcal{O}(T|Z|^3)$ can be found in [409, 410], which will have higher efficiency for larger $|Z|$. It should be noted that the K shortest path problem can be solved by the ranked assignment algorithm, but for the δ-GLMB prediction, the K shortest path algorithm is more effective. However, the ranking of association maps can't be described as the K shortest path problem due to the constraint that each target can only generate at most one measurement.

B. Calculation of updated parameters

Next, the cost matrix $\boldsymbol{C}_Z^{(I,\vartheta)}$(7.3.36) of the ranked assignment problem and the updated parameters $\eta_Z^{(\vartheta,\theta)}(\ell)$ and $p^{(\vartheta,\theta)}(\cdot,\ell|Z)$ of the δ-GLMB components are calculated.

(1) Gaussian mixture implementation: for the linear Gaussian multi-target model, $p_D(\boldsymbol{x},\ell) = p_D$, $g(z|\boldsymbol{x},\ell) = \mathcal{N}(z;\boldsymbol{H}\boldsymbol{x},\boldsymbol{R})$, where $\boldsymbol{H}$ is the measurement matrix and $\boldsymbol{R}$ is the measurement noise covariance. The Gaussian mixture expression provides the most general configuration for the linear Gaussian model. Assume that each single-target density $p^{(\vartheta)}(\cdot,\ell)$ is in the Gaussian mixture form

$$p^{(\vartheta)}(\cdot,\ell) = \sum_{n=1}^{J^{(\vartheta)}(\ell)} \omega_n^{(\vartheta)}(\ell)\mathcal{N}(\boldsymbol{x};\boldsymbol{m}_n^{(\vartheta)}(\ell),\boldsymbol{P}_n^{(\vartheta)}(\ell)) \tag{7.3.40}$$

then we have

$$c_{i,j} = -\ln\left[\frac{p_D\sum_{n=1}^{J^{(\vartheta)}(\ell_i)}\omega_n^{(\vartheta)}(\ell_i)q_n^{(\vartheta)}(z_j;\ell_i)}{(1-p_D)\kappa(z_j)}\right] \tag{7.3.41}$$

In addition, for the updated association history (ϑ, θ), we have

$$\eta_Z^{(\vartheta,\theta)}(\ell) = \sum_{n=1}^{J^{(\vartheta)}(\ell)} w_{Z,n}^{(\vartheta,\theta)}(\ell) \tag{7.3.42}$$

$$p^{(\vartheta,\theta)}(\boldsymbol{x}, \ell|Z) = \sum_{n=1}^{J^{(\vartheta)}(\ell)} \frac{w_{Z,n}^{(\vartheta,\theta)}(\ell)}{\eta_Z^{(\vartheta,\theta)}(\ell)} \mathcal{N}(\boldsymbol{x}; \boldsymbol{m}_{Z,n}^{(\vartheta,\theta)}(\ell), \boldsymbol{P}_n^{(\vartheta,\theta)}(\ell)) \tag{7.3.43}$$

where

$$w_{Z,n}^{(\vartheta,\theta)}(\ell) = \omega_n^{(\vartheta)}(\ell) \times \begin{cases} p_D q_n^{(\vartheta)}(z_{\theta(\ell)}; \ell) \big/ \kappa(z_{\theta(\ell)}) & \text{If } \theta(\ell) > 0 \\ (1 - p_D) & \text{If } \theta(\ell) = 0 \end{cases} \tag{7.3.44}$$

$$q_n^{(\vartheta)}(z; \ell) = \mathcal{N}(z; \boldsymbol{H}\boldsymbol{m}_n^{(\vartheta)}(\ell), \boldsymbol{H}\boldsymbol{P}_n^{(\vartheta)}(\ell)\boldsymbol{H}^{\mathrm{T}} + \boldsymbol{R}) \tag{7.3.45}$$

$$\boldsymbol{m}_{Z,n}^{(\vartheta,\theta)}(\ell) = \begin{cases} \boldsymbol{m}_n^{(\vartheta)}(\ell) + \boldsymbol{G}_n^{(\vartheta,\theta)}(\ell)(z_{\theta(\ell)} - \boldsymbol{H}\boldsymbol{m}_n^{(\vartheta)}(\ell)) & \text{If } \theta(\ell) > 0 \\ \boldsymbol{m}_n^{(\vartheta)}(\ell) & \text{If } \theta(\ell) = 0 \end{cases} \tag{7.3.46}$$

$$\boldsymbol{P}_n^{(\vartheta,\theta)}(\ell) = (\boldsymbol{I} - \boldsymbol{G}_n^{(\vartheta,\theta)}(\ell)\boldsymbol{H})\boldsymbol{P}_n^{(\vartheta)}(\ell) \tag{7.3.47}$$

$$\boldsymbol{G}_n^{(\vartheta,\theta)}(\ell) = \begin{cases} \boldsymbol{P}_n^{(\vartheta)}(\ell)\boldsymbol{H}^{\mathrm{T}}(\boldsymbol{H}\boldsymbol{P}_n^{(\vartheta)}(\ell)\boldsymbol{H}^{\mathrm{T}} + \boldsymbol{R})^{-1} & \text{If } \theta(\ell) > 0 \\ \boldsymbol{0} & \text{If } \theta(\ell) = 0 \end{cases} \tag{7.3.48}$$

When the parameters of the measurement model are dependent on label ℓ, it is only necessary to substitute $p_D = p_D(\ell)$, $\boldsymbol{H} = \boldsymbol{H}(\ell)$, and $\boldsymbol{R} = \boldsymbol{R}(\ell)$ into the above equations.

(2) Sequential Monte Carlo implementation: for the sequential Monte Carlo approximation, assume that each single-target density $p^{(\vartheta)}(\cdot, \ell)$ is represented by a set of weighted samples $\{\omega_n^{(\vartheta)}(\ell), \boldsymbol{x}_n^{(\vartheta)}(\ell)\}_{n=1}^{J^{(\vartheta)}(\ell)}$, then we have

$$c_{i,j} = -\ln\left[\frac{\sum_{n=1}^{J^{(\vartheta)}(\ell_i)} \omega_n^{(\vartheta)}(\ell_i) p_D(\boldsymbol{x}_n^{(\vartheta)}(\ell_i), \ell_i) g(z_j|\boldsymbol{x}_n^{(\vartheta)}(\ell_i), \ell_i)}{\sum_{n=1}^{J^{(\vartheta)}(\ell_i)} \omega_n^{(\vartheta)}(\ell_i)(1 - p_D(\boldsymbol{x}_n^{(\vartheta)}(\ell_i), \ell_i))\kappa(z_j)}\right] \tag{7.3.49}$$

In addition, for a given updated association history (ϑ, θ), we have

$$\eta_Z^{(\vartheta,\theta)}(\ell) = \sum_{n=1}^{J^{(\vartheta)}(\ell)} \omega_n^{(\vartheta)}(\ell) \varphi_Z(\boldsymbol{x}_n^{(\vartheta)}(\ell), \ell; \theta) \tag{7.3.50}$$

and $p^{(\vartheta,\theta)}(\cdot, \ell|Z)$ is expressed by the following weighted sample set

$$\left\{ \frac{\varphi_Z(\boldsymbol{x}_n^{(\vartheta)}(\ell), \ell; \theta)\omega_n^{(\vartheta)}(\ell)}{\eta_Z^{(\vartheta,\theta)}(\ell)}, \boldsymbol{x}_n^{(\vartheta)}(\ell) \right\}_{n=1}^{J^{(\vartheta)}(\ell)} \tag{7.3.51}$$

C. Pruning of filtering densities

Given a δ-GLMB predicted density with numerated parameter set $\{(I^{(h)}, \vartheta^{(h)}, w^{(h)}, p^{(h)})\}_{h=1}^{H}$, the δ-GLMB filtering density (7.3.8) can be written as

$$\pi(\mathrm{X}|\mathrm{Z}) = \sum_{h=1}^{H} \pi^{(h)}(\mathrm{X}|\mathrm{Z}) \tag{7.3.52}$$

where

$$\pi^{(h)}(\mathrm{X}|\mathrm{Z}) = \Delta(\mathrm{X}) \sum_{j=1}^{|\Theta(I^{(h)})|} w^{(h,j)} \delta_{I^{(h)}}(\mathcal{L}(\mathrm{X}))[p^{(h,j)}]^{\mathrm{X}} \tag{7.3.53}$$

$$w^{(h,j)} \triangleq w^{(I^{(h)},\vartheta^{(h)},\theta^{(h,j)})}(Z) \tag{7.3.54}$$

$$p^{(h,j)} \triangleq p^{(I^{(h)},\vartheta^{(h)},\theta^{(h,j)})}(\cdot|\mathrm{Z}) \tag{7.3.55}$$

Thus, each δ-GLMB predicted component with index h generates $|\Theta(I^{(h)})|$ δ-GLMB filtering density components.

In order to prune the δ-GLMB filtering density (7.3.52), a simple and highly parallel strategy is to prune $\pi^{(h)}(\cdot|Z)$. For each $h = 1, \ldots, H$, solving the ranked assignment problem with cost matrix $\boldsymbol{C}_Z^{(I^{(h)},\vartheta^{(h)})}$ yields $T^{(h)}$ hypotheses $\theta^{(h,j)}$, $j = 1, \ldots, T^{(h)}$ with the highest weights in non-increasing order, as shown in Fig. 7.3. In the figure, the prior component h generates a large number of posterior components, and the ranked assignment algorithm determines the $T^{(h)}$ components with the maximum weights $w^{(h,1)} \geq w^{(h,2)} \geq \ldots \geq w^{(h,T^{(h)})}$. As a result, the pruned version of $\pi^{(h)}(\cdot|\mathrm{Z})$ is

$$\hat{\pi}^{(h)}(\mathrm{X}|\mathrm{Z}) = \Delta(\mathrm{X}) \sum_{j=1}^{T^{(h)}} w^{(h,j)} \delta_{I^{(h)}}(\mathcal{L}(\mathrm{X}))[p^{(h,j)}]^{\mathrm{X}} \tag{7.3.56}$$

The pruned δ-GLMB density has a total of $T = \sum_{h=1}^{H} T^{(h)}$ components, and it is obtained through the normalization of the sum of weights. Table 7.3 summarizes the pseudocode of the update step. Note that both internal and external for loops are parallel.

The concrete value of the required number of components $T^{(h)}$ is generally specified by users or is application specific. A general strategy is to select $T^{(h)} = \lceil w^{(h)} J_{\max} \rceil$, where $J_{\max}$ is the expected total number of hypotheses. Another strategy

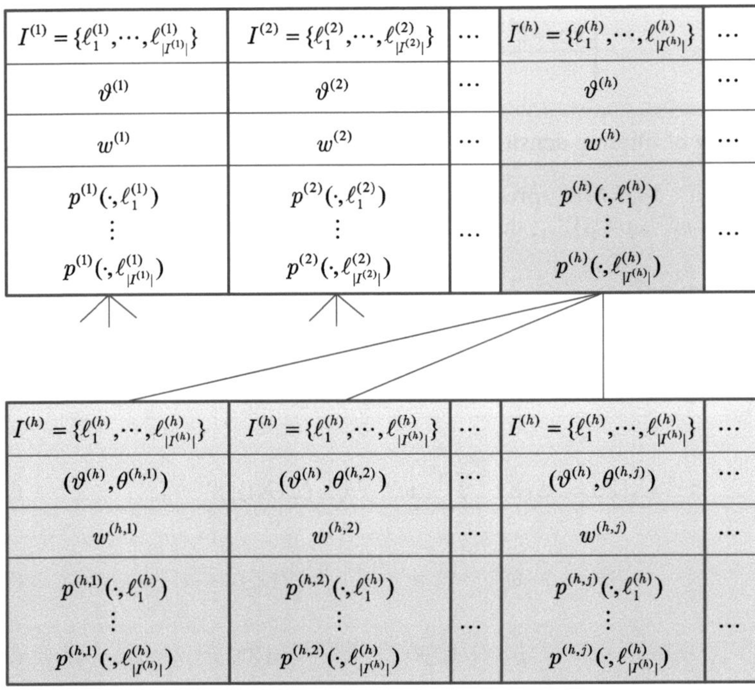

Fig. 7.3 Schematic diagram of the δ-GLMB update

Table 7.3 Pseudo code of the δ-GLMB update

• Input: $\{(I^{(h)}, \vartheta^{(h)}, w^{(h)}, p^{(h)}, T^{(h)})\}_{h=1}^{H}$, Z
• Output: $\{(I^{(h,j)}, \vartheta^{(h,j)}, w^{(h,j)}, p^{(h,j)})\}_{(h,j)=(1,1)}^{(H,T^{(h)})}$
for $h = 1 : H$
Calculate $\boldsymbol{C}_Z^{(h)} := \boldsymbol{C}_Z^{(I^{(h)}, \xi^{(h)})}$ according to (7.3.36), (7.3.41) or (7.3.49)
$\{\theta^{(h,j)}\}_{j=1}^{T^{(h)}}$:= ranked_assignment(Z, $I^{(h)}$, $\boldsymbol{C}_Z^{(h)}$, $T^{(h)}$)
for $j = 1 : T^{(h)}$
Calculate $\eta_Z^{(h,j)} := \eta_Z^{(\vartheta^{(h)}, \theta^{(h,j)})}$ according to (7.3.42) or (7.3.50)
Calculate $p^{(h,j)} := p^{(\vartheta^{(h)}, \theta^{(h,j)})}(\cdot \vert Z)$ according to (7.3.43) or (7.3.51)
$w^{(h,j)} := w^{(h)}[\eta_Z^{(h,j)}]^{I^{(h)}}$
$I^{(h,j)} := I^{(h)}$
$\vartheta^{(h,j)} := (\vartheta^{(h)}, \theta^{(h,j)})$
end
end
Normalize weights $\{w^{(h,j)}\}_{(h,j)=(1,1)}^{(H,T^{(h)})}$

is to always keep $T = J_{\max}$ maximum weight components of $\pi(\cdot|Z)$. However, in addition to increasing the dimensionality of the ranked assignmen problem by H times, this strategy also loses the parallelism.

As mentioned earlier, the actual value of the association history $\vartheta^{(h)}$ is not required in the prediction and update calculations and it is only used as an index for track density $p^{(\vartheta^{(h)})}$. Since the track density is equivalently indexed by h, i.e., $p^{(h)} \triangleq p^{(\vartheta^{(h)})}$, in fact, it is not necessary to transfer $\vartheta^{(h)}$. However, $\vartheta^{(h)}$ is still kept in the pseudocodes of the prediction and update for convenience of presentation.

7.3.2.3 Multi-target State Estimation

Given the multi-target filtering density, many multi-target state estimators can be used to obtain state estimates. There are two types of Bayesian optimal estimators [14, 392]. The first multi-target state estimator is called the marginal multi-target (MaM) estimator, which only considers the cardinality information in the FISST density. The second one is called the joint multi-target (JoM) estimator, which takes into account the cardinality information and spatial distribution information related to the multi-target state in the FISST density. Both estimators are the Bayesian optimal, that is, the corresponding Bayesian risk function is minimized [392]. The JoM estimator simultaneously minimizes the cardinality and spatial differences between the true RFS and its estimate, while the MaM estimator first minimizes the cardinality difference, and then extracts the MAP estimate according to the relevant FISST posterior probability density. Therefore, compared with the MaM estimator, the JoM estimator is more suitable to obtain the multi-target state estimation, especially when the cardinality estimation is related to the spatial information in the FISST probability density, e.g., the target state under low observable conditions [127].

However, although the JoM estimator is Bayesian optimal, it is difficult to calculate [14]. For the δ-GLMB density, a simple and intuitive multi-target estimator is the MB estimator, which chooses the set $L \subseteq \mathbb{L}$ of tracks or labels with existence probabilities above a certain threshold (existence probability of track ℓ is the sum of weights of all hypotheses containing track ℓ, i.e., $\sum_{(I,\vartheta)\in\mathcal{F}(\mathbb{L})\times\Xi} w^{(I,\vartheta)} 1_I(\ell)$ [69]), and then estimates the track states according to the MAP or EAP of density $p^{(\vartheta)}(\cdot,\ell)$, $\ell \in L$. Moreover, the tractable MaM estimator can also be obtained as follows [69]. First, find the MAP cardinality estimate according to the following cardinality distribution [69]

$$\rho(n) = \sum_{(I,\vartheta)\in\mathcal{F}_n(\mathbb{L})\times\Xi} w^{(I,\vartheta)} \tag{7.3.57}$$

where $\mathcal{F}_n(\mathbb{L})$ represents a class of finite subsets of $\mathbb{L}$ with n elements. Then, the labels and state means are found from the highest weighted components with the same cardinality as the MAP cardinality estimate. Table 7.4 gives the pseudocode for the multi-target state estimation.

Table 7.4 Multi-target state estimation

1:	• Input:$N_{\max}$, $\{(I^{(h,j)}, \vartheta^{(h,j)}, w^{(h,j)}, p^{(h,j)})\}_{(h,j)=(1,1)}^{(H,T^{(h)})}$
2:	• Output:$\hat{\mathrm{X}}$
3:	$\rho(n) := \sum_{h=1}^{H} \sum_{j=1}^{T^{(h)}} w^{(h,j)} \delta_n(\lvert I^{(h,j)} \rvert); \ n = 0, \ldots, N_{\max}$
4:	$\hat{N} := \arg\max \rho$
5:	$(\hat{h}, \hat{j}) := \arg\max_{(h,j)} w^{(h,j)} \delta_{\hat{N}}(\lvert I^{(h,j)} \rvert)$
6:	$\hat{\mathrm{X}} := \{(\hat{\boldsymbol{x}}, \ell) : \ell \in I^{(\hat{h},\hat{j})}, \hat{\boldsymbol{x}} = \int \boldsymbol{x} p^{(\hat{h},\hat{j})}(\boldsymbol{x}, \ell) d\boldsymbol{x}\}$

Table 7.5 δ-GLMB filter

1:	for $k = 1 : K$
2:	Prediction
3:	Update
4:	Multi-target state estimation
5:	end

The pseudocode in Table 7.5 summarizes the algorithm of the complete δ-GLMB filter.

7.4 Labeled Multi-Bernoulli (LMB) Filter

The multi-Bernoulli filter approximates the posterior as a multi-Bernoulli (MB) RFS and is an approximate version of the multi-target Bayesian filter. In order to estimate target tracks, the track label is assigned to each component by post-processing (see Sect. 6.3.3). Although the time prediction step of the MB filter is exact, the posterior is approximated by using probability generating functional (PGFl) twice during the process of measurement update, so that the MB filter will exhibit a significant cardinality bias problem in the environment of low signal-to-noise ratio. Whereas in the δ-GLMB filter, the δ-GLMB density is closed under the multi-target prediction and update operations [69] and provides the track information. Although the δ-GLMB filter is superior to the CPHD and MB filters, its computational burden is significantly increased. See [69, 70] for details.

The basic idea of the labeled MB (LMB) filter is to approximate the predicted and posterior multi-target densities as the LMB process, and propagate the LMB multi-target posterior density forward over time, which is an approximation of the δ-GLMB filter. With the help of the LMB RFS, the intuitive mathematical structure of the MB RFS can be called, but there is no disadvantage of the MB filter: it can't provide the track estimation formally and exhibits the cardinality bias problem. The reason is that the LMB filter doesn't approximate the multi-target PGFl but adopts the LMB approximation that can accurately match the first-order posterior moment, thus eliminating the cardinality bias problem. However, this tracking filter also suffers from the disadvantage that the number of posterior components grows exponentially. For the purpose of pruning the predicted and posterior densities, like the δ-GLMB filter, the K shortest path and ranked assignment algorithms can be used, respectively.

The LMB recursion is composed of such two steps as prediction and update. The exact time prediction step and approximate measurement update step of the LMB filter are introduced as follows, which will be used to describe the complete implementation of the LMB filter below.

7.4.1 LMB Prediction

As mentioned earlier, although the number of terms in the predicted GLMB expression grows exponentially, the GLMB (and δ-GLMB) density is closed in the prediction step [69]. Although the LMB is a special case of the GLMB (with only one term), whether the predicted LMB density is still a LMB density needs to be strictly confirmed. In fact, assume that both the prior and birth target distributions are the LMB distributions, i.e.,

$$\pi(\mathrm{X}) = \Delta(\mathrm{X}) w(\mathcal{L}(\mathrm{X})) p^{\mathrm{X}} \tag{7.4.1}$$

$$\pi_\gamma(\mathrm{X}) = \Delta(\mathrm{X}) w_\gamma(\mathcal{L}(\mathrm{X})) p_\gamma^{\mathrm{X}} \tag{7.4.2}$$

where $\Delta(\cdot)$ is the distinct label indicator (DLI) defined in (3.3.15), and

$$w(L) = \prod_{i \in \mathbb{L}} (1 - \varepsilon^{(i)}) \prod_{\ell \in L} \frac{1_{\mathbb{L}}(\ell) \varepsilon^{(\ell)}}{1 - \varepsilon^{(\ell)}} \tag{7.4.3}$$

$$w_\gamma(L) = \prod_{i \in \mathbb{B}} (1 - \varepsilon_\gamma^{(i)}) \prod_{\ell \in L} \frac{1_{\mathbb{B}}(\ell) \varepsilon_\gamma^{(\ell)}}{1 - \varepsilon_\gamma^{(\ell)}} \tag{7.4.4}$$

$$p(\boldsymbol{x}, \ell) = p^{(\ell)}(\boldsymbol{x}) \tag{7.4.5}$$

$$p_\gamma(\boldsymbol{x}, \ell) = p_\gamma^{(\ell)}(\boldsymbol{x}) \tag{7.4.6}$$

Then, according to Proposition 15, the prediction of the LMB is a GLMB with state space $\mathbb{L}$ and (finite) label space $\mathbb{L}_+ = \mathbb{B} \cup \mathbb{L}$ ($\mathbb{B} \cap \mathbb{L} = \emptyset$), i.e.,

$$\pi_+(\mathrm{X}_+) = \Delta(\mathrm{X}_+) w_+(\mathcal{L}(\mathrm{X}_+)) p_+^{\mathrm{X}_+} \tag{7.4.7}$$

where

$$w_+(I_+) = w_S(I_+ \cap \mathbb{L}) w_\gamma(I_+ \cap \mathbb{B}) \tag{7.4.8}$$

$$p_+(\boldsymbol{x}, \ell) = 1_{\mathbb{L}}(\ell) p_{+,S}(\boldsymbol{x}, \ell) + 1_{\mathbb{B}}(\ell) p_\gamma(\boldsymbol{x}, \ell) \tag{7.4.9}$$

$$p_{+,S}(\boldsymbol{x}, \ell) = \langle p_S(\cdot, \ell)\phi(\boldsymbol{x}|\cdot, \ell), p(\cdot, \ell)\rangle \big/ \eta_S(\ell) \tag{7.4.10}$$

$$\eta_S(\ell) = \langle p_S(\cdot, \ell), p(\cdot, \ell)\rangle \tag{7.4.11}$$

$$w_S(L) = \eta_S^L \sum_{I \supseteq L} [1 - \eta_S]^{I-L} w(I) \tag{7.4.12}$$

In the above equations, $p_S(\boldsymbol{x}, \ell)$ is the state-dependent survival probability, $\eta_S(\ell)$ is the survival probability of track ℓ, and $\phi(\boldsymbol{x}|\boldsymbol{\xi}, \ell)$ is the single-target transition density of track ℓ.

Note that $w_\gamma(L)$ is the weight of the LMB, however, the weight $w_+(I_+)$ of the predicted density in (7.4.8) doesn't seem to be the weight of the LMB, because $w_S(L)$ in (7.4.12) results from the sum over the superset of L instead of the product (7.4.3) over L. However, the sum of (7.4.12) can be decomposed into the product form and rewritten as the LMB weight by applying Lemma 6 in Appendix I, leading to the following proposition (see Appendix I for proof).

Proposition 19 *Suppose that the multi-target posterior density is an LMB with space-state $\mathbb{X}$, (infinite) label space $\mathbb{L}$, and parameter set $\pi = \{\varepsilon^{(\ell)}, p^{(\ell)}\}_{\ell \in \mathbb{L}}$, and the multi-target birth model is an LMB with state-space $\mathbb{X}$, (finite) label space $\mathbb{B}$, and parameter set $\pi_\gamma = \{\varepsilon_\gamma^{(\ell)}, p_\gamma^{(\ell)}\}_{\ell \in \mathbb{B}}$, then the multi-target predicted density is also an LMB with state-space $\mathbb{X}$, (finite) label space $\mathbb{L}_+ = \mathbb{B} \cup \mathbb{L}$ ($\mathbb{B} \cap \mathbb{L} = \emptyset$), and parameter set π_+, whose concrete expression is*

$$\pi_+ = \{\varepsilon_\gamma^{(\ell)}, p_\gamma^{(\ell)}\}_{\ell \in \mathbb{B}} \cup \{\varepsilon_{+,S}^{(\ell)}, p_{+,S}^{(\ell)}\}_{\ell \in \mathbb{L}} \tag{7.4.13}$$

where, the first LMB represents the LMB birth component, which can be specified by the prior. For the birth tracks, label $\ell \in \mathbb{B}$ is a new label that is different from each other. The second LMB represents the previous survival label Bernoulli track given by (7.4.14) and (7.4.15). For the survival tracks, the predicted label is the same as the previous label and the predicted existence probability and spatial distribution are re-weighted by the survival probability and the transition density, respectively, i.e.,

$$\varepsilon_{+,S}^{(\ell)} = \eta_S(\ell)\varepsilon^{(\ell)} \tag{7.4.14}$$

$$p_{+,S}^{(\ell)}(\boldsymbol{x}) = \langle p_S(\cdot, \ell)\phi(\boldsymbol{x}|\cdot, \ell), p(\cdot, \ell)\rangle \big/ \eta_S(\ell) \tag{7.4.15}$$

with $\eta_S(\ell)$ being given in (7.4.11).

Equation (7.4.13) shows that the predicted LMB is composed of the union of predicted surviving tracks and birth tracks. Compared with the GLMB prediction, the LMB prediction is less computationally expensive because it does not involve the summation over the subset of $\mathbb{L}$ in (7.2.3).

For the LMB, the multi-target prediction is actually consistent with the prediction of the (unlabeled or label-free) MB filter, and the component indices of the MB filter can be interpreted as track labels. As a result, in order to perform the prediction of the LMB filter, it is only required to predict parameters forward according to (7.4.13), which is the same as the prediction of the MB filter. This conclusion can be applied to its concrete implementation.

7.4.2 LMB Update

Although the LMB family is closed under the prediction step, it is no longer closed under the update operation. In other words, the multi-target posterior density $\pi(\cdot|Z)$ is generally no longer the LMB but the GLMB. Drawing on the MB filter [4], an LMB approximation that matches the first-order moment of the multi-target posterior density can be sought. Compared with the update step of the MB filter, which requires approximating the probability generating functional (PGFl) of the multi-target posterior twice, the important advantage of the LMB update is that, it doesn't involve the approximation of the posterior PGFl, but directly approximates the posterior multi-target density through the accurate moment matching, so only one approximation of the multi-target posterior density is involved. Therefore, in addition to the capability to generate target tracks, the performance of the LMB filter is also better than that of the MB filter.

Proposition 20 *Assuming that the multi-target predicted density is an LMB with state-space $\mathbb{X}$, (finite) label space $\mathbb{L}_+$, and parameter set $\pi_+ = \{\varepsilon_+^{(\ell)}, p_+^{(\ell)}\}_{\ell \in \mathbb{L}_+}$, the LMB that exactly matches the first-order moment of the multi-target posterior density is $\pi(\cdot|Z) = \{\varepsilon^{(\ell)}, p^{(\ell)}(\cdot)\}_{\ell \in \mathbb{L}_+}$, where*

$$\varepsilon^{(\ell)} = \sum_{(I_+,\theta)\in\mathcal{F}(\mathbb{L}_+)\times\Theta(I_+)} 1_{I_+}(\ell) w^{(I_+,\theta)}(Z) \tag{7.4.16}$$

$$p^{(\ell)}(\boldsymbol{x}) = \frac{1}{\varepsilon^{(\ell)}} \sum_{(I_+,\theta)\in\mathcal{F}(\mathbb{L}_+)\times\Theta(I_+)} 1_{I_+}(\ell) w^{(I_+,\theta)}(Z) p^{(\theta)}(\boldsymbol{x}, \ell|Z) \tag{7.4.17}$$

$$w^{(I_+,\theta)}(Z) \propto [\eta_Z^{(\theta)}]^{I_+} w_+(I_+) \tag{7.4.18}$$

$$\begin{aligned} w_+(I_+) &= \prod_{\ell \in \mathbb{L}_+} (1 - \varepsilon_+^{(\ell)}) \prod_{\ell' \in I_+} \frac{1_{\mathbb{L}_+}(\ell')\varepsilon_+^{(\ell')}}{1 - \varepsilon_+^{(\ell')}} \\ &= \prod_{\ell' \in \mathbb{L}_+ - I_+} (1 - \varepsilon_+^{(\ell')}) \prod_{\ell \in I_+} 1_{\mathbb{L}}(\ell)\varepsilon_+^{(\ell)} \end{aligned} \tag{7.4.19}$$

$$p^{(\theta)}(\boldsymbol{x}, \ell|Z) = p_+(\boldsymbol{x}, \ell)\varphi_Z(\boldsymbol{x}, \ell; \theta) \Big/ \eta_Z^{(\theta)}(\ell) \tag{7.4.20}$$

$$\eta_Z^{(\theta)}(\ell) = \langle p_+(\cdot, \ell), \varphi_Z(\cdot, \ell; \theta)\rangle \tag{7.4.21}$$

in which, $\Theta(I_+)$ is the space of map $\theta : I_+ \to \{0, 1, \ldots, |Z|\}$, which satisfies $\ell = \ell'$ when $\theta(\ell) = \theta(\ell') > 0$, $\varphi_Z(\boldsymbol{x}, \ell; \theta)$ is defined by (3.4.21). The physical meaning of (7.4.17) is that summing up all the spatial distributions containing a single track label obtains the spatial distribution of the track.

In order to prove Proposition 20, the LMB predicted density is written as the following δ-GLMB form

$$\pi_+(\mathrm{X}) = \Delta(\mathrm{X}) \sum_{I_+ \in \mathcal{F}(\mathbb{L}_+)} \delta_{I_+}(\mathcal{L}(\mathrm{X}))w_+(I_+)p_+^{\mathrm{X}} \tag{7.4.22}$$

Using the conclusion in [69], the multi-target posterior is a δ-GLMB with state-space $\mathbb{X}$ and (finite) label space $\mathbb{L}_+$ (and $\Xi = \Theta(I_+)$), i.e.,

$$\pi(\mathrm{X}|Z) = \Delta(\mathrm{X}) \sum_{(I_+,\theta) \in \mathcal{F}(\mathbb{L}_+)\times\Theta(I_+)} \delta_{I_+}(\mathcal{L}(\mathrm{X}))w^{(I_+,\theta)}(Z)[p^{(\theta)}(\cdot|Z)]^{\mathrm{X}} \tag{7.4.23}$$

According to (3.3.57), the (label-free) probability hypothesis density (PHD) of the full posterior is

$$v(\boldsymbol{x}) = \sum_{(I_+,\theta) \in \mathcal{F}(\mathbb{L}_+)\times\Theta(I_+)} w^{(I_+,\theta)}(Z) \sum_{\ell \in I_+} p^{(\theta)}(\boldsymbol{x}, \ell|Z) \tag{7.4.24}$$

Equation (7.4.24) can be interpreted as the weighted sum of the PHDs of all independent tracks. Therefore, the full posterior (7.4.23) can be interpreted as a distribution of tracks weighted by their probability of existence. For an independent track label, the summation of all component weights gives the existence probability of that track. Similarly, summing over all spatial distributions containing an independent track label yields the spatial distribution of this independent track.

Substituting (7.4.16) and (7.4.17) into (3.3.11) yields the label-free PHD of the LMB approximation. Comparing the obtained result with the label-free PHD of the full posterior (7.4.24), it is easy to know that the two are the same. Therefore, from the perspective of the decomposition into independent tracks and label-free first order moment (i.e., the PHD), the above LMB approximation matches the original posterior. Since the PHD mass provides the average number of targets (mean cardinality), the mean cardinality of the LMB approximation is equal to that of the full posterior. However, the cardinality distributions are not the same, because the cardinality distribution of the LMB approximation follows that of the MB RFS (3.3.10), while the cardinality distribution of the full posterior is given by (3.3.59). Thus, although the LMB approximation matches the PHD (i.e., average cardinality) of the full GLMB distribution, it fails to match the full cardinality distribution. The reason is that the LMB model has smaller degrees of freedom than the GLMB one, and therefore, its cardinality distribution form is more limited. Since the LMB filter perform the approximation after each update, it leads to an accumulative error in the cardinality distribution, which makes the estimation error larger than that of the GLMB filter. However, the update step of the δ-GLMB filter needs to propagate a large number of multi-object exponential sums, while the update of the LMB filter only needs to propagate one component that approximates the multi-object exponential sum.

Although there are other possible options for approximating the posterior, the particular approximation presented above can be intuitively interpreted. From the perspective of preserving the spatial density to be estimated for each track and accurately matching the first-order moment, this option is the best approximation of the original distribution.

Based on the above LMB update, the concrete implementation step is given as follows.

(1) Represent the predicted LMB as δ-GLMB

Since the predicted multi-target density is the LMB, the predicted density needs to be expressed in the δ-GLMB form in order to perform the measurement update. For track label set $\mathbb{L}_+$, the predicted δ-GLMB $\pi_+(\mathrm{X}) = \Delta(\mathrm{X}) \sum_{I_+ \in \mathcal{F}(\mathbb{L}_+)} \delta_{I_+}(\mathcal{L}(\mathrm{X})) w_+(I_+) p_+^{\mathrm{X}}$ is given by (7.4.22), where $w_+(I_+)$ is obtained from (7.4.19). As a result, the predicted δ-GLMB is determined separately by each predicted component or track.

A brute force method to enumerate the summation term in (7.4.22) is to generate all possible combinations of label set $\mathbb{L}_+$ and cardinality $n = 0, 1, \ldots, |\mathbb{L}_+|$. The number of combinations for each cardinality is determined by the binomial coefficient $C(|\mathbb{L}_+|, n) = |\mathbb{L}_+|! / (n!(|\mathbb{L}_+| - n)!)$, while the number of combinations for the track label set is $2^{|\mathbb{L}_+|}$. Therefore, the explicit computation of all combinations is only feasible when $|\mathbb{L}_+|$ is small. For larger $|\mathbb{L}_+|$, the summation can be approximate to its K most significant terms by utilizing the K shortest path algorithm, thus eliminating the need to enumerate all possible terms. Therefore, I_+ contains only the most important hypotheses. Another solution to enumerating and producing the K most significant terms is to utilize the (random) sampling method, which may be computationally faster for larger number of targets or components. In this method, by sampling an independent identically distributed (IID) random number $a^{(\ell)} \sim \mathcal{U}([0, 1])$ from a uniform distribution $\mathcal{U}(\cdot)$, and testing the acceptance of each labeled Bernoulli track according to $I_+ = \{\ell | a^{(\ell)} < \varepsilon_+^{(\ell)}, \forall \ell \in \mathbb{L}_+\}$, the desired number of unique label samples I_+ are obtained.

(2) δ-GLMB Update

After the measurements Z are obtained, the δ-GLMB is updated according to (7.4.23). Because of the combinatorial nature of the update, the number of components or hypotheses increases exponentially with the number $|\mathbb{L}_+|$ of track labels. Therefore, for larger $|\mathbb{L}_+|$, it is necessary to prune the posterior distribution (7.4.23), which can be achieved using the ranked assignment algorithm. This algorithm only needs to compute the M most significant hypotheses, thus eliminating the need to compute all possible solutions [69, 70].

(C) Approximate the updated δ-GLMB to the LMB

After the measurement update, it is also necessary to convert the δ-GLMB form back to the LMB form

$$\pi(\cdot|Z) \approx \{(\varepsilon^{(\ell)}, p^{(\ell)})\}_{\ell\in\mathbb{L}_+} \tag{7.4.25}$$

where $\varepsilon^{(\ell)}$ and $p^{(\ell)}(\boldsymbol{x})$ are obtained from (7.4.16) and (7.4.17), respectively.

7.4.3 Multi-target State Extraction

Since the updated track is represented by the LMB, track pruning can be realized by deleting the tracks whose existence probabilities are lower than a specified smaller threshold, and then the multi-target state can be extracted by following the steps listed in Table 7.6.

Another solution to track extraction is to pick out all tracks whose existence probabilities exceed a given larger threshold, i.e.,

$$\hat{\mathrm{X}} = \{(\hat{\boldsymbol{x}}, \ell) : \varepsilon^{(\ell)} > \varepsilon_{\mathrm{Th}}\} \tag{7.4.26}$$

where $\hat{\boldsymbol{x}} = \arg_{\boldsymbol{x}} \max p^{(\ell)}(\boldsymbol{x})$. On the one hand, choosing a higher $\varepsilon_{\mathrm{Th}}$ will significantly reduce the number of cluttered tracks at the expense of delayed detection of birth tracks. On the other hand, a lower $\varepsilon_{\mathrm{Th}}$ can quickly report birth tracks, but at the cost of producing more cluttered tracks. Moreover, when a higher $\varepsilon_{\mathrm{Th}}$ is selected, attention shall also be paid to the problem of missed detections. In the case of $p_D \approx 1$, missed detections significantly reduce the existence probability, perhaps thereby suppressing the output of the previous confirmed tracks with $\varepsilon^{(\ell)} \approx 1$.

Table 7.6 LMB estimation extraction

1:	• Input: $n_{\max}, \pi = \{(\varepsilon^{(\ell)}, p^{(\ell)})\}_{\ell\in\mathbb{L}_-}$
2:	• Output: $\hat{\mathrm{X}}$
3:	$\rho(n) = \sum_{I\in\mathcal{F}(\mathbb{L}),\, |I|=n} w(I), \quad n = 1, \ldots, n_{\max}$
4:	$\hat{n} = \arg_n \max \rho(n), \widehat{\mathbb{L}} = \emptyset$
5:	$\hat{\mathbb{L}} = \hat{\mathbb{L}} \cup \arg\max_{\ell\in\mathbb{L}\setminus\hat{\mathbb{L}}} \varepsilon^{(\ell)}, \quad n = 1, \ldots, \hat{n}$
6:	$\hat{\mathrm{X}} := \{(\hat{\boldsymbol{x}}, \hat{\ell}) : \hat{\ell} \in \hat{\mathbb{L}}, \quad \hat{\boldsymbol{x}} = \arg_{\boldsymbol{x}} \max p^{(\hat{\ell})}(\boldsymbol{x})\}$

In order to alleviate this problem, the following hysteresis mechanism can be adopted: only if the maximum existence probability $\varepsilon_{\max}^{(\ell)}$ of the track exceeds a certain larger threshold ε_{U} and the current existence probability $\varepsilon^{(\ell)}$ is greater than a certain lower threshold ε_{L}, then the track is output, i.e.,

$$\hat{\mathrm{X}} = \{(\hat{\boldsymbol{x}}, \ell) : \varepsilon_{\max}^{(\ell)} > \varepsilon_{\mathrm{U}} \text{ and } \varepsilon^{(\ell)} > \varepsilon_{\mathrm{L}}\} \tag{7.4.27}$$

A more efficient and practical implementation of the LMB filter considering grouping and adaptive birth distribution can be found in [152].

7.5 Marginalized δ-GLMB (Mδ-GLMB) Filter

The marginalized δ-GLMB (Mδ-GLMB) filter is based on an approximation of the GLMB proposed in [152], which is called the Mδ-GLMB filter because the result can be interpreted as performing the marginalization on the association history. There are two important conclusions regarding this filter, i.e., it is an efficient approximation of the Bayesian optimal δ-GLMB filter, easy to perform the multi-sensor update, and the Mδ-GLMB approximation exactly matches the (labeled) PHD and cardinality distribution of the δ-GLMB filtering density.

7.5.1 Mδ-GLMB Approximation

As mentioned earlier, one of the major factors influencing the computational complexity of the δ-GLMB filter [70] is that, updating the prior generates an explicit sum over association history variables, whereby the number of hypotheses grows exponentially. Furthermore, in the multi-sensor scenario, the number of association histories will further increase due to the continuous update steps. The basic idea of the Mδ-GLMB filter is to establish a principled approximation $\hat{\pi}(\cdot)$ of the GLMB posterior density $\pi(\cdot)$, which results in the marginalization on the association histories, thus significantly reducing the number of components required to represent the posterior (or filtering) density.

Definition 4: Given a labeled multi-target density π defined on $\mathcal{F}(\mathbb{X} \times \mathbb{L})$, for any positive integer n, the joint existence probability of the label set $\{\ell_1, \ldots, \ell_n\}$ and the joint probability density of $\boldsymbol{x}_1, \ldots, \boldsymbol{x}_n$ on $\mathbb{X}^n$ conditioned on $\{\ell_1, \ldots, \ell_n\}$ are respectively defined as

$$w(\{\ell_1, \ldots, \ell_n\}) \triangleq \int_{\mathbb{X}^n} \pi(\{(\boldsymbol{x}_1, \ell_1), \ldots, (\boldsymbol{x}_n, \ell_n)\}) d(\boldsymbol{x}_1, \ldots, \boldsymbol{x}_n) \tag{7.5.1}$$

$$p(\{(\boldsymbol{x}_1, \ell_1), \ldots, (\boldsymbol{x}_n, \ell_n)\}) \triangleq \frac{\pi(\{(\boldsymbol{x}_1, \ell_1), \ldots, (\boldsymbol{x}_n, \ell_n)\})}{w(\{\ell_1, \ldots, \ell_n\})} \tag{7.5.2}$$

For $n = 0$, $w(\theta) = \pi(\theta)$ and $p(\theta) = 1$ are specified. By default, as long as $w(\mathcal{L}(\mathrm{X}))$ is 0, $p(\mathrm{X})$ is equal to 0. Thus, the labeled multi-target density can be expressed as

$$\pi(\mathrm{X}) = w(\mathcal{L}(\mathrm{X}))p(\mathrm{X}) \tag{7.5.3}$$

Note that $\sum_{L\in\mathcal{F}(\mathbb{L})} w(L) = 1$. Moreover, according to Lemma 7, since π is symmetric with respect to its parameters, and thus, $w(\cdot)$ is also symmetric with respect to $\ell_1, \ldots, \ell_n$. Therefore, $w(\cdot)$ is actually a probability distribution over $\mathcal{F}(\mathbb{L})$.

The Mδ-GLMB density is a tractable approximation of the GLMB density, which is used for approximating the arbitrary labeled multi-target density π given by (7.5.3). The Mδ-GLMB density is calculated numerically via the δ-GLMB form, which involves the explicit enumeration of the label set. Due to the lack of general information on the form of π, a natural option is the δ-GLMB class of the form

$$\hat{\pi}(\mathrm{X}) = \Delta(\mathrm{X}) \sum_{I\in\mathcal{F}(\mathbb{L})} \delta_I(\mathcal{L}(\mathrm{X}))\hat{w}^{(I)}[\hat{p}^{(I)}]^{\mathrm{X}} \tag{7.5.4}$$

where each $\hat{p}^{(I)}(\cdot, \ell)$ is the density on $\mathbb{X}$ and each non-negative weight $\hat{w}^{(I)}$ satisfies $\sum_{I\subseteq\mathbb{L}} \hat{w}^{(I)} = 1$. In fact, the above δ-GLMB class has better properties.

Proposition 21 *(see Appendix J for proof): given an arbitrary labeled multi-target density π (7.5.3), the δ-GLMB density that preserves the cardinality distribution and PHD of π and has the minimum Kullback–Leibler divergence (KLD) of π is given by* (7.5.4)*, where, the parameters of $\hat{\pi}$ and π have the following relationships*

$$\hat{w}^{(I)} = w(I) \tag{7.5.5}$$

$$\hat{p}^{(I)}(\boldsymbol{x}, \ell) = p_{I-\{\ell\}}(\boldsymbol{x}, \ell) \tag{7.5.6}$$

where $w(I)$ is the joint existence probability defined by (7.5.1), and

$$p_{\{\ell_1,\ldots,\ell_n\}}(\boldsymbol{x}, \ell) = \int p(\{(\boldsymbol{x}, \ell), (\boldsymbol{x}_1, \ell_1), \ldots, (\boldsymbol{x}_n, \ell_n)\})d(\boldsymbol{x}_1, \ldots, \boldsymbol{x}_n) \tag{7.5.7}$$

with $p(\mathrm{X})$ being the joint probability density defined by (7.5.2).

Note that according to the definition of $\hat{p}^{(I)}(\boldsymbol{x}, \ell)$ in (7.5.6), we have

$$\hat{p}^{(\{\ell,\ell_1,\ldots,\ell_n\})}(\boldsymbol{x}, \ell) = \int p(\{(\boldsymbol{x}, \ell), (\boldsymbol{x}_1, \ell_1), \ldots, (\boldsymbol{x}_n, \ell_n)\})d(\boldsymbol{x}_1, \ldots, \boldsymbol{x}_n) \tag{7.5.8}$$

Therefore, $\hat{p}^{(\{\ell_1,\ldots,\ell_n\})}(\cdot, \ell_i),\ \ i = 1, \ldots, n$ defined by (7.5.6) is the marginalized probability of the label-conditioned joint density $p(\{(\cdot, \ell_1), \ldots, (\cdot, \ell_n)\})$ of π given in (7.5.2). As a result, the δ-GLMB density of the form (7.5.4) is referred to as the marginalized δ-GLMB (Mδ-GLMB) density. Note that this δ-GLMB is fully characterized by the parameter set $\{\hat{w}^{(I)}, \hat{p}^{(I)}\}_{I\in\mathcal{F}(\mathbb{L})}$.

Proposition 21 states that, replacing the label-conditioned joint density $p(\{(\cdot,\ell_1),\ldots,(\cdot,\ell_n)\})$ of the labeled multi-target density π with the product of their marginal probabilities $\hat{p}^{(\{\ell_1,\ldots,\ell_n\})}(\cdot,\ell_i)$ yields the δ-GLMB of the form (7.5.4), which matches the PHD and cardinality distribution of π and has the minimum KLD with π.

The above strategy for matching the PHD and cardinality distribution borrows from Mahler's IIDC approximation strategy in the CPHD filter. This strategy can be easily extended to the arbitrary labeled multi-target density (also known as labeled RFS mixture density) of the form

$$\pi(\mathrm{X}) = \Delta(\mathrm{X})\sum_{c\in\mathbb{C}} w^{(c)}(\mathcal{L}(\mathrm{X}))p^{(c)}(\mathrm{X}) \tag{7.5.9}$$

where $p^{(c)}(\mathrm{X})$ is symmetric with respect to the elements of X, $p^{(c)}(\{(\cdot,\ell_1),\ldots,(\cdot,\ell_n)\})$ is the joint PDF on $\mathbb{X}^n$, and weight $w^{(c)}(\cdot)$ and density $p^{(c)}(\cdot)$ satisfy the following relations, respectively

$$\sum_{L\subseteq\mathbb{L}}\sum_{c\in\mathbb{C}} w^{(c)}(L) = 1 \tag{7.5.10}$$

$$\int p^{(c)}(\{(\boldsymbol{x}_1,\ell_1),\ldots,(\boldsymbol{x}_n,\ell_n)\})d(\boldsymbol{x}_1,\ldots,\boldsymbol{x}_n) = 1 \tag{7.5.11}$$

However, for such a quite general of the form (7.5.9), it is difficult to draw a conclusion about the KLD. Nevertheless, following the proof process of Proposition 21, it is easy to obtain the following proposition.

Proposition 22 *Given the arbitrary labeled multi-target density of the form* (7.5.9), *the* δ*-GLMB that preserves the cardinality distribution and the PHD of* π *is*

$$\hat{\pi}(\mathrm{X}) = \Delta(\mathrm{X})\sum_{(I,c)\in\mathcal{F}(\mathbb{L})\times\mathbb{C}} \delta_I(\mathcal{L}(\mathrm{X}))\hat{w}^{(I,c)}[\hat{p}^{(I,c)}]^{\mathrm{X}} \tag{7.5.12}$$

where

$$\hat{w}^{(I,c)} = w^{(c)}(I) \tag{7.5.13}$$

$$\hat{p}^{(I,c)}(\boldsymbol{x},\ell) = 1_I(\ell)p^{(c)}_{I-\{\ell\}}(\boldsymbol{x},\ell) \tag{7.5.14}$$

$$p^{(c)}_{\{\ell_1,\ldots,\ell_n\}}(\boldsymbol{x},\ell) = \int p^{(c)}(\{(\boldsymbol{x},\ell),(\boldsymbol{x}_1,\ell_1),\ldots,(\boldsymbol{x}_n,\ell_n)\})d(\boldsymbol{x}_1,\ldots,\boldsymbol{x}_n) \tag{7.5.15}$$

For the δ-GLMB density given in (3.3.55), the corresponding Mδ-GLMB density is given by the following proposition.

Proposition 23 *(see Appendix J for proof): the M δ-GLMB density $\pi(\cdot)$ that matches the PHD and cardinality distribution of the δ-GLMB density given in* (3.3.55) *is*

$$\pi(\mathrm{X}) = \Delta(\mathrm{X}) \sum_{I \in \mathcal{F}(\mathbb{L})} \delta_I(\mathcal{L}(\mathrm{X})) w^{(I)} [p^{(I)}]^{\mathrm{X}} \tag{7.5.16}$$

where

$$w^{(I)} = \sum_{\vartheta \in \Xi} w^{(I,\vartheta)} \tag{7.5.17}$$

$$p^{(I)}(\boldsymbol{x}, \ell) = \frac{1_I(\ell)}{\sum_{\vartheta \in \Xi} w^{(I,\vartheta)}} \sum_{\vartheta \in \Xi} w^{(I,\vartheta)} p^{(\vartheta)}(\boldsymbol{x}, \ell) = \frac{1_I(\ell)}{w^{(I)}} \sum_{\vartheta \in \Xi} w^{(I,\vartheta)} p^{(\vartheta)}(\boldsymbol{x}, \ell) \tag{7.5.18}$$

Note that the Mδ-GLMB density in (7.5.16) can also be rewritten into the following equivalent form

$$\pi(\mathrm{X}) = \Delta(\mathrm{X}) w^{(\mathcal{L}(\mathrm{X}))} [p^{(\mathcal{L}(\mathrm{X}))}]^{\mathrm{X}} \tag{7.5.19}$$

7.5.2 Mδ-GLMB Recursion

By using the δ-GLMB prediction equation for forward prediction and computing the Mδ-GLMB approximation after the δ-GLMB update, the resulting Mδ-GLMB density can be used to construct an efficient recursive multi-target tracking filter.

7.5.2.1 Mδ-GLMB Prediction

If the current multi-target prior density is the Mδ-GLMB of the form (7.5.16), according to the δ-GLMB prediction equation in Proposition 17, the multi-target predicted density is also an Mδ-GLMB, which is [329]

$$\pi_+(\mathrm{X}_+) = \Delta(\mathrm{X}_+) \sum_{I \in \mathcal{F}(\mathbb{L}_+)} \delta_I(\mathcal{L}(\mathrm{X}_+)) w_+^{(I)} [p_+^{(I)}]^{\mathrm{X}_+} \tag{7.5.20}$$

where

$$w_+^{(I)} = w_\gamma(I \cap \mathbb{B}) w_S^{(I)}(I \cap \mathbb{L}) \tag{7.5.21}$$

$$\begin{aligned} w_S^{(I)}(L) &= [\eta_S^{(I)}]^L \sum_{J \subseteq \mathbb{L}} 1_J(L)[1-\eta_S^{(I)}]^{J-L} w^{(J)} \\ &= [\eta_S^{(I)}]^L \sum_{J \supseteq L} [1-\eta_S^{(I)}]^{J-L} w^{(J)} \end{aligned} \tag{7.5.22}$$

$$p_+^{(I)}(\boldsymbol{x}, \ell) = 1_{\mathbb{L}}(\ell) p_{+,S}^{(I)}(\boldsymbol{x}, \ell) + 1_{\mathbb{B}}(\ell) p_\gamma(\boldsymbol{x}, \ell) \tag{7.5.23}$$

$$p_{+,S}^{(I)}(\boldsymbol{x}, \ell) = \langle p_S(\cdot, \ell)\phi(\boldsymbol{x}|\cdot, \ell), p^{(I)}(\cdot, \ell)\rangle \Big/ \eta_S^{(I)}(\ell) \tag{7.5.24}$$

$$\eta_S^{(I)}(\ell) = \langle p_S(\cdot, \ell), p^{(I)}(\cdot, \ell)\rangle \tag{7.5.25}$$

In the above equations, $p_S(\boldsymbol{x}, \ell)$ is the state-dependent survival probability, $\phi(\boldsymbol{x}|\boldsymbol{\xi}, \ell)$ is the single-target transition density of track ℓ, and w_γ and $p_\gamma^{(\ell)}$ are the parameters related to the birth density.

Equations (7.5.20)–(7.5.25) explicitly describe the process of calculating the parameters of the predicted multi-target density according to the parameters of the multi-target density at the previous time [70]. In essence, these equations have a corresponding relationship with the δ-GLMB prediction (7.3.1). Nevertheless, due to the marginalization operation of (7.5.17)–(7.5.18), the previous association history is not involved here (i.e., $\Xi = \theta$), and the superscript (ϑ) is replaced by (I_+).

Note: the number of components $(w_+^{(I)}, p_+^{(I)})$ obtained from the Mδ-GLMB prediction (7.5.20) is $|\mathcal{F}(\mathbb{L}_+)|$, while for components $(\omega_+^{(I_+,\vartheta)}, p_+^{(\vartheta)})$ obtained from the δ-GLMB prediction (7.3.1), the numbers of $\omega_+^{(I_+,\vartheta)}$ and $p_+^{(\vartheta)}$ are $|\mathcal{F}(\mathbb{L}_+) \times \Xi|$ and $|\Xi|$, respectively. As a result, the number of weights $w_+^{(I)}$ of the Mδ-GLMB is significantly lower than that of $\omega_+^{(I_+,\vartheta)}$ of the δ-GLMB. In addition, due to the growth of association history $\vartheta \in \Xi$, the growth rate of the number of the δ-GLMB predicted components $p_+^{(\vartheta)}$ has a super-exponential relationship with time [69, 70], while that of the number $|\mathcal{F}(\mathbb{L}_+)|$ of the Mδ-GLMB predicted components $p_+^{(I)}$ is well constrained.

7.5.2.2 Mδ-GLMB Update

Assume that the multi-target predicted density is the Mδ-GLMB of the form (7.5.16), under the action of the likelihood function defined by (3.4.19), the multi-target posterior density is usually no longer the Mδ-GLMB but the δ-GLMB as follows

$$\pi(\mathrm{X}|Z) = \Delta(\mathrm{X}) \sum_{(I) \in \mathcal{F}(\mathbb{L})} \sum_{\theta \in \Theta(I)} w^{(I,\theta)}(Z) \delta_I(\mathcal{L}(\mathrm{X}))[p^{(I,\theta)}(\cdot|Z)]^{\mathrm{X}} \tag{7.5.26}$$

where

$$w^{(I,\theta)}(Z) \propto [\eta_Z^{(I,\theta)}]^I w^{(I)} \tag{7.5.27}$$

$$p^{(I,\theta)}(\boldsymbol{x},\ell|Z) = p^{(I)}(\boldsymbol{x},\ell)\varphi_Z(\boldsymbol{x},\ell;\theta)\Big/\eta_Z^{(I,\theta)}(\ell) \tag{7.5.28}$$

$$\eta_Z^{(I,\theta)}(\ell) = \langle p^{(I)}(\cdot,\ell), \varphi_Z(\cdot,\ell;\theta)\rangle \tag{7.5.29}$$

in which, $\Theta(I)$ is the space of map $\theta : I \to \{0, 1, \ldots, |Z|\}$ satisfying $\ell = \ell'$ when $\theta(\ell) = \theta(\ell') > 0$, $\varphi_Z(\boldsymbol{x},\ell;\theta)$ is defined by (3.4.21), and $p_D(\boldsymbol{x},\ell)$ is the detection probability at $(\boldsymbol{x},\ell)$.

According to (7.5.17)–(7.5.18), the Mδ-GLMB density corresponding to the δ-GLMB density in (7.5.26) is a probability density of the form (7.5.16). In this case,

$$w^{(I)} = \sum_{\theta\in\Theta(I)} w^{(I,\theta)}(Z) \tag{7.5.30}$$

$$p^{(I)}(\boldsymbol{x},\ell) = \frac{1_I(\ell)}{w^{(I)}} \sum_{\theta\in\Theta(I)} w^{(I,\theta)}(Z)p^{(I,\theta)}(\boldsymbol{x},\ell|Z) \tag{7.5.31}$$

The Mδ-GLMB density provided by (7.5.30)–(7.5.31) preserves the PHD and cardinality distribution of the original δ-GLMB density (7.5.26).

Remark 6: for the δ-GLMB posterior, each hypothesis $I \in \mathcal{F}(\mathbb{L})$ generates $|\Theta(I)|$ new measurement-track association maps. For hypothesis $(w^{(I,\vartheta,\theta)}, p^{(\vartheta,\theta)})$ after the δ-GLMB update, the numbers of $w^{(I,\vartheta,\theta)}$ and $p^{(\vartheta,\theta)}$ are respectively $|\mathcal{F}(\mathbb{L})\times\Xi|\times\sum_{I\in\mathcal{F}(\mathbb{L})}|\Theta(I)|$ and $|\Xi|\cdot\sum_{I\in\mathcal{F}(\mathbb{L})}|\Theta(I)|$, while the number of components $(w^{(I,\theta)}, p^{(I,\theta)})$ stored/calculated after performing the δ-GLMB update (7.5.26) is only $|\mathcal{F}(\mathbb{L})| \times \sum_{I\in\mathcal{F}(\mathbb{L})}|\Theta(I)|$. Further, after the marginalization step (7.5.30)–(7.5.31), since all new contributions from association map $\Theta(I)$ are accumulated in a single component, the number of hypotheses is only $|\mathcal{F}(\mathbb{L})|$. Note that $|\mathcal{F}(\mathbb{L})|$ is the number of hypotheses generated by the prediction step (7.5.20). Therefore, the prediction step determines the upper bound on the total number of hypotheses retained in each full Mδ-GLMB update step.

Therefore, the number of hypotheses remaining after the Mδ-GLMB update is always kept at $|\mathcal{F}(\mathbb{L})|$, while the number of hypotheses for the δ-GLMB increases exponentially. From the perspective of information storage and computational burden, it is easy to know that the Mδ-GLMB is more popular than the δ-GLMB. This advantage is more prominent especially in the multi-sensor fusion scenario, because the number $|\mathcal{F}(\mathbb{L})|$ of hypotheses after the Mδ-GLMB update has nothing to do with the number of measurements collected by multiple sensors, while the δ-GLMB is seriously influenced by the number of measurements (sensors). Compared with the δ-GLMB, the Mδ-GLMB greatly reduces the need for pruning hypotheses and obtains a principled approximation. In particular, in the multi-sensor scenario with low signal-to-noise ratio (such as high clutter intensity, low detection probability, etc.) and limited storage/computing power, the pruning operation of the δ-GLMB

Table 7.7 Mδ-GLMB estimate extraction

1:	• Input: $n_{\max}$, π
2:	• Output: $\hat{X}$
3:	$\rho(n) = \sum_{I \in \mathcal{F}(\mathbb{L}),\ \lvert I \rvert = n} w^{(I)},\ n = 1, \ldots, n_{\max}$
4:	$\hat{n} = \arg_n \max \rho(n)$
5:	$\hat{I} = \arg_{I \in \mathcal{F}_{\hat{n}}(\mathbb{L})} \max w^{(I)}$
6:	$\hat{X} := \{(\hat{x}, \hat{\ell}) : \hat{\ell} \in \hat{I}, \hat{x} = \arg_x \max p^{(\hat{I})}(x, \hat{\ell})\}$

may lead to a poor performance. This is because, if some sensors fail to detect one or more targets, the hypotheses related to the true tracks may be deleted due to the pruning.

In conclusion, the Mδ-GLMB approximation significantly reduces the number of hypotheses in the posterior density, while still preserves the characteristics of the posterior PHD and cardinality distribution [152]. Furthermore, the Mδ-GLMB is well suitable for efficient and treatable information fusion (e.g., multi-sensor processing).

After performing the Mδ-GLMB update, the multi-target state estimation can be extracted by following the steps listed in Table 7.7.

The concrete implementation of the Mδ-GLMB can directly adopt the implementation method of the δ-GLMB filter. Specifically, for the linear Gaussian multi-target model, it is assumed that: (1) the single-target transition density, likelihood and birth intensity obey the Gaussian distribution; (2) the survival probability and detection probability are state independent; (3) and each single-target density can be expressed as a Gaussian mixture. Using the standard Gaussian mixture prediction and update equations of the Kalman filter, the corresponding Gaussian mixture predicted and updated densities can be calculated. In the case of the non-linear single-target transition density or likelihood, the well-known extended or unscented Kalman filter can be resorted to. On the other hand, for the non-linear non-Gaussian multi-target model (with state-dependent survival and detection probabilities), each single-target density can be represented as a set of weighted particles, and the corresponding predicted and updated densities are calculated based on the standard particle (or SMC) filter.

7.6 Simulation and Comparison

The following makes a comprehensive comparison among the PHD (Chap. 4), CPHD (Chap. 5), CBMeMBer (Chap. 6), δ-GLMB (Sect. 7.3, abbreviated as the GLMB in the subsequent simulation diagrams), single-step Gibbs implementation of the δ-GLMB joint prediction and update [146] (abbreviated as the JointGLMB in the follow-up simulation diagrams), LMB (Sect. 7.4), single-step Gibbs implementation of the LMB joint prediction and update (abbreviated as the JointLMB) in terms of the tracking performance (mainly reflected in such indicators as the CPEP (9.3.43), the mean and standard deviation of cardinality estimation, and OSPA) and operational efficiency (mainly reflected in the running time) under the linear Gaussian and nonlinear Gaussian conditions, respectively. The Matlab source code of each filter can be downloaded from Vo's personal website: http://ba-tuong.vo-au.com/codes.html.

7.6.1 Simulation and Comparison Under Linear Gaussian Case

Suppose that a sensor is located at the origin and obtains the positional measurements of both targets and clutters. The monitoring area of the sensor is $[-1000, 1000](\text{m}) \times [-1000, 1000](\text{m})$, with a sampling interval of $T_s = 1$. Within 100 sampling times of the sensor, a total of $N = 10$ targets appear, and each target moves in a straight line at a uniform speed. The initial state of each target and its start time and end time are shown in Table 7.8, and the true track of each target is shown in Fig. 7.4. The clutter obeys a uniform distribution in the monitoring area, and the number of clutters obeys a Poisson distribution with a mean of $\lambda_C = 30$.

In all filters, it is assumed that each target is born at such four fixed positions as (0, 0), (400, −600), (−800,−200), and (−200, 800), the velocity component of the birth target is set as 0, the corresponding covariance matrix is set to $\text{diag}([10^2, 10^2, 0^2, 0^2])$, and the weight is set as 0.03 for the Gaussian mixture birth distribution; For the LMB birth distribution, the existence probability is set as 0.03; the target survival probability is set to 0.99, the sensor detection probability is 0.98, the state transition matrix and the process noise covariance matrix are respectively

Table 7.8 Initial state of each target and its start time and end time

Target No.	Start time	End time	Initial state $[\,x_0\ y_0\ \dot{x}_0\ \dot{y}_0\,]^{\mathrm{T}}$
1	1	70	(0, 0, 0, − 10)
2	1	100	(400, − 600, − 10, 5)
3	1	70	(− 800, − 200, 20, − 5)
4	20	100	(400, − 600, − 7, − 4)
5	20	100	(400, − 600, − 2.5, 10)
6	20	100	(0, 0, 7.5, − 5)
7	40	100	(− 800, − 200, 12, 7)
8	40	100	(− 200, 800, 15, − 10)
9	60	100	(− 800, − 200, 3, 15)
10	60	100	(− 200, 800, − 3, − 15)

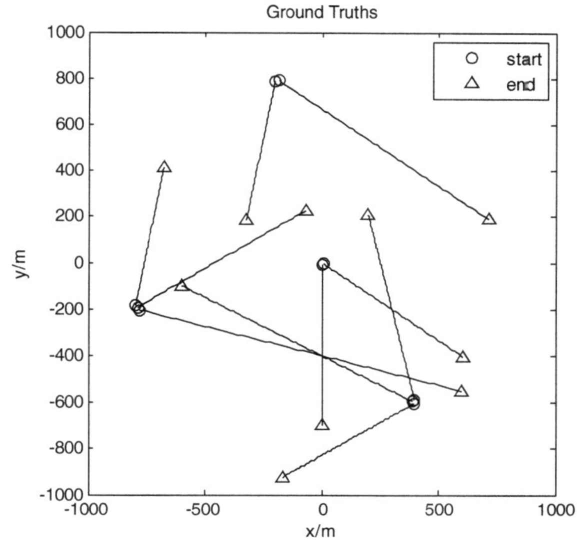

Fig. 7.4 True motion tracks of multiple targets

$$F = \begin{bmatrix} 1 & 0 & T_s & 0 \\ 0 & 0 & 1 & 0 \\ 0 & 1 & 0 & T_s \\ 0 & 0 & 0 & 1 \end{bmatrix}, \quad Q = 5^2 \begin{bmatrix} T_s^2/2 & 0 \\ 0 & T_s^2/2 \\ T_s & 0 \\ 0 & T_s \end{bmatrix} \begin{bmatrix} T_s^2/2 & 0 \\ 0 & T_s^2/2 \\ T_s & 0 \\ 0 & T_s \end{bmatrix}^{\mathrm{T}}$$

and the observation matrix and observation noise covariance matrix are respectively

$$H = \begin{bmatrix} 1 & 0 & 0 & 0 \\ 0 & 1 & 0 & 0 \end{bmatrix}, \quad R = \begin{bmatrix} 100 & 0 \\ 0 & 100 \end{bmatrix}$$

In the GM-PHD filter, the merging parameters used in the process of state extraction (see Table 4.2) include: weight threshold $T = 10^{-5}$, merging threshold $U = 4$m, and the maximum number of Gaussian components $J_{\max} = 100$.

In the GM-CPHD filter, the relevant parameters of state extraction are the same as those of the GM-PHD filter. Moreover, the capping parameter for the cardinality distribution is set to $N_{\max} = 100$ [64].

For subsequent CBMeMBer, LMB and δ-GLMB filters, in each hypothetical track, the maximum number of Gaussian components is 10, and the pruning and merging thresholds of Gaussian components are set as $T = 10^{-5}$ and $U = 4$m, respectively.

In the GM-CBMeMBer filter, the maximum number of hypothetical tracks is set as 100 and the pruning threshold for tracks is set as 10^{-3}.

In the GM-LMB filter, the relevant parameters are the same as those of the GM-CBMeMBer filter. In addition, in the course of conversion from the LMB to the GLMB before the update step, the numbers of corresponding birth hypotheses, survival hypotheses and hypotheses after the GLMB update are set as 5, 1000 and 1000, respectively.

In the GM-Joint-LMB filter, the relevant parameters are the same as those of the GM-CBMeMBer filter. Furthermore, the number of corresponding hypotheses after the GLMB update is set to 1000.

In the GM-δ-GLMB filter, the relevant parameters are the same as those of the GM-LMB filter. Moreover, the upper limit of the number of hypotheses is set as 1000 and the pruning threshold for hypotheses is 10^{-15}.

In the GM-Joint-δ-GLMB filter, the relevant parameters are the same as those of the GM-Joint-LMB filter. In addition, the pruning threshold for hypotheses is 10^{-15}.

The relevant parameters of performance indicators are set as follows: the CPEP radius is set as $r = 20$ m [see the definition of (9.3.43)], and the order parameter and threshold parameter of the OSPA are set as $p = 1$ and $c = 100$ m, respectively.

In the above simulation configuration, typical tracking results through a single run are shown in Fig. 7.5. Since the estimation results of the CPHD and CBMeMBer filters are similar to the result of the PHD filter, and the estimation results of the Joint-LMB, δ-GLMB and Joint-δ-GLMB filters are similar to the result of the LMB filter, only state estimation results of the PHD and LMB filters are shown in the figure to avoid repetition. As can be seen intuitively from the figure, these filters all give accurate state estimates, especially the label RFS filters, which also give correct track estimates due to the inclusion of label information.

In order to compare the advantages and disadvantages of each filter, the cardinality estimation, OSPA and CPEP performances are statistically calculated through 100 Monte Carlo simulations. As seen from Fig. 7.6a, the CBMeMBer filter has obvious cardinality overestimation, while the GLMB filter has a certain cardinality underestimation; though performances of the PHD and CPHD filters are relatively closed to each other, the CPHD filter is generally better than the PHD filter, but it has

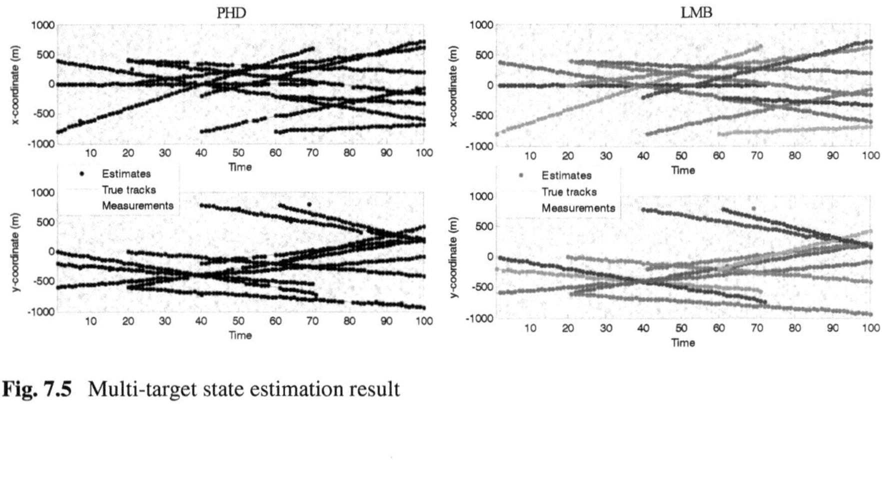

Fig. 7.5 Multi-target state estimation result

a slower response to the target birth process than the PHD filter. The performances of the LMB, JointLMB and JointGLMB filters are similar, and they can estimate the number of targets most accurately and give fast response to the target births (the PHD and CBMeMBer filters have the fastest response speed). The above phenomena show that the LMB, JointLMB and JointGLMB filters have the most accurate cardinality estimation and respond quickly to the birth targets; the CPHD filter is slightly less accurate and has a slower response to target births; Although the PHD filter responds quickly, its accuracy is poor, while the accuracies of the GLMB and CBMeMBer filters are relatively the worst.

In terms of the standard deviation of cardinality estimation, it can be seen from Fig. 7.6b that the PHD and CBMeMBer filters are relatively close and have the largest standard deviation, followed by the CPHD filter; the LMB, JointLMB and JointGLMB filters are close to each other and have the minimum standard deviation. However, the GLMB filter shows an unexpected behavior: its performance is close to that of the three filters when $k < 40$, but its standard deviation is significantly increased and even becomes worse than that of the CPHD filter since then, which is mainly due to the problem of missed target tracking of the GLMB filter (though the simulation shows that this filter is free from this problem most of the time), as seen from Fig. 7.6a, c, d. The above phenomenon shows that the LMB, JointLMB and JointGLMB filters are the most stable in terms of cardinality estimation, followed by the GLMB and CPHD filters, while the PHD and CBMeMBer filters are relatively fluctuating.

In terms of the OSPA, as seen from Fig. 7.6, the LMB, JointLMB and JointGLMB filters are the most accurate and have an OSPA distance (OSPA Dist) up to about 10 m; the GLMB filter is close to these three filters when $k < 40$ in terms of the OSPA Dist, but its OSPA Dist gradually increases subsequently and is kept around 14 m, close to that of the CPHD filter; while the PHD and CBMeMBer filters have the worst accuracy and their OSPA Dist is about 17 m.

In terms of the CPEP, since the CPEP measures the missed tracking performance, the smaller the value is, the better the performance is. The maximum CPEP value is 1, indicating that the filter cannot track the targets, while the minimum CPEP value is 0, implying that the filter can track the target completely. As seen from Fig. 7.6d, the LMB, JointLMB and JointGLMB filters are close to each other and have the best performance. The performances of the CBMeMBer, CPHD and PHD filters deteriorate in turn; and the GLMB filter is comparable to the LMB, JointLMB and JointGLMB filters when $k < 40$, after which its performance gradually degrades.

Figure 7.7 compares the computation time of each filter. In the figure, the vertical axis represents the 100 times Monte Carlo average of the total running time (100 sampling intervals). The average operation times of the PHD, CPHD, CBMeMBer, LMB, JointLMB, GLMB and JointGLMB filters are 0.58 s, 9.69 s, 0.87 s, 81.53 s, 11.60 s, 54.83 s and 14.63 s, respectively. It can be seen that the PHD and CBMeMBer filters have the lowest computational complexity, the LMB filter has the lowest operational efficiency, reaching an unacceptable 81.53 s, and the GLMB filter seems a slight decrease, but it is also difficult to meet the real-time performance. However, the

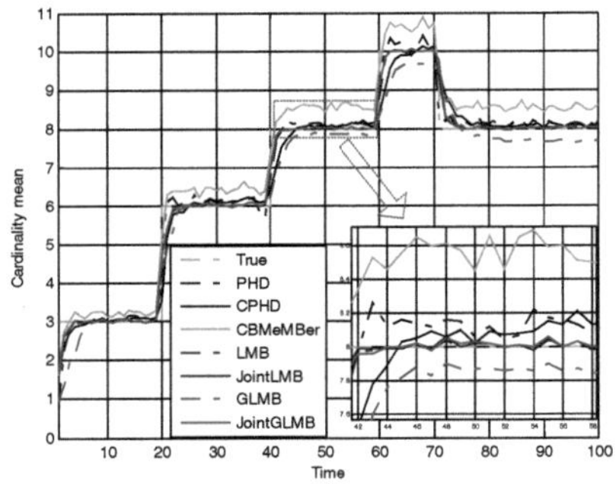

(a) Mean of cardinality estimates

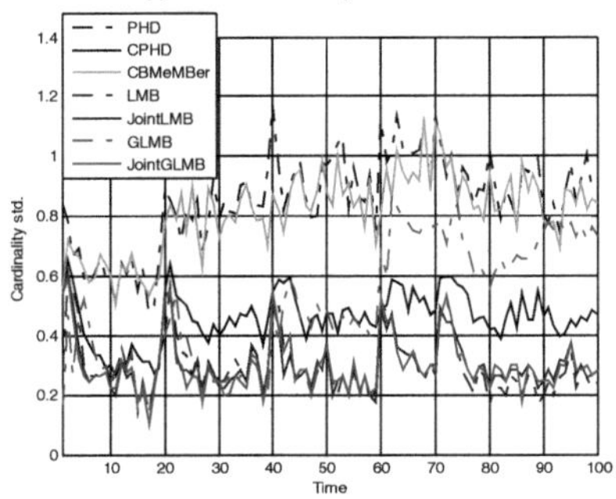

(b) Standard deviation of cardinality estimates (cardinality std.)

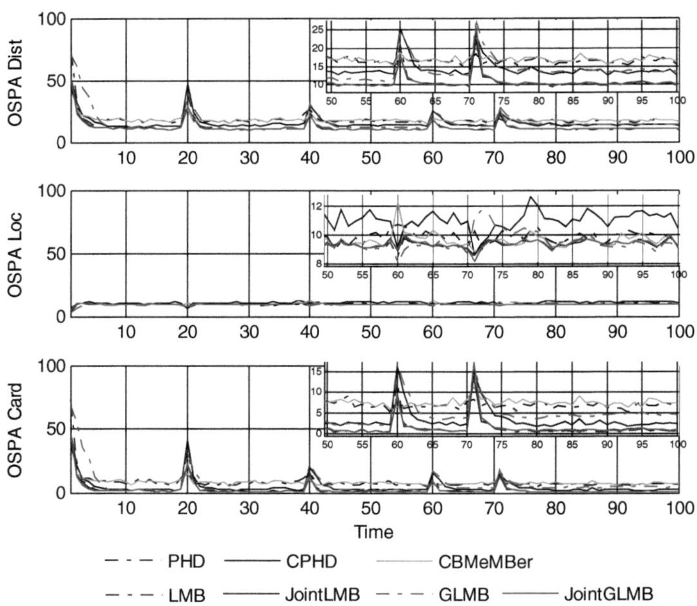

(c) OSPA

Fig. 7.6 Statistical performance of multi-target state estimation for each filter **a** Mean of cardinality estimates **b** standard deviation of cardinality estimates (cardinality std.) **c** OSPA **d** CPEP

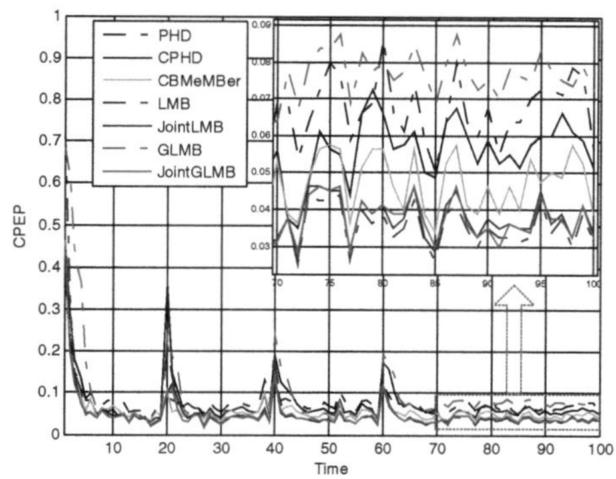

(d) CPEP

Fig. 7.6 (continued)

joint Gibbs implementation of the two filters significantly improves the operational efficiency and achieves comparable performance to the CPHD filter.

To sum up, under the linear Gaussian condition, the JointLMB and JointGLMB filters achieve the best performance in the tracking performance such as the estimation accuracy of the number of targets, stability and tracking accuracy at the cost of increasing a certain amount of computation. Although the PHD and CBMeMBer filters have the highest operational efficiency, their tracking performances are not satisfactory. The CPHD filter achieves a compromise between the operational efficiency and the tracking performance. Though the LMB filter is similar to the

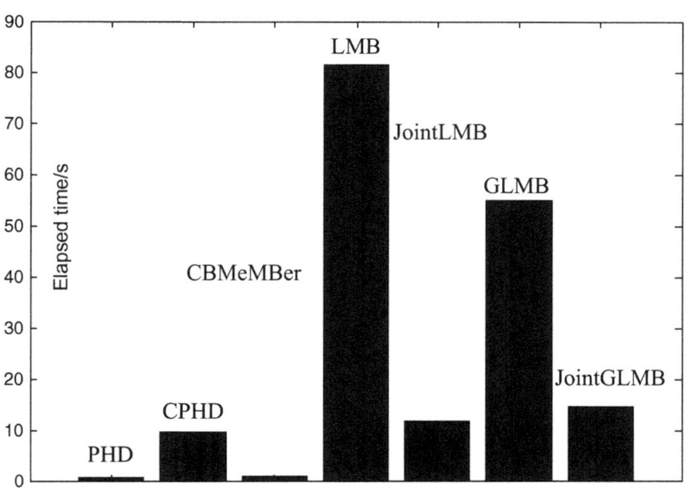

Fig. 7.7 Comparison of the running time of each filter

JointLMB and JointGLMB filters in terms of the tracking performance, its operational efficiency is too low. The GLMB filter is not only inefficient, but also has the problem of missed target tracking. Therefore, under the linear Gaussian condition, the JointLMB and JointGLMB filters are preferred.

7.6.2 *Simulation and Comparison Under Nonlinear Gaussian Condition*

Assume that a sensor is located at the origin, and obtains slant range r (m) and azimuth a (rad) of both targets and clutters. The monitoring area of the sensor is $r \in [0, 2000]$, $a \in [-\pi/2, \pi/2]$, and the sampling interval is $T_s = 1$. During the 100 sampling times of the sensor, there are a total of $N = 10$ targets, each of which makes a uniform turning motion with a turning angular velocity of ω (rad/s). See Table 7.9 for the initial state of each target and its starting time and end time, and see Fig. 7.8 for true target tracks. The clutter obeys a uniform distribution in the monitoring area, and the number of clutters obeys the Poisson distribution with mean $\lambda_C = 10$.

In each filter, it is assumed that the target is generated at such four fixed positions as (− 1500, 250), (− 250, 1000), (250, 750), and (1000, 1500), both the velocity component and the turning angular velocity of each birth target are set as 0, the corresponding covariance matrix is $\mathrm{diag}([50^2, 50^2, 50^2, 50^2, (\pi/3)^2])$, the weight for the Gaussian mixture birth distribution is set as 0.03, and the existence probability for the LMB birth distribution is set as 0.03. The target survival probability is 0.99, the sensor detection probability is set as $p_{D,k}(\boldsymbol{x}) = 0.98\mathcal{N}([x_k, y_k]^{\mathrm{T}}; \boldsymbol{0}, 2000^2\boldsymbol{I}_2)$, and the state transition matrix and process noise covariance matrix are respectively

Table 7.9 Initial state of each target and its starting time and end time

Target No.	Starting time	End time	Initial state $[\, x_0 \;\; y_0 \;\; \dot{x}_0 \;\; \dot{y}_0 \;\; \omega_0 \,]^{\mathrm{T}}$
1	1	100	(1000, 1500, − 10, − 10, $\pi/720$)
2	10	100	(− 250, 1000, 20, 3, $-\pi/270$)
3	10	100	(− 1500, 250, 11, 10, $-\pi/180$)
4	20	66	(− 1500, 250, 43, 0, 0)
5	20	80	(250, 750, 11, 5, $\pi/360$)
6	40	100	(− 250, 1000, − 12, − 12, $\pi/180$)
7	40	100	(1000, 1500, 0, − 10, $\pi/360$)
8	40	80	(250, 750, − 50, 0, $-\pi/360$)
9	60	100	(1000, 1500, − 50, 0, $-\pi/360$)
10	60	100	(250, 750, − 40, 25, $-\pi/360$)

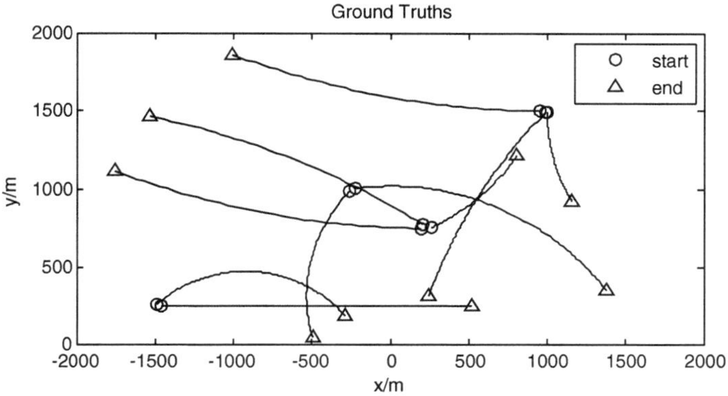

Fig. 7.8 True motion tracks of multiple targets

$$
\boldsymbol{F} = \begin{bmatrix} 1 & 0 & \sin(\omega T_s)/\omega & -(1-\cos(\omega T_s))/\omega & 0 \\ 0 & 0 & \cos(\omega T_s) & -\sin(\omega T_s) & 0 \\ 0 & 1 & (1-\cos(\omega T_s))/\omega & \sin(\omega T_s)/\omega & 0 \\ 0 & 0 & \sin(\omega T_s)/\omega & \cos(\omega T_s) & 0 \\ 0 & 0 & 0 & 0 & 1 \end{bmatrix},
$$

$$
\boldsymbol{Q} = \sigma_v^2 \begin{bmatrix} T_s^2/2 & 0 & 0 \\ 0 & T_s^2/2 & 0 \\ T_s & 0 & 0 \\ 0 & T_s & 0 \\ 0 & 0 & \sigma_\omega T_s/\sigma_v \end{bmatrix} \begin{bmatrix} T_s^2/2 & 0 & 0 \\ 0 & T_s^2/2 & 0 \\ T_s & 0 & 0 \\ 0 & T_s & 0 \\ 0 & 0 & \sigma_\omega T_s/\sigma_v \end{bmatrix}^{\mathrm{T}}
$$

where $\sigma_v = 5$ m/s^2 and $\sigma_\omega = \pi/180$ rad/s are the standard deviations of the process noise and turning angular velocity, respectively. The nonlinear observation function is

$$
z_k = \begin{bmatrix} \sqrt{x_k^2 + y_k^2} \\ \operatorname{atan2}(y_k, x_k) \end{bmatrix} + \boldsymbol{n}_k
$$

where $\boldsymbol{n}_k$ is the zero mean Gaussian noise, and the corresponding observation noise covariance matrix is $\boldsymbol{R} = \operatorname{diag}([10^2, (\pi/90)^2])$.

In the SMC-PHD filter, the maximum number of particles is 10^5, and the number of particles corresponding to each target is set to 10^3.

In the SMC-CPHD filter, the relevant parameters are the same as those of the SMC-PHD filter. Moreover, the capping parameter for the cardinality distribution is set as $N_{\max} = 100$.

In the SMC-CBMeMBer filter, the maximum number of hypothetical tracks is 100 and the pruning threshold for tracks is set as 10^{-3}. For each hypothetical track, the minimum and maximum numbers of particles are set to 300 and 1000, respectively.

In the SMC-LMB filter, the maximum number of hypothetical tracks is set as 100, and 1000 particles are allocated to each track. The pruning threshold for tracks is set as 10^{-3}. In addition, during the conversion from the LMB to the GLMB before the update step, the corresponding numbers for birth hypotheses, survival hypotheses and the hypotheses after the GLMB update are set as 5, 1000 and 1000, respectively, and the maximum number of retained posterior hypotheses is 1000. The threshold for the number of effective particles when resampling is set to 500. The threshold for pruning hypotheses is 10^{-15}.

In the SMC-Joint-LMB filter, the maximum number of hypothetical tracks is 100, and 1000 particles are allocated to each track. The pruning threshold for tracks is set as 10^{-3}. Moreover, the corresponding number of hypotheses after the GLMB update is 1000. The threshold for the number of effective particles when resampling is set to 500.

In the SMC-δ-GLMB filter, the corresponding numbers of the birth hypotheses, survival hypotheses and the hypotheses after the GLMB update are set as 5, 1000 and 1000, respectively, and the maximum number of retained posterior hypotheses is 1000. 1000 particles are allocated to each track, and the threshold for the number of effective particles during resampling is set to 500. The threshold for pruning hypotheses is 10^{-15}.

In the SMC-Joint-δ-GLMB filter, the corresponding number of hypotheses after the GLMB update is set as 1000 and the maximum number of posterior hypotheses is 1000. 1000 particles are allocated to each track, and the threshold for the number of effective particles during resampling is set to 500. The threshold for pruning hypotheses is 10^{-15}.

The parameters related to the performance indicators are set as follows: the CPEP radius is set as $r = 20\,\text{m}$, and the order parameter and threshold parameter for the OSPA are set as $p = 1$ and $c = 100$ m, respectively.

In the above simulation configuration, typical tracking results from a single run are shown in Fig. 7.9. Since the estimation result of the CPHD filter is similar to that of the PHD filter, and the estimation results of the Joint-LMB, δ-GLMB and Joint-δ-GLMB filters are similar to the estimation result of the LMB filter, to avoid repetition, only the state estimation results of the PHD, CBMeMBer and LMB filters are shown in the figure. It can be seen intuitively from the figure that, the CBMeMBer filter has a better tracking accuracy than the PHD/CPHD filter due to avoidance of the clustering operation when extracting the state estimation; while labeled RFS filters such as the LMB filter are superior to unlabeled RFS filters such as the CBMeMBer filter in terms of both estimation accuracy of the number of targets and tracking accuracy. In addition, they also provide the accurate track estimation because they contain the label information.

In order to compare the advantages and disadvantages of all filters, the cardinality estimation, OSPA and CPEP performances of each filter are calculated through 100 Monte Carlo simulations. As can be seen from Fig. 7.10a, on the whole, these filters can accurately estimate the number of targets. Due to the relatively low clutter density, the CBMeMBer filter doesn't show the obvious cardinality overestimation problem, and the accuracy of estimating the number of targets is better than those of the PHD

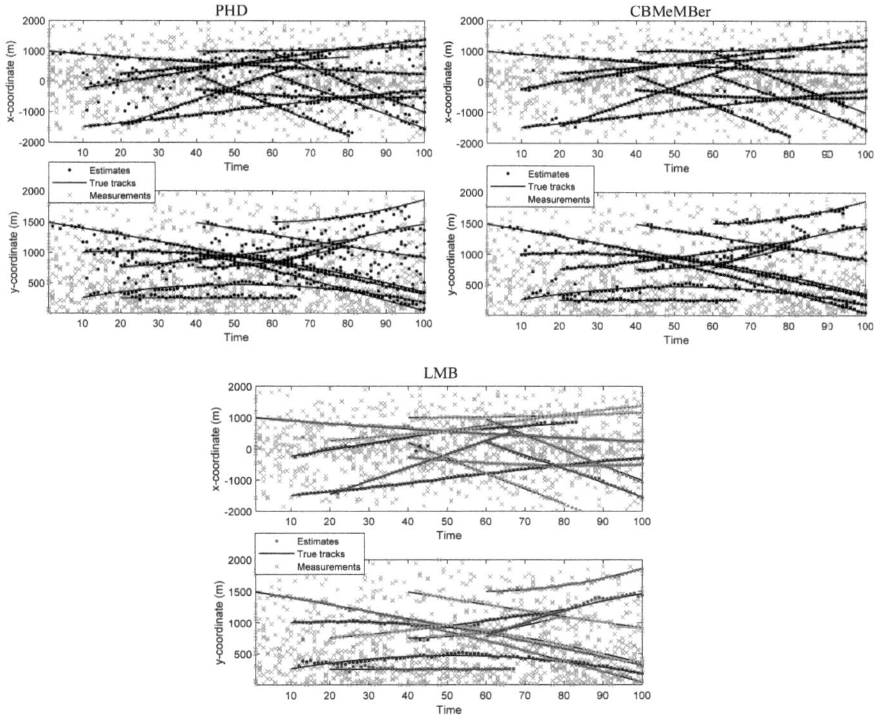

Fig. 7.9 Multi-target state estimation

and CPHD filters. Note that the CPHD filter with the cardinality distribution doesn't play an advantage, while the LMB, JointLMB, GLMB and JointGLMB filters have similar performances, better than the PHD and CPHD filters, and they all have smooth curves, which are free from the burr phenomenon of the CBMeMBer filter.

In terms of the standard deviation of cardinality estimation, it can be seen from Fig. 7.10b that the performances of the LMB, JointLMB, GLMB and JointGLMB filters are close to each other and are the best, while the performances of the CPHD, CBMeMBer and PHD filters deteriorate in turn.

In terms of the OSPA, it can be seen from Fig. 7.10c that the LMB, JointLMB, GLMB and JointGLMB filters have the highest accuracy, followed by the CBMeMBer filter, which is significantly better than the PHD and CPHD filters. This shows the advantage of the CBMeMBer filter in extracting target states without clustering operation.

In terms of the CPEP, as seen from Fig. 7.10d, the LMB, JointLMB, GLMB and JointGLMB filters have the best performance, and the CBMeMBer filter is quite close to them and obviously better than the PHD and CPHD filters.

Figure 7.11 compares the computation time of each filter. In the figure, the vertical axis represents the 100-time Monte Carlo average of the total running time (100 sampling intervals). It can be seen from the figure that the average operation times

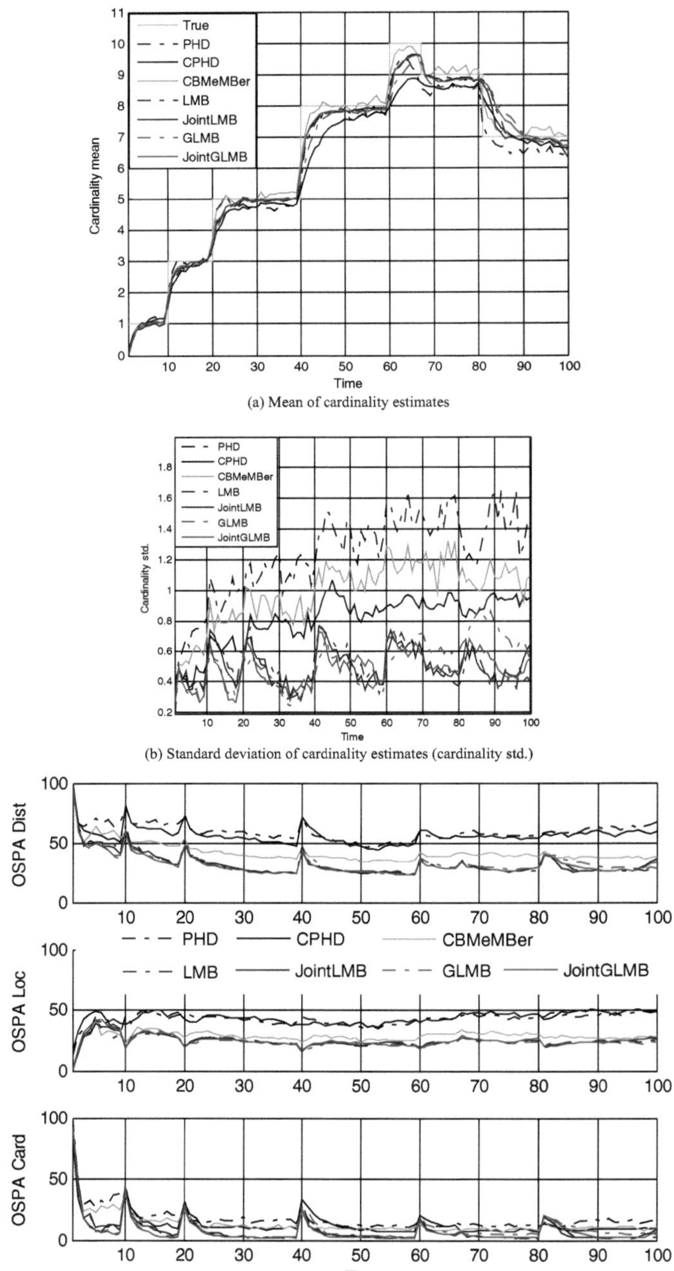

(a) Mean of cardinality estimates

(b) Standard deviation of cardinality estimates (cardinality std.)

(c) OSPA

Fig. 7.10 Statistical performance of multi-target state estimation of each filter **a** Mean of cardinality estimates **b** standard deviation of cardinality estimates (cardinality std.) **c** OSPA **d** CPEP

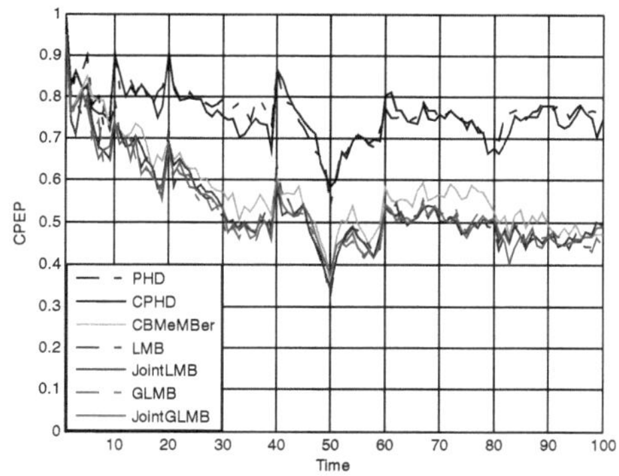

(d) CPEP

Fig. 7.10 (continued)

of the PHD, CPHD, CBMeMBer, LMB, JointLMB, GLMB and JointGLMB filters are 6.02 s, 81.30 s, 3.29 s, 3876.71 s, 69.69 s, 77.32 s and 24.48 s, respectively. As a result, the LMB filter has the lowest operational efficiency, reaching an unbearable 3876.71 s. The right plot shows the comparison among other filters after removal of the LMB filter. It can be seen that the PHD and CBMeMBer filters have the lowest computational complexity, and the GLMB, JointLMB and JointGLMB filters have a lower computational complexity than the CPHD filter at this time. In particular, the time consumption of the JointGLMB filter is greatly reduced.

To sum up, under the non-linear Gaussian condition, the GLMB, JointLMB and JointGLMB filters achieve the best tracking performance in terms of estimation accuracy of the number of targets, stability, tracking accuracy at the expense of increasing the amount of computation. Among them, the JointGLMB filter has the lowest computational burden. Although the LMB filter is similar to these three filters in terms of the tracking performance, its operational efficiency is too low. For the unlabeled RFS filters, the CBMeMBer filter not only has lower computational complexity,

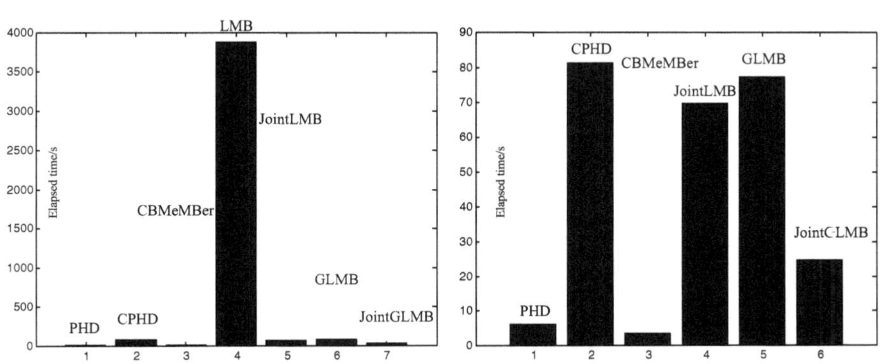

Fig. 7.11 Comparison of running times of different filters

but also has significantly better tracking performance than the PHD and CPHD filters. However, it should be noted that the theoretical analysis shows that the CBMeMBer filter has the problem of cardinality overestimation under the condition of low signal-to-noise ratio. Therefore, under the nonlinear Gaussian condition, the JointGLMB filter can be considered as the first option.

7.7 Summary

Based on the introduction of the general GLMB filter, this chapter gives the δ-GLMB, LMB and Mδ-GLMB filters successively, and elaborates the specific implementation process. These labeled RFS filters are true multi-target trackers, which can effectively provide the estimation information of target tracks. Among them, the δ-GLMB filter is the closed form solution of the full multi-target Bayesian recursion; while both the LMB filter and the Mδ-GLMB filter are the approximate solutions of the δ-GLMB filter in order to reduce the computational burden. Recently, in order to reduce the amount of computation without losing the accuracy of the δ-GLMB filter, Vo et al. developed a single-step implementation scheme of the δ-GLMB filter, which combines the original prediction and update steps into a single step. In addition, the Gibbs sampling was also proposed to replace the ranked assignment algorithm, thus reducing the calculation amount of the δ-GLMB filter by two orders of magnitude (relative to the number of measurements). According to the simulation and comparison, the Gibbs sampling implementation of the joint δ-GLMB filter shows the excellent performance under both linear Gaussian and nonlinear Gaussian conditions, and can be regarded as the first choice of the multi-target tracking filters.

Chapter 8
Maneuvering Target Tracking

8.1 Introduction

The system model of target tracking is composed of a measurement model and a motion model. As mentioned in the introduction of Chap. 2, the measurement uncertainty is a major challenge faced by target tracking. In fact, there is another important problem: the uncertainty of target dynamics. For a practical tracker, the choice of the target motion model exerts direct impact on its performance. The motion model that matches the actual motion of the target will improve the tracking performance; conversely, it will deteriorate the tracking performance and even lead to the filter divergence, because the actual error may fall outside the predicted range of the error covariance. Since the practical tracking system lacks the prior information about non-cooperative target motion model, that is, the so-called target motion equation is of uncertainty, the performance of the tracker is seriously restricted by the motion model. Hence, it is necessary to use the maneuvering target tracking algorithm to tackle this problem.

Among maneuvering target tracking algorithms, the multiple model (MM) filter is a well-known solution for dealing with non-cooperative maneuvering targets. For example, in the vehicle environment perception, the tracking algorithm is required to track all relevant objects (e.g., cars and pedestrians) around the vehicle. Since these objects usually have different behavioral characteristics, the multi-target tracking algorithm needs to use multiple models to obtain the optimal estimation of target states. The MM filter is often based on the jump Markov (JM) model [411]. In this approach, a target switches between a model set in a Markov fashion, where the transition between target motion models follows the Markov chain probability rule. From the point of view that the target motion can be modeled using different motion models, the multi-maneuvering-target system can be viewed as a JM system.

This chapter mainly introduces the MM-based maneuvering target tracking algorithms under the RFS framework. On the basis of introducing the JM system, it describes the MM-PHD, MM-CBMeMber, and MM-δ-GLMB filters, and provides the specific implementation process.

W. Wu et al., *Target Tracking with Random Finite Sets*,
https://doi.org/10.1007/978-981-19-9815-7_8

8.2 Jump Markov Systems

8.2.1 Nonlinear Jump Markov System

A typical example of using jump Markov (JM) system model to represent the maneuvering target is the aircraft dynamics. The aircraft can use the approximate constant velocity model, acceleration/deceleration model, and coordinate turning model [12] for flight, and in this case, it is often difficult for a single model to represent behaviors of a maneuvering target at all times [412, 380]. The JM system is described by a set of parameterized state-space models whose parameters evolve over time according to finite-state Markov chains. Specifically, under the framework of the JM system, a target that moves with a certain motion model at any time is assumed to follow the same motion model with a certain probability, or switch to a different motion model (which belong to the preselected motion model set) with a certain probability at the next moment. This system has been applied to many fields of signal processing and it provides a natural method for maneuvering target modeling.

In the JM system model, the target motion model is switched according to a Markov chain. Consider the discrete and finite set $\mathcal{M}$ of motion models (also referred to as modes). $\mu_k \in \mathcal{M}$ is assumed to be model index parameter and it is governed by a potential Markov process. Its corresponding model transition probability is $\tau_{k|k-1}(\mu_k|\mu_{k-1})$, which represents the probability of transition from model μ_{k-1} to model μ_k.

In a linear JM system, the transition probability among models μ_k is constant, which forms the well-known model transition matrix $\mathcal{T}_k = [t_k^{(m,n)}]_{1\leq m,n\leq|\mathcal{M}|}$, in which each element represents the probability of transition between two models, i.e.,

$$t_k^{(m,n)} = \tau_{k|k-1}(\mu_k = n|\mu_{k-1} = m) \tag{8.2.1}$$

The Markov transition probability matrix describes the probability of changing/maintaining the motion model at the next moment when the motion model of a specific target is given at the current time. In the Markov transition probability matrix, when the current model is given, the sum of conditional probabilities of all possible motion models at the next moment is 1, i.e.,

$$\sum_{\mu_k\in\mathcal{M}} \tau_{k|k-1}(\mu_k|\mu_{k-1}) = 1 \tag{8.2.2}$$

Let $\boldsymbol{\xi}_k \in \mathbb{R}^{n_\xi}$ and $z_k \in \mathbb{R}^{n_z}$ indicate the kinetic state (such as the target's position and speed) and the measurement at time k, respectively. For the general non-linear model, the target dynamics can be described by the following general parameterized non-linear equation

$$\boldsymbol{\xi}_k = f_{k|k-1}(\boldsymbol{\xi}_{k-1}, \boldsymbol{v}_{k-1}, \mu_k) \tag{8.2.3}$$

where $f_{k|k-1}(\cdot)$ is a general non-linear function, and $\boldsymbol{v}_{k-1}$ is the model-dependent process noise vector whose statistical characteristic is assumed to be known.

The measurements are generated by targets or clutter. Let the measurement received by the sensor from the target be

$$z_k = h_k(\boldsymbol{\xi}_k, \boldsymbol{n}_k, \mu_k) \tag{8.2.4}$$

where $h_k(\cdot)$ is a general non-linear measurement function, and $\boldsymbol{n}_k$ is a model-dependent measurement noise vector with a known statistical characteristic. Some targets may miss detections at time k, and the detection probability of a target with state $\boldsymbol{\xi}_k$ at time k is denoted as $p_{D,k}(\boldsymbol{\xi}_k)$. Besides, the sensor may also receive measurements from clutter. The measurement set Z_k at time k consists of target-originated measurements and clutter measurements.

In order to achieve the MM-based tracking by using the JM method, the random state variable is generalized to the augmented state variable containing the corresponding motion model, i.e., $\boldsymbol{x} = (\boldsymbol{\xi}, \mu)$. Note: by defining the augmented system state as $\boldsymbol{x} = (\boldsymbol{\xi}, \mu)$, the JM system model can be re-written as a standard state-space model. Through the parameterization of discrete model variables, the probability density function (PDF) of an augmented random state has the following relationship with that of the base state

$$p(\boldsymbol{\xi}) = \int_{\mathcal{M}} p(\boldsymbol{\xi}, \mu) d\mu = \sum_{\mu} p(\mu) p(\boldsymbol{\xi}|\mu) \tag{8.2.5}$$

The JM method in which the random state is extended to include the target motion model has changed the conventional target random model, and the augmented state transition model and measurement likelihood model have become related to motion model μ, i.e.,

$$\begin{aligned} \phi_{k|k-1}(\boldsymbol{x}_k|\boldsymbol{x}_{k-1}) &= \phi_{k|k-1}(\boldsymbol{\xi}_k, \mu_k|\boldsymbol{\xi}_{k-1}, \mu_{k-1}) \\ &= \tilde{\phi}_{k|k-1}(\boldsymbol{\xi}_k|\boldsymbol{\xi}_{k-1}, \mu_k)\tau_{k|k-1}(\mu_k|\mu_{k-1}) \end{aligned} \tag{8.2.6}$$

$$g_k(z_k|\boldsymbol{x}_k) = g_k(z_k|\boldsymbol{\xi}_k, \mu_k) \tag{8.2.7}$$

We can notice that, although measurement likelihood function (8.2.7) is generally dependent on the JM variable, it is usually independent of the target motion model. Therefore, for simplicity, it can be assumed to be independent of the motion model, i.e., $g_k(z_k|\boldsymbol{x}_k) = g_k(z_k|\boldsymbol{\xi}_k)$.

Maneuvering target tracking means estimating the target kinematic state $\boldsymbol{\xi}_k$ or augmented state $\boldsymbol{x}_k$ at time k based on a series of measurements $Z_{1:k} = \{Z_1, \cdots, Z_k\}$. The JM system (or MM method) has been shown highly effective for maneuvering target tracking. Moreover, the JM system model is not only used for maneuvering target tracking but also applicable to the estimation of unknown clutter parameters [60, 173].

8.2.2 Linear Gaussian Jump Markov System

The linear Gaussian jump Markov (LGJM) system is a JM system under the linear Gaussian model, i.e., conditioned on model μ_k, the state transition density and measurement likelihood are respectively

$$\tilde{\phi}_{k|k-1}(\boldsymbol{\xi}_k|\boldsymbol{\xi}_{k-1},\mu_k) = \mathcal{N}(\boldsymbol{\xi}_k; \boldsymbol{F}_{k-1}(\mu_k)\boldsymbol{\xi}_{k-1}, \boldsymbol{Q}_{k-1}(\mu_k)) \tag{8.2.8}$$

$$g_k(z_k|\boldsymbol{\xi}_k,\mu_k) = \mathcal{N}(z_k; \boldsymbol{H}_k(\mu_k)\boldsymbol{\xi}_k, \boldsymbol{R}_k(\mu_k)) \tag{8.2.9}$$

where $\boldsymbol{F}_{k-1}(\mu_k)$ and $\boldsymbol{H}_k(\mu_k)$ denote the state transition and measurement matrices conditioned on model μ_k, respectively, and $\boldsymbol{Q}_{k-1}(\mu_k)$ and $\boldsymbol{R}_k(\mu_k)$ denote the covariance matrices of process noise and measurement noise conditioned on model μ_k, respectively.

8.3 Multiple Model PHD (MM-PHD) Filter

In this section, the sequential Monte Carlo (SMC) and Gaussian mixture (GM) implementations of the multiple model PHD (MM-PHD) filter are descried and they are referred as the SMC multiple model PHD (SMC-MM-PHD) filter and the GM multiple model PHD (GM-MM-PHD) filter, respectively.

8.3.1 Sequential Monte Carlo MM-PHD Filter

This part firstly gives the MM-PHD recursion and then describes the detailed SMC implementation steps.

8.3.1.1 MM-PHD Recursion

The MM method adopted in the MM-PHD algorithm has a similar structure to the IMM estimator at the stage of "mixing and integration". When $|\mathcal{M}|$ models are used to describe the target dynamics, only $|\mathcal{M}|$ PHD filters are required to work in parallel for the MM-PHD filter. In addition, the MM-PHD filter also conducts soft switching between models, without the maneuvering detection decision. The MM method herein is different from the IMM estimator in that: at the stage of mixing and integration, the latter only uses the first order and second order statistics of the target density. This technique shall not be used for integration of model-dependent PHD filter outputs, because the density is not necessarily Gaussian distributed. In fact, since the density representing multiple targets may be multi-mode, it is not reasonable to use the first order and second order statistics for approximation. As a result, the MM-PHD filter uses the branched true density at the stage of mixing and update, i.e., the full density conditioned on each model. This approach can be used to handle the MM target density at the expense of some increase in computational burden compared to the IMM estimator. The MM-PHD recursion is composed of the following two steps.

(1) Prediction

The prediction step of the MM-PHD filter involves the model prediction and the state prediction. Based on Markov model transition probability $\tau_{k|k-1}(\mu_k|\mu_{k-1})$ and model-dependent prior intensity $v_{k-1}(\boldsymbol{\xi}_{k-1}, \mu_{k-1}|Z_{1:k-1})$, the initial intensity $\tilde{v}_{k|k-1}(\boldsymbol{\xi}_{k-1}, \mu_k|Z_{1:k-1})$ that is fed back to the PHD filter and matched to model μ_k can be calculated as

$$\tilde{v}_{k|k-1}(\boldsymbol{\xi}_{k-1}, \mu_k|Z_{1:k-1}) = \sum_{\mu_{k-1}} v_{k-1}(\boldsymbol{\xi}_{k-1}, \mu_{k-1}|Z_{1:k-1})\tau_{k|k-1}(\mu_k|\mu_{k-1}) \quad (8.3.1)$$

The above equation corresponds to the model prediction, and it doesn't consider the target spawning, birth and death processes which are taken into consideration in the state prediction step given in (8.3.2). In [188], the step corresponding to (8.3.1) is referred to as the mixing step. Intensities $\tilde{v}_{k|k-1}(\cdot)$ in (8.3.1) and $v_{k-1}(\cdot)$ are similar to the probability densities, but their integrals are not 1. The mixing described in (8.3.1) is analogous to the total probability theorem.

Once the initial intensity of the PHD filter that matches model μ_k is calculated, the model-dependent predicted intensity can be obtained as follows (i.e., state prediction)

$$\begin{aligned} & v_{k|k-1}(\boldsymbol{\xi}_k, \mu_k|Z_{1:k-1}) \\ & = \int p_{S,k}(\boldsymbol{\xi}_{k-1})\phi_{k|k-1}(\boldsymbol{\xi}_k|\boldsymbol{\xi}_{k-1}, \mu_k)\tilde{v}_{k|k-1}(\boldsymbol{\xi}_{k-1}, \mu_k|Z_{1:k-1})d\boldsymbol{\xi}_{k-1} \\ & + \int v_{\beta,k}(\boldsymbol{\xi}_k|\boldsymbol{\xi}_{k-1}, \mu_k)\tilde{v}_{k|k-1}(\boldsymbol{\xi}_{k-1}, \mu_k|Z_{1:k-1})d\boldsymbol{\xi}_{k-1} \\ & + v_{\gamma,k}(\boldsymbol{\xi}_k, \mu_k) \end{aligned} \quad (8.3.2)$$

where $\phi_{k|k-1}(\boldsymbol{\xi}_k|\boldsymbol{\xi}_{k-1}, \mu_k)$ is the single-target Markov transition density conditioned on model μ_k. The above equation shows that the model-dependent PHD prediction is not just applicable to model-matched single-model PHD prediction. In fact, the birth target PHD $v_{\gamma,k}(\cdot)$ and the spawning target PHD $v_{\beta,k}(\cdot|\cdot)$ are also model-dependent. Assuming that all targets follow the target dynamics described by model μ_k, the integral of model-dependent PHD $v_{k|k-1}(\boldsymbol{\xi}_k, \mu_k|Z_{1:k-1})$ within a certain region gives the expected (predicted) number of targets in that region.

(2) Update

With the measurements obtained at time k, the model-matched updated intensity can be calculated as follows

$$v_k(\boldsymbol{\xi}_k, \mu_k|Z_{1:k}) = [1 - p_{D,k}(\boldsymbol{\xi}_k)]v_{k|k-1}(\boldsymbol{\xi}_k, \mu_k|Z_{1:k-1}) + \sum_{z \in Z_k} \frac{p_{D,k}(\boldsymbol{\xi}_k)g_k(z|\boldsymbol{\xi}_k, \mu_k)v_{k|k-1}(\boldsymbol{\xi}_k, \mu_k|Z_{1:k-1})}{\kappa_k(z) + \int p_{D,k}(\boldsymbol{\xi}_k)g_k(z|\boldsymbol{\xi}_k, \mu_k)v_{k|k-1}(\boldsymbol{\xi}_k, \mu_k|Z_{1:k-1})d\boldsymbol{\xi}_k} \tag{8.3.3}$$

Although there is no explicit model probability update in the MM-PHD filter, the update step implicitly updates the model probability. However, it is not necessary to calculate the value of the updated model probability in the recursive MM-PHD filter, since the model-dependent intensity as a whole is fed back to the filter in the mixing stage.

8.3.1.2 SMC Implementation

The SMC implementation of the MM-PHD filter utilizes a set of random samples or particles to represent the posterior MM-PHD and these particles are composed of the state and model information with weights. The core idea of this implementation is to incorporate a model index parameter in the weighted sample set to represent the model-dependent posterior density. In the MM-PHD filter, multiple PHD filters run in parallel and each one matches different target dynamics. The model index parameter in the sample guides the MM-PHD filter to select the PHD filter that matches the corresponding target model. Additionally, for different models, the number of samples used by each PHD filter is not necessarily the same. Sine the model probabilities are updated in update step of the filter, the PHD filter that matches the target motion will contain more samples. The calculation is more efficient than using an equal number of samples to represent the PHD filter matching each model.

In essence, since the SMC implementation of the MM-PHD filter can be considered as a special case of the SMC-PHD filter with appropriate modifications where the target state is augmented with a model index parameter. Therefore, the prediction and update operations of the SMC-PHD filter need to be modified to include the uncertainty of the target motion model.

Let the posterior MM-PHD $v_{k-1}(\boldsymbol{\xi}_{k-1}, \mu_{k-1}|Z_{1:k-1})$ be represented by the particle set $\{w_{k-1}^{(i)}, \boldsymbol{\xi}_{k-1}^{(i)}, \mu_{k-1}^{(i)}\}_{i=1}^{L_{k-1}}$, i.e.,

$$v_{k-1}(\boldsymbol{\xi}_{k-1}, \mu_{k-1}|Z_{1:k-1})=\sum_{i=1}^{L_{k-1}} w_{k-1}^{(i)}\delta_{\boldsymbol{\xi}_{k-1}^{(i)},\mu_{k-1}^{(i)}}(\boldsymbol{\xi}_{k-1}, \mu_{k-1}) \tag{8.3.4}$$

where δ is the Dirac delta function.

(1) Prediction: as mentioned earlier, the prediction step of the MM-PHD filter involves the model prediction as well as the state prediction. For survival targets, the model prediction is conducted based on model transition probability $\tau_{k|k-1}(\mu_k|\mu_{k-1})$. The model samples $\left\{\mu_{k|k-1}^{(i)}\right\}_{i=1}^{L_{k-1}}$ from the MM-PHD $\tilde{v}_{k|k-1}(\boldsymbol{\xi}_{k-1}, \mu_k|Z_{1:k-1})$ after the model prediction are generated by importance sampling according to the proposal distribution $q_{\mu,k}(\cdot|\mu_{k-1}^{(i)})$. The independent and identically distributed (IID) model samples $\{\ \mu_{k|k-1}^{(i)}\}_{i=L_{k-1}+1}^{L_{k-1}+J_k}$ corresponding to the birth targets are generated according to another proposal density $b_{\mu,k}(\cdot)$, i.e.,

$$\mu_{k|k-1}^{(i)} \sim \begin{cases} q_{\mu,k}(\cdot|\mu_{k-1}^{(i)}), \ i = 1, \ldots, L_{k-1} \\ b_{\mu,k}(\cdot), \qquad\quad i = L_{k-1}+1, \ldots, L_{k-1}+J_k \end{cases} \tag{8.3.5}$$

where $q_{\mu,k}(\cdot|\mu_{k-1}^{(i)})$ and $b_{\mu,k}(\cdot)$ are the probability mass functions (PMF). Therefore, the discrete weighted approximation of MM-PHD $\tilde{v}_{k|k-1}(\boldsymbol{\xi}_{k-1}, \mu_k|Z_{1:k-1})$ after the model prediction is

$$\tilde{v}_{k|k-1}(\boldsymbol{\xi}_{k-1}, \mu_k|Z_{1:k-1}) = \sum_{i=1}^{L_{k-1}+J_k} \tilde{w}_{k|k-1}^{(i)}\delta_{\boldsymbol{\xi}_{k-1}^{(i)},\mu_{k|k-1}^{(i)}}(\boldsymbol{\xi}_{k-1}, \mu_k) \tag{8.3.6}$$

with

$$\tilde{w}_{k|k-1}^{(i)} = \begin{cases} \frac{\tau_{k|k-1}(\mu_{k|k-1}^{(i)}|\mu_{k-1}^{(i)})w_{k-1}^{(i)}}{q_{\mu,k}(\mu_{k|k-1}^{(i)}|\mu_{k-1}^{(i)})}, \ i = 1, \ldots, L_{k-1} \\ \frac{p_{\mu,k}(\mu_{k|k-1}^{(i)})}{J_k b_{\mu,k}(\mu_{k|k-1}^{(i)})}, \qquad\quad i = L_{k-1}+1, \ldots, L_{k-1}+J_k \end{cases} \tag{8.3.7}$$

where PMF $p_{\mu,k}(\cdot)$ indicates the model distribution of birth targets at time k, the number J_k of birth particles can be a function of time k in order to accommodate the time-varying number of targets.

Then, the importance sampling is applied to generate state samples of approximate predicted MM-PHD $v_{k|k-1}(\boldsymbol{\xi}_k, \mu_k|Z_{1:k-1})$. State samples $\left\{\boldsymbol{\xi}_{k|k-1}^{(i)}\right\}_{i=1}^{L_{k-1}}$ are generated based on a proposal density $q_{\boldsymbol{\xi},k}(\cdot|\boldsymbol{\xi}_{k-1}^{(i)}, \mu_{k|k-1}^{(i)}, Z_k)$, and IID state samples $\left\{\boldsymbol{\xi}_{k|k-1}^{(i)}\right\}_{i=L_{k-1}+1}^{L_{k-1}+J_k}$ corresponding to the birth targets are generated according to another proposal density $b_{\boldsymbol{\xi},k}(\cdot|\mu_{k|k-1}^{(i)}, Z_k)$, i.e.,

$$\boldsymbol{\xi}_{k|k-1}^{(i)} = \begin{cases} q_{\xi,k}(\cdot|\boldsymbol{\xi}_{k-1}^{(i)}, \mu_{k|k-1}^{(i)}, Z_k), & i = 1, \ldots, L_{k-1} \\ b_{\xi,k}(\cdot|\mu_{k|k-1}^{(i)}, Z_k), & i = L_{k-1}+1, \ldots, L_{k-1}+J_k \end{cases} \tag{8.3.8}$$

Then, the weighted approximation of the predicted MM-PHD is

$$v_{k|k-1}(\boldsymbol{\xi}_k, \mu_k|Z_{1:k-1}) = \sum_{i=1}^{L_{k-1}+J_k} w_{k|k-1}^{(i)} \delta_{\boldsymbol{\xi}_{k|k-1}^{(i)}, \mu_{k|k-1}^{(i)}}(\boldsymbol{\xi}_k, \mu_k) \tag{8.3.9}$$

where

$$w_{k|k-1}^{(i)} = \begin{cases} \dfrac{p_{S,k}(\boldsymbol{\xi}_{k|k-1}^{(i)})\phi_{k|k-1}(\boldsymbol{\xi}_{k|k-1}^{(i)}|\boldsymbol{\xi}_{k-1}^{(i)}, \mu_{k|k-1}^{(i)}) + v_{\beta,k}(\boldsymbol{\xi}_{k|k-1}^{(i)}|\boldsymbol{\xi}_{k-1}^{(i)}, \mu_{k|k-1}^{(i)})}{q_{\xi,k}(\boldsymbol{\xi}_{k|k-1}^{(i)}|\boldsymbol{\xi}_{k-1}^{(i)}, \mu_{k|k-1}^{(i)}, Z_k)} \tilde{w}_{k|k-1}^{(i)}, & i = 1, \ldots, L_{k-1} \\ \dfrac{v_{\gamma,k}(\boldsymbol{\xi}_{k|k-1}^{(i)}|\mu_{k|k-1}^{(i)})}{b_{\xi,k}(\boldsymbol{\xi}_{k|k-1}^{(i)}|\mu_{k|k-1}^{(i)}, Z_k)} \tilde{w}_{k|k-1}^{(i)}, & i = L_{k-1}+1, \ldots, L_{k-1}+J_k \end{cases} \tag{8.3.10}$$

In (8.3.10), the functions describing the Markov target transition density $\phi_{k|k-1}(\cdot)$, target spawning $v_{\beta,k}(\cdot|\cdot)$ and birth target $v_{\gamma,k}(\cdot)$ are all conditioned on a specific motion model. As a consequence, although the model is not explicitly shown by the equations, there are actually $|\mathcal{M}|$ PHD filters running in parallel when implemented.

(2) Update: using the measurement set Z_k at time k, the updated particle weight is calculated as follows

$$\begin{aligned} \tilde{w}_k^{(i)} = & \Big[(1 - p_{D,k}(\boldsymbol{\xi}_{k|k-1}^{(i)})) \\ & + \sum_{z \in Z_k} \frac{\psi_{k,z}(\boldsymbol{\xi}_{k|k-1}^{(i)}, \mu_{k|k-1}^{(i)})}{\kappa_k(z) + \sum_{i=1}^{L_{k-1}+J_k} \psi_{k,z}(\boldsymbol{\xi}_{k|k-1}^{(i)}, \mu_{k|k-1}^{(i)}) w_{k|k-1}^{(i)}}\Big] w_{k|k-1}^{(i)} \end{aligned} \tag{8.3.11}$$

where

$$\psi_{k,z}(\boldsymbol{\xi}_{k|k-1}^{(i)}, \mu_{k|k-1}^{(i)}) = p_{D,k}(\boldsymbol{\xi}_{k|k-1}^{(i)}) g_k(z|\boldsymbol{\xi}_{k|k-1}^{(i)}, \mu_{k|k-1}^{(i)}) \tag{8.3.12}$$

In (8.3.12), the measurement likelihood function $g_k(\cdot)$ is written in a model-dependent form, considering the general case that the measurement function can also be model-dependent.

(3) Re-sampling: since the weights are not normalized to 1 in the PHD filter, the expected number of targets is calculated by summing all weights in order to implement re-sampling, i.e.,

$$\hat{N}_k = \sum_{i=1}^{L_{k-1}+J_k} \tilde{w}_k^{(i)} \tag{8.3.13}$$

Then, the updated particle set $\{\tilde{w}_k^{(i)} \big/ \hat{N}_k, \boldsymbol{\xi}_{k|k-1}^{(i)}, \mu_{k|k-1}^{(i)}\}_{i=1}^{L_{k-1}+J_k}$ is re-sampled to obtain $\{w_k^{(i)} \big/ \hat{N}_k, \boldsymbol{\xi}_k^{(i)}, \mu_k^{(i)}\}_{i=1}^{L_k}$, and finally the weight is multiplied by $\hat{N}_k$ to get $\{w_k^{(i)}, \boldsymbol{\xi}_k^{(i)}, \mu_k^{(i)}\}_{i=1}^{L_k}$, which keeps the total weight after re-sampling unchanged at $\hat{N}_k$. At this point, the discrete approximation of the updated posterior MM-PHD at time k is

$$v_{k|k}(\boldsymbol{\xi}_k, \mu_k|Z_{1:k}) = \sum_{i=1}^{L_k} w_k^{(i)} \delta_{\boldsymbol{\xi}_k^{(i)}, \mu_k^{(i)}}(\boldsymbol{\xi}_k, \mu_k) \tag{8.3.14}$$

The SMC implementation of the MM-PHD filter is structurally similar to the sampling importance resampling (SIR) of the particle filter [26]. Unlike the particle filter, the sum $\sum_{i=1}^{L_{k-1}} w_{k-1}^{(i)}$ of all weights is not equal to 1. Instead, it gives the expected number of targets at time $k-1$. Furthermore, it is easy to distinguish model-dependent posterior PHDs by grouping particles based on model index parameters, and the posterior model probability corresponding to a specific target is approximately proportional to the total sample weights of each model corresponding to that target in the index set $\left\{\mu_{k-1}^{(i)}\right\}_{i=1}^{L_{k-1}}$.

8.3.2 *Gaussian Mixture MM-PHD Filter*

The Gaussian mixture multiple model PHD (GM-MM-PHD) filter is a closed form solution to the MM-PHD recursion and it is applicable to target switching between linear Gaussian models. First, the LGJM system multi-target model is described. On this basis, the corresponding MM-PHD recursion is given. Finally, the generalized closed form solution to the MM-PHD recursion under the general condition is provided.

8.3.2.1 Multi-target Model for Linear Gaussian Jump Markov System

In addition to the three assumptions (A.1–A.3) in Sect. 8.4.2, the multi-target model for the LGJM system also contain the LGJM system model of each target in Sect. 8.2.2, and the models for the survival probability and detection probability independent of the kinetic state, and the models for target birth and target spawning. First, focus will be paid to the modeling process of target birth and spawning models. Similar to the motion model, birth and spawning models can be naturally described in terms of kinetic states. However, although the distribution of the augmented state can be viewed as the product of the model distribution and the distribution of the model-conditioned kinetic state, i.e., $p(\boldsymbol{\xi}, \mu) = p(\mu)p(\boldsymbol{\xi}|\mu)$, this rationality can not be extended to the birth and spawning intensities. In other words, the intensity of the augmented state is not necessarily the product of the model intensity and the intensity of the model-conditioned kinetic state.

In order to specify the birth and spawning models of kinetic states and modes so that they produce the effective birth intensity and spawning intensity in the augmented state, we borrow from the well-known conclusion from point process theory, i.e., the Campbell's theory for marked point process [394]. The target birth intensity and spawning intensity are modeled using the Campbell's theory [394]. Specifically, this theory considers that, the Cartesian product of a point process with intensity $\tilde{v}$ in the kinetic state-space $\mathbb{R}^{n_\xi}$ and a point process in the mode space $\mathcal{M}$ is the intensity of the point process in space $\mathbb{R}^{n_\xi} \times \mathcal{M}$, i.e.,

$$v(\boldsymbol{\xi}, \mu) = p(\mu|\boldsymbol{\xi})\tilde{v}(\boldsymbol{\xi}) \tag{8.3.15}$$

where $p(\cdot|\boldsymbol{\xi})$ is the model distribution of the point with kinetic state $\boldsymbol{\xi}$ for a given product point process. Moreover, if the point process in $\mathbb{R}^{n_\xi}$ is of Poisson type, the product point process in $\mathbb{R}^{n_\xi} \times \mathcal{M}$ is also of Poisson type [413].

(1) Birth model 1. In the context of the multi-target birth model, the birth intensity of the augmented state at time k is

$$v_{\gamma,k}(\boldsymbol{\xi}, \mu) = p_{\gamma,k}(\mu|\boldsymbol{\xi})\tilde{v}_{\gamma,k}(\boldsymbol{\xi}) \tag{8.3.16}$$

where $\tilde{v}_{\gamma,k}$ is the birth intensity of the kinetic state at time k, and $p_{\gamma,k}(\cdot|\boldsymbol{\xi})$ is the probability distribution of the birth mode given the kinetic state $\boldsymbol{\xi}$ at time k. According to the assumptions of the LGJM system, mode transition probability $\tau_{k|k-1}$ is not a function of the kinetic state, and the multi-target model of the LGJM system also assumes that the birth mode distribution does not depend on the kinetic state, i.e., $p_{\gamma,k}(\mu|\boldsymbol{\xi}) = p_{\gamma,k}(\mu)$. Moreover, birth intensity $\tilde{v}_{\gamma,k}$ of the kinetic state is assumed to be in the following Gaussian mixture form

$$\tilde{v}_{\gamma,k}(\boldsymbol{\xi}) = \sum_{i=1}^{J_{\gamma,k}} w_{\gamma,k}^{(i)} \mathcal{N}(\boldsymbol{\xi}; \boldsymbol{m}_{\gamma,k}^{(i)}, \boldsymbol{P}_{\gamma,k}^{(i)}) \tag{8.3.17}$$

where $J_{\gamma,k}$, $w_{\gamma,k}^{(i)}$, $\boldsymbol{m}_{\gamma,k}^{(i)}$, and $\boldsymbol{P}_{\gamma,k}^{(i)}$, $i = 1, 2, \cdots, J_{\gamma,k}$ are given model parameters.

Similarly, the intensity of augmented state $[\boldsymbol{\xi}^{\mathrm{T}}, \mu]^{\mathrm{T}}$ at time k derived from the target with augmented state $[\boldsymbol{\xi}'^{\mathrm{T}}, \mu']^{\mathrm{T}}$ at time $k-1$ is

$$v_{\beta,k|k-1}(\boldsymbol{\xi}, \mu|\boldsymbol{\xi}', \mu') = p_{\beta,k|k-1}(\mu|\boldsymbol{\xi}, \boldsymbol{\xi}', \mu')\tilde{v}_{\beta,k|k-1}(\boldsymbol{\xi}|\boldsymbol{\xi}', \mu') \tag{8.3.18}$$

where $\tilde{v}_{\beta,k|k-1}(\cdot|\boldsymbol{\xi}', \mu')$ denotes the intensity of the kinetic state at time k derived from state $[\boldsymbol{\xi}'^{\mathrm{T}}, \mu']^{\mathrm{T}}$, and $p_{\beta,k|k-1}(\cdot|\boldsymbol{\xi}, \boldsymbol{\xi}', \mu')$ is the probability distribution of the mode given the kinetic state $\boldsymbol{\xi}$ at time k derived from state $[\boldsymbol{\xi}'^{\mathrm{T}}, \mu']^{\mathrm{T}}$. Consistent with the assumptions of the LGJM system, the multi-target model of the LGJM system assumes that the mode distribution of a spawning target is uncorrelated with neither its kinetic state nor its parent kinetic state, i.e., $p_{\beta,k|k-1}(\mu|\boldsymbol{\xi}, \boldsymbol{\xi}', \mu') = p_{\beta,k|k-1}(u|\mu')$, and it also assumes that the intensity $\tilde{v}_{\beta,k|k-1}(\cdot|\boldsymbol{\xi}', \mu')$ of the spawning kinetic state is in the form of Gaussian mixtures

$$\tilde{v}_{\beta,k|k-1}(\boldsymbol{\xi}|\boldsymbol{\xi}', \mu') = \sum_{i=1}^{J_{\beta,k}(\mu')} w_{\beta,k}^{(i)}(\mu')\mathcal{N}(\boldsymbol{\xi}; \boldsymbol{F}_{\beta,k-1}^{(i)}(\mu')\boldsymbol{\xi}' + \boldsymbol{d}_{\beta,k-1}^{(i)}(\mu'), \boldsymbol{P}_{\beta,k-1}^{(i)}(\mu')) \tag{8.3.19}$$

where $J_{\beta,k}(\mu')$, $w_{\beta,k}^{(i)}(\mu')$, $\boldsymbol{F}_{\beta,k-1}^{(i)}(\mu')$, $\boldsymbol{d}_{\beta,k-1}^{(i)}(\mu')$ and $\boldsymbol{P}_{\beta,k-1}^{(i)}(\mu')$, $i = 1, 2, \cdots, J_{\beta,k}(\mu')$ are given model parameters. Typically, a spawning target is modeled in the vicinity of its parent target.

(2) Birth model 2: by exchanging the variables in the kinetic state-space and the mode space in the PHD prediction equation, another consensus model can also be derived for the birth targets and spawning targets. At this point, the birth intensity of the augmented state at time k is

$$v_{\gamma,k}(\boldsymbol{\xi}, \mu) = p_{\gamma,k}(\boldsymbol{\xi}|\mu)\tilde{v}_{\gamma,k}(\mu) \tag{8.3.20}$$

where $\tilde{v}_{\gamma,k}$ is now the intensity for the birth model at time k, and $p_{\gamma,k}(\cdot|\mu)$ is the distribution of the birth kinetic state given model μ. We can notice that, the birth model intensity is not a function of the kinetic state. In the multi-target model of the LGJM system, the birth kinetic state is assumed to be Gaussian mixture distributed, i.e.,

$$p_{\gamma,k}(\boldsymbol{\xi}|\mu) = \sum_{i=1}^{J_{\gamma,k}(\mu)} w_{\gamma,k}^{(i)}(\mu)\mathcal{N}(\boldsymbol{\xi}; \boldsymbol{m}_{\gamma,k}^{(i)}(\mu), \boldsymbol{P}_{\gamma,k}^{(i)}(\mu)) \tag{8.3.21}$$

where $J_{\gamma,k}(\mu)$, $w_{\gamma,k}^{(i)}(\mu)$, $\boldsymbol{m}_{\gamma,k}^{(i)}(\mu)$, and $\boldsymbol{P}_{\gamma,k}^{(i)}(\mu)$, $i = 1, 2, \cdots, J_{\gamma,k}(\mu)$ are given model parameters. Note that they are different from those of birth model 1 in that, these model parameters are now related to model μ.

Similarly, the intensity of augmented state $[\boldsymbol{\xi}^{\mathrm{T}}, \mu]^{\mathrm{T}}$ at time k derived from the augmented state $[\boldsymbol{\xi}'^{\mathrm{T}}, \mu']^{\mathrm{T}}$ at time $k-1$ is

$$v_{\beta,k|k-1}(\boldsymbol{\xi}, \mu|\boldsymbol{\xi}', \mu') = p_{\beta,k|k-1}(\boldsymbol{\xi}|\mu, \boldsymbol{\xi}', \mu')\tilde{v}_{\beta,k|k-1}(\mu|\boldsymbol{\xi}', \mu') \tag{8.3.22}$$

where $\tilde{v}_{\beta,k|k-1}(\cdot|\boldsymbol{\xi}', \mu')$ is the mode spawning intensity, and $p_{\beta,k|k-1}(\cdot|\mu, \boldsymbol{\xi}', \mu')$ is the distribution of the spawning kinetic state given model μ. The multi-target model of the LGJM system assumes that, the spawning mode intensity is unrelated to its parent kinetic state, i.e., $\tilde{v}_{\beta,k|k-1}(\mu|\boldsymbol{\xi}', \mu') = \tilde{v}_{\beta,k|k-1}(\mu|\mu')$, and the distribution $p_{\beta,k|k-1}(\cdot|\mu, \boldsymbol{\xi}', \mu')$ of the spawning kinetic state is in the form of Gaussian mixture, i.e.,

$$p_{\beta,k|k-1}(\boldsymbol{\xi}|\mu, \boldsymbol{\xi}', \mu')$$

$$= \sum_{i=1}^{J_{\beta,k}(\mu,\mu')} w_{\beta,k}^{(i)}(\mu,\mu')\mathcal{N}(\boldsymbol{\xi};\boldsymbol{F}_{\beta,k-1}^{(i)}(\mu,\mu')\boldsymbol{\xi}'+\boldsymbol{d}_{\beta,k-1}^{(i)}(\mu,\mu'),\boldsymbol{P}_{\beta,k-1}^{(i)}(\mu,\mu')) \tag{8.3.23}$$

where $J_{\beta,k}(\mu,\mu')$, $w_{\beta,k}^{(i)}(\mu,\mu')$, $\boldsymbol{F}_{\beta,k-1}^{(i)}(\mu,\mu')$, $\boldsymbol{d}_{\beta,k-1}^{(i)}(\mu,\mu')$ and $\boldsymbol{P}_{\beta,k-1}^{(i)}(\mu,\mu')$, $i = 1, 2, \cdots, J_{\beta,k}(\mu,\mu')$ are given model parameters and are dependent on current mode μ and previous mode μ'. We can notice that, in birth model 1, these model parameters are merely related to model μ'.

From the perspective of modeling and application, model 1 is different from model 2. However, from the point of view of algorithm or computation, model 1 can be considered as a special case of model 2. Therefore, model 2 is adopted here.

In conclusion, in addition to the three assumptions (A.1–A.3) in Sect. 8.4.2, the LGJM system multi-target model also involves the following assumptions.

A.8: Each target follows the LGJM system model, i.e., the dynamic model and the measurement model of the augmented state have the following form

$$\phi_{k|k-1}(\boldsymbol{\xi},\mu|\boldsymbol{\xi}',\mu') = \mathcal{N}(\boldsymbol{\xi};\boldsymbol{F}_{S,k-1}(\mu)\boldsymbol{\xi}',\boldsymbol{Q}_{S,k-1}(\mu))\tau_{k|k-1}(\mu|\mu') \tag{8.3.24}$$

$$g_k(\boldsymbol{z}|\boldsymbol{\xi},\mu) = \mathcal{N}(\boldsymbol{z};\boldsymbol{H}_k(\mu)\boldsymbol{\xi},\boldsymbol{R}_k(\mu)) \tag{8.3.25}$$

where $\boldsymbol{F}_{S,k-1}(\mu)$ and $\boldsymbol{Q}_{S,k-1}(\mu)$ denote the linear target dynamic model parameters conditioned on mode μ, $\boldsymbol{H}_k(\mu)$ and $\boldsymbol{R}_k(\mu)$ indicate the linear measurement model parameters conditioned on mode μ, and $\tau_{k|k-1}(\mu|\mu')$ is the model transition probability. Specifically, conditioned on model μ, $\boldsymbol{F}_{S,k-1}(\mu)$ is the state transition matrix, $\boldsymbol{Q}_{S,k-1}(\mu)$ is the process noise covariance matrix, $\boldsymbol{H}_k(\mu)$ is the measurement matrix and $\boldsymbol{R}_k(\mu)$ is the measurement noise covariance matrix.

A.9: The target survival probability and target detection probability are independent of the kinetic state, i.e.,

$$p_{S,k|k-1}(\boldsymbol{\xi}',\mu') = p_{S,k|k-1}(\mu') \tag{8.3.26}$$

$$p_{D,k}(\boldsymbol{\xi},\mu) = p_{D,k}(\mu) \tag{8.3.27}$$

Assumptions A. 8 and A. 9 are commonly adopted in maneuvering target tracking algorithms, such as [9, 54].

A.10: the birth and spawning RFS intensities can be expressed in the form of Gaussian mixtures

$$v_{\gamma,k}(\boldsymbol{\xi},\mu) = \tilde{v}_{\gamma,k}(\mu)\sum_{i=1}^{J_{\gamma,k}(\mu)} w_{\gamma,k}^{(i)}(\mu)\mathcal{N}(\boldsymbol{\xi};\boldsymbol{m}_{\gamma,k}^{(i)}(\mu),\boldsymbol{P}_{\gamma,k}^{(i)}(\mu)) \tag{8.3.28}$$

$$v_{\beta,k|k-1}(\boldsymbol{\xi},\mu|\boldsymbol{\xi}',\mu') = \tilde{v}_{\beta,k|k-1}(\mu|\mu')$$

$$\times \sum_{i=1}^{J_{\beta,k}(\mu,\mu')} w_{\beta,k}^{(i)}(\mu,\mu')\mathcal{N}(\boldsymbol{\xi}; \boldsymbol{F}_{\beta,k-1}^{(i)}(\mu,\mu')\boldsymbol{\xi}' + \boldsymbol{d}_{\beta,k-1}^{(i)}(\mu,\mu'), \boldsymbol{P}_{\beta,k-1}^{(i)}(\mu,\mu')) \tag{8.3.29}$$

where $J_{\gamma,k}(\mu)$, $w_{\gamma,k}^{(i)}(\mu)$, $\boldsymbol{m}_{\gamma,k}^{(i)}(\mu)$ and $\boldsymbol{P}_{\gamma,k}^{(i)}(\mu)$, $i = 1, 2, \cdots, J_{\gamma,k}(\mu)$ are the parameters of the Gaussian mixture density of the birth target kinetic state at time k conditioned on mode μ, and $\tilde{v}_{\gamma,k}(\mu)$ is the birth mode intensity at time k. Similarly, $J_{\beta,k}(\mu,\mu')$, $w_{\beta,k}^{(i)}(\mu,\mu')$, $\boldsymbol{F}_{\beta,k-1}^{(i)}(\mu,\mu')$, $\boldsymbol{d}_{\beta,k-1}^{(i)}(\mu,\mu')$ and $\boldsymbol{P}_{\beta,k-1}^{(i)}(\mu,\mu')$, $i = 1, 2, \cdots, J_{\beta,k}(\mu,\mu')$ are, conditioned on mode μ, the parameters of the Gaussian mixture density of the kinetic state at time k of the target derived from augmented state $[\boldsymbol{\xi}'^{\mathrm{T}}, \mu']^{\mathrm{T}}$ at time $k-1$, and $\tilde{v}_{\beta,k|k-1}(\cdot|\mu')$ is the mode intensity at time k derived from the target with mode μ' at time $k-1$.

Compared with the model adopted by the standard multi-target tracking algorithm, the multi-target model of the LGJM system is more general. Most of existing algorithms don't consider the birth and spawning processes, whereas the multi-target model herein involves both. When the birth and spawning processes may vary from mode to mode, the birth model is compatible with both birth and spawning models with different intensities for a given mode μ. Similarly, the multi-target model of the LGJM system also involves the target death (survival) and target detection models in a given model μ. For such a general model, the traditional multi-target filtering techniques are computationally intractable.

8.3.2.2 Gaussian Mixture Implementation

By using two lemmas in Appendix A, the closed-form PHD recursion of the multi-target model of the LGJM system can be derived.

Proposition 24 *For the LGJM multi-target model, if posterior intensity v_{k-1} at time $k-1$ is a Gaussian mixture*

$$v_{k-1}(\boldsymbol{\xi}', \mu') = \sum_{i=1}^{J_{k-1}(\mu')} w_{k-1}^{(i)}(\mu')\mathcal{N}(\boldsymbol{\xi}'; \boldsymbol{m}_{k-1}^{(i)}(\mu'), \boldsymbol{P}_{k-1}^{(i)}(\mu')) \tag{8.3.30}$$

then predicted intensity $v_{k|k-1}$ at time k is also in the form of Gaussian mixture, which is

$$v_{k|k-1}(\boldsymbol{\xi}, \mu) = v_{S,k|k-1}(\boldsymbol{\xi}, \mu) + v_{\beta,k|k-1}(\boldsymbol{\xi}, \mu) + v_{\gamma,k}(\boldsymbol{\xi}, \mu) \tag{8.3.31}$$

where $v_{\gamma,k}(\cdot)$ is the birth intensity at time k and is given by (8.3.28), and

$$v_{S,k|k-1}(\boldsymbol{\xi},\mu)=\sum_{\mu'}\sum_{j=1}^{J_{k-1}(\mu')} w_{S,k|k-1}^{(j)}(\mu,\mu')\mathcal{N}(\boldsymbol{\xi};\boldsymbol{m}_{S,k|k-1}^{(j)}(\mu,\mu'),\boldsymbol{P}_{S,k|k-1}^{(j)}(\mu,\mu')) \tag{8.3.32}$$

$$w_{S,k|k-1}^{(j)}(\mu,\mu')=\tau_{k|k-1}(\mu|\mu')p_{S,k}(\mu')w_{k-1}^{(j)}(\mu') \tag{8.3.33}$$

$$\boldsymbol{m}_{S,k|k-1}^{(j)}(\mu,\mu')=\boldsymbol{F}_{S,k-1}(\mu)\boldsymbol{m}_{k-1}^{(j)}(\mu') \tag{8.3.34}$$

$$\boldsymbol{P}_{S,k|k-1}^{(j)}(\mu,\mu')=\boldsymbol{F}_{S,k-1}(\mu)\boldsymbol{P}_{k-1}^{(j)}(\mu')\boldsymbol{F}_{S,k-1}^{\mathrm{T}}(\mu)+\boldsymbol{Q}_{S,k-1}(\mu) \tag{8.3.35}$$

$$\begin{aligned}&v_{\beta,k|k-1}(\boldsymbol{\xi},\mu)\\&=\sum_{\mu'}\sum_{i=1}^{J_{k-1}(\mu')}\sum_{j=1}^{J_{\beta,k}(\mu,\mu')} w_{\beta,k|k-1}^{(i,j)}(\mu,\mu')\mathcal{N}(\boldsymbol{\xi};\boldsymbol{m}_{\beta,k|k-1}^{(i,j)}(\mu,\mu'),\boldsymbol{P}_{\beta,k|k-1}^{(i,j)}(\mu,\mu'))\end{aligned} \tag{8.3.36}$$

$$w_{\beta,k|k-1}^{(i,j)}(\mu,\mu')=\tilde{v}_{\beta,k|k-1}(\mu|\mu')w_{k-1}^{(i)}(\mu')w_{\beta,k}^{(j)}(\mu,\mu') \tag{8.3.37}$$

$$\boldsymbol{m}_{\beta,k|k-1}^{(i,j)}(\mu,\mu')=\boldsymbol{F}_{\beta,k-1}^{(j)}(\mu,\mu')\boldsymbol{m}_{k-1}^{(i)}(\mu')+\boldsymbol{d}_{\beta,k-1}^{(j)}(\mu,\mu') \tag{8.3.38}$$

$$\boldsymbol{P}_{\beta,k|k-1}^{(i,j)}(\mu,\mu')=\boldsymbol{F}_{\beta,k-1}^{(j)}(\mu,\mu')\boldsymbol{P}_{k-1}^{(i)}(\mu')(\boldsymbol{F}_{\beta,k-1}^{(j)}(\mu,\mu'))^{\mathrm{T}}+\boldsymbol{P}_{\beta,k-1}^{(j)}(\mu,\mu') \tag{8.3.39}$$

Proof According to (4.2.1), due to the target birth, spawning and motion processes, the predicted intensity is composed of such 3 terms as $v_{\gamma,k}$, $v_{\beta,k|k-1}$ and $v_{S,k|k-1}$. $v_{\gamma,k}$ has been given by the multi-target model. For $v_{\beta,k|k-1}$, by substituting (8.3.29) and (8.3.30) into $\int v_{\beta,k|k-1}(\boldsymbol{x}|\boldsymbol{x}')v_{k-1}(\boldsymbol{x}')d\boldsymbol{x}'$, exchanging the order of summation and integration and applying Lemma 4 in Appendix A to each term, we can obtain (8.3.36). For $v_{S,k|k-1}$, by substituting (8.3.24) and (8.3.30) into $\int \phi_{k|k-1}(\boldsymbol{x}|\boldsymbol{x}')v_{k-1}(\boldsymbol{x}')d\boldsymbol{x}'$, exchanging the order of summation and integration and applying Lemma 4 in Appendix A to each term, we can obtain (8.3.32).

Proposition 25 *For the LGJM multi-target model, if predicted intensity $v_{k|k-1}$ at time k is a Gaussian mixture*

$$v_{k|k-1}(\boldsymbol{\xi},\mu)=\sum_{i=1}^{J_{k|k-1}(\mu)} w_{k|k-1}^{(i)}(\mu)\mathcal{N}(\boldsymbol{\xi};\boldsymbol{m}_{k|k-1}^{(i)}(\mu),\boldsymbol{P}_{k|k-1}^{(i)}(\mu)) \tag{8.3.40}$$

then posterior (updated) intensity v_k at time k is also a Gaussian mixture, i.e.,

$$v_k(\boldsymbol{\xi}, \mu) = (1 - p_{D,k}(\mu))v_{k|k-1}(\boldsymbol{\xi}, \mu) + \sum_{z \in Z_k} v_{D,k}(\boldsymbol{\xi}, \mu; z)$$
$$= v_{k|k-1}(\boldsymbol{\xi}, \mu) - v_{D,k}(\boldsymbol{\xi}, \mu) + \sum_{z \in Z_k} v_{D,k}(\boldsymbol{\xi}, \mu; z) \tag{8.3.41}$$

where

$$v_{D,k}(\boldsymbol{\xi}, \mu; z) = \sum_{j=1}^{J_{k|k-1}(\mu)} w_k^{(j)}(\mu; z)\mathcal{N}(\boldsymbol{\xi}; \boldsymbol{m}_{k|k}^{(j)}(\mu; z), \boldsymbol{P}_{k|k}^{(j)}(\mu)) \tag{8.3.42}$$

$$w_k^{(j)}(\mu; z) = \frac{p_{D,k}(\mu)w_{k|k-1}^{(j)}(\mu)q_k^{(j)}(\mu; z)}{\kappa_k(z) + \sum_\mu p_{D,k}(\mu)\sum_{i=1}^{J_{k|k-1}(\mu)} w_{k|k-1}^{(i)}(\mu)q_k^{(i)}(\mu; z)} \tag{8.3.43}$$

$$q_k^{(j)}(\mu; z) = \mathcal{N}(z; \boldsymbol{\eta}_{k|k-1}^{(j)}(\mu), \boldsymbol{S}_{k|k-1}^{(j)}(\mu)) \tag{8.3.44}$$

$$\boldsymbol{m}_{k|k}^{(j)}(\mu; z) = \boldsymbol{m}_{k|k-1}^{(j)}(\mu) + \boldsymbol{G}_k^{(j)}(\mu)(z - \boldsymbol{\eta}_{k|k-1}^{(j)}(\mu)) \tag{8.3.45}$$

$$\boldsymbol{P}_{k|k}^{(j)}(\mu) = (\boldsymbol{I} - \boldsymbol{G}_k^{(j)}(\mu)\boldsymbol{H}_k(\mu))\boldsymbol{P}_{k|k-1}^{(j)}(\mu) \tag{8.3.46}$$

$$\boldsymbol{\eta}_{k|k-1}^{(j)}(\mu) = \boldsymbol{H}_k(\mu)\boldsymbol{m}_{k|k-1}^{(j)}(\mu) \tag{8.3.47}$$

$$\boldsymbol{S}_{k|k-1}^{(j)}(\mu) = \boldsymbol{H}_k(\mu)\boldsymbol{P}_{k|k-1}^{(j)}(\mu)\boldsymbol{H}_k^{\mathrm{T}}(\mu) + \boldsymbol{R}_k(\mu) \tag{8.3.48}$$

$$\boldsymbol{G}_k^{(j)}(\mu) = \boldsymbol{P}_{k|k-1}^{(j)}(\mu)\boldsymbol{H}_k^{\mathrm{T}}(\mu)(\boldsymbol{S}_{k|k-1}^{(j)}(\mu))^{-1} \tag{8.3.49}$$

Proof According to (4.2.3), the updated intensity is composed of 3 terms. The first term is the given predicted intensity $v_{k|k-1}(\boldsymbol{\xi}, \mu)$, the second term is the product $p_{D,k}(\mu)v_{k|k-1}(\boldsymbol{\xi}, \mu)$ and denoted as $v_{D,k}(\boldsymbol{\xi}, \mu)$, the third term is the sum $\sum_{z \in Z_k} v_{D,k}(\boldsymbol{\xi}, \mu; z)$, where

$$v_{D,k}(\boldsymbol{\xi}, \mu; z) = v_{D,k}(\boldsymbol{x}; z) = \frac{p_{D,k}(\boldsymbol{x})g_k(z|\boldsymbol{x})v_{k|k-1}(\boldsymbol{x})}{\kappa_k(z) + \int p_{D,k}(\boldsymbol{x}')g_k(z|\boldsymbol{x}')v_{k|k-1}(\boldsymbol{x}')d\boldsymbol{x}'} \tag{8.3.50}$$

For $v_{D,k}(\boldsymbol{\xi}, \mu; z)$, first, substituting (8.3.25) and (8.3.40) into the numerator in (8.3.50) and applying Lemma 5 in Appendix A yields the sum of weighted Gaussians. Then, applying Lemma 4 in Appendix A to the integral in the denominator in (8.3.50) gives the (double) summations in the denominator in (8.3.43). Finally, combining the results of the numerator and denominator in (8.3.50) yields (8.3.42).

Propositions 24 and 25 show the process of analytical propagation of intensities $v_{k|k-1}$ and v_k over time under the assumption of the LGJM system multi-target model.

The recursions of means and covariances of $v_{S,k|k-1}(\cdot)$ and $v_{\beta,k|k-1}(\cdot)$ correspond to the Kalman prediction, and those of $v_{D,k}(\cdot)$ correspond to the Kalman update. The complexity of the PHD filter is $\mathcal{O}(J_{k-1}|Z_k|)$, where J_{k-1} is the number of Gaussian components used to express v_{k-1} under a given model μ' at time $k-1$, and $|Z_k|$ represents the number of measurements at time k.

The above propositions also show that the number of components of predicted and posterior intensities increases with time and needs to be processed by applying a pruning step similar to the GM-PHD filter.

According to Propositions 24 and 25, it is easy to draw the following two corollary.

Corollary 3 *According to Proposition* 24, *the expected number of predicted targets is*

$$\hat{N}_{k|k-1} = \hat{N}_{S,k|k-1} + \hat{N}_{\beta,k|k-1} + \hat{N}_{\gamma,k} \tag{8.3.51}$$

where

$$\hat{N}_{\gamma,k} = \sum_{\mu} \sum_{i=1}^{J_{\gamma,k}(\mu)} \tilde{v}_{\gamma,k}(\mu) w_{\gamma,k}^{(i)}(\mu) \tag{8.3.52}$$

$$\hat{N}_{\beta,k|k-1} = \sum_{\mu} \sum_{\mu'} \sum_{i=1}^{J_{k-1}(\mu')} \sum_{j=1}^{J_{\beta,k}(\mu,\mu')} \tilde{v}_{\beta,k|k-1}(\mu|\mu') w_{k-1}^{(i)}(\mu') w_{\beta,k}^{(j)}(\mu,\mu') \tag{8.3.53}$$

$$\hat{N}_{S,k|k-1} = \sum_{\mu} \sum_{\mu'} \sum_{i=1}^{J_{k-1}(\mu')} p_{S,k}(\mu') \tau_{k|k-1}(\mu|\mu') w_{k-1}^{(i)}(\mu') \tag{8.3.54}$$

Corollary 4 *Under the premise of Proposition* 25, *the expected number of targets is*

$$\hat{N}_k = \sum_{\mu} (1 - p_{D,k}(\mu)) \sum_{i=1}^{J_{k|k-1}(\mu)} w_{k|k-1}^{(i)}(\mu) + \sum_{z \in Z_k} \sum_{\mu} \sum_{i=1}^{J_{k|k-1}(\mu)} w_k^{(i)}(\mu; z) \tag{8.3.55}$$

For the posterior intensity at time k

$$v_k(\boldsymbol{\xi}, \mu) = \sum_{i=1}^{J_k(\mu)} w_k^{(i)}(\mu) \mathcal{N}(\boldsymbol{\xi}; \boldsymbol{m}_k^{(i)}(\mu), \boldsymbol{P}_k^{(i)}(\mu)) \tag{8.3.56}$$

the peaks of the intensity correspond to the points where the expected number of targets have the strongest local concentration. In order to extract target states from posterior intensity v_k at time k, it is necessary to estimate the number $\hat{N}_k$ of targets, which can be obtained by rounding off $\sum_{i=1}^{J_k(\mu)} w_k^{(i)}(\mu)$. The estimation of the multi-target state is the set of $\hat{N}_k$ ordered pairs of mean and mode $(\boldsymbol{m}_k^{(i)}(\mu), \mu)$ with the largest weights $w_k^{(i)}(\mu)$; $\mu \in \mathcal{M}, i = 1, \ldots, J_k(\mu)$.

8.3.2.3 A Generalized Solution to the MM-PHD Recursion

In addition to the linear Gaussian multi-target model and the multi-target model of the LGJM system, the closed form solution to the MM-PHD recursion under a more general condition can also be obtained.

Proposition 26 *Let the parameters of multi-target state transition model of Eqs.* (8.3.26) *and* (8.3.24) *be relaxed to*

$$p_{S,k|k-1}(\boldsymbol{\xi}', \mu') = w_{S,k|k-1}^{(0)}(\mu') + \sum_{j=1}^{J_{S,k|k-1}(\mu')} w_{S,k|k-1}^{(j)}(\mu')\mathcal{N}(\boldsymbol{\xi}'; \boldsymbol{m}_{S,k|k-1}^{(j)}(\mu'), \boldsymbol{P}_{S,k|k-1}^{(j)}(\mu')) \tag{8.3.57}$$

$$\phi_{k|k-1}(\boldsymbol{\xi}, \mu|\boldsymbol{\xi}', \mu') = \sum_{j=1}^{J_{\phi,k|k-1}(\mu,\mu')} w_{\phi,k|k-1}^{(j)}(\mu, \mu')\mathcal{N}(\boldsymbol{\xi}; \boldsymbol{F}_{S,k-1}^{(j)}(\mu, \mu')\boldsymbol{\xi}', \boldsymbol{Q}_{S,k-1}^{(j)}(\mu, \mu')) \tag{8.3.58}$$

If posterior intensity v_{k-1} at time $k-1$ is in the Gaussian mixture form given in (8.3.30), the predicted intensity $v_{k|k-1}$ at time k is also in the following Gaussian mixture form

$$v_{k|k-1}(\boldsymbol{\xi}, \mu) = v_{S,k|k-1}(\boldsymbol{\xi}, \mu) + v_{\beta,k|k-1}(\boldsymbol{\xi}, \mu) + v_{\gamma,k}(\boldsymbol{\xi}, \mu) \tag{8.3.59}$$

where $v_{\gamma,k}(\cdot)$ is the birth intensity at time k given by (8.3.28), $v_{\beta,k|k-1}(\boldsymbol{\xi}, \mu)$ is the target spawning intensity given by (8.3.36), and

$$v_{S,k|k-1}(\boldsymbol{\xi}, \mu) = \sum_{\mu'} \sum_{i=1}^{J_{k-1}(\mu')} \sum_{s=0}^{J_{S,k|k-1}(\mu')} \sum_{j=1}^{J_{\phi,k|k-1}(\mu,\mu')} w_{S,k|k-1}^{(i,j,s)}(\mu, \mu')\mathcal{N}(\boldsymbol{\xi}; \boldsymbol{m}_{S,k|k-1}^{(i,j,s)}(\mu, \mu'), \boldsymbol{P}_{S,k|k-1}^{(i,j,s)}(\mu, \mu')) \tag{8.3.60}$$

$$w_{S,k|k-1}^{(i,j,s)}(\mu, \mu') = w_{\phi,k|k-1}^{(j)}(\mu, \mu')w_{S,k|k-1}^{(s)}(\mu')w_{k-1}^{(i)}(\mu')q_{S,k|k-1}^{(i,s)}(\mu') \tag{8.3.61}$$

$$q_{S,k|k-1}^{(i,s)}(\mu') = \mathcal{N}(\boldsymbol{m}_{S,k|k-1}^{(s)}(\mu'); \boldsymbol{m}_{k-1}^{(i)}(\mu'), \boldsymbol{P}_{S,k|k-1}^{(s)}(\mu') + \boldsymbol{P}_{k-1}^{(i)}(\mu')), \quad q_{S,k|k-1}^{(i,0)}(\mu') = 1 \tag{8.3.62}$$

$$\boldsymbol{m}_{S,k|k-1}^{(i,j,s)}(\mu, \mu') = \boldsymbol{F}_{S,k-1}^{(j)}(\mu, \mu')\boldsymbol{m}_{S,k|k-1}^{(i,s)}(\mu') \tag{8.3.63}$$

$$\boldsymbol{m}_{S,k|k-1}^{(i,s)}(\mu') = \boldsymbol{m}_{k-1}^{(i)}(\mu') + \boldsymbol{G}_{k-1}^{(i,s)}(\mu')(\boldsymbol{m}_{S,k|k-1}^{(s)}(\mu') - \boldsymbol{m}_{k-1}^{(i)}(\mu')),\ \boldsymbol{m}_{S,k|k-1}^{(i,0)}(\mu') = \boldsymbol{m}_{k-1}^{(i)}(\mu') \tag{8.3.64}$$

$$\boldsymbol{P}_{S,k|k-1}^{(i,j,s)}(\mu,\mu') = \boldsymbol{F}_{S,k-1}^{(j)}(\mu,\mu')\boldsymbol{P}_{S,k|k-1}^{(i,s)}(\mu')[\boldsymbol{F}_{S,k-1}^{(j)}(\mu,\mu')]^{\mathrm{T}} + \boldsymbol{Q}_{S,k-1}^{(j)}(\mu,\mu') \tag{8.3.65}$$

$$\boldsymbol{P}_{S,k|k-1}^{(i,s)}(\mu') = (\boldsymbol{I} - \boldsymbol{G}_{k-1}^{(i,s)}(\mu'))\boldsymbol{P}_{k-1}^{(i)}(\mu'),\ \ \boldsymbol{P}_{S,k|k-1}^{(i,0)}(\mu') = \boldsymbol{P}_{k-1}^{(i)}(\mu') \tag{8.3.66}$$

$$\boldsymbol{G}_{k-1}^{(i,s)}(\mu') = \boldsymbol{P}_{k-1}^{(i)}(\mu')(\boldsymbol{P}_{k-1}^{(i)}(\mu') + \boldsymbol{P}_{S,k|k-1}^{(s)}(\mu'))^{-1} \tag{8.3.67}$$

Proof For $v_{S,k|k-1}(\boldsymbol{\xi},\mu)$, firstly substituting Eqs. (8.3.57) and (8.3.30) into $p_{S,k|k-1}(\boldsymbol{x}')v_{k-1}(\boldsymbol{x}')$, and applying Lemma 5 in Appendix A, we can obtain the (double) sums of weighted Gaussians. Then, by substituting the resulting Gaussian mixture and (8.3.58) into $\int p_{S,k|k-1}(\boldsymbol{x}')\phi_{k|k-1}(\boldsymbol{x}|\boldsymbol{x}')v_{k-1}(\boldsymbol{x}')d\boldsymbol{x}'$, exchanging the order of summation and integration and applying Lemma 4 in Appendix A to each term, we can obtain (8.3.60).

Proposition 27 *Let the parameters of multi-target measurement model of Eqs. (8.3.27) and (8.3.25) be relaxed to*

$$p_{D,k}(\boldsymbol{\xi},\mu) = w_{D,k}^{(0)}(\mu) + \sum_{j=1}^{J_{D,k}(\mu)} w_{D,k}^{(j)}(\mu)\mathcal{N}(\boldsymbol{\xi};\boldsymbol{m}_{D,k}^{(j)}(\mu),\boldsymbol{P}_{D,k}^{(j)}(\mu)) \tag{8.3.68}$$

$$g_k(z|\boldsymbol{\xi},\mu) = \sum_{j=1}^{J_{g,k}(\mu)} w_{g,k}^{(j)}(\mu)\mathcal{N}(z;\boldsymbol{H}_k^{(j)}(\mu)\boldsymbol{\xi},\boldsymbol{R}_k^{(j)}(\mu)) \tag{8.3.69}$$

If predicted intensity $v_{k|k-1}$ at time k is in the Gaussian mixture form given in (8.3.40), the posterior (or updated) intensity v_k at time k is also in the Gaussian mixture form as follows

$$v_k(\boldsymbol{\xi},\mu) = v_{k|k-1}(\boldsymbol{\xi},\mu) - v_{D,k}(\boldsymbol{\xi},\mu) + \sum_{z\in Z_k} v_{D,k}(\boldsymbol{\xi},\mu;z) \tag{8.3.70}$$

where

$$v_{D,k}(\boldsymbol{\xi},\mu) = \sum_{i=1}^{J_{k|k-1}(\mu)}\sum_{j=0}^{J_{D,k}(\mu)} w_{k|k-1}^{(i,j)}(\mu)\mathcal{N}(\boldsymbol{\xi};\boldsymbol{m}_{k|k-1}^{(i,j)}(\mu),\boldsymbol{P}_{k|k-1}^{(i,j)}(\mu)) \tag{8.3.71}$$

$$w_{k|k-1}^{(i,j)}(\mu) = w_{D,k}^{(j)}(\mu)w_{k|k-1}^{(i)}(\mu)q_{k|k-1}^{(i,j)}(\mu) \tag{8.3.72}$$

$$q_{k|k-1}^{(i,j)}(\mu) = \mathcal{N}(\boldsymbol{m}_{D,k}^{(j)}(\mu); \boldsymbol{m}_{k|k-1}^{(i)}(\mu), \boldsymbol{P}_{D,k}^{(j)}(\mu) + \boldsymbol{P}_{k|k-1}^{(i)}(\mu)), \quad q_{k|k-1}^{(i,0)}(\mu) = 1 \tag{8.3.73}$$

$$\begin{aligned}\boldsymbol{m}_{k|k-1}^{(i,j)}(\mu) = \boldsymbol{m}_{k|k-1}^{(i)}(\mu) + \boldsymbol{G}_{k|k-1}^{(i,j)}(\mu)(\boldsymbol{m}_{D,k}^{(j)}(\mu) \\ - \boldsymbol{m}_{k|k-1}^{(i)}(\mu)), \quad \boldsymbol{m}_{k|k-1}^{(i,0)}(\mu) = \boldsymbol{m}_{k|k-1}^{(i)}(\mu)\end{aligned} \tag{8.3.74}$$

$$\boldsymbol{P}_{k|k-1}^{(i,j)}(\mu) = [\boldsymbol{I} - \boldsymbol{G}_{k|k-1}^{(i,j)}(\mu)]\boldsymbol{P}_{k|k-1}^{(i)}(\mu), \quad \boldsymbol{P}_{k|k-1}^{(i,0)}(\mu) = \boldsymbol{P}_{k|k-1}^{(i)}(\mu) \tag{8.3.75}$$

$$\boldsymbol{G}_{k|k-1}^{(i,j)}(\mu) = \boldsymbol{P}_{k|k-1}^{(i)}(\mu)(\boldsymbol{P}_{k|k-1}^{(i)}(\mu) + \boldsymbol{P}_{D,k}^{(j)}(\mu))^{-1} \tag{8.3.76}$$

$$v_{D,k}(\boldsymbol{\xi}, \mu; \boldsymbol{z}) = \sum_{i=1}^{J_{k|k-1}(\mu)} \sum_{j=0}^{J_{D,k}(\mu)} \sum_{l=1}^{J_{g,k}(\mu)} w_k^{(i,l,j)}(\mu; \boldsymbol{z})\mathcal{N}(\boldsymbol{\xi}; \boldsymbol{m}_{k|k}^{(i,l,j)}(\mu; \boldsymbol{z}), \boldsymbol{P}_{k|k}^{(i,l,j)}(\mu)) \tag{8.3.77}$$

$$w_k^{(i,l,j)}(\mu; \boldsymbol{z}) = \frac{w_{k|k-1}^{(i,j)}(\mu)w_{g,k}^{(l)}(\mu)q_k^{(i,l,j)}(\mu; \boldsymbol{z})}{\kappa_k(\boldsymbol{z}) + \sum_{r=1}^{J_{k|k-1}(\mu)} \sum_{s=0}^{J_{D,k}(\mu)} \sum_{t=1}^{J_{g,k}(\mu)} w_{k|k-1}^{(r,s)}(\mu)w_{g,k}^{(t)}(\mu)q_k^{(r,s,t)}(\mu; \boldsymbol{z})} \tag{8.3.78}$$

$$q_k^{(i,l,j)}(\mu; \boldsymbol{z}) = \mathcal{N}(\boldsymbol{z}; \boldsymbol{\eta}_{k|k-1}^{(i,l,j)}(\mu), \boldsymbol{S}_{k|k-1}^{(i,l,j)}(\mu)) \tag{8.3.79}$$

$$\boldsymbol{m}_{k|k}^{(i,l,j)}(\mu; \boldsymbol{z}) = \boldsymbol{m}_{k|k-1}^{(i,j)}(\mu) + \boldsymbol{G}_k^{(i,l,j)}(\mu)(\boldsymbol{z} - \boldsymbol{\eta}_{k|k-1}^{(i,l,j)}(\mu)) \tag{8.3.80}$$

$$\boldsymbol{P}_{k|k}^{(i,l,j)}(\mu) = (\boldsymbol{I} - \boldsymbol{G}_k^{(i,l,j)}(\mu)\boldsymbol{H}_k^{(l)}(\mu))\boldsymbol{P}_{k|k-1}^{(i,j)}(\mu) \tag{8.3.81}$$

$$\boldsymbol{\eta}_{k|k-1}^{(i,l,j)}(\mu) = \boldsymbol{H}_k^{(l)}(\mu)\boldsymbol{m}_{k|k-1}^{(i,j)}(\mu) \tag{8.3.82}$$

$$\boldsymbol{S}_{k|k-1}^{(i,l,j)}(\mu) = \boldsymbol{H}_k^{(l)}(\mu)\boldsymbol{P}_{k|k-1}^{(i,j)}(\mu)[\boldsymbol{H}_k^{(l)}(\mu)]^{\mathrm{T}} + \boldsymbol{R}_k^{(l)}(\mu) \tag{8.3.83}$$

$$\boldsymbol{G}_k^{(i,l,j)}(\mu) = \boldsymbol{P}_{k|k-1}^{(i,j)}(\mu)[\boldsymbol{H}_k^{(l)}(\mu)]^{\mathrm{T}}[\boldsymbol{S}_{k|k-1}^{(i,j)}(\mu)]^{-1} \tag{8.3.84}$$

Proof For $v_{D,k}(\boldsymbol{\xi}, \mu)$, substituting Eqs. (8.3.68) and (8.3.40) into $p_{D,k}(\boldsymbol{x})v_{k|k-1}(\boldsymbol{x})$, and applying Lemma 5 in Appendix A to each term, we can obtain (8.3.71). For $v_{D,k}(\boldsymbol{\xi}, \mu; \boldsymbol{z})$, firstly substituting Eqs. (8.3.69) and (8.3.71) into the numerator in (8.3.50), and applying Lemma 5 in Appendix A, we can obtain the (triple) summations of weighted Gaussians. Then, applying Lemma 4 in Appendix A to the integral in the denominator of (8.3.50) gives the (triple) summations in the denominator in (8.3.78). Finally, combining the numerator and denominator in (8.3.50) yields (8.3.77).

8.4 Multiple Model CBMeMBer (MM-CBMeMBer) Filter

In this section, the multiple model CBMeMBer (MM-CBMeMBer) recursion is given first. On this basis, the sequential Monte Carlo (SMC) implementation and Gaussian mixture (GM) implementation of the MM-CBMeMBer recursion are introduced respectively, and the two are respectively called are sequential Monte Carlo MM-CBMeMBer (SMC-MM-CBMeMBer) and Gaussian mixture MM-CBMeMBer (GM-MM-CBMeMBer) filters.

8.4.1 MM-CBMeMBer Recursion

Using the technology similar to the MM-PHD filter described in Sect. 8.3, the CBMeMBer filter can be extended to the multiple model case, which is achieved by extending the traditional Bernoulli RFS state variable to include the discrete random variable representing the motion model. Let the dimension-expanded RFS be

$$\begin{aligned}(X, \mathcal{M}) &= \{(\boldsymbol{\xi}, \mu)^{(1)}, \cdots, (\boldsymbol{\xi}, \mu)^{(M)}\} \\ &= \bigcup_{m=1}^{M} \{(\boldsymbol{\xi}, \mu)^{(m)}\} = \bigcup_{m=1}^{M} (X, \mathcal{M})^{(m)}\end{aligned} \tag{8.4.1}$$

where M is the number of the MB components, each single element $(X, \mathcal{M})^{(m)}$ is expressed as the Bernoulli set $(\varepsilon^{(m)}, p^{(m)}(\boldsymbol{\xi}, \mu))$ (referred to as the MM-BRFS), and its corresponding union set is referred to as the MM-MBRFS.

Then, the multi-target PDF at time k has the following MM-MBRFS form

$$\pi_k(X, \mathcal{M}) \sim \{(\varepsilon_k^{(i)}, p_k^{(i)}(\boldsymbol{\xi}, \mu))\}_{i=1}^{M_k} \tag{8.4.2}$$

Similar to the standard CBMeMBer filter, the MM-MBRFS forms a recursion filter through the prediction and update and it is called the MM-CBMeMber filter.

8.4.1.1 Prediction of MM-CBMeMber Filter

The prediction step of the MM-CBMeMBer filter is quite similar to that of the standard CBMeMBer filter. The predicted multi-target PDF at time $k-1$ is in the form of MM-MBRFS $\pi_{k|k-1}(X, \mathcal{M})$, and it can be approximated as the union of the following two MM-MBRFS sets

$$\pi_{k|k-1}(X, \mathcal{M}) \approx \{(\varepsilon_{S,k|k-1}^{(i)}, p_{S,k|k-1}^{(i)}(\boldsymbol{\xi}, \mu))\}_{i=1}^{M_{k-1}} \cup \{(\varepsilon_{\gamma,k}^{(i)}, p_{\gamma,k}^{(i)}(\boldsymbol{\xi}, \mu))\}_{i=1}^{M_{\gamma,k}} \tag{8.4.3}$$

where $\{(\varepsilon_{\gamma,k}^{(i)}, p_{\gamma,k}^{(i)}(\boldsymbol{\xi}, \mu))\}_{i=1}^{M_{\gamma,k}}$ is the parameters of the birth MM-MBRFS at time k, and note that it takes into account the initial model state. Similar to the birth state intensity equation $v_{\gamma,k}(\boldsymbol{\xi}, \mu) = p_{\gamma,k}(\boldsymbol{\xi}|\mu)\tilde{v}_{\gamma,k}(\mu)$ of multiple model PHD (see (8.3.20)), the dimension-extended birth state distribution including the initial model state is

$$p_{\gamma,k}(\boldsymbol{\xi}, \mu) = p_{\gamma,k}(\boldsymbol{\xi}|\mu)p_{\mu,k}(\mu) \tag{8.4.4}$$

The predicted MM-MBRFS can be calculated by using equations similar to prediction Eqs. (6.2.3)–(6.2.4) of the basic CBMeMBer filter, but it needs to be extended to include multiple model variables. The single model random equation is replaced by the model-dependent transition equation and the new MM-CBMeMBer prediction equation is

$$\varepsilon_{S,k|k-1}^{(i)} = \varepsilon_{k-1}^{(i)}\langle p_{k-1}^{(i)}(\cdot, \cdot), p_{S,k}(\cdot, \cdot)\rangle \tag{8.4.5}$$

$$p_{S,k|k-1}^{(i)}(\boldsymbol{\xi}, \mu) = \frac{\langle \phi_{k|k-1}(\boldsymbol{\xi}, \mu|\cdot, \cdot), p_{k-1}^{(i)}(\cdot, \cdot)p_{S,k}(\cdot, \cdot)\rangle}{\langle p_{k-1}^{(i)}(\cdot, \cdot), p_{S,k}(\cdot, \cdot)\rangle} \tag{8.4.6}$$

where it shall be emphasized that, the symbol $\langle \cdot, \cdot \rangle$ indicates the integral over full state variables (including the state variable and the mode variable), i.e.,

$$\begin{aligned}\langle \alpha, \beta \rangle &= \int_{(X,\mathcal{M})} \alpha(\boldsymbol{\xi}, \mu)\beta(\boldsymbol{\xi}, \mu)d\boldsymbol{\xi}d\mu \\ &= \int_X \int_{\mathcal{M}} \alpha(\boldsymbol{\xi}, \mu)\beta(\boldsymbol{\xi}, \mu)d\mu d\boldsymbol{\xi} \\ &= \sum_{\mu} \int_X \alpha(\boldsymbol{\xi}|\mu)\beta(\boldsymbol{\xi}|\mu)d\boldsymbol{\xi}\end{aligned} \tag{8.4.7}$$

8.4.1.2 Update of MM-CBMeMber Filter

The update operation of the MM-CBMeMBer filter is also quite similar to that of the standard CBMeMBer filter. The MM-MBRFS posterior multi-target PDF $\pi_k(X, \mathcal{M})$ at time k given by (8.4.2) consists of the union of the following two MM-MBRFS sets

$$\pi_k(X, \mathcal{M}) \approx \{(\varepsilon_{L,k}^{(i)}, p_{L,k}^{(i)}(\boldsymbol{\xi}, \mu))\}_{i=1}^{M_{k|k-1}} \cup \{(\varepsilon_{U,k}(z), p_{U,k}(\boldsymbol{\xi}, \mu; z))\}_{z\in Z_k} \tag{8.4.8}$$

The MM-MBRFS update equation in (8.4.8) is similar to the standard CBMeMBer update equation given by (6.3.8), but it is generalized to incorporate the multiple model parameter. Similar to Eqs. (6.2.7)–(6.2.8), the (legacy) update equation for missed detections is

$$\varepsilon_{L,k}^{(i)} = \varepsilon_{k|k-1}^{(i)} \frac{1 - \langle p_{k|k-1}^{(i)}(\cdot,\cdot), p_{D,k}(\cdot,\cdot)\rangle}{1 - \varepsilon_{k|k-1}^{(i)} \langle p_{k|k-1}^{(i)}(\cdot,\cdot), p_{D,k}(\cdot,\cdot)\rangle} \tag{8.4.9}$$

$$p_{L,k}^{(i)}(\boldsymbol{\xi},\mu) = p_{k|k-1}^{(i)}(\boldsymbol{\xi},\mu) \frac{1 - p_{D,k}(\boldsymbol{\xi},\mu)}{1 - \langle p_{k|k-1}^{(i)}(\cdot,\cdot), p_{D,k}(\cdot,\cdot)\rangle} \tag{8.4.10}$$

Finally, the measurement update MM-MBRFS equation in (8.4.8) is also similar to the standard measurement update equation in (6.2.21) and (6.3.9). However, the single-target measurement likelihood function is replaced by the model-dependent measurement likelihood function $g_k(\mathbf{z}|\boldsymbol{\xi},\mu)$, i.e.,

$$\varepsilon_{U,k}(\mathbf{z}) = \frac{\sum_{i=1}^{M_{k|k-1}} \frac{\varepsilon_{k|k-1}^{(i)} \langle p_{k|k-1}^{(i)}(\cdot,\cdot), \psi_{k,\mathbf{z}}(\cdot,\cdot)\rangle}{1-\varepsilon_{k|k-1}^{(i)} \langle p_{k|k-1}^{(i)}(\cdot,\cdot), p_{D,k}(\cdot,\cdot)\rangle}}{\kappa_k(\mathbf{z}) + \sum_{i=1}^{M_{k|k-1}} \frac{\varepsilon_{k|k-1}^{(i)} \langle p_{k|k-1}^{(i)}(\cdot,\cdot), \psi_{k,\mathbf{z}}(\cdot,\cdot)\rangle}{1-\varepsilon_{k|k-1}^{(i)} \langle p_{k|k-1}^{(i)}(\cdot,\cdot), p_{D,k}(\cdot,\cdot)\rangle}} \tag{8.4.11}$$

$$p_{U,k}(\boldsymbol{\xi},\mu;\mathbf{z}) = \frac{\sum_{i=1}^{M_{k|k-1}} \frac{\varepsilon_{k|k-1}^{(i)} p_{k|k-1}^{(i)}(\boldsymbol{\xi},\mu) \psi_{k,\mathbf{z}}(\boldsymbol{\xi},\mu)}{1-\varepsilon_{k|k-1}^{(i)} \langle p_{k|k-1}^{(i)}(\cdot,\cdot), p_{D,k}(\cdot,\cdot)\rangle}}{\sum_{i=1}^{M_{k|k-1}} \frac{\varepsilon_{k|k-1}^{(i)} \langle p_{k|k-1}^{(i)}(\cdot,\cdot), \psi_{k,\mathbf{z}}(\cdot,\cdot)\rangle}{1-\varepsilon_{k|k-1}^{(i)} \langle p_{k|k-1}^{(i)}(\cdot,\cdot), p_{D,k}(\cdot,\cdot)\rangle}} \tag{8.4.12}$$

$$\psi_{k,\mathbf{z}}(\boldsymbol{\xi},\mu) = p_{D,k}(\boldsymbol{\xi},\mu) g_k(\mathbf{z}|\boldsymbol{\xi},\mu) \tag{8.4.13}$$

It shall be noted that, the MBRFS model state transition is not involved in the update step.

The above MM-CBMeMBer equation is based on the standard CBMeMBer equation. Compared with the standard CBMeMBer equation, the key change is the addition of the model state variable. On this basis, the motion transition and measurement likelihood functions in the prediction and update steps are appropriately extended to capture and estimate the uncertainty of the target motion. The above prediction and update steps form the MM-CBMeMBer filter, based on which the SMC and GM implementations can be obtained, respectively.

8.4.2 Sequential Monte Carlo MM-CBMeMBer Filter

Like the SMC approximation of the standard CBMeMBer filter, the MM-CBMeMBer filter also can be implemented using the particle approximation. This implementation involves incorporating a model identification variable into the state-weight pair (two-tuple) of each particle, and this model parameter is a discrete representation of the current particle state model. The dimension-extended particle set is described in detail as follows

$$
\begin{aligned}
(X_k, \mathcal{M}_k) &= \{(\varepsilon_k^{(i)}, p_k^{(i)}(\boldsymbol{\xi}, \mu))\}_{i=1}^{M_k} \\
&\approx \{(\varepsilon_k^{(i)}, \{(\boldsymbol{\xi}_k^{(i,j)}, \mu_k^{(i,j)}, w_k^{(i,j)})\}_{j=1}^{L_k^{(i)}})\}_{i=1}^{M_k}
\end{aligned}
\tag{8.4.14}
$$

Similar to the SMC approximation, this forms the PDF approximation of the multiple model extended BRFS in (8.4.2), i.e.,

$$
p_k^{(i)}(\boldsymbol{\xi}, \mu) = \sum_{j=1}^{L_k^{(i)}} w_k^{(i,j)} \delta_{\boldsymbol{\xi}_k^{(i,j)}, \mu_k^{(i,j)}}(\boldsymbol{\xi}, \mu) \tag{8.4.15}
$$

8.4.2.1 Prediction of SMC-MM-CBMeMBer Filter

The transition of the dimension-extended particle set described by Eq. (8.4.14) can be forward predicted using the MM-CBMeMBer prediction equation in the SMC implementation. These equations are similar to those of the standard SMC-CBMeMBer, but they account for additional model-conditioned particle states.

Assume that the posterior MB multi-target density at time $k-1$ is $\pi_{k-1} = \{(\varepsilon_{k-1}^{(i)}, p_{k-1}^{(i)}(\boldsymbol{\xi}, \mu))\}_{i=1}^{M_{k-1}}$ and each of $p_{k-1}^{(i)}(\boldsymbol{\xi}, \mu), i = 1, \cdots, M_{k-1}$ is composed of a set of weighted samples $\{(w_{k-1}^{(i,j)}, \boldsymbol{\xi}_{k-1}^{(i,j)}, \mu_{k-1}^{(i,j)})\}_{j=1}^{L_{k-1}^{(i)}}$, i.e.,

$$
p_{k-1}^{(i)}(\boldsymbol{\xi}, \mu) = \sum_{j=1}^{L_{k-1}^{(i)}} w_{k-1}^{(i,j)} \delta_{\boldsymbol{\xi}_{k-1}^{(i,j)}, \mu_{k-1}^{(i,j)}}(\boldsymbol{\xi}, \mu) \tag{8.4.16}
$$

Then, the predicted multiple model (MB) multi-target density $\pi_{k|k-1} = \{(\varepsilon_{S,k|k-1}^{(i)}, p_{S,k|k-1}^{(i)})\}_{i=1}^{M_{k-1}} \cup \{(\varepsilon_{\gamma,k}^{(i)}, p_{\gamma,k}^{(i)})\}_{i=1}^{M_{\gamma,k}}$ can be calculated as follows

$$
\varepsilon_{S,k|k-1}^{(i)} = \varepsilon_{k-1}^{(i)} \sum_{j=1}^{L_{k-1}^{(i)}} w_{k-1}^{(i,j)} p_{S,k}(\boldsymbol{\xi}_{k-1}^{(i,j)}, \mu_{k-1}^{(i,j)}) \tag{8.4.17}
$$

$$
p_{S,k|k-1}^{(i)}(\boldsymbol{\xi}, \mu) = \sum_{j=1}^{L_{k-1}^{(i)}} \tilde{w}_{S,k|k-1}^{(i,j)} \delta_{\boldsymbol{\xi}_{S,k|k-1}^{(i,j)}, \mu_{S,k|k-1}^{(i,j)}}(\boldsymbol{\xi}, \mu) \tag{8.4.18}
$$

$$
\varepsilon_{\gamma,k}^{(i)} = \int_{(X,\mathcal{M})} b_{\boldsymbol{\xi},k}^{(i)}(\boldsymbol{\xi}|\mu, Z_k) b_{\mu,k}^{(i)}(\mu) d\boldsymbol{\xi} d\mu \tag{8.4.19}
$$

$$p_{\gamma,k}^{(i)}(\boldsymbol{\xi},\mu)=\sum_{j=1}^{L_{\gamma,k}^{(i)}}\tilde{w}_{\gamma,k}^{(i,j)}\delta_{\boldsymbol{\xi}_{\gamma,k}^{(i,j)},\mu_{\gamma,k}^{(i,j)}}(\boldsymbol{\xi},\mu) \tag{8.4.20}$$

where

$$\mu_{S,k|k-1}^{(i,j)}\sim q_{\mu,k}^{(i)}(\cdot|\mu_{k-1}^{(i,j)}),\ j=1,\cdots,L_{k-1}^{(i)} \tag{8.4.21}$$

$$\boldsymbol{\xi}_{S,k|k-1}^{(i,j)}\sim q_{\boldsymbol{\xi},k}^{(i)}(\cdot|\boldsymbol{\xi}_{k-1}^{(i,j)},\mu_{S,k|k-1}^{(i,j)},Z_k),\ j=1,\cdots,L_{k-1}^{(i)} \tag{8.4.22}$$

$$\tilde{w}_{S,k|k-1}^{(i,j)}=w_{S,k|k-1}^{(i,j)}\Big/\sum_{j=1}^{L_{k-1}^{(i)}}w_{S,k|k-1}^{(i,j)} \tag{8.4.23}$$

$$w_{S,k|k-1}^{(i,j)}=w_{k-1}^{(i,j)}\frac{\tilde{\phi}_{k|k-1}(\boldsymbol{\xi}_{S,k|k-1}^{(i,j)}|\boldsymbol{\xi}_{k-1}^{(i,j)},\mu_{S,k|k-1}^{(i,j)})\tau(\mu_{S,k|k-1}^{(i,j)}|\mu_{k-1}^{(i,j)})p_{S,k}(\boldsymbol{\xi}_{k-1}^{(i,j)},\mu_{k-1}^{(i,j)})}{q_{\boldsymbol{\xi},k}^{(i)}(\boldsymbol{\xi}_{S,k|k-1}^{(i,j)}|\boldsymbol{\xi}_{k-1}^{(i,j)},\mu_{S,k|k-1}^{(i,j)},Z_k)q_{\mu,k}^{(i)}(\mu_{S,k|k-1}^{(i,j)}|\mu_{k-1}^{(i,j)})} \tag{8.4.24}$$

$$\mu_{\gamma,k}^{(i,j)}\sim b_{\mu,k}^{(i)}(\cdot),\ j=1,\ldots,L_{\gamma,k}^{(i)} \tag{8.4.25}$$

$$\boldsymbol{\xi}_{\gamma,k}^{(i,j)}\sim b_{\boldsymbol{\xi},k}^{(i)}(\cdot|\mu_{\gamma,k}^{(i,j)},Z_k),\ j=1,\cdots,L_{\gamma,k}^{(i)} \tag{8.4.26}$$

$$\tilde{w}_{\gamma,k}^{(i,j)}=w_{\gamma,k}^{(i,j)}\Big/\sum_{j=1}^{L_{\gamma,k}^{(i)}}w_{\gamma,k}^{(i,j)} \tag{8.4.27}$$

$$w_{\gamma,k}^{(i,j)}=\frac{p_{\gamma,k}(\boldsymbol{\xi}_{\gamma,k}^{(i,j)},\mu_{\gamma,k}^{(i,j)})}{b_k^{(i)}(\boldsymbol{\xi}_{\gamma,k}^{(i,j)},\mu_{\gamma,k}^{(i,j)}|Z_k)}=\frac{p_{\gamma,k}(\boldsymbol{\xi}_{\gamma,k}^{(i,j)}|\mu_{\gamma,k}^{(i,j)})p_{\mu,k}(\mu_{\gamma,k}^{(i,j)})}{b_{\boldsymbol{\xi},k}^{(i)}(\boldsymbol{\xi}_{\gamma,k}^{(i,j)}|\mu_{\gamma,k}^{(i,j)},Z_k)b_{\mu,k}^{(i)}(\mu_{\gamma,k}^{(i,j)})} \tag{8.4.28}$$

with $q_{\mu,k}^{(i)}$ and $q_{\boldsymbol{\xi},k}^{(i)}$ being the importance sampling distributions of the motion model and kinetic state, respectively. The core difference from the standard CBMeMBer filter is that the motion model is sampled in (8.4.21) and (8.4.22), which is applied to the calculation of the weights (8.4.24). Probability mass function (PMF) $p_{\mu,k}(\cdot)$ indicates the model distribution of the birth target at time k, $p_{\gamma,k}(\cdot|\mu)$ is the state distribution conditioned on model μ, and $b_{\mu,k}^{(i)}$ and $b_{\boldsymbol{\xi},k}^{(i)}$ correspond to the given importance (or suggested) densities for sampling model variables (8.4.25) and state variables (8.4.26), respectively.

8.4.2.2 Update of SMC-MM-CBMeMBer Filter

The update of the SMC-MM-CBMeMBer filter is similar to that of the standard SMC-CBMeMBer filter. However, the single-target measurement likelihood function is replaced by a model-dependent measurement likelihood function. Operations in the update step are similar to those of the basic particle filtering algorithm but they are conducted in the RFS context.

Assume that the predicted multiple model (MB) multi-target density at time k is $\pi_{k|k-1} = \{(\varepsilon_{k|k-1}^{(i)}, p_{k|k-1}^{(i)}(\boldsymbol{\xi}, \mu))\}_{i=1}^{M_{k|k-1}}$, and each of $p_{k|k-1}^{(i)}(\boldsymbol{\xi}, \mu), i = 1, \cdots, M_{k|k-1}$ is composed of a set of weighted samples $\{(w_{k|k-1}^{(i,j)}, \boldsymbol{\xi}_{k|k-1}^{(i,j)}, \mu_{k|k-1}^{(i,j)})\}_{j=1}^{L_{k|k-1}^{(i)}}$, i.e.,

$$p_{k|k-1}^{(i)}(\boldsymbol{\xi}, \mu) = \sum_{j=1}^{L_{k|k-1}^{(i)}} w_{k|k-1}^{(i,j)} \delta_{\boldsymbol{\xi}_{k|k-1}^{(i,j)}, \mu_{k|k-1}^{(i,j)}}(\boldsymbol{\xi}, \mu) \tag{8.4.29}$$

Then, the MB approximation of the updated multi-target density $\pi_k = \{(\varepsilon_{L,k}^{(i)}, p_{L,k}^{(i)}(\boldsymbol{\xi}, \mu))\}_{i=1}^{M_{k|k-1}} \cup \{(\varepsilon_{U,k}^{*}(z), p_{U,k}^{*}(\boldsymbol{\xi}, \mu; z))\}_{z \in Z_k}$ can be calculated as follows

$$\varepsilon_{L,k}^{(i)} = \varepsilon_{k|k-1}^{(i)} \frac{1 - \eta_{L,k}^{(i)}}{1 - \varepsilon_{k|k-1}^{(i)} \eta_{L,k}^{(i)}} \tag{8.4.30}$$

$$p_{L,k}^{(i)}(\boldsymbol{\xi}, \mu) = \sum_{j=1}^{L_{k|k-1}^{(i)}} \tilde{w}_{L,k}^{(i,j)} \delta_{\boldsymbol{\xi}_{k|k-1}^{(i,j)}, \mu_{k|k-1}^{(i,j)}}(\boldsymbol{\xi}, \mu) \tag{8.4.31}$$

$$\varepsilon_{U,k}^{*}(z) = \frac{\sum_{i=1}^{M_{k|k-1}} \frac{\varepsilon_{k|k-1}^{(i)}(1-\varepsilon_{k|k-1}^{(i)})\eta_{U,k}^{(i)}(z)}{\left(1-\varepsilon_{k|k-1}^{(i)}\eta_{L,k}^{(i)}\right)^2}}{\kappa_k(z) + \sum_{i=1}^{M_{k|k-1}} \frac{\varepsilon_{k|k-1}^{(i)}\eta_{U,k}^{(i)}(z)}{1-\varepsilon_{k|k-1}^{(i)}\eta_{L,k}^{(i)}}} \tag{8.4.32}$$

$$p_{U,k}^{*}(\boldsymbol{\xi}, \mu; z) = \sum_{i=1}^{M_{k|k-1}} \sum_{j=1}^{L_{k|k-1}^{(i)}} \tilde{w}_{U,k}^{*(i,j)}(z) \delta_{\boldsymbol{\xi}_{k|k-1}^{(i,j)}, \mu_{k|k-1}^{(i,j)}}(\boldsymbol{\xi}, \mu) \tag{8.4.33}$$

where

$$\eta_{L,k}^{(i)} = \sum_{j=1}^{L_{k|k-1}^{(i)}} w_{k|k-1}^{(i,j)} p_{D,k}(\boldsymbol{\xi}_{k|k-1}^{(i,j)}, \mu_{k|k-1}^{(i,j)}) \tag{8.4.34}$$

$$\tilde{w}_{L,k}^{(i,j)} = w_{L,k}^{(i,j)} \Big/ \sum_{j=1}^{L_{k|k-1}^{(i)}} w_{L,k}^{(i,j)} \tag{8.4.35}$$

$$w_{L,k}^{(i,j)} = w_{k|k-1}^{(i,j)}(1 - p_{D,k}(\boldsymbol{\xi}_{k|k-1}^{(i,j)}, \mu_{k|k-1}^{(i,j)})) \tag{8.4.36}$$

$$\eta_{U,k}^{(i)}(z) = \sum_{j=1}^{L_{k|k-1}^{(i)}} w_{k|k-1}^{(i,j)} \psi_{k,z}(\boldsymbol{\xi}_{k|k-1}^{(i,j)}, \mu_{k|k-1}^{(i,j)}) \tag{8.4.37}$$

$$\tilde{w}_{U,k}^{*(i,j)}(z) = w_{U,k}^{*(i,j)}(z) \Bigg/ \sum_{i=1}^{M_{k|k-1}} \sum_{j=1}^{L_{k|k-1}^{(i)}} w_{U,k}^{*(i,j)}(z) \tag{8.4.38}$$

$$w_{U,k}^{*(i,j)}(z) = w_{k|k-1}^{(i,j)} \frac{\varepsilon_{k|k-1}^{(i)}}{1 - \varepsilon_{k|k-1}^{(i)}} \psi_{k,z}(\boldsymbol{\xi}_{k|k-1}^{(i,j)}, \mu_{k|k-1}^{(i,j)}) \tag{8.4.39}$$

$$\psi_{k,z}(\boldsymbol{\xi}, \mu) = g_k(z|\boldsymbol{\xi}, \mu) p_{D,k}(\boldsymbol{\xi}, \mu) \tag{8.4.40}$$

The key difference from the standard SMC-CBMeMBer filter is that, the particle-expressed detection probability is applied to the kinetic state and motion model, and the model-conditioned target likelihood function is used in the weight update calculation of (8.4.37) and (8.4.39).

8.4.3 Gaussian Mixture MM-CBMeMBer Filter

The GM-CBMeMBer filter can also be extended to multiple model filtering. Similarly, each GM component is extended to incorporate the model parameter and the linear Gaussian jump Markov (LGJM) model is used to model switch parameters in a way similar to extension of the GM-PHD filter to the LGJM model (see Sect. 8.3.2 for details). The GM-MM-MBRFS can be described as follows

$$\begin{aligned}(X_k, \mathcal{M}_k) &= \{(\varepsilon_k^{(i)}, p_k^{(i)}(\boldsymbol{\xi}, \mu))\}_{i=1}^{M_k} \\ &\approx \{(\varepsilon_k^{(i)}, \{(w_k^{(i,j)}, \boldsymbol{m}_k^{(i,j)}, \boldsymbol{P}_k^{(i,j)}, \mu_k^{(i,j)})\}_{j=1}^{J_k^{(i)}})\}_{i=1}^{M_k}\end{aligned} \tag{8.4.41}$$

where

$$p_k^{(i)}(\boldsymbol{\xi}, \mu) = \sum_{j=1}^{J_k^{(i)}} w_k^{(i,j)}(\mu) \mathcal{N}(\boldsymbol{\xi}; \boldsymbol{m}_k^{(i,j)}(\mu), \boldsymbol{P}_k^{(i,j)}(\mu)) \tag{8.4.42}$$

8.4.3.1 Prediction of GM-MM-CBMeMBer Filter

Given the posterior multiple model MB multi-target density $\pi_{k-1} = \{(\varepsilon_{k-1}^{(i)}, p_{k-1}^{(i)}(\boldsymbol{\xi}, \mu))\}_{i=1}^{M_{k-1}}$ at time $k-1$, assuming that each of $p_{k-1}^{(i)}(\boldsymbol{\xi}, \mu), i = 1, \cdots, M_{k-1}$ is in the Gaussian mixture form

$$p_{k-1}^{(i)}(\boldsymbol{\xi}, \mu) = \sum_{j=1}^{J_{k-1}^{(i)}} w_{k-1}^{(i,j)}(\mu)\mathcal{N}(\boldsymbol{\xi}; \boldsymbol{m}_{k-1}^{(i,j)}(\mu), \boldsymbol{P}_{k-1}^{(i,j)}(\mu)) \tag{8.4.43}$$

the predicted multiple model (MB) multi-target density $\pi_{k|k-1} = \{(\varepsilon_{S,k|k-1}^{(i)}, p_{S,k|k-1}^{(i)}(\boldsymbol{\xi}, \mu))\}_{i=1}^{M_{k-1}} \cup \{(\varepsilon_{\gamma,k}^{(i)}, p_{\gamma,k}^{(i)}(\boldsymbol{\xi}, \mu))\}_{i=1}^{M_{\gamma,k}}$ can be calculated as follows

$$\varepsilon_{S,k|k-1}^{(i)} = \varepsilon_{k-1}^{(i)} \sum_{\mu'} \sum_{j=1}^{J_{k-1}^{(i)}} p_{S,k}(\mu') w_{k-1}^{(i,j)}(\mu') \tag{8.4.44}$$

$$p_{S,k|k-1}^{(i)}(\boldsymbol{\xi}, \mu) = \sum_{\mu'} \sum_{j=1}^{J_{k-1}^{(i)}} w_{S,k|k-1}^{(i,j)}(\mu|\mu')\mathcal{N}(\boldsymbol{\xi}; \boldsymbol{m}_{S,k|k-1}^{(i,j)}(\mu|\mu'), \boldsymbol{P}_{S,k|k-1}^{(i,j)}(\mu|\mu')) \tag{8.4.45}$$

with

$$w_{S,k|k-1}^{(i,j)}(\mu|\mu') = p_{S,k}(\mu')\tau_{k|k-1}(\mu|\mu')w_{k-1}^{(i,j)}(\mu') \tag{8.4.46}$$

$$\boldsymbol{m}_{S,k|k-1}^{(i,j)}(\mu|\mu') = \boldsymbol{F}_{S,k-1}(\mu)\boldsymbol{m}_{k-1}^{(i,j)}(\mu') \tag{8.4.47}$$

$$\boldsymbol{P}_{S,k|k-1}^{(i,j)}(\mu|\mu') = \boldsymbol{F}_{S,k-1}(\mu)\boldsymbol{P}_{k-1}^{(i,j)}(\mu')\boldsymbol{F}_{S,k-1}^{\mathrm{T}}(\mu) + \boldsymbol{Q}_{k-1}(\mu) \tag{8.4.48}$$

where $\boldsymbol{F}_{S,k-1}(\mu)$ is the linear state transition matrix conditioned on motion model μ, and $\boldsymbol{Q}_{k-1}(\mu)$ is the process noise matrix under the same motion model. Moreover, $p_{S,k}(\mu)$ indicates the model-dependent survival probability. However, in most cases, this parameter is model independent, i.e., $p_{S,k}(\mu) = p_{S,k}$. Compared with the standard GM-CBMeMBer filter, an important extension of the GM-MM-CBMeMBer filter is to incorporate the model parameter in each target motion parameter, in particular, the model transition probability $\tau_{k|k-1}(\mu|\mu')$ is included in (8.4.46).

For the birth MBRFS set $\{(\varepsilon_{\gamma,k}^{(i)}, p_{\gamma,k}^{(i)}(\boldsymbol{\xi}, \mu))\}_{i=1}^{M_{\gamma,k}}$, a multiple model Gaussian mixture is used to approximate the PDF of the birth set, i.e.,

$$p_{\gamma,k}^{(i)}(\boldsymbol{\xi}, \mu) = \sum_{j=1}^{J_{\gamma,k}^{(i)}} w_{\gamma,k}^{(i,j)}(\mu)\mathcal{N}(\boldsymbol{\xi}; \boldsymbol{m}_{\gamma,k}^{(i,j)}(\mu), \boldsymbol{P}_{\gamma,k}^{(i,j)}(\mu)) \tag{8.4.49}$$

where $\boldsymbol{m}_{\gamma,k}^{(i,j)}(\mu)$, $\boldsymbol{P}_{\gamma,k}^{(i,j)}(\mu)$, and $w_{\gamma,k}^{(i,j)}(\mu)$ represent the Gaussian mixture approximation of the PDF of multiple model birth RFS, and $\varepsilon_{\gamma,k}^{(i)}$ is the model setup parameter.

8.4.3.2 Update of GM-MM-CBMeMBer Filter

Assume that the predicted multiple model (MB) multi-target density at time k is $\pi_{k|k-1} = \{(\varepsilon_{k|k-1}^{(i)}, p_{k|k-1}^{(i)}(\boldsymbol{\xi}, \mu))\}_{i=1}^{M_{k|k-1}}$, and each of $p_{k|k-1}^{(i)}(\boldsymbol{\xi}, \mu)$, $i = 1, \cdots, M_{k|k-1}$ is in the following Gaussian mixture form

$$p_{k|k-1}^{(i)}(\boldsymbol{\xi}, \mu) = \sum_{j=1}^{J_{k|k-1}^{(i)}} w_{k|k-1}^{(i,j)}(\mu)\mathcal{N}(\boldsymbol{\xi}; \boldsymbol{m}_{k|k-1}^{(i,j)}(\mu), \boldsymbol{P}_{k|k-1}^{(i,j)}(\mu)) \tag{8.4.50}$$

Then, the MB approximation $\pi_k = \{(\varepsilon_{L,k}^{(i)}, p_{L,k}^{(i)}(\boldsymbol{\xi}, \mu))\}_{i=1}^{M_{k|k-1}} \cup \{(\varepsilon_{U,k}^{*}(\boldsymbol{z}), p_{U,k}^{*}(\boldsymbol{\xi}, \mu, \boldsymbol{z}))\}_{\boldsymbol{z}\in Z_k}$ of the updated multiple model multi-target density can be calculated as follows

$$\varepsilon_{L,k}^{(i)} = \varepsilon_{k|k-1}^{(i)} \frac{1 - p_{D,k}(\mu)}{1 - \varepsilon_{k|k-1}^{(i)} p_{D,k}(\mu)} \tag{8.4.51}$$

$$p_{L,k}^{(i)}(\boldsymbol{\xi}, \mu) = p_{k|k-1}^{(i)}(\boldsymbol{\xi}, \mu) \tag{8.4.52}$$

$$\varepsilon_{U,k}^{*}(\boldsymbol{z}) = \frac{\sum_{i=1}^{M_{k|k-1}} \frac{\varepsilon_{k|k-1}^{(i)}(1-\varepsilon_{k|k-1}^{(i)})\eta_{U,k}^{(i)}(\boldsymbol{z})}{(1-\varepsilon_{k|k-1}^{(i)} p_{D,k}(\mu))^2}}{\kappa_k(\boldsymbol{z}) + \sum_{i=1}^{M_{k|k-1}} \frac{\varepsilon_{k|k-1}^{(i)}\eta_{U,k}^{(i)}(\boldsymbol{z})}{1-\varepsilon_{k|k-1}^{(i)} p_{D,k}(\mu)}} \tag{8.4.53}$$

$$p_{U,k}^{*}(\boldsymbol{\xi}, \mu; \boldsymbol{z}) = \frac{\sum_{i=1}^{M_{k|k-1}} \sum_{j=1}^{J_{k|k-1}^{(i)}} w_{U,k}^{(i,j)}(\mu; \boldsymbol{z})\mathcal{N}(\boldsymbol{\xi}; \boldsymbol{m}_{U,k}^{(i,j)}(\mu; \boldsymbol{z}), \boldsymbol{P}_{U,k}^{(i,j)}(\mu))}{\sum_{i=1}^{M_{k|k-1}} \sum_{j=1}^{J_{k|k-1}^{(i)}} w_{U,k}^{(i,j)}(\mu; \boldsymbol{z})} \tag{8.4.54}$$

where

$$\eta_{U,k}^{(i)}(\boldsymbol{z}) = p_{D,k}(\mu) \sum_{j=1}^{J_{k|k-1}^{(i)}} w_{k|k-1}^{(i,j)}(\mu) q_k^{(i,j)}(\mu; \boldsymbol{z}) \tag{8.4.55}$$

$$q_k^{(i,j)}(\mu; \boldsymbol{z}) = \mathcal{N}(\boldsymbol{z}; \boldsymbol{H}_k(\mu)\boldsymbol{m}_{k|k-1}^{(i,j)}(\mu), \boldsymbol{S}_k^{(i,j)}(\mu)) \tag{8.4.56}$$

$$w_{U,k}^{(i,j)}(\mu;z) = p_{D,k}(\mu) w_{k|k-1}^{(i,j)}(\mu) \frac{\varepsilon_{k|k-1}^{(i)}}{1-\varepsilon_{k|k-1}^{(i)}} q_k^{(i,j)}(\mu;z) \tag{8.4.57}$$

$$\boldsymbol{m}_{U,k}^{(i,j)}(\mu;z) = \boldsymbol{m}_{k|k-1}^{(i,j)}(\mu) + \boldsymbol{G}_{U,k}^{(i,j)}(\mu)(z - \boldsymbol{H}_k(\mu)\boldsymbol{m}_{k|k-1}^{(i,j)}(\mu)) \tag{8.4.58}$$

$$\boldsymbol{P}_{U,k}^{(i,j)}(\mu) = (\boldsymbol{I} - \boldsymbol{G}_{U,k}^{(i,j)}(\mu)\boldsymbol{H}_k(\mu))\boldsymbol{P}_{k|k-1}^{(i,j)}(\mu) \tag{8.4.59}$$

$$\boldsymbol{G}_{U,k}^{(i,j)}(\mu) = \boldsymbol{P}_{k|k-1}^{(i,j)}(\mu)\boldsymbol{H}_k^{\mathrm{T}}(\mu)(\boldsymbol{S}_k^{(i,j)}(\mu))^{-1} \tag{8.4.60}$$

$$\boldsymbol{S}_k^{(i,j)}(\mu) = \boldsymbol{H}_k(\mu)\boldsymbol{P}_{k|k-1}^{(i,j)}(\mu)\boldsymbol{H}_k^{\mathrm{T}}(\mu) + \boldsymbol{R}_k(\mu) \tag{8.4.61}$$

The main difference between the GM-MM-CBMeMBer update equation and the standard GM-CBMeMBer update equation is the use of model-dependent detection probabilities. However, in most cases, the detection probability is independent of the model, i.e., $p_{D,k}(\mu) = p_{D,k}$. Furthermore, the measurement likelihood model is also extended to include the model parameter.

8.5 Multiple Model GLMB Filter

In the above two sections, the joint transition of the state and motion models is of the following form

$$\phi(\boldsymbol{\xi},\mu|\boldsymbol{\xi}',\mu') = \tilde{\phi}(\boldsymbol{\xi}|\boldsymbol{\xi}',\mu)\tau(\mu|\mu') \tag{8.5.1}$$

where μ and μ' are the current and previous model variables, respectively, and $\tilde{\phi}(\boldsymbol{\xi}|\boldsymbol{\xi}',\mu)$ indicates the state transition density from previous state $\boldsymbol{\xi}'$ to current state $\boldsymbol{\xi}$. Moreover, since the measurement function may also depend on model μ, the likelihood of state $\boldsymbol{\xi}$ generating measurement z is denoted as $g(z|\boldsymbol{\xi},\mu)$.

In order to extend the GLMB filter to the multiple model version, the state transition and measurement likelihood models need to consider the label variable ℓ in addition to the model variable μ. Let the (labeled) state of the maneuvering target consist of kinetic state $\boldsymbol{\xi}$, mode index μ and label ℓ, i.e., $\mathbf{x} = (\boldsymbol{x},\ell) = (\boldsymbol{\xi},\mu,\ell)$, and it is modeled as a jump Markov (JM) system. Note that although the target label is a part of the state vector, it is assumed to remain unchanged throughout its lifetime. Therefore, the JM system state equation for the target with the label ℓ is indexed by ℓ, i.e., $\tilde{\phi}^{(\ell)}(\boldsymbol{\xi}|\boldsymbol{\xi}',\mu)$ and $g^{(\ell)}(z|\boldsymbol{\xi},\mu)$. The new state of a surviving target is jointly dominated by the survival probability, the target transition probability from the previous model to this model, and the associated state transition function. As a result, the joint transition and likelihood functions for the state and model index become

$$\phi(\boldsymbol{\xi},\mu|\boldsymbol{\xi}',\mu',\ell) = \tilde{\phi}^{(\ell)}(\boldsymbol{\xi}|\boldsymbol{\xi}',\mu)\tau(\mu|\mu') \tag{8.5.2}$$

$$g(z|\boldsymbol{\xi}, \mu, \ell) = g^{(\ell)}(z|\boldsymbol{\xi}, \mu) \tag{8.5.3}$$

Notice that, for any function $f(\cdot)$, due to $\boldsymbol{x} = (\boldsymbol{\xi}, \mu)$, then we have

$$\int f(\boldsymbol{x})d\boldsymbol{x} = \sum_{\mu\in\mathcal{M}} \int f(\boldsymbol{\xi}, \mu)d\boldsymbol{\xi} \tag{8.5.4}$$

Substituting Eqs. (8.5.2) and (8.5.3) into the GLMB prediction and update equations yields a GLMB filter suitable for maneuvering targets.

For simplicity, the target birth model, motion model and measurement model are assumed to be linear Gaussian models. Given posterior density (3.3.38) with state $\mathbf{x} = (\boldsymbol{\xi}, \mu, \ell)$ at time $k-1$, the prediction equation of the GLMB filter can be written as follows

$$\pi_+(\mathrm{X}_+) = \Delta(\mathrm{X}_+)\sum_{c\in\mathbb{C}} w_+^{(c)}(\mathcal{L}(\mathrm{X}_+))[p_+^{(c)}]^{\mathrm{X}_+} \tag{8.5.5}$$

with

$$w_+^{(c)}(L) = w_S^{(c)}(L\cap\mathbb{L})w_\gamma(L-\mathbb{L}) \tag{8.5.6}$$

$$p_+^{(c)}(\boldsymbol{\xi}, \mu, \ell) = 1_{\mathbb{L}}(\ell)p_{+,S}^{(c)}(\boldsymbol{\xi}, \mu, \ell) + (1-1_{\mathbb{L}}(\ell))p_\gamma(\boldsymbol{\xi}, \mu, \ell) \tag{8.5.7}$$

$$w_S^{(c)}(L) = [\eta_S^{(c)}]^L\sum_{I\supseteq L}[1-\eta_S^{(c)}]^{I-L}w^{(c)}(I) \tag{8.5.8}$$

$$p_{+,S}^{(c)}(\boldsymbol{\xi}, \mu, \ell) = \sum_{\mu'\in\mathcal{M}} \langle p_S(\cdot, \mu', \ell)\phi(\boldsymbol{\xi}, \mu|\cdot, \mu', \ell), p^{(c)}(\cdot, \mu', \ell)\rangle/\eta_S^{(c)}(\ell) \tag{8.5.9}$$

$$\eta_S^{(c)}(\ell) = \sum_{\mu'\in\mathcal{M}} \langle p_S(\cdot, \mu', \ell), p^{(c)}(\cdot, \mu', \ell)\rangle \tag{8.5.10}$$

$$\phi(\boldsymbol{\xi}, \mu|\boldsymbol{\xi}', \mu', \ell) = \mathcal{N}(\boldsymbol{\xi}; \boldsymbol{F}_S(\mu)\boldsymbol{\xi}', \boldsymbol{Q}_S(\mu))\tau(\mu|\mu') \tag{8.5.11}$$

$$p_\gamma(\boldsymbol{\xi}, \mu, \ell) = \sum_{i=1}^{J_\gamma^{(\ell)}} \omega_\gamma^{(i)}(\mu)\mathcal{N}(\boldsymbol{\xi}; \boldsymbol{m}_\gamma^{(i)}(\mu), \boldsymbol{P}_\gamma^{(i)}(\mu)) \tag{8.5.12}$$

where $p_S(\boldsymbol{\xi}', \mu', \ell)$ is the probability that the target with the previous label state $(\boldsymbol{\xi}', \mu', \ell)$ survives at time k, $\boldsymbol{F}_S(\mu)$ is the state transition matrix conditioned on motion model μ, $\boldsymbol{Q}_S(\mu)$ is the corresponding covariance matrix, $w_\gamma(L)$ is the probability that birth targets have the label set L, and $\omega_\gamma^{(i)}(\mu)$, $\boldsymbol{m}_\gamma^{(i)}(\mu)$ and $\boldsymbol{P}_\gamma^{(i)}(\mu)$ are the weight, mean and covariance parameters determining the shape of the birth distribution, respectively.

If the prediction of the GLMB filter is given by (3.3.38), the GLMB update equation can be written as follows

$$\pi(\mathrm{X}|Z) = \Delta(\mathrm{X}) \sum_{c\in\mathbb{C}} \sum_{\theta} w_Z^{(c,\theta)}(\mathcal{L}(\mathrm{X}))[p^{(c,\theta)}(\cdot|Z)]^{\mathrm{X}} \tag{8.5.13}$$

with

$$w_Z^{(c,\theta)}(L) = \frac{\delta_{\theta^{-1}(\{0:|Z|\})}(L)[\eta_Z^{(c,\theta)}]^L w^{(c)}(L)}{\sum_{c\in\mathbb{C}} \sum_{\theta\in\Theta} \sum_{J\subseteq\mathbb{L}} \delta_{\theta^{-1}(\{0:|Z|\})}(J)[\eta_Z^{(c,\theta)}]^J w^{(c)}(J)} \tag{8.5.14}$$

$$p^{(c,\theta)}(\boldsymbol{\xi}, \mu, \ell|Z) = p^{(c)}(\boldsymbol{\xi}, \mu, \ell)\varphi_Z(\boldsymbol{\xi}, \mu, \ell; \theta)/\eta_Z^{(c,\theta)}(\ell) \tag{8.5.15}$$

$$\eta_Z^{(c,\theta)}(\ell) = \sum_{\mu\in\mathcal{M}} \langle \varphi_Z(\cdot, \mu, \ell; \theta), p^{(c)}(\cdot, \mu, \ell)\rangle \tag{8.5.16}$$

$$\varphi_Z(\boldsymbol{\xi}, \mu, \ell; \theta) = \begin{cases} p_D(\boldsymbol{\xi}, \mu, \ell)g(z_{\theta(\ell)}|\boldsymbol{\xi}, \mu, \ell)/\kappa(z_{\theta(\ell)}), & if\ \theta(\ell) > 0 \\ 1 - p_D(\boldsymbol{\xi}, \mu, \ell), & if\ \theta(\ell) = 0 \end{cases} \tag{8.5.17}$$

$$g(z|\boldsymbol{\xi}, \mu, \ell) = \mathcal{N}(z; \boldsymbol{H}(\mu)\boldsymbol{\xi}, \boldsymbol{R}(\mu)) \tag{8.5.18}$$

where Θ is the space of association map $\theta : \mathbb{L} \to \{0 : |Z|\} \triangleq \{0, 1, \cdots, |Z|\}$, such that $i = i'$ when $\theta(i) = \theta(i') > 0$, $p_D(\boldsymbol{\xi}, \mu, \ell)$ is the detection probability of the target with state $(\boldsymbol{\xi}, \mu, \ell)$, $\kappa(\cdot)$ is the Poisson clutter intensity, and $\boldsymbol{H}(\mu)$ and $\boldsymbol{R}(\mu)$ are respectively the measurement matrix and measurement covariance matrix conditioned on motion model μ.

In terms of the state extraction, it is similar to the single-model system. In order to estimate the motion model for each label, we select the motion model that maximizes the marginal probability of the model over the full density of that label, i.e., for the label ℓ with component $\boldsymbol{\xi}$, the estimated motion model $\hat{\mu}$ is

$$\hat{\mu} = \arg_{\mu} \max \int p^{(c)}(\boldsymbol{\xi}, \mu, \ell)d\boldsymbol{\xi} \tag{8.5.19}$$

In the above solution, the posterior density of each track is in the form of the Gaussian mixture and each mixture component is related to the current motion model. For a particular track, at each new moment, the posterior densities are forwardly predicted for all motion models in the system, resulting in a new Gaussian mixture. The weight of each new component is the product of the weight of the parent component and the switching probability of the corresponding motion model. Thus, the number of mixture components grows exponentially. Therefore, after the update step, in order to keep the computation feasible, additional pruning and merging must be performed on each track in each GLMB hypothesis.

For moderately nonlinear motion and measurement models, the UKF can be used to predict and update each mixture Gaussian component. The particle filter can also be called to represent the posterior density of each track in the hypothesis. Unlike the Gaussian mixture, the density is represented by a set of particles in the particle filter; as in the Gaussian mixture case, the number of particles in the posterior density grows exponentially in each forward prediction process. Therefore, it is necessary to perform resampling to discard secondary particles, so as to keep the total number of particles under control.

8.6 Summary

Aiming at the problem of maneuvering target tracking, this chapter introduces the multiple model PHD, multiple model CBMeMBer, and multiple model GLMB filters under the RFS framework based on the jump Markov system. In essence, since these multiple model filters can be considered as special cases of the corresponding filters after appropriate modification. The target state is augmented with the model index parameter. Therefore, it is necessary to modify the prediction and update operations of the corresponding filtering algorithms, so as to include the uncertainty of the target motion model.

It shall be noted that, for moderately non-maneuvering target, the use of the multiple model approach may weaken the tracker performance and increase the computational burden. However, for highly maneuvering targets, the multiple model method is required. Judging whether the multiple model method is needed can be based on the maneuvering index, which quantifies the maneuverability of a target in terms of the process noise, sensor measurement noise, and sensor sampling interval. Reference [414] compares the IMM estimator and the Kalman filter based on the maneuvering index, and interested readers can further refer to this reference.

Chapter 9
Target Tracking for Doppler Radars

9.1 Introduction

The previous chapter extended the standard RFS-based target tracking algorithms from the perspective of the motion model, and from the beginning of this chapter to Chap. 12, the standard RFS-based target tracking algorithms will be extended from the perspective of the measurement model. Specifically, Chap. 10 will focus on the measurement model with the amplitude information, Chap. 11 will focus on the non-standard measurement model, and Chap. 12 will focus on the multi-sensor measurement model.

This chapter focuses on the measurement model with the Doppler information. The Doppler radar is widely used for airborne early warning, navigation and guidance, satellite tracking and battle reconnaissance, etc. The Doppler radar detects moving targets by using the Doppler effect. Compared with conventional radars, its most notable feature is that it can obtain the Doppler measurement (also known as the radial velocity) in addition to the positional measurement (such as slant range, azimuth, and elevation) provided by conventional radars. Therefore, the primary purpose of target tracking with the Doppler radar is how to make full use of this one-dimensional incremental information to improve the target tracking performance. However, influenced by the inherent Doppler blind zone (DBZ) of Doppler radar, target tracking with Doppler radar faces serious problems such as tracks being intermittent, short, heavy, and reinitialized.

In order to give full play to the performance of the Doppler radar under the RFS framework, this chapter firstly studies the gain of Doppler information to improve the performance, and presents a GM-CPHD filter with the Doppler measurement. Then, in order to alleviate the track intermittent problem of target tracking caused by the DBZ, a GM-PHD filter with the minimum detectable velocity (MDV) and Doppler information is presented by using the MDV information related to the DBZ. Finally, considering the potential systematic error when Doppler radars are networked, an augmented state GM-PHD filter with the registration error is presented.

W. Wu et al., *Target Tracking with Random Finite Sets*,
https://doi.org/10.1007/978-981-19-9815-7_9

9.2 GM-CPHD Filter with Doppler Measurement

This section takes the CPHD filter as an example to introduce the method of introducing the Doppler information, and describes the multi-target tracking algorithm of airborne Doppler radar based on the GM-CPHD filter, which is abbreviated as the GM-CPHD filter with the Doppler measurement (GM-CPHDwD). This algorithm is based on the standard GM-CPHD algorithm, which firstly uses the positional measurement to update the states, and then utilizes the Doppler measurement for sequential update to obtain a more accurate state estimation and likelihood function. The simulation results verify the effectiveness of this algorithm and show that, the introduction of Doppler information can effectively suppress clutter and significantly improve the multi-target tracking performance under clutter conditions.

9.2.1 Doppler Measurement Model

The model assumption of the GM-CPHDwD filter is basically the same as that of the GM-CPHD filter in Sect. 5.4. The only difference lies in the measurement model. Therefore, only the measurement model of the Doppler radar is described here.

Denote the set of target states at time k as $X_k = \{\boldsymbol{x}_{k,1}, \boldsymbol{x}_{k,2}, \ldots, \boldsymbol{x}_{k,N_k}\}$, N_k is the number of targets, the set of measurements is $Z_k = \{z_{k,1}, z_{k,2}, \ldots, z_{k,M_k}\}$, and M_k is the number of measurements. The measurement equation for measurement $z_{k,j}$ from state $\boldsymbol{x}_{k,i}$ in the two-dimensional coordinate system is

$$z_{k,j} = \left[\boldsymbol{y}_{k,j}^c;\, y_{k,j}^d\right] = h_k(\boldsymbol{x}_{k,i}) + \boldsymbol{n}_{k,j} = \begin{bmatrix} x_{k,j}^m \\ y_{k,j}^m \\ y_{k,j}^d \end{bmatrix} = \begin{bmatrix} x_{k,i} \\ y_{k,i} \\ \dot{r}_{k,i} \end{bmatrix} + \begin{bmatrix} n_{k,j}^x \\ n_{k,j}^y \\ n_{k,j}^d \end{bmatrix} \tag{9.2.1}$$

$$\dot{r}_{k,i} = h_{d,k}\left(\boldsymbol{x}_{k,i}\right) = \frac{x_{k,i}\dot{x}_{k,i} + y_{k,i}\dot{y}_{k,i}}{\sqrt{x_{k,i}^2 + y_{k,i}^2}} \tag{9.2.2}$$

where $\boldsymbol{x}_{k,i} = [x_{k,i}\ y_{k,i}\ \dot{x}_{k,i}\ \dot{y}_{k,i}]^{\mathrm{T}}$, $\boldsymbol{n}_{k,j} \sim \mathcal{N}(\boldsymbol{n}_{k,j}; \boldsymbol{0}, \boldsymbol{R}_k)$ is the Gaussian white noise with zero mean and covariance $\boldsymbol{R}_k = \mathrm{diag}(\sigma_x^2, \sigma_y^2, \sigma_d^2) = \mathrm{diag}(\boldsymbol{R}_{c,k}, \sigma_d^2)$, and $\boldsymbol{y}_{k,j}^c$ and $y_{k,j}^d$ represent the positional measurement and Doppler measurement, respectively, and their likelihood functions are

$$g_k\left(\boldsymbol{y}_{k,j}^c|\boldsymbol{x}_{k,i}\right) = \mathcal{N}\left(\boldsymbol{y}_{k,j}^c; \boldsymbol{H}_{c,k}\boldsymbol{x}_{k,i}, \boldsymbol{R}_{c,k}\right) \tag{9.2.3}$$

$$g_k\left(y_{k,j}^d|\boldsymbol{x}_{k,i}\right) = \mathcal{N}\left(y_{k,j}^d; h_{d,k}(\boldsymbol{x}_{k,i}), \sigma_d^2\right) \tag{9.2.4}$$

where $\boldsymbol{H}_{c,k} = \text{diag}(\boldsymbol{I}_2, \boldsymbol{0}_2)$ is the positional measurement matrix, $\boldsymbol{I}_n$ and $\boldsymbol{0}_n$ represent the identity matrix and all-zero matrix of size $n \times n$, respectively, $\boldsymbol{R}_{c,k}$ is the positional measurement noise covariance, $h_{d,k}(\cdot)$ is the non-linear Doppler measurement function, and σ_d^2 is the Doppler measurement noise variance. The above equations assume that the positional measurement $\boldsymbol{y}_{k,j}^c$ and the Doppler measurement $y_{k,j}^d$ are uncorrelated and that the positional measurement function is linear. For non-linear polar coordinates, the measurement matrix $\boldsymbol{H}_{c,k}$ and the corresponding covariance $\boldsymbol{R}_{c,k}$ can be obtained through measurement transformation [20, 415].

9.2.2 *Sequential GM-CPHD Filter with Doppler Measurement*

Since only the measurement model is different, the prediction step of the GM-CPHDwD filter is the same as that of the standard GM-CPHD filter in Sect. 5.4.1, which will not be repeated here. Therefore, only the update step will be given as follows.

Assume that the predicted intensity $v_{k|k-1}$ and the predicted cardinality distribution $\rho_{k|k-1}$ at time k are given, and $v_{k|k-1}$ is the following Gaussian mixture

$$v_{k|k-1}(\boldsymbol{x}) = \sum_{i=1}^{J_{k|k-1}} w_{k|k-1}^{(i)} \mathcal{N}\left(\boldsymbol{x}; \boldsymbol{m}_{k|k-1}^{(i)}, \boldsymbol{P}_{k|k-1}^{(i)}\right) \tag{9.2.5}$$

Then, the updated cardinality distribution ρ_k is given by (5.4.6), and the updated intensity v_k is also a Gaussian mixture, which is [64]

$$\begin{aligned} v_k(\boldsymbol{x}) = & \frac{\left\langle \Upsilon_k^{(1)}[w_{k|k-1}, Z_k], \rho_{k|k-1} \right\rangle}{\left\langle \Upsilon_k^{(0)}[w_{k|k-1}, Z_k], \rho_{k|k-1} \right\rangle} (1 - p_{D,k}) v_{k|k-1}(\boldsymbol{x}) \\ & + \sum_{z_{k,m} \in Z_k} \sum_{j=1}^{J_{k|k-1}} w_k^{(j)}(z_{k,m}) \mathcal{N}\left(\boldsymbol{x}; \boldsymbol{m}_{k|k}^{(j)}(z_{k,m}), \boldsymbol{P}_{k|k}^{(j)}(z_{k,m})\right) \end{aligned} \tag{9.2.6}$$

where $\langle \cdot, \cdot \rangle$ represents the inner product and $\Upsilon_k^{(u)}[w, Z](n)$ is given in (9.2.24), $\boldsymbol{m}_{k|k}^{(j)}(z_{k,m})$ and $\boldsymbol{P}_{k|k}^{(j)}(z_{k,m})$ in (9.2.6) are obtained from (9.2.12) and (9.2.13), respectively, and $w_k^{(j)}(z_{k,m})$ is given by (9.2.23).

In order to effectively utilize the Doppler information in the update step of the GM-CPHD filter, a sequential filtering method is employed. That is, the positional measurement is used to update the states first, then the Doppler measurement is used to further update the states, and after a more accurate state estimate and likelihood function are obtained, the weight is finally calculated using the position and Doppler measurement.

- Step 1: Update the target state with the positional measurement $\boldsymbol{y}_{k,m}^{c}$

$$\boldsymbol{m}_{k|k}^{(j)}\left(\boldsymbol{y}_{k,m}^{c}\right) = \left[\, x_{k|k}^{(m,j)} \; y_{k|k}^{(m,j)} \; \dot{x}_{k|k}^{(m,j)} \; \dot{y}_{k|k}^{(m,j)} \,\right]^{\mathrm{T}} = \boldsymbol{m}_{k|k-1}^{(j)} + \boldsymbol{G}_{c,k}^{(j)} \tilde{\boldsymbol{y}}_{k,m,j}^{c} \tag{9.2.7}$$

$$\boldsymbol{P}_{k|k}^{(j)}\left(\boldsymbol{y}_{k,m}^{c}\right) = \left[\boldsymbol{I} - \boldsymbol{G}_{c,k}^{(j)} \boldsymbol{H}_{c,k}\right] \boldsymbol{P}_{k|k-1}^{(j)} \tag{9.2.8}$$

where

$$\tilde{\boldsymbol{y}}_{k,m,j}^{c} = \boldsymbol{y}_{k,m}^{c} - \boldsymbol{H}_{c,k} \boldsymbol{m}_{k|k-1}^{(j)} \tag{9.2.9}$$

$$\boldsymbol{G}_{c,k}^{(j)} = \boldsymbol{P}_{k|k-1}^{(j)} \boldsymbol{H}_{c,k}^{\mathrm{T}} \left[\boldsymbol{S}_{c,k}^{(j)}\right]^{-1} \tag{9.2.10}$$

$$\boldsymbol{S}_{c,k}^{(j)} = \boldsymbol{H}_{c,k} \boldsymbol{P}_{k|k-1}^{(j)} \boldsymbol{H}_{c,k}^{\mathrm{T}} + \boldsymbol{R}_{c,k} \tag{9.2.11}$$

- Step 2: Sequentially update the target state using the Doppler measurement $y_{k,m}^{d}$

$$\boldsymbol{m}_{k|k}^{(j)}\left(\boldsymbol{z}_{k,m}\right) = \boldsymbol{m}_{k|k}^{(j)}\left(\boldsymbol{y}_{k,m}^{c}\right) + \boldsymbol{G}_{d,k}^{(j)}\left(\boldsymbol{y}_{k,m}^{c}\right) \tilde{y}_{k,m,j}^{d} \tag{9.2.12}$$

$$\boldsymbol{P}_{k|k}^{(j)}\left(\boldsymbol{z}_{k,m}\right) = \left[\boldsymbol{I} - \boldsymbol{G}_{d,k}^{(j)}\left(\boldsymbol{y}_{k,m}^{c}\right) \boldsymbol{H}_{d,k}\left(\boldsymbol{y}_{k,m}^{c}\right)\right] \boldsymbol{P}_{k|k}^{(j)}\left(\boldsymbol{y}_{k,m}^{c}\right) \tag{9.2.13}$$

where the residual of Doppler measurement is

$$\tilde{y}_{k,m,j}^{d} = y_{k,m}^{d} - \hat{y}_{k,m,j}^{d} \tag{9.2.14}$$

the predicted Doppler measurement is

$$\hat{y}_{k,m,j}^{d} = \dot{r}_{k,j}\left(\boldsymbol{m}_{k|k}^{(j)}(\boldsymbol{y}_{k,m}^{c})\right) \tag{9.2.15}$$

the gain and innovation covariance about the Doppler measurement respectively are

$$\boldsymbol{G}_{d,k}^{(j)}\left(\boldsymbol{y}_{k,m}^{c}\right) = \boldsymbol{P}_{k|k}^{(j)}\left(\boldsymbol{y}_{k,m}^{c}\right) \boldsymbol{H}_{d,k}^{\mathrm{T}}\left(\boldsymbol{y}_{k,m}^{c}\right) \left[S_{d,k}^{(m,j)}\right]^{-1} \tag{9.2.16}$$

$$S_{d,k}^{(m,j)} = \boldsymbol{H}_{d,k}\left(\boldsymbol{y}_{k,m}^{c}\right) \boldsymbol{P}_{k|k}^{(j)}\left(\boldsymbol{y}_{k,m}^{c}\right) \boldsymbol{H}_{d,k}^{\mathrm{T}}\left(\boldsymbol{y}_{k,m}^{c}\right) + \sigma_{d}^{2} \tag{9.2.17}$$

and the Jacobian matrix about the Doppler measurement is

$$\boldsymbol{H}_{d,k}\left(\boldsymbol{y}_{k,m}^{c}\right) = \nabla \dot{r}_{k,j}\left(\boldsymbol{y}_{k,m}^{c}\right) = \left[\, h_1 \; h_2 \; h_3 \; h_4 \,\right] \tag{9.2.18}$$

In the above equation,

$$h_1 = \frac{\dot{x}_{k|k}^{(m,j)}}{\left[\left(x_{k|k}^{(m,j)}\right)^2 + \left(y_{k|k}^{(m,j)}\right)^2\right]^{1/2}} - x_{k|k}^{(m,j)} \frac{x_{k|k}^{(m,j)} \dot{x}_{k|k}^{(m,j)} + y_{k|k}^{(m,j)} \dot{y}_{k|k}^{(m,j)}}{\left[\left(x_{k|k}^{(m,j)}\right)^2 + \left(y_{k|k}^{(m,j)}\right)^2\right]^{3/2}}$$
$$= \left(\dot{x}_{k|k}^{(m,j)} - \hat{y}_{k,m,j}^{d} \cos \hat{a}_{k,m,j}\right) / \hat{r}_{k,m,j} \tag{9.2.19}$$

$$h_2 = \frac{\dot{y}_{k|k}^{(m,j)}}{\left[\left(x_{k|k}^{(m,j)}\right)^2 + \left(y_{k|k}^{(m,j)}\right)^2\right]^{1/2}} - y_{k|k}^{(m,j)} \frac{x_{k|k}^{(m,j)} \dot{x}_{k|k}^{(m,j)} + y_{k|k}^{(m,j)} \dot{y}_{k|k}^{(m,j)}}{\left[\left(x_{k|k}^{(m,j)}\right)^2 + \left(y_{k|k}^{(m,j)}\right)^2\right]^{3/2}}$$
$$= \left(\dot{y}_{k|k}^{(m,j)} - \hat{y}_{k,m,j}^{d} \sin \hat{a}_{k,m,j}\right) / \hat{r}_{k,m,j} \tag{9.2.20}$$

$$h_3 = x_{k|k}^{(m,j)} / \left[\left(x_{k|k}^{(m,j)}\right)^2 + \left(y_{k|k}^{(m,j)}\right)^2\right]^{1/2} \triangleq \cos \hat{a}_{k,m,j} \tag{9.2.21}$$

$$h_4 = y_{k|k}^{(m,j)} / \left[\left(x_{k|k}^{(m,j)}\right)^2 + \left(y_{k|k}^{(m,j)}\right)^2\right]^{1/2} \triangleq \sin \hat{a}_{k,m,j} \tag{9.2.22}$$

with $\hat{r}_{k,m,j} \triangleq [(x_{k|k}^{(m,j)})^2 + (y_{k|k}^{(m,j)})^2]^{1/2}$.

- Step 3: Calculate the weight using the position and Doppler measurement $z_{k,m} = [\boldsymbol{y}_{k,m}^{c}; y_{k,m}^{d}]$

$$w_k^{(j)}(z_{k,m}) = p_{D,k} w_{k|k-1}^{(j)} q_k^{(j)}(z_{k,m}) \frac{\left\langle 1, \kappa_k^c \kappa_k^d \right\rangle}{\kappa_k^c\left(\boldsymbol{y}_{k,m}^c\right) \kappa_k^d\left(y_{k,m}^d\right)} \times$$
$$\frac{\left\langle \Upsilon_k^{(1)}\left[w_{k|k-1}, Z_k - \{z_{k,m}\}\right], \rho_{k|k-1} \right\rangle}{\left\langle \Upsilon_k^{(0)}\left[w_{k|k-1}, Z_k\right], \rho_{k|k-1} \right\rangle} \tag{9.2.23}$$

where

$$\Upsilon_k^{(u)}[w, Z](n) = \sum_{j=0}^{\min(|Z|,n)} (|Z| - j)! \rho_{C,k}(|Z| - j) P_{j+u}^{n} \times$$
$$\frac{(1 - p_{D,k})^{n-(j+u)}}{\langle 1, w \rangle^{j+u}} e_j(\Lambda_k(w, Z)) \tag{9.2.24}$$

$$\Lambda_k(w, Z) = \left\{ \frac{\left\langle 1, \kappa_k^c \kappa_k^d \right\rangle}{\kappa_k^c\left(\boldsymbol{y}_{k,m}^c\right) \kappa_k^d\left(y_{k,m}^d\right)} p_{D,k} w^{\mathrm{T}} q_k(z_{k,m}) : z_{k,m} \in Z_k \right\} \tag{9.2.25}$$

the predicted weight vector is

$$w_{k|k-1} = \left[w_{k|k-1}^{(1)}, \ldots, w_{k|k-1}^{(J_{k|k-1})}\right]^{\mathrm{T}} \tag{9.2.26}$$

the likelihood function vector is

$$q_k(z_{k,m}) = \left[q_k^{(1)}(z_{k,m}), \ldots, q_k^{(J_{k|k-1})}(z_{k,m})\right]^{\mathrm{T}} \tag{9.2.27}$$

in which the likelihood function of the jth component is

$$q_k^{(j)}(z_{k,m}) = q_{c,k}^{(j)}(y_{k,m}^c) q_{d,k}^{(j)}(y_{k,m}^d) \tag{9.2.28}$$

and the likelihood functions corresponding to the position component and the Doppler component are respectively

$$q_{c,k}^{(j)}\left(y_{k,m}^c\right) = \mathcal{N}\left(y_{k,m}^c; \boldsymbol{H}_{c,k} \boldsymbol{m}_{k|k-1}^{(j)}, \boldsymbol{S}_{c,k}^{(j)}\right) \tag{9.2.29}$$

$$q_{d,k}^{(j)}\left(y_{k,m}^d\right) = \mathcal{N}\left(y_{k,m}^d; \hat{y}_{k,m,j}^d, S_{d,k}^{(m,j)}\right) \tag{9.2.30}$$

where $|\cdot|$ represents the cardinality of a set (the number of elements in the set), $P_j^n = n!/(n-j)!$ is the permutation coefficient, $\rho_{C,k}(\cdot)$ represents the clutter cardinality distribution, $\kappa_k^c(y_{k,m}^c) = \lambda_c \cdot V \cdot u(y_{k,m}^c)$ is the clutter intensity on the positional measurement component, λ_c is the clutter density, V is the volume of the monitoring area, and $u(\cdot)$ is the uniform density. Assume that the clutter velocity is uniformly distributed in the range of $[-v_{\max}, v_{\max}]$, then the Doppler clutter intensity is $\kappa_k^d(y_{k,m}^d) = 1/(2v_{\max})$, where $v_{\max}$ is the maximum velocity that the sensor can detect [223]. $e_j(Z)$ is the elementary symmetric function (ESF) of order j defined on a finite set of real numbers Z, and its specific expression is shown in (5.2.9).

9.2.3 Simulation Analysis

Consider the multi-target tracking in a scenario where the targets move in a straight line at a uniform speed in the x–y plane. The initial states and the starting and end times of targets are shown in Table 9.1. Assume that the sensor monitoring area is $[-1000\ 1000]\,(\mathrm{m}) \times [-1000\ 1000]\,(\mathrm{m})$, $V = 4\times10^6\ \mathrm{m}^2$, the detection probability is $p_{D,k} = 0.98$, $\boldsymbol{R}_{c,k} = \sigma_c^2 \boldsymbol{I}_2$, $\sigma_c = 10$ m, $\sigma_d = 0.5$ m/s, and $v_{\max} = 35$ m/s. In the GM-PHD and GM-CPHD filters, $p_{S,k} = 0.99$, and

$$\boldsymbol{F}_{k-1} = \begin{bmatrix} \boldsymbol{I}_2 & \Delta\cdot \boldsymbol{I}_2 \\ \boldsymbol{0}_2 & \boldsymbol{I}_2 \end{bmatrix}, \ \boldsymbol{Q}_{k-1} = \sigma_v^2 \begin{bmatrix} \Delta^4 \boldsymbol{I}_2/4 & \Delta^3 \boldsymbol{I}_2/2 \\ \Delta^3 \boldsymbol{I}_2/2 & \Delta^2 \boldsymbol{I}_2 \end{bmatrix}$$

where $\Delta = 1$ s, $\sigma_v = 5$ m/s. The birth intensity is set to $\boldsymbol{m}_{\gamma,k}^{(1)} = [0\ \mathrm{m}, 0\ \mathrm{m}, 0\ \mathrm{m/s}, 0\ \mathrm{m/s}]^{\mathrm{T}}$, $\boldsymbol{m}_{\gamma,k}^{(2)} = [400\ \mathrm{m}, -600\ \mathrm{m}, 0\ \mathrm{m/s}, 0\ \mathrm{m/s}]^{\mathrm{T}}$, $\boldsymbol{m}_{\gamma,k}^{(3)} = [-200\ \mathrm{m}, 800\ \mathrm{m}, 0\ \mathrm{m/s}, 0\ \mathrm{m/s}]^{\mathrm{T}}$, $\boldsymbol{m}_{\gamma,k}^{(4)} = [-800\ \mathrm{m}, -200\ \mathrm{m}, 0\ \mathrm{m/s}, 0\ \mathrm{m/s}]^{\mathrm{T}}$, $J_{\gamma,k} = 4$, $w_{\gamma,k}^{(i)} = 0.03$,

Table 9.1 Initial state and starting and end times of each target

Target No.	Starting time	End time	Initial state $[\, x_0 \;\; y_0 \;\; \dot{x}_0 \;\; \dot{y}_0 \,]^{\mathrm{T}}$
1	1	70	(− 800, − 200, 20, − 5)
2	40	100	(− 800, − 200, 12.5, 7)
3	60	100	(− 800, − 200, 2.5, 10.5)
4	40	100	(− 200, 800, 16, − 9.7)
5	60	100	(− 200, 800, − 2.5, − 14.6)
6	80	100	(− 200, 800, 17.5, − 5)
7	1	70	(0, 0, 0, − 10)
8	20	100	(0, 0, 7.5, − 5)
9	80	100	(0, 0, − 20, − 15)
10	1	100	(400, − 600, − 10.5, 5)
11	20	100	(400, − 600, − 2.5, 10.2)
12	20	100	(400, − 600, − 7.5, − 4.5)

$\boldsymbol{P}_{\gamma,k}^{(i)} = \mathrm{blkdiag}\left(\sqrt{10}\,\mathrm{m},\ \sqrt{10}\,\mathrm{m},\ \sqrt{10}\,\mathrm{m/s},\ \sqrt{10}\,\mathrm{m/s}\right)^2, i = 1, 2, 3, 4$. The merging parameters used in the state extraction [63] include: weight threshold $T = 10^{-5}$, merging threshold $U = 4\,\mathrm{m}$, and the maximum number of Gaussian components $J_{\max} = 100$. The capping parameter used in the approximation of the cardinality distribution in the GM-CPHD filter is $N_{\max} = 100$ [64]. The order and threshold parameters in the OSPA (see Sect. 3.8.3) are set to $p = 2$ and $c = 20\,\mathrm{m}$, respectively.

Under the condition of $\lambda_c = 12.5 \times 10^{-6}\,\mathrm{m}^{-2}$, Fig. 9.1 visually shows the typical multi-target tracking result of the GM-CPHD algorithm with Doppler information (i.e., GM-CPHDwD). In the figure, the black dotted line represents the true target track and the red circle is the estimated target state. It can be seen that the GM-CPHDwD algorithm can well realize multi-target tracking under the clutter condition.

Figure 9.2 shows the estimated number of targets and the tracking performance in terms of the OSPA obtained by this algorithm after 100 Monte Carlo statistics. The dashed line in Fig. 9.2a represents the standard deviation of the estimation. It can be seen that the algorithm estimates the number of targets accurately and stably. Figure 9.2b shows that, when the number of targets is stable, the OSPA distance can be maintained at around 20 m, while during the change of the number of targets, the OSPA distance fluctuates sharply.

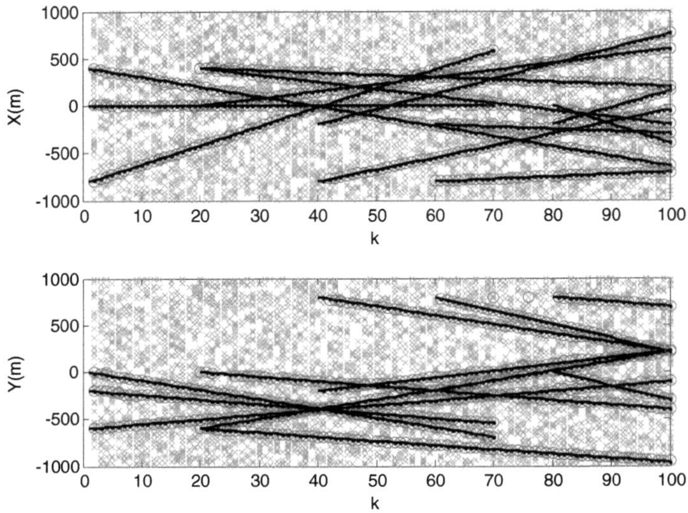

Fig. 9.1 Typical multi-target tracking results of the GM-CPHDwD algorithm

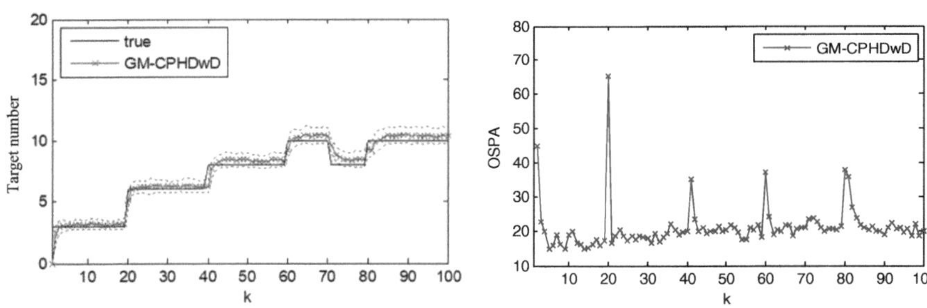

(a) Estimated number of targets (b) OSPA

Fig. 9.2 Tracking performance of the GM-CPHDwD algorithm at different times

In order to validate the capability of introducing the Doppler information into the GM-CPHDwD for clutter suppression and performance improvement, the standard GM-PHD [63] and GM-CPHD [64] algorithms without Doppler information are respectively denoted as the GM-PHDwoD and GM-CPHDwoD, and the GM-PHD algorithm with Doppler information proposed in [223] is recorded as the GM-PHDwD. Figure 9.3 shows the OSPAs of the four algorithms at different clutter densities, as well as the cardinality estimation error in the OSPA (OSPA Card) and the localization error in the OSPA (OSPA Loc) (see Sect. 3.8.3). In general, the introduction of the Doppler information significantly improves the performances of both the GM-PHD and the GM-CPHD, and the higher the clutter density, the more significant the performance improvement, which indicates that the use of Doppler

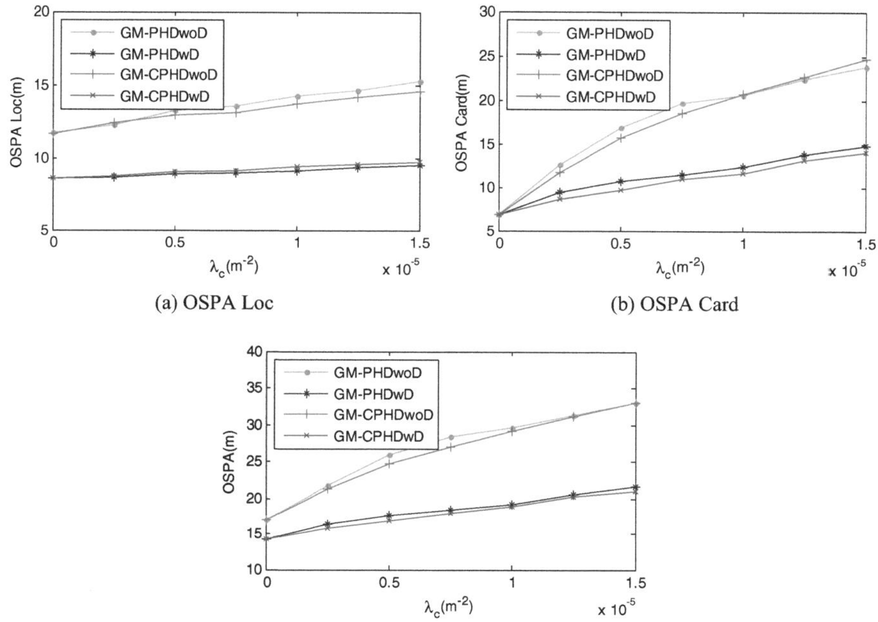

(a) OSPA Loc (b) OSPA Card (c) OSPA

Fig. 9.3 Comparison of tracking performances of all algorithms at different clutter densities

information can effectively suppress the clutter interference. Compared with the GM-PHDwD, Fig. 9.3a shows that the GM-CPHDwD is slightly inferior in the localization accuracy, while Fig. 9.3b shows that it is better than the GM-PHDwD in terms of the performance of cardinality estimation. As a result, from the perspective of the overall OSPA, the GM-CPHDwD is better than the GM-PHDwD, as shown in Fig. 9.3c.

9.3 GM-PHD Filter in the Presence of DBZ

The existence of the Doppler blind zone (DBZ) complicates the problem of the multi-target tracking with Doppler radars, because the resulting successive missed detections will severely deteriorate the tracking performance. In general, the width of the DBZ is determined by the minimum detectable velocity (MDV), which is an important tracking parameter. In this section, a GM-PHD filter is used to track multiple targets in the presence of the DBZ. Based on the detection probability model incorporating into the MDV, by substituting it into the update equation of the standard GM-PHD filter, the GM-PHD update equation in the presence of the DBZ is derived, and finally the detailed implementation steps of the new GM-PHD update are provided. This algorithm makes full use of the MDV and Doppler information. The performance of the algorithm is verified by comparing with the Monte Carlo experiment of the conventional GM-PHD algorithm with only Doppler measurement, which proves that the algorithm can improve the tracking performance, especially when the MDV value is small.

9.3.1 Detection Probability Model Incorporating MDV

Let $\boldsymbol{x}_k = [x_k\ y_k\ z_k\ \dot{x}_k\ \dot{y}_k\ \dot{z}_k]^{\mathrm{T}}$ denote the target state at time k, where (x_k, y_k, z_k) is the position component and $(\dot{x}_k, \dot{y}_k, \dot{z}_k)$ is the velocity component. Similarly, Let $\boldsymbol{x}_k^s = [x_k^s\ y_k^s\ z_k^s\ \dot{x}_k^s\ \dot{y}_k^s\ \dot{z}_k^s]^{\mathrm{T}}$ represent the state of sensor s. Let $\dot{r}_k$ be the Doppler velocity, and $\dot{r}_{c,k}$ be the corresponding background clutter Doppler at the position of the state, then we have

$$\dot{r}_k = h_d(\boldsymbol{x}_k) = \frac{(x_k - x_k^s)(\dot{x}_k - \dot{x}_k^s) + (y_k - y_k^s)(\dot{y}_k - \dot{y}_k^s) + (z_k - z_k^s)(\dot{z}_k - \dot{z}_k^s)}{\sqrt{(x_k - x_k^s)^2 + (y_k - y_k^s)^2 + (z_k - z_k^s)^2}} \tag{9.3.1}$$

$$\dot{r}_{c,k} = -\frac{\dot{x}_k^s(x_k - x_k^s) + \dot{y}_k^s(y_k - y_k^s) + \dot{z}_k^s(z_k - z_k^s)}{\sqrt{(x_k - x_k^s)^2 + (y_k - y_k^s)^2 + (z_k - z_k^s)^2}} \tag{9.3.2}$$

Following [85], the difference between the target Doppler velocity and the surrounding clutter Doppler velocity is referred to as the clutter notch function, which is denoted as n_c and defined as

$$n_c = n_c(\boldsymbol{x}_k) \triangleq \dot{r}_k - \dot{r}_{c,k} = \frac{\dot{x}_k(x_k - x_k^s) + \dot{y}_k(y_k - y_k^s) + \dot{z}_k(z_k - z_k^s)}{\sqrt{(x_k - x_k^s)^2 + (y_k - y_k^s)^2 + (z_k - z_k^s)^2}} \tag{9.3.3}$$

The clutter notch mainly suppresses the clutter, but also affects the detection of targets with lower Doppler velocities. In the standard GM-PHD filter of Sect. 4.4.1, (4.4.4) assumes that the detection probability is independent of the target state. However, the detection probability $p_{D,k}(\boldsymbol{x})$ is generally a function of the target state. In particular, for a GMTI radar, due to the DBZ of the sensor, the detection probability is strongly influenced by the presence of the DBZ. Specifically, when n_c falls within the DBZ, i.e., $|n_c| <$ MDV, $p_{D,k}(\boldsymbol{x})$ will reduce to 0; when it is far from the clutter notch, i.e., $|n_c| \gg$ MDV, $p_{D,k}(\boldsymbol{x})$ reaches its saturation value p_D, where p_D is the conventional detection probability when the target Doppler velocity is outside the DBZ and the influence of antenna pattern and propagation is considered. To this end, the detection probability is modeled as [215]

$$p_{D,k}(\boldsymbol{x}) \approx p_D\big[1 - \exp\big(-(n_c(\boldsymbol{x}_k)/\mathrm{MDV})^2 \log 2\big)\big] \tag{9.3.4}$$

The detection probability in (9.3.4) is in exponential form. In order to facilitate to be called under the subsequent GM-PHD framework, it is required to be transformed into a Gaussian form. For this reason, the first-order Taylor expansion of non-linear clutter notch function n_c is carried out around the predicted value $\boldsymbol{x}_{k|k-1} = [\hat{x}_{k|k-1}\ \hat{y}_{k|k-1}\ \hat{z}_{k|k-1}\ \hat{\dot{x}}_{k|k-1}\ \hat{\dot{y}}_{k|k-1}\ \hat{\dot{z}}_{k|k-1}]^{\mathrm{T}}$ to obtain

$$
\begin{aligned}
n_c(\boldsymbol{x}_k) &\approx n_c(\hat{\boldsymbol{x}}_{k|k-1}) + \frac{\partial n_c}{\partial \boldsymbol{x}_k}\Big|_{\boldsymbol{x}_k=\hat{\boldsymbol{x}}_{k|k-1}} (\boldsymbol{x}_k - \hat{\boldsymbol{x}}_{k|k-1}) \\
&= n_c(\hat{\boldsymbol{x}}_{k|k-1}) - \frac{\partial n_c}{\partial \boldsymbol{x}_k}\Big|_{\boldsymbol{x}_k=\hat{\boldsymbol{x}}_{k|k-1}} \hat{\boldsymbol{x}}_{k|k-1} + \frac{\partial n_c}{\partial \boldsymbol{x}_k}\Big|_{\boldsymbol{x}_k=\hat{\boldsymbol{x}}_{k|k-1}} \boldsymbol{x}_k \\
&= y_f(\hat{\boldsymbol{x}}_{k|k-1}) - \boldsymbol{H}_f(\hat{\boldsymbol{x}}_{k|k-1})\boldsymbol{x}_k
\end{aligned} \tag{9.3.5}
$$

where the pseudo-measurement function is

$$
y_f = y_f(\hat{\boldsymbol{x}}_{k|k-1}) = n_c(\hat{\boldsymbol{x}}_{k|k-1}) + \boldsymbol{H}_f(\hat{\boldsymbol{x}}_{k|k-1})\hat{\boldsymbol{x}}_{k|k-1} \tag{9.3.6}
$$

and the pseudo-measurement matrix is

$$
\boldsymbol{H}_f(\hat{\boldsymbol{x}}_{k|k-1}) = -\frac{\partial n_c}{\partial \boldsymbol{x}_k}\Big|_{\boldsymbol{x}_k=\hat{\boldsymbol{x}}_{k|k-1}} = \begin{bmatrix} n_1\ n_2\ n_3\ n_4\ n_5\ n_6 \end{bmatrix}^{\mathrm{T}} \tag{9.3.7}
$$

In the above equation, we have

$$
n_1 = -\left(\hat{\dot{x}}_{k|k-1} - \hat{n}_c \cos\hat{a}_k \cos\hat{e}_k\right)/\hat{r}_k, \quad n_2 = -\left(\hat{\dot{y}}_{k|k-1} - \hat{n}_c \sin\hat{a}_k \cos\hat{e}_k\right)/\hat{r}_k,
$$

$$
\begin{aligned}
n_3 &= -\left(\hat{\dot{z}}_{k|k-1} - \hat{n}_c \sin\hat{e}_k\right)/\hat{r}_k, \quad n_4 = -\cos\hat{a}_k \cos\hat{e}_k, \\
n_5 &= -\sin\hat{a}_k \cos\hat{e}_k, \quad n_6 = -\sin\hat{e}_k,
\end{aligned}
$$

$$
\begin{aligned}
\hat{r}_k &\triangleq \sqrt{(\hat{x}_{k|k-1} - x_k^s)^2 + (\hat{y}_{k|k-1} - y_k^s)^2 + (\hat{z}_{k|k-1} - z_k^s)^2}, \\
\hat{a}_k &\triangleq \mathrm{a}\tan 2\left[(\hat{y}_{k|k-1} - y_k^s), (\hat{x}_{k|k-1} - x_k^s)\right],
\end{aligned}
$$

$$
\hat{e}_k \triangleq \arctan\left[(\hat{z}_{k|k-1} - z_k^s)/\sqrt{(\hat{x}_{k|k-1} - x_k^s)^2 + (\hat{y}_{k|k-1} - y_k^s)^2}\right],
$$

$$
n_c = n_c(\hat{\boldsymbol{x}}_{k|k-1}) = \frac{\hat{\dot{x}}_{k|k-1}(\hat{x}_{k|k-1} - x_k^s) + \hat{\dot{y}}_{k|k-1}(\hat{y}_{k|k-1} - y_k^s) + \hat{\dot{z}}_{k|k-1}(\hat{z}_{k|k-1} - z_k^s)}{\sqrt{(\hat{x}_{k|k-1} - x_k^s)^2 + (\hat{y}_{k|k-1} - y_k^s)^2 + (\hat{z}_{k|k-1} - z_k^s)^2}}.
$$

Substituting the above approximate $n_c(\boldsymbol{x})$ into (9.3.4) yields

$$
p_{D,k}(\boldsymbol{x}) = p_D \cdot \left[1 - c_f \mathcal{N}\left(y_f(\hat{\boldsymbol{x}}_{k|k-1}); \boldsymbol{H}_f(\hat{\boldsymbol{x}}_{k|k-1})\boldsymbol{x}, R_f\right)\right] \tag{9.3.8}
$$

where $c_f = \mathrm{MDV}\sqrt{\pi/\log 2}$ is the normalization factor and $R_f = \mathrm{MDV}^2/(2\log 2)$ is the variance of the pseudo-measurement in the Doppler domain. Therefore, the clutter notch information in (9.3.8) acts as a pseudo-measurement, and the MDV plays the role of the standard deviation of the pseudo-measurement.

9.3.2 GM-PHD Filter with MDV and Doppler Measurements

From (9.2.3) and (9.2.4), it can be known that for a Doppler radar, the likelihood function $g_k(z|\boldsymbol{x})$ can be modeled as

$$g_k(z|\boldsymbol{x}) = \mathcal{N}\big(\boldsymbol{y}_c; \boldsymbol{H}_{c,k}\boldsymbol{x}, \boldsymbol{R}_{c,k}\big)\mathcal{N}\big(y_d; h_d(\boldsymbol{x}), \sigma_d^2\big) \tag{9.3.9}$$

where z is the conventional detection, which includes the positional measurement component $\boldsymbol{y}_c$ and the Doppler measurement component y_d, $\boldsymbol{H}_{c,k}$ is the positional measurement matrix, $\boldsymbol{R}_{c,k}$ is the positional measurement noise covariance, $h_d(\cdot)$ is the non-linear Doppler measurement function, and σ_d is the standard deviation of the Doppler measurement noise.

According to the PHD recursion (4.2.1) and (4.2.2), the detection probability and likelihood function only influence the updated intensity, but have no effect on the predicted intensity. As a consequence, for the Gaussian mixture implementation of the PHD filter, the relevant prediction equation is exactly the same as the prediction step in Sect. 4.4.2, hence will not be repeated here. The next focus is on the derivation of the updated intensity incorporating the MDV and Doppler information.

Assume that the predicted intensity at time k is in the following Gaussian mixture form

$$v_{k|k-1}(\boldsymbol{x}) = \sum_{j=1}^{J_{k|k-1}} w_{k|k-1}^{(j)}\mathcal{N}\Big(\boldsymbol{x}; \boldsymbol{m}_{k|k-1}^{(j)}, \boldsymbol{P}_{k|k-1}^{(j)}\Big) \tag{9.3.10}$$

where $w_{k|k-1}^{(j)}$, $\boldsymbol{m}_{k|k-1}^{(j)}$, and $\boldsymbol{P}_{k|k-1}^{(j)}$ are the weight, mean, and covariance of the predicted component, respectively. Substituting the detection probability in (9.3.8) and the likelihood function in (9.3.9) into (4.2.2), successively applying the product equation of normal densities in Lemma 5, and reorganizing the obtained results yields the following updated intensity

$$\begin{aligned} v_k(\boldsymbol{x}) &= \sum_{j=1}^{J_{k|k}} w_{k|k}^{(j)}\mathcal{N}\Big(\boldsymbol{x}; \boldsymbol{m}_{k|k}^{(j)}, \boldsymbol{P}_{k|k}^{(j)}\Big) \\ &= \sum_{j=1}^{J_{k|k-1}} w_{k|k,0}^{(j)}\mathcal{N}\Big(\boldsymbol{x}; \boldsymbol{m}_{k|k,0}^{(j)}, \boldsymbol{P}_{k|k,0}^{(j)}\Big) + \sum_{z\in Z_k}\sum_{j=1}^{J_{k|k-1}} w_{k|k}^{(j)}(z)\mathcal{N}\Big(\boldsymbol{x}; \boldsymbol{m}_{k|k}^{(j)}(z), \boldsymbol{P}_{k|k}^{(j)}(z)\Big) \\ &\quad + \sum_{j=1}^{J_{k|k-1}} w_{k|k,f}^{(j)}\mathcal{N}\Big(\boldsymbol{x}; \boldsymbol{m}_{k|k,f}^{(j)}, \boldsymbol{P}_{k|k,f}^{(j)}\Big) \end{aligned}$$

$$+\sum_{z\in Z_k}\sum_{j=1}^{J_{k|k-1}} w_{k|k}^{(j)}(z_f)\mathcal{N}\Big(x;\, m_{k|k}^{(j)}(z_f),\, P_{k|k}^{(j)}(z_f)\Big) \tag{9.3.11}$$

where z_f represents the augmented measurement, which includes the conventional measurement component z and the pseudo-measurement component y_f. The components $\{w_{k|k,0}^{(j)}, m_{k|k,0}^{(j)}, P_{k|k,0}^{(j)}\}$, $\{w_{k|k}^{(j)}(z), m_{k|k}^{(j)}(z), P_{k|k}^{(j)}(z)\}$, $\{w_{k|k,f}^{(j)}, m_{k|k,f}^{(j)}, P_{k|k,f}^{(j)}\}$, and $\{w_{k|k}^{(j)}(z_f), m_{k|k}^{(j)}(z_f), P_{k|k}^{(j)}(z_f)\}$, are updated with the conventional missed measurement, conventional measurement, pseudo measurement and augmented measurement, respectively.

Component $\{w_{k|k,0}^{(j)}, m_{k|k,0}^{(j)}, P_{k|k,0}^{(j)}\}$ is given as follows

$$w_{k|k,0}^{(j)} = (1-p_D)w_{k|k-1}^{(j)} \tag{9.3.12}$$

$$m_{k|k,0}^{(j)} = m_{k|k-1}^{(j)},\; P_{k|k,0}^{(j)} = P_{k|k-1}^{(j)} \tag{9.3.13}$$

Component $\{w_{k|k}^{(j)}(z), m_{k|k}^{(j)}(z), P_{k|k}^{(j)}(z)\}$ is obtained by sequentially processing the positional measurement and Doppler measurement, namely

$$w_{k|k}^{(j)}(z) = \frac{p_D w_{k|k-1}^{(j)} q_k^{(j)}(z)}{\kappa_k(z) + w_{\text{sum}}} \tag{9.3.14}$$

$$m_{k|k}^{(j)}(z) = m_{k|k}^{(j)}(y_c) + G_{d,k}^{(j)}(y_c)\Big(y_d - h_d\Big(m_{k|k}^{(j)}(y_c)\Big)\Big) \tag{9.3.15}$$

$$P_{k|k}^{(j)}(z) = \Big[I - G_{d,k}^{(j)}(y_c)H_d\Big(m_{k|k}^{(j)}(y_c)\Big)\Big]P_{k|k}^{(j)}(y_c) \tag{9.3.16}$$

where $\kappa_k(z)$, w_{sum} and $q_k^{(j)}(z)$ are given by (9.3.35), (9.3.36) and (9.3.37), respectively, and $m_{k|k}^{(j)}(z)$ and $P_{k|k}^{(j)}(z)$ are respectively the updated mean and covariance with the measurement z (including position measurement y_c and Doppler measurement y_d). Similarly, $m_{k|k}^{(j)}(y_c)$ and $P_{k|k}^{(j)}(y_c)$ are respectively the updated mean and covariance with positional measurement y_c, i.e.,

$$m_{k|k}^{(j)}(y_c) = m_{k|k-1}^{(j)} + G_{c,k}^{(j)}\Big(y_c - H_{c,k}m_{k|k-1}^{(j)}\Big) \tag{9.3.17}$$

$$P_{k|k}^{(j)}(y_c) = \Big[I - G_{c,k}^{(j)}H_{c,k}\Big]P_{k|k-1}^{(j)} \tag{9.3.18}$$

where the positional measurement gain and the innovation covariance are respectively

$$G_{c,k}^{(j)} = P_{k|k-1}^{(j)}H_{c,k}\Big[S_{c,k}^{(j)}\Big]^{-1} \tag{9.3.19}$$

$$S_{c,k}^{(j)} = H_{c,k} P_{k|k-1}^{(j)} H_{c,k}^{\mathrm{T}} + R_{c,k} \tag{9.3.20}$$

The Doppler measurement gain $G_{d,k}^{(j)}(y_c)$ in (9.3.15) and (9.3.16) is

$$G_{d,k}^{(j)}(y_c) = P_{k|k}^{(j)}(y_c) H_d\left(m_{k|k}^{(j)}(y_c)\right)\left[S_{d,k}^{(j)}(y_c)\right]^{-1} \tag{9.3.21}$$

where the Doppler measurement innovation covariance is

$$S_{d,k}^{(j)}(y_c) = H_d\left(m_{k|k}^{(j)}(y_c)\right) P_{k|k}^{(j)}(y_c)\left[H_d(m_{k|k}^{(j)}(y_c))\right]^{\mathrm{T}} + \sigma_d^2 \tag{9.3.22}$$

In (9.3.16), (9.3.21) and (9.3.22), $H_d(m_{k|k}^{(j)}(y_c))$ is the Jacobian matrix of $\dot{r}_k$ with respect to x_k calculated at $m_{k|k}^{(j)}(y_c)$ (for each component j), i.e.,

$$H_d\left(m_{k|k}^{(j)}(y_c)\right) = \frac{\partial \dot{r}_k}{\partial x_k}\bigg|_{x_k = m_{k|k}^{(j)}(y_c)} = \left[h_1^{(j)}\ h_2^{(j)}\ h_3^{(j)}\ h_4^{(j)}\ h_5^{(j)}\ h_6^{(j)}\right] \tag{9.3.23}$$

where

$$\begin{aligned}
m_{k|k}^{(j)}(y_c) &= \left[\hat{x}_{k|k}^{(j)}\ \hat{y}_{k|k}^{(j)}\ \hat{z}_{k|k}^{(j)}\ \hat{\dot{x}}_{k|k}^{(j)}\ \hat{\dot{y}}_{k|k}^{(j)}\ \hat{\dot{z}}_{k|k}^{(j)}\right]^{\mathrm{T}}, \\
h_1^{(j)} &= \left[(\hat{\dot{x}}_{k|k}^{(j)} - \dot{x}_k^s) - h_d(m_{k|k}^{(j)}(y_c)) \cos \hat{a}_k^{(j)} \cos \hat{e}_k^{(j)}\right] / \hat{r}_k^{(j)}, \\
h_2^{(j)} &= \left[(\hat{\dot{y}}_{k|k}^{(j)} - \dot{y}_k^s) - h_d(m_{k|k}^{(j)}(y_c)) \sin \hat{a}_k^{(j)} \cos \hat{e}_k^{(j)}\right] / \hat{r}_k^{(j)}, \\
h_3^{(j)} &= \left[(\hat{\dot{z}}_{k|k}^{(j)} - \dot{z}_k^s) - h_d(m_{k|k}^{(j)}(y_c)) \sin \hat{e}_k^{(j)}\right] / \hat{r}_k^{(j)}, \\
h_4^{(j)} &= \cos \hat{a}_k^{(j)} \cos \hat{e}_k^{(j)},\ h_5^{(j)} = \sin \hat{a}_k^{(j)} \cos \hat{e}_k^{(j)},\ h_6^{(j)} = \sin \hat{e}_k^{(j)}, \\
\hat{r}_k^{(j)} &= \sqrt{(\hat{x}_{k|k}^{(j)} - x_k^s)^2 + (\hat{y}_{k|k}^{(j)} - y_k^s)^2 + (\hat{z}_{k|k}^{(j)} - z_k^s)^2}, \\
\hat{a}_k^{(j)} &= \mathrm{a}\tan 2\left[(\hat{y}_{k|k}^{(j)} - y_k^s), (\hat{x}_{k|k}^{(j)} - x_k^s)\right], \\
\hat{e}_k^{(j)} &= \arctan\left[(\hat{z}_{k|k}^{(j)} - z_k^s) / \left((\hat{x}_{k|k}^{(j)} - x_k^s)^2 + (\hat{y}_{k|k}^{(j)} - y_k^s)^2\right)^{1/2}\right].
\end{aligned}$$

Note that since the Doppler measurement is a non-linear function, similar to the EKF update [416], the following equation is used in the derivation of (9.3.15) and (9.3.16)

$$\begin{aligned}
&\mathcal{N}(y_d; h_d(x), \sigma_d^2)\mathcal{N}\left(x; m_{k|k}^{(j)}(y_c), P_{k|k}^{(j)}(y_c)\right) \\
&\approx \mathcal{N}\left(y_d; h_d(m_{k|k}^{(j)}(y_c)), S_{d,k}^{(j)}(y_c)\right)\mathcal{N}\left(x; m_{k|k}^{(j)}(z), P_{k|k}^{(j)}(z)\right)
\end{aligned} \tag{9.3.24}$$

Component $\{w_{k|k,f}^{(j)}, m_{k|k,f}^{(j)}, P_{k|k,f}^{(j)}\}$ is updated with the pseudo measurement, which is

$$w_{k|k,f}^{(j)} = \frac{p_D}{1 - p_D} c_f \mathcal{N}\left(y_f(m_{k|k-1}^{(j)}); H_f(m_{k|k-1}^{(j)}) m_{k|k-1}^{(j)}, S_{f,k}^{(j)}\right) w_{k|k,0}^{(j)} \tag{9.3.25}$$

$$\boldsymbol{m}_{k|k,f}^{(j)} = \boldsymbol{m}_{k|k-1}^{(j)} + \boldsymbol{K}_{f,k}^{(j)}\left(y_f(\boldsymbol{m}_{k|k-1}^{(j)}) - \boldsymbol{H}_f(\boldsymbol{m}_{k|k-1}^{(j)})\boldsymbol{m}_{k|k-1}^{(j)}\right) \tag{9.3.26}$$

$$\boldsymbol{P}_{k|k,f}^{(j)} = \left[\boldsymbol{I} - \boldsymbol{K}_{f,k}^{(j)}\boldsymbol{H}_f(\boldsymbol{m}_{k|k-1}^{(j)})\right]\boldsymbol{P}_{k|k-1}^{(j)} \tag{9.3.27}$$

where the pseudo measurement gain $\boldsymbol{K}_{f,k}^{(j)}$ is given by

$$\boldsymbol{K}_{f,k}^{(j)} = \boldsymbol{P}_{k|k-1}^{(j)}\boldsymbol{H}_f\left(\boldsymbol{m}_{k|k-1}^{(j)}\right)\left[S_{f,k}^{(j)}\right]^{-1} \tag{9.3.28}$$

with the pseudo measurement innovation covariance being

$$S_{f,k}^{(j)} = \boldsymbol{H}_f(\boldsymbol{m}_{k|k-1}^{(j)})\boldsymbol{P}_{k|k-1}^{(j)}\boldsymbol{H}_f^{\mathrm{T}}(\boldsymbol{m}_{k|k-1}^{(j)}) + R_f \tag{9.3.29}$$

Based on component $\{w_{k|k}^{(j)}(z), \boldsymbol{m}_{k|k}^{(j)}(z), \boldsymbol{P}_{k|k}^{(j)}(z)\}$, component $\{w_{k|k}^{(j)}(z_f), \boldsymbol{m}_{k|k}^{(j)}(z_f), \boldsymbol{P}_{k|k}^{(j)}(z_f)\}$ is further updated with the pseudo measurement, given by

$$w_{k|k}^{(j)}(z_f) = -c_f\mathcal{N}\left(y_f(\boldsymbol{m}_{k|k}^{(j)}(z)); \boldsymbol{H}_f(\boldsymbol{m}_{k|k}^{(j)}(z))\boldsymbol{m}_{k|k}^{(j)}(z), S_{f,k}^{(j)}(z)\right)w_{k|k}^{(j)}(z) \tag{9.3.30}$$

$$\boldsymbol{m}_{k|k}^{(j)}(z_f) = \boldsymbol{m}_{k|k}^{(j)}(z) + \boldsymbol{G}_{f,k}^{(j)}(z)\left(y_f(\boldsymbol{m}_{k|k}^{(j)}(z)) - \boldsymbol{H}_f(\boldsymbol{m}_{k|k}^{(j)}(z))\boldsymbol{m}_{k|k}^{(j)}(z)\right) \tag{9.3.31}$$

$$\boldsymbol{P}_{k|k}^{(j)}(z_f) = \left[\boldsymbol{I} - \boldsymbol{G}_{f,k}^{(j)}(z)\boldsymbol{H}_f(\boldsymbol{m}_{k|k}^{(j)}(z))\right]\boldsymbol{P}_{k|k}^{(j)}(z) \tag{9.3.32}$$

where the corresponding pseudo-measurement gain $\boldsymbol{G}_{f,k}^{(j)}(z)$ is

$$\boldsymbol{G}_{f,k}^{(j)}(z) = \boldsymbol{P}_{k|k}^{(j)}(z)\boldsymbol{H}_f\left(\boldsymbol{m}_{k|k}^{(j)}(z)\right)\left[S_{f,k}^{(j)}(z)\right]^{-1} \tag{9.3.33}$$

and the corresponding pseudo-measurement innovation covariance is

$$S_{f,k}^{(j)}(z) = \boldsymbol{H}_f\left(\boldsymbol{m}_{k|k}^{(j)}(z)\right)\boldsymbol{P}_{k|k}^{(j)}(z)\boldsymbol{H}_f^{\mathrm{T}}\left(\boldsymbol{m}_{k|k}^{(j)}(z)\right) + R_f \tag{9.3.34}$$

In the denominator of (9.3.14), the clutter intensity is modeled as

$$\kappa_k(z) = \kappa_{c,k}(\boldsymbol{y}_c)\kappa_{d,k}(y_d) \tag{9.3.35}$$

where $\kappa_{c,k}(\boldsymbol{y}_c)$ and $\kappa_{d,k}(y_d)$ are the clutter intensities of the position and Doppler components, respectively, and w_{sum} is given by

$$w_{\text{sum}} = p_D \sum_{j=1}^{J_{k|k-1}} w_{k|k-1}^{(j)} q_k^{(j)}(z) - c_f p_D \sum_{j=1}^{J_{k|k-1}} w_{k|k-1}^{(j)} q_k^{(j)}(z_f) \tag{9.3.36}$$

where

$$q_k^{(j)}(z) = \mathcal{N}\Big(\boldsymbol{y}_c; \boldsymbol{H}_{c,k}\boldsymbol{m}_{k|k-1}^{(j)}, \boldsymbol{S}_{c,k}^{(j)}\Big)\mathcal{N}\Big(y_d; h_d(\boldsymbol{m}_{k|k}^{(j)}(\boldsymbol{y}_c)), S_{d,k}^{(j)}(\boldsymbol{y}_c)\Big) \tag{9.3.37}$$

$$q_k^{(j)}(z_f) = q_k^{(j)}(z)\mathcal{N}\Big(y_f(\boldsymbol{m}_{k|k}^{(j)}(z)); \boldsymbol{H}_f(\boldsymbol{m}_{k|k}^{(j)}(z))\boldsymbol{m}_{k|k}^{(j)}(z), S_{f,k}^{(j)}(z)\Big) \tag{9.3.38}$$

Since the number of Gaussian components increases with time, like the standard GM-PHD filter, the pruning and merging steps need to be performed after the measurement update step. In the subsequent simulation, the parameter settings for the steps and multi-target state extraction are the same as those in [63].

The GM-PHD filter with the Doppler and MDV information introduced above is abbreviated as the GM-PHD-D&MDV filter. It is worth pointing out that, when MDV $= 0$, then $c_f = 0$. Therefore, the GM-PHD-D&MDV filter will degrade into the GM-PHD filter with Doppler information (abbreviated as GM-PHD-D) proposed in [223]. By comparing with the GM-PHD-D filter, it can be seen that since the detection probability in (9.3.8) consists of two parts, each component is split into two. Therefore, substituting the detection probability into the updated intensity Eq. (4.2.2) leads to the multiplying number of Gaussian components. Specifically, in addition to $J_{k|k-1}$ components $\{w_{k|k,0}^{(j)}, \boldsymbol{m}_{k|k,0}^{(j)}, \boldsymbol{P}_{k|k,0}^{(j)}\}$ and $|Z_k|J_{k|k-1}$ components $\{w_{k|k}^{(j)}(z), \boldsymbol{m}_{k|k}^{(j)}(z), \boldsymbol{P}_{k|k}^{(j)}(z)\}$, the same number of additional components are also obtained, including $J_{k|k-1}$ components $\{w_{k|k,f}^{(j)}, \boldsymbol{m}_{k|k,f}^{(j)}, \boldsymbol{P}_{k|k,f}^{(j)}\}$ and $|Z_k|J_{k|k-1}$ components $\{w_{k|k}^{(j)}(z_f), \boldsymbol{m}_{k|k}^{(j)}(z_f), \boldsymbol{P}_{k|k}^{(j)}(z_f)\}$. Therefore, the incorporation of the DBZ information in the GM-PHD filter results in additional Gaussian mixture components [85]. Since more components are obtained in the update step of the filter, the GM-PHD-D&MDV filter can interpret the filter behavior in a more complex way than the GM-PHD-D filter. Specifically, component $\{w_{k|k,0}^{(j)}, \boldsymbol{m}_{k|k,0}^{(j)}, \boldsymbol{P}_{k|k,0}^{(j)}\}$ deals with the conventional missed detection case, $\{w_{k|k}^{(j)}(z), \boldsymbol{m}_{k|k}^{(j)}(z), \boldsymbol{P}_{k|k}^{(j)}(z)\}$ tackles the conventional measurement case, $\{w_{k|k,f}^{(j)}, \boldsymbol{m}_{k|k,f}^{(j)}, \boldsymbol{P}_{k|k,f}^{(j)}\}$ copes with the missed detection of the target masked by the DBZ, that is, the pseudo-measurement case, and $\{w_{k|k}^{(j)}(z_f), \boldsymbol{m}_{k|k}^{(j)}(z_f), \boldsymbol{P}_{k|k}^{(j)}(z_f)\}$ deals with the augmented measurement case. For the third case, when a target enters the DBZ, the corresponding component will have a larger weight according to (9.3.25), so as to avoid the corresponding track being deleted by the pruning threshold. Therefore, the missed detection can provide the valuable information. For the last case, the weight $w_{k|k}^{(j)}(z_f)$ of the corresponding component is negative (see (9.3.30)). In the specific implementation process, this component will be deleted in the pruning and merging step, because the negative weight is necessarily less than the positive pruning threshold. However, considering that the sum of all weights represents the average cardinality, this weight is assigned to the corresponding component with positive weight, i.e., $w_{k|k}^{(j)}(z) = w_{k|k}^{(j)}(z) + w_{k|k}^{(j)}(z_f)$.

Moreover, Eq. (9.3.11) shows that each component is split into two, regardless of whether the state falls in the DBZ. Actually, in order to retain the components corresponding to targets in the DBZ, it is only required to split the corresponding component when the target state falls into the clutter notch, i.e., the component whose state $(\boldsymbol{m}_{k|k-1}^{(i)}, \boldsymbol{P}_{k|k-1}^{(i)})$ satisfies the condition $n_c(\boldsymbol{m}_{k|k-1}^{(j)}) \le \text{MDV} + \sqrt{S_{d,k}^{(j)}}$ (where $S_{d,k}^{(j)} = \boldsymbol{H}_d(\boldsymbol{m}_{k|k-1}^{(j)})\boldsymbol{P}_{k|k-1}^{(j)}\boldsymbol{H}_d^{\mathrm{T}}(\boldsymbol{m}_{k|k-1}^{(j)}) + \sigma_d^2$) is split, and vice versa not split. Hereinafter, this approximation filter is referred to as the GM-PHD-D&MDV1 for simplicity. For completeness, Table 9.2 summarizes the key update steps of the GM-PHD-D&MDV1 filter.

9.3.3 Simulation Analysis

In order to verify the effectiveness of the GM-PHD-D&MDV1 filter, it is compared with the GM-PHD-D filter. In addition, the performances of the GM-PHD filter (without Doppler measurement) and the original GM-PHD-D&MDV filter are used as a reference.

For illustration purpose, each target follows a linear Gaussian dynamic in the x–y plane

$$\boldsymbol{x}_k = \boldsymbol{F}_{k-1}\boldsymbol{x}_{k-1} + \boldsymbol{v}_{k-1} \tag{9.3.39}$$

with $\boldsymbol{x}_k = [\,x_k\ y_k\ \dot{x}_k\ \dot{y}_k\,]^{\mathrm{T}}$ being the target state at time k, and the transition matrix being

$$\boldsymbol{F}_{k-1} = \begin{bmatrix} 1 & \tau_k \\ 0 & 1 \end{bmatrix} \otimes \boldsymbol{I}_2 \tag{9.3.40}$$

where $\boldsymbol{I}_n$ is the $n \times n$ identity matrix, $\tau_k = 1$ s is the time interval, $\otimes$ denotes the Kronecker product, $\boldsymbol{v}_{k-1}$ is a zero-mean white Gaussian process noise and its covariance is

$$\boldsymbol{Q}_{k-1} = \begin{bmatrix} \tau_k^4/4 & \tau_k^3/2 \\ \tau_k^3/2 & \tau_k^2 \end{bmatrix} \otimes \text{blkdiag}\left(\sigma_x^2, \sigma_y^2\right) \tag{9.3.41}$$

In the above equation, σ_x^2 and σ_y^2 are the variances of the accelerate process noise in x axis and y axis directions, respectively, and blkdiag represents the block diagonal matrix.

Each target has a survival probability of $p_{S,k} = 0.99$ and the detection probability is $p_D = 0.98$. The observation follows the measurement model given in (9.3.9),

Table 9.2 Pseudo code of the GM-PHD-D&MDV1 update

Input: $\{w_{k\|k-1}^{(j)}, \boldsymbol{m}_{k\|k-1}^{(j)}, \boldsymbol{P}_{k\|k-1}^{(j)}\}_{j=1}^{J_{k\|k-1}}$, measurement set Z_k
Output: $\{w_{k\|k}^{(j)}, \boldsymbol{m}_{k\|k}^{(j)}, \boldsymbol{P}_{k\|k}^{(j)}\}_{j=1}^{J_{k\|k}}$
for $j = 1, \ldots, J_{k\|k-1}$
$\quad \boldsymbol{\eta}_{k\|k-1}^{(j)} = \boldsymbol{H}_{c,k}\boldsymbol{m}_{k\|k-1}^{(j)}$, $\quad$ calculate $\boldsymbol{G}_{c,k}^{(j)}$ and $\boldsymbol{S}_{c,k}^{(j)}$ according to (9.3.19) and (9.3.20), $\quad \boldsymbol{P}_{k}^{(j)} = [\boldsymbol{I} - \boldsymbol{G}_{c,k}^{(j)}\boldsymbol{H}_{c,k}]\boldsymbol{P}_{k\|k-1}^{(j)}$,
end
for $j = 1, \ldots, J_{k\|k-1}$
$\quad w_{k\|k}^{(j)} = (1 - p_D)w_{k\|k-1}^{(j)}, \boldsymbol{m}_{k\|k}^{(j)} = \boldsymbol{m}_{k\|k-1}^{(j)}, \boldsymbol{P}_{k\|k}^{(j)} = \boldsymbol{P}_{k\|k-1}^{(j)}, S_{d,k}^{(j)} = \boldsymbol{H}_d(\boldsymbol{m}_{k\|k-1}^{(j)})$ $\quad \boldsymbol{P}_{k\|k-1}^{(j)}\boldsymbol{H}_d^{\mathrm{T}}(\boldsymbol{m}_{k\|k-1}^{(j)}) + \sigma_d^2$,
$\quad$ if $n_c(\boldsymbol{m}_{k\|k-1}^{(j)}) \leq \mathrm{MDV} + \sqrt{S_{d,k}^{(j)}}$
$\quad\quad$ Calculate $\boldsymbol{K}_{f,k}^{(j)}$ and $S_{f,k}^{(j)}$ according to (9.3.28) and (9.3.29),
$\quad\quad \boldsymbol{m}_{k\|k}^{(J_{k\|k-1}+j)} = \boldsymbol{m}_{k\|k-1}^{(j)} + \boldsymbol{K}_{f,k}^{(j)}(y_f(\boldsymbol{m}_{k\|k-1}^{(j)}) - \boldsymbol{H}_f(\boldsymbol{m}_{k\|k-1}^{(j)})\boldsymbol{m}_{k\|k-1}^{(j)})$, $\quad\quad \boldsymbol{P}_{k\|k}^{(J_{k\|k-1}+j)} = [\boldsymbol{I} - \boldsymbol{K}_{f,k}^{(j)}\boldsymbol{H}_f(\boldsymbol{m}_{k\|k-1}^{(j)})] \cdot \boldsymbol{P}_{k\|k-1}^{(j)}$,
$\quad\quad w_{k\|k}^{(J_{k\|k-1}+j)} = \frac{p_D}{1-p_D}c_f\mathcal{N}(y_f(\boldsymbol{m}_{k\|k-1}^{(j)}); \boldsymbol{H}_f(\boldsymbol{m}_{k\|k-1}^{(j)})\boldsymbol{m}_{k\|k-1}^{(j)}, S_{f,k}^{(j)})w_{k\|k}^{(j)}$,
$\quad$ else $(w_{k\|k}^{(J_{k\|k-1}+j)} = 0, \boldsymbol{m}_{k\|k}^{(J_{k\|k-1}+j)} = \boldsymbol{m}_{k\|k-1}^{(j)}, \boldsymbol{P}_{k\|k}^{(J_{k\|k-1}+j)} = \boldsymbol{P}_{k\|k-1}^{(j)})$
$\quad$ end
end
$l = 0$
for $\boldsymbol{z} = [\boldsymbol{y}_c; y_d] \in Z_k$
$\quad l \leftarrow l + 1$,
$\quad$ for $j = 1, \ldots, J_{k\|k-1}$
$\quad\quad \boldsymbol{m}_{k\|k}^{(l \cdot 2J_{k\|k-1}+j)} = \boldsymbol{m}_{k\|k-1}^{(j)} + \boldsymbol{G}_{c,k}^{(j)}(\boldsymbol{y}_c - \boldsymbol{\eta}_{k\|k-1}^{(j)}), \boldsymbol{P}_{k\|k}^{(l \cdot 2J_{k\|k-1}+j)} = \boldsymbol{P}_{k}^{(j)}$,
$\quad\quad \boldsymbol{H}_d(\boldsymbol{m}_{k\|k}^{(l \cdot 2J_{k\|k-1}+j)}) = \frac{\partial \dot{r}_k}{\partial \boldsymbol{x}_k}\Big\|_{\boldsymbol{x}_k = \boldsymbol{m}_{k\|k}^{(l \cdot 2J_{k\|k-1}+j)}}, \hat{y}_d^{(j)} = h_d(\boldsymbol{m}_{k\|k}^{(l \cdot 2J_{k\|k-1}+j)})$
$\quad\quad S_d = \boldsymbol{H}_d(\boldsymbol{m}_{k\|k}^{(l \cdot 2J_{k\|k-1}+j)})\boldsymbol{P}_{k\|k}^{(l \cdot 2J_{k\|k-1}+j)}\boldsymbol{H}_d^{\mathrm{T}}(\boldsymbol{m}_{k\|k}^{(l \cdot 2J_{k\|k-1}+j)}) + \sigma_d^2$, $\quad\quad \boldsymbol{G}_d = \boldsymbol{P}_{k\|k}^{(l \cdot 2J_{k\|k-1}+j)}\boldsymbol{H}_d(\boldsymbol{m}_{k\|k}^{(l \cdot 2J_{k\|k-1}+j)})S_d^{-1}$,
$\quad\quad \boldsymbol{m}_{k\|k}^{(l \cdot 2J_{k\|k-1}+j)} = \boldsymbol{m}_{k\|k}^{(l \cdot 2J_{k\|k-1}+j)} + \boldsymbol{G}_d(y_d - \hat{y}_d^{(j)})$, $\quad\quad \boldsymbol{P}_{k\|k}^{(l \cdot 2J_{k\|k-1}+j)} = [\boldsymbol{I} - \boldsymbol{G}_d\boldsymbol{H}_d(\boldsymbol{m}_{k\|k}^{(l \cdot 2J_{k\|k-1}+j)})]\boldsymbol{P}_{k\|k}^{(l \cdot 2J_{k\|k-1}+j)}$,
$\quad\quad w_{k\|k}^{(l \cdot 2J_{k\|k-1}+j)} = p_D w_{k\|k-1}^{(j)}\mathcal{N}(\boldsymbol{y}_c; \boldsymbol{\eta}_{k\|k-1}^{(j)}, \boldsymbol{S}_{c,k}^{(j)})\mathcal{N}(y_d; \hat{y}_d^{(j)}, S_d)$,

(continued)

Table 9.2 (continued)

if $n_c(\boldsymbol{m}_{k\mid k-1}^{(j)}) \le \mathrm{MDV} + \sqrt{S_d}$
$S_f = \boldsymbol{H}_f(\boldsymbol{m}_{k\mid k}^{(l\cdot 2J_{k\mid k-1}+j)})\boldsymbol{P}_{k\mid k}^{(l\cdot 2J_{k\mid k-1}+j)}\boldsymbol{H}_f^{\mathrm{T}}(\boldsymbol{m}_{k\mid k}^{(l\cdot 2J_{k\mid k-1}+j)}) + R_f,$
$\boldsymbol{G}_f = \boldsymbol{P}_{k\mid k}^{(l\cdot 2J_{k\mid k-1}+j)}\boldsymbol{H}_f(\boldsymbol{m}_{k\mid k}^{(l\cdot 2J_{k\mid k-1}+j)})S_f^{-1},$
$\boldsymbol{m}_{k\mid k}^{(l\cdot 2J_{k\mid k-1}+J_{k\mid k-1}+j)} = \boldsymbol{m}_{k\mid k}^{(l\cdot 2J_{k\mid k-1}+j)} + \boldsymbol{G}_f(y_f(\boldsymbol{m}_{k\mid k}^{(l\cdot 2J_{k\mid k-1}+j)})$ $-\boldsymbol{H}_f(\boldsymbol{m}_{k\mid k}^{(l\cdot 2J_{k\mid k-1}+j)})\boldsymbol{m}_{k\mid k}^{(l\cdot 2J_{k\mid k-1}+j)}),$
$\boldsymbol{P}_{k\mid k}^{(l\cdot 2J_{k\mid k-1}+J_{k\mid k-1}+j)} = [\boldsymbol{I} - \boldsymbol{G}_f\boldsymbol{H}_f(\boldsymbol{m}_{k\mid k}^{(l\cdot 2J_{k\mid k-1}+j)})]\boldsymbol{P}_{k\mid k}^{(l\cdot 2J_{k\mid k-1}+j)},$
$w_{k\mid k}^{(l\cdot 2J_{k\mid k-1}+J_{k\mid k-1}+j)} = -c_f w_{k\mid k}^{(l\cdot 2J_{k\mid k-1}+j)}\mathcal{N}(y_f(\boldsymbol{m}_{k\mid k}^{(l\cdot 2J_{k\mid k-1}+j)});$ $\boldsymbol{H}_f(\boldsymbol{m}_{k\mid k}^{(l\cdot 2J_{k\mid k-1}+j)})\boldsymbol{m}_{k\mid k}^{(l\cdot 2J_{k\mid k-1}+j)}, S_f),$
else $(\boldsymbol{m}_{k\mid k}^{(l\cdot 2J_{k\mid k-1}+J_{k\mid k-1}+j)} = \boldsymbol{m}_{k\mid k}^{(l\cdot 2J_{k\mid k-1}+j)},$ $\boldsymbol{P}_{k\mid k}^{(l\cdot 2J_{k\mid k-1}+J_{k\mid k-1}+j)} = \boldsymbol{P}_{k\mid k}^{(l\cdot 2J_{k\mid k-1}+j)}, w_{k\mid k}^{(l\cdot 2J_{k\mid k-1}+J_{k\mid k-1}+j)} = 0)$
end
end
$w_{k\mid k}^{(l\cdot 2J_{k\mid k-1}+j)} = w_{k\mid k}^{(l\cdot 2J_{k\mid k-1}+j)}/[\kappa_{c,k}(\boldsymbol{y}_c)\kappa_{d,k}(y_d) + \sum_{i=1}^{2J_{k\mid k-1}} w_{k\mid k}^{(l\cdot 2J_{k\mid k-1}+i)}],$ $j = 1, 2, \ldots, 2\cdot J_{k\mid k-1}$
end
for $j = 1, \ldots, J_{k\mid k-1}$
$w_{k\mid k}^{(l\cdot 2J_{k\mid k-1}+j)} \leftarrow w_{k\mid k}^{(l\cdot 2J_{k\mid k-1}+j)} + w_{k\mid k}^{(l\cdot 2J_{k\mid k-1}+J_{k\mid k-1}+j)}, w_{k\mid k}^{(l\cdot 2J_{k\mid k-1}+J_{k\mid k-1}+j)} \leftarrow 0$
end
$J_{k\mid k} = 2(l+1)J_{k\mid k-1}$

where $\boldsymbol{H}_{c,k} = [\,\boldsymbol{I}_2\ \boldsymbol{0}_2\,]$, $\boldsymbol{R}_{c,k} = \sigma_c^2\boldsymbol{I}_2$, $\sigma_d = 0.5$ m/s, and $\sigma_c = 10$ m is the standard deviation of the positional measurement noise.

The clutter is uniformly distributed in the surveillance area [− 1000, 1000] (m) × [− 1000, 1000] (m). For the positional component, $\kappa_{c,k}(\cdot) = \lambda_c \cdot V \cdot u(\cdot)$; for the Doppler component, $\kappa_{d,k} = 1/(2\upsilon_{\max})$, where λ_c is the average number of clutters in the unit volume, V is the volume of the surveillance area, $u(\cdot)$ is the uniform density of the surveillance area, and $\upsilon_{\max} = 35$ m/s is the maximum velocity that can be detected by the sensor.

For the purpose of comparison, all filters are assumed to have a fixed initial position. The pruning parameter is $T = 10^{-5}$, the merging threshold is $U = 4$ and the maximum number of Gaussian components is $J_{\max} = 100$. Besides, the multi-target extraction threshold is 0.5.

9.3.3.1 Two-Dimensional Stationary Sensor Scenario

Consider first the example of two targets being tracked by a sensor located at the origin in the x–y plane. The initial states of the targets are set to $\boldsymbol{x}_{0,1} = \left[-500\,\text{m}\ 200\,\text{m}\ 10\,\text{m/s}\ 0\,\text{m/s}\right]^{\text{T}}$ and $\boldsymbol{x}_{0,2} = \left[-500\,\text{m}\ -200\,\text{m}\ 10\,\text{m/s}\ 0\,\text{m/s}\right]^{\text{T}}$, respectively, and $\sigma_x = \sigma_y = 5\,\text{m/s}^2$. The birth target RFS follows the Poisson distribution with an intensity of

$$v_{\gamma,k}(\boldsymbol{x}) = 0.1 \times \sum_{j=1}^{2} \mathcal{N}\left(\boldsymbol{x}; \boldsymbol{m}_{\gamma,k}^{(j)}, \boldsymbol{P}_{\gamma,k}^{(j)}\right) \tag{9.3.42}$$

where $\boldsymbol{m}_{\gamma,k}^{(1)} = [-500\,\text{m}, 200\,\text{m}, 0\,\text{m/s}, 0\,\text{m/s}]^{\text{T}}$, $\boldsymbol{m}_{\gamma,k}^{(2)} = [-500\,\text{m}, -200\,\text{m}, 0\,\text{m/s}, 0\,\text{m/s}]^{\text{T}}$, $\boldsymbol{P}_{\gamma,k}^{(j)} = \text{blkdiag}(100\,\text{m}, 100\,\text{m}, 25\,\text{m/s}, 25\,\text{m/s})^2$, $j = 1, 2$.

Figure 9.4 shows the sensor position, the true target tracks and the clutter. Figure 9.5 shows the true Doppler curves of the two targets and the minimum detectable velocity (MDV) versus time in the above configuration. It can be seen that these two targets fly tangentially at time $k = 50$. Therefore, their Dopplers are close to 0 at that time, and the larger the MDV value, the longer the target passes through the DBZ, and the more missed detections.

Figure 9.6 shows the true target tracks and estimation results of different algorithms when MDV = 1 m/s. Before the targets fall into the DBZ (Stage 1), all filters can successfully track both targets. Nevertheless, the GM-PHD filter has the worst performance, while the other three have similar performances. However, when the targets are both within the DBZ (Stage 2), all filters fail to track these targets due to missed detections caused by the MDV. Note that all filters are assumed to have fixed birth positions, so the GM-PHD and GM-PHD-D filters cannot track any target again due to a series of missed detections. Nevertheless, once the targets fly out of the DBZ (Stage 3), both the GM-PHD-D&MDV and GM-PHD-D&MDV1 can track them again, because they preserve the Gaussian components corresponding to the targets

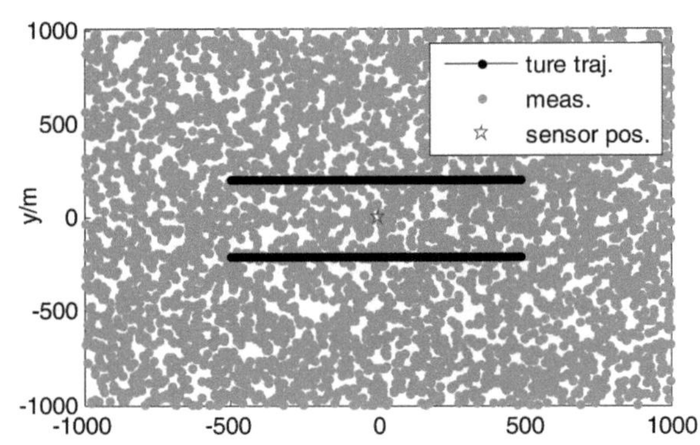

Fig. 9.4 Sensor/target geometry and clutter distribution at a clutter ratio of 12.5×10^{-6}

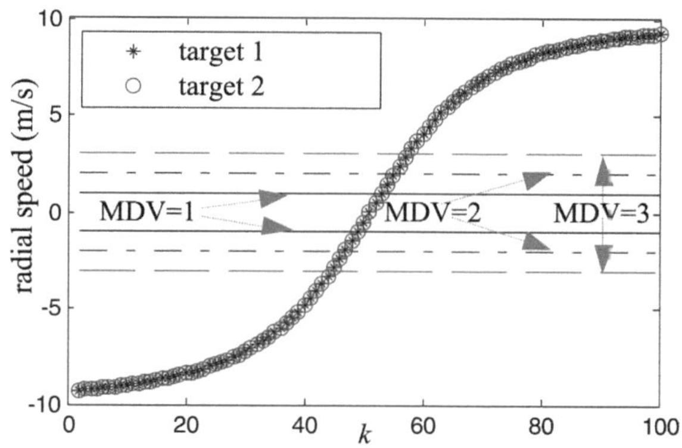

Fig. 9.5 Doppler curves of two targets and different MDVs versus time

inside the DBZ. As a result, these two algorithms can reasonably handle missed detections by incorporating the additional MDV information, and thus maintain the target tracks.

To evaluate the performance, the following track loss performance, i.e., circular positional error probability (CPEP), is studied [63]

$$\text{CPEP}_k(r) = \frac{1}{|X_k|} \sum_{\boldsymbol{x} \in X_k} \alpha_k(\boldsymbol{x}, r) \tag{9.3.43}$$

where for a certain radius r of the position error, X_k is the set of the true target states, $\alpha_k(\boldsymbol{x}, r) = \Pr\{\left\| \boldsymbol{H}\hat{\boldsymbol{x}} - \boldsymbol{H}\boldsymbol{x} \right\|_2 > r \text{ for all } \hat{\boldsymbol{x}} \in \hat{X}_k\}$, $\Pr(\cdot)$ represents the event

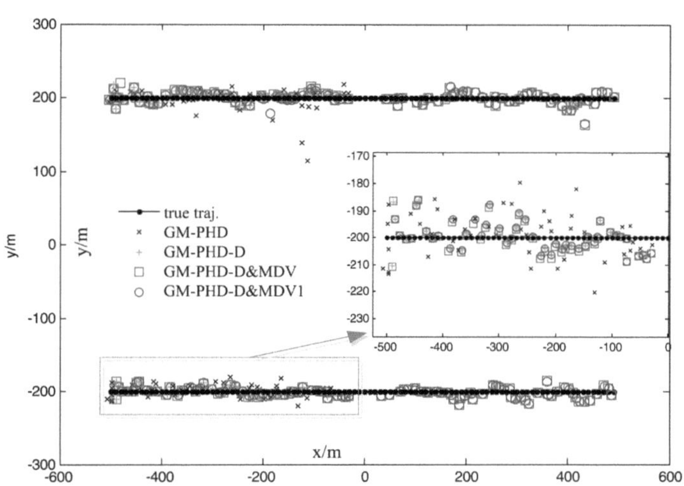

Fig. 9.6 True tracks and estimations from different algorithms (MDV = 1 m/s)

probability, $\boldsymbol{H} = [\, \boldsymbol{I}_2 \; \boldsymbol{0}_2 \,]$, and $\|\cdot\|_2$ is the 2-norm. Moreover, the optimal sub-pattern assignment (OSPA) performance is also investigated. The CPEP radius is $r = 20$ m, and the order parameter and threshold parameter of the OSPA are set as $p = 2$ and $c = 20$ m, respectively.

Through 1000 Monte Carlo runs, Figs. 9.7, 9.8 and 9.9 show the tracking performances of different algorithms at different MDV values. In Fig. 9.7 where MDV = 1 m/s, it can be seen that during the first stage (Stage 1), due to the introduction of Doppler information, these filters that incorporate the Doppler information such as the GM-PHD-D, GM-PHD-D&MDV and GM-PHD-D&MDV1 filters, have similar performances in terms of CPEP and OSPA, and they outperform the GM-PHD filter without Doppler information. In the second stage (Stage 2), since the weight of Gaussian component is less than the extraction threshold so that no state is extracted, all filters can't track targets obscured by the DBZ. However, the filters incorporating the MDV information such as the GM-PHD-D&MDV and GM-PHD-D&MDV1 filters are fundamentally different from the GM-PHD-D and GM-PHD filters that do not contain the MDV information. For the GM-PHD-D&MDV and GM-PHD-D&MDV1 filters, it is possible to obtain a higher weight than the pruning threshold according to (9.3.25), thus avoiding track deletion. However, for the GM-PHD-D and GM-PHD filters, when targets are obscured by the DBZ, they do not have a similar mechanism to effectively preserve the components corresponding to the targets, and thus these components will be deleted. Therefore, when a target flies out of the DBZ, the GM-PHD-D&MDV and GM-PHD-D&MDV1 filters are able to track the target again, while the GM-PHD-D and GM-PHD filters can no longer track the target.

The similar improvement trend can be seen from Figs. 9.8 and 9.9. In addition, according to the comparative analysis of Figs. 9.7, 9.8 and 9.9, for the GM-PHD-D&MDV and GM-PHD-D&MDV1 filters, the following points are worth noting. First, the performance of the approximate GM-PHD-D&MDV1 filter is quite similar to that of the original GM-PHD-D&MDV filter. On the other hand, it can be seen from Fig. 9.10 that the former performs faster than the latter. In the figure, the absolute time refers to the average time of 1000 Monte Carlo runs when 100 steps are executed each run. For example, when MDV = 1 m/s, the GM-PHD-D&MDV1 filter requires 11.05 s while the GM-PHD-D&MDV filter takes 22.87 s. Relatively speaking, the time consumed by the former is only 48.29% of that of the latter. Therefore, the approximation method can effectively reduce the computational complexity, without significant performance loss. Second, with the increase of the MDV value, the CPEP value after the DBZ masking doesn't drop to the original level in the first stage, and the CPEP gap between the first stage and the third stage becomes larger, because it is more difficult to retain components after more missed detections are generated, thus resulting in an increase in the probability that targets masked by the DBZ cannot be tracked. Third, as the MDV increases, it becomes easier for the component weight updated by the pseudo-measurement to be larger than the extraction threshold, so more false tracks will be extracted. Furthermore, the OSPA cardinality will increase before the targets enter the DBZ. In other words, when the MDV is large, the performance improvement of the proposed filter comes at the expense of increasing false tracks. Nevertheless, when the MDV value was small, the improvement in the tracking

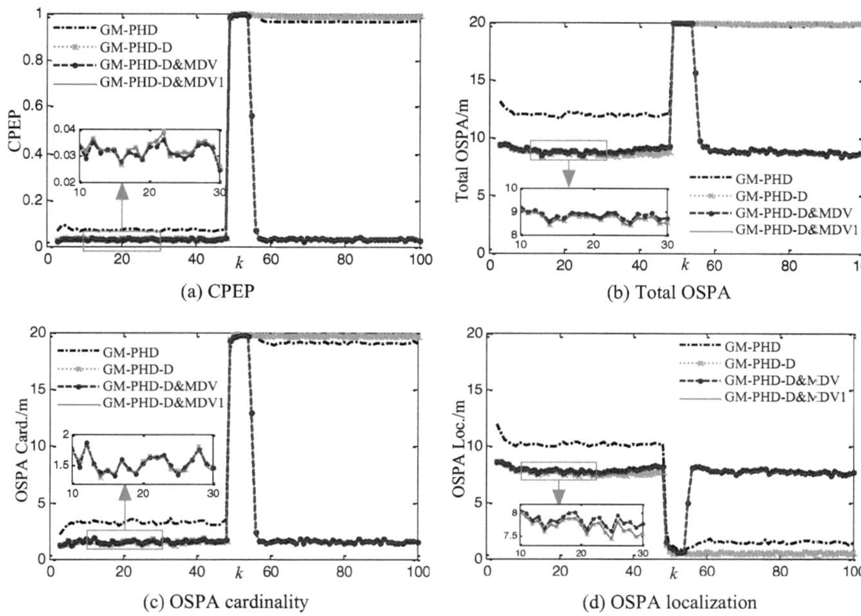

(a) CPEP (b) Total OSPA

(c) OSPA cardinality (d) OSPA localization

Fig. 9.7 Tracking performances of different algorithms (MDV = 1 m/s)

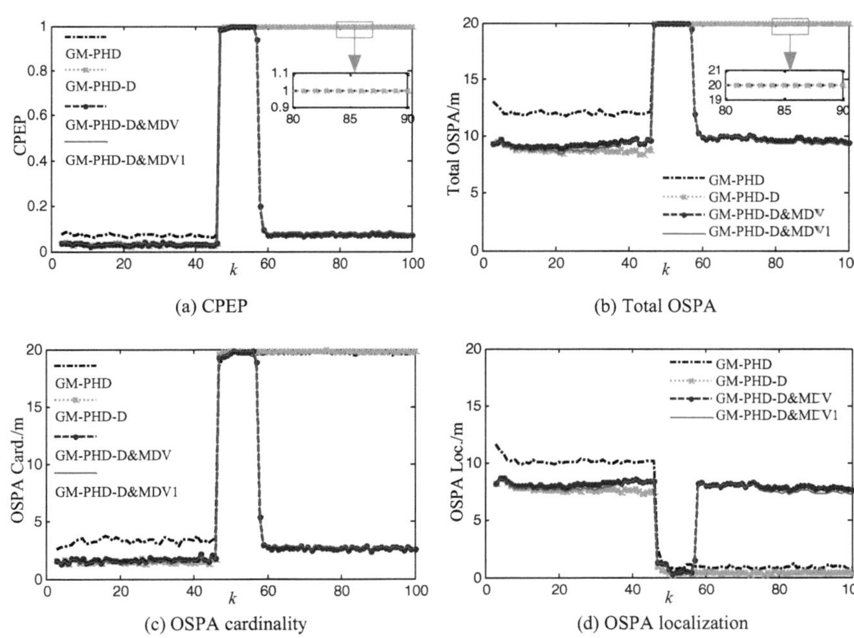

(a) CPEP (b) Total OSPA

(c) OSPA cardinality (d) OSPA localization

Fig. 9.8 Tracking performances of different algorithms (MDV = 2 m/s)

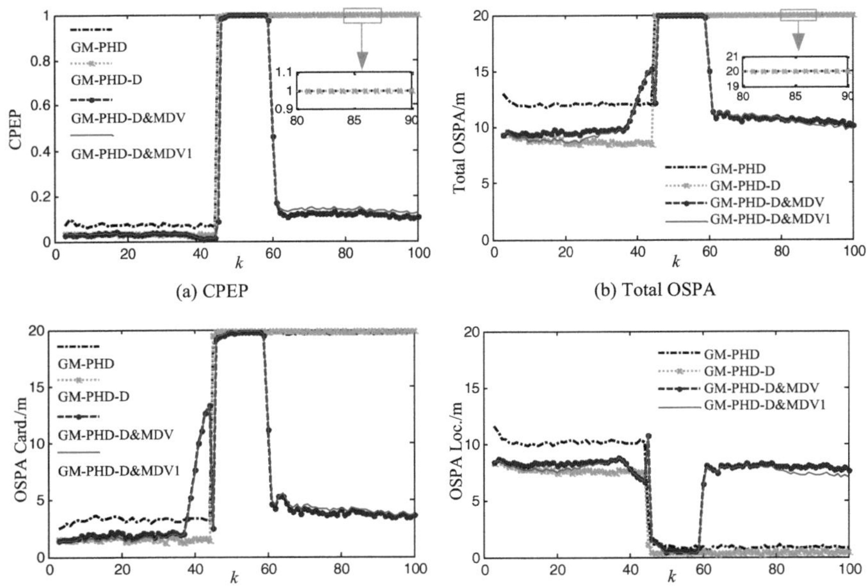

(a) CPEP (b) Total OSPA

(c) OSPA cardinality (d) OSPA localization

Fig. 9.9 Tracking performances of different algorithms (MDV = 3 m/s)

performance doesn't increase the number of false tracks. In addition, in Stage 3, the GM-PHD and GM-PHD-D filters seem to be better than the GM-PHD-D&MDV and GM-PHD-D&MDV1 filters in terms of the OSPA localization performance. In fact, this phenomenon is caused by the corresponding OSPA cardinality performance approaching the cutoff threshold. Therefore, the GM-PHD-D and GM-PHD filters are still inferior to the GM-PHD-D&MDV and GM-PHD-D&MDV1 filters in terms of the overall OSPA performance.

Fig. 9.10 Comparison of different methods in terms of absolute time and relative time

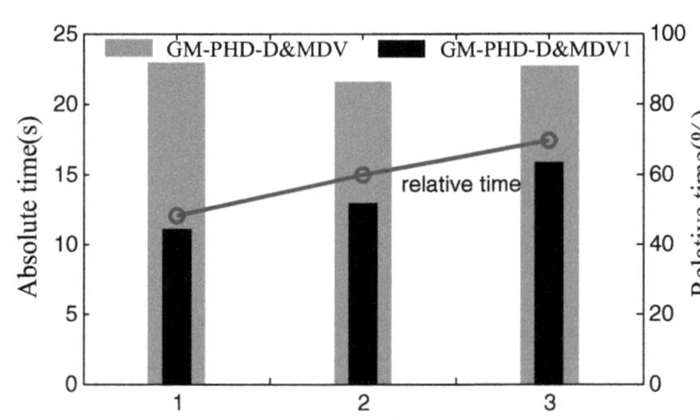

9.3.3.2 Three-Dimensional Moving Sensor Scenario

For a more general case, consider the example of 1 moving sensor tracking 2 targets. The sensor makes the uniform circular motion in the plane of z = 100 m, with an initial state of $\left[600\,\text{m}\ -150\,\text{m}\ 10\,\text{m/s}\ 0\,\text{m/s}\right]^{\text{T}}$ and turn rate of 0.063 rad/s. The initial states of the two targets are set as $\boldsymbol{x}_{0,1} = [0\text{ m } 0\text{ m } 5\text{ m/s } 5\text{ m/s}]^{\text{T}}$ and $\boldsymbol{x}_{0,2} = [0\text{ m } 0\text{ m } 5\text{ m/s } -5\text{ m/s}]^{\text{T}}$, respectively. The standard deviations of the process noise are set to $\sigma_x = \sigma_y = 2\,\text{m/s}^2$. The birth target RFS follows the Poisson distribution with an intensity of

$$v_{\gamma,k}(\boldsymbol{x}) = 0.1\mathcal{N}\left(\boldsymbol{x}; \boldsymbol{m}_{\gamma,k}, \boldsymbol{P}_{\gamma,k}\right) \tag{9.3.42}$$

where $\boldsymbol{m}_{\gamma,k} = [0\,\text{m}, 0\,\text{m}, 0\,\text{m/s}, 0\,\text{m/s}]^{\text{T}}$ and $\boldsymbol{P}_{\gamma,k} = \text{blkdiag}(100\,\text{m}, 100\,\text{m}, 25\,\text{m/s}, 25\,\text{m/s})^2$.

Figure 9.11 shows the sensor track and target tracks. Figure 9.12 shows the relationship between the true tracks of the two targets and the corresponding DBZ versus time under this configuration. From the figure, it can be seen that the time periods of target 1 and target 2 in the DBZ are 43–53 s and 61–65 s, respectively.

Through 1000 Monte Carlo runs, Fig. 9.13 shows the tracking performances of different algorithms under the condition of MDV = 1 m/s. As can be seen from this figure, all the sensors fail to track target 2 but can track target 1 during the time period of 43–53 s when target 2 is masked by the DBZ. As a consequence, the CPEP increases to around 0.5 and the total OSPA increases to a higher level. Once target 2 leaves the DBZ, the performances of both the GM-PHD-D&MDV and GM-PHD-D&MDV1 filters will be improved, while the performances of the GM-PHD and GM-PHD-D filters remain unchanged, which indicates that the first 2 filters can track target 2 again while the last 2 filters can't. Similarly, the CPEPs and the total OSPAs of the GM-PHD and GM-PHD-D filters increase to around 1 m and 20 m, respectively, during the time period of 61–65 s when target 1 is masked by the DBZ, indicating that they can't track target 1 and target 2. During this time period, the CPEPs and the total OSPAs of the GM-PHD-D&MDV and GM-PHD-D&MDV1 filters recover to the level at 43–53 s, because target 1 can not be tracked but target 2

Fig. 9.11 Sensor-target relative geometry

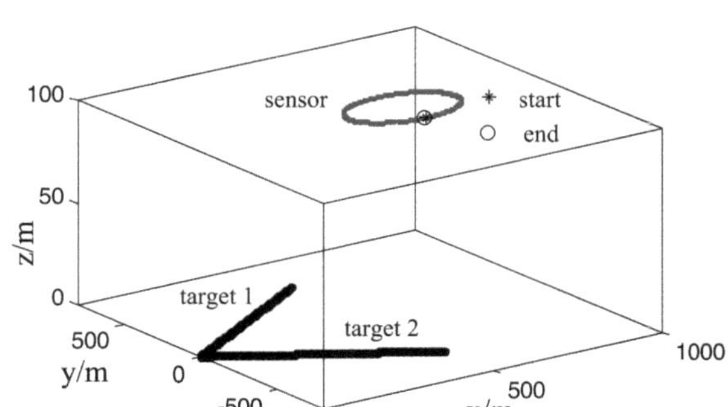

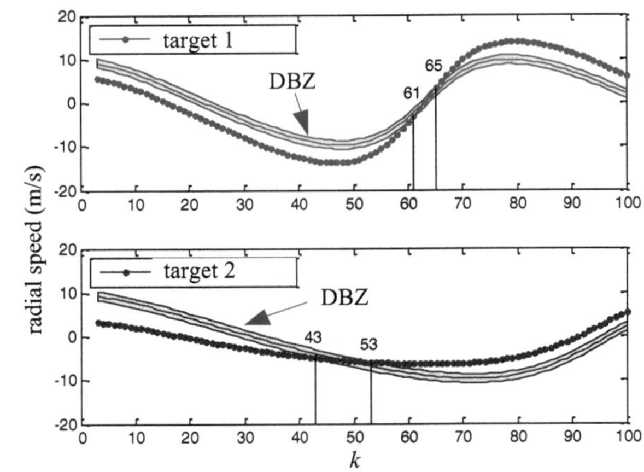

Fig. 9.12 Doppler curves of two targets and the corresponding DBZs versus time (MDV = 1 m/s)

is still tracked. After 66 s, when both target 1 and target 2 are outside the DBZ, they can be successfully tracked by the GM-PHD-D&MDV and GM-PHD-D&MDV1 filters. However, the GM-PHD and GM-PHD-D filters can't track them.

In conclusion, due to the incorporation of the Doppler measurement, the GM-PHD-D filter is superior to the GM-PHD filter before the targets enter the DBZ. Similarly, when the GM-PHD-D filter incorporates the MDV information, the resulting

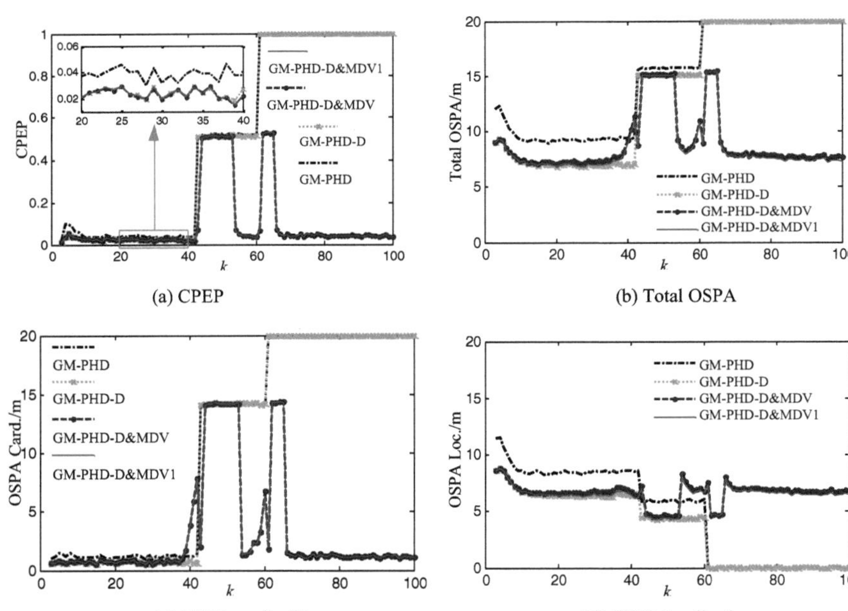

Fig. 9.13 Tracking performance of different algorithms (MDV = 1 m/s)

GM-PHD-D&MDV filter is superior to the GM-PHD-D filter. Through reasonable processing of additional components, the approximate GM-PHD-D&MDV1 filter is further proposed, which has similar performance to the original GM-PHD-D&MDV filter, but greatly reduces the computational burden.

9.4 GM-PHD Filter with Registration Error for Netted Doppler Radars

Doppler radar networking can effectively improve the multi-target tracking performance. However, when multi-sensor fusion is performed, data association and sensor registration are two difficult problems. Traditional methods deal with the two separately. In fact, they affect each other. Aiming at the problem that the systematic bias affects the multi-target tracking performance when the Doppler radar is networked, an augmented state GM-PHD filter with the registration error is proposed based on the PHD filter that avoids data association. The augmented state is composed of the target state and the sensor bias. First, the linear Gaussian dynamic and measurement models of the augmented state are constructed, and then the relevant formulas of the standard GM-PHD filter applied to the augmented state system are derived. In order to effectively utilize the Doppler measurement in the augmented state GM-PHD filter, a sequential processing method is employed, that is, first update the target state and sensor bias with polar coordinate measurements, then update the target state with Doppler measurements, and finally calculate weights with polar coordinate and Doppler measurements. Monte Carlo simulation results verify the effectiveness of the proposed filter.

9.4.1 Problem Formulation

Let $\boldsymbol{x}_{k,i} = [x_{k,i}, y_{k,i}, \dot{x}_{k,i}, \dot{y}_{k,i}]^{\mathrm{T}}$ be the standard target state of target i at time k in the common coordinate system, where $(x_{k,i}, y_{k,i})$ is the position of target i and $(\dot{x}_{k,i}, \dot{y}_{k,i})$ is the velocity. The dynamics of each target is modeled as

$$\boldsymbol{x}_{k,i} = \boldsymbol{F}_{k-1}\boldsymbol{x}_{k-1,i} + \boldsymbol{v}_{k-1} \tag{9.4.1}$$

with the transition matrix being

$$\boldsymbol{F}_{k-1} = \begin{bmatrix} 1 & \tau_k \\ 0 & 1 \end{bmatrix} \otimes \boldsymbol{I}_2 \tag{9.4.2}$$

where $\boldsymbol{I}_n$ is the identity matrix, τ_k is the time interval, $\otimes$ is the Kronecker product, and $\boldsymbol{v}_{k-1}$ is the zero-mean white Gaussian process noise and its covariance is

$$\boldsymbol{Q}_{k-1} = \begin{bmatrix} \tau_k^4/4 & \tau_k^3/2 \\ \tau_k^3/2 & \tau_k^2 \end{bmatrix} \otimes \text{blkdiag}(\sigma_x^2, \sigma_y^2) \tag{9.4.3}$$

with σ_x^2 and σ_y^2 being the variances of the acceleration process noise in x and y directions, respectively.

Assume that there are S mismatched sensors with independent biases, and the dynamics of each bias can be modeled as a first-order Gauss–Markov process [339], i.e.,

$$\boldsymbol{\theta}_{s,k} = \boldsymbol{\theta}_{s,k-1} + \boldsymbol{w}_{s,k-1}, s = 1, \ldots, S \tag{9.4.4}$$

where $\boldsymbol{w}_{s,k-1}$ is the white Gaussian process noise of sensor s with zero mean and covariance $\boldsymbol{B}_{s,k-1}$.

Let $\bar{\boldsymbol{x}}_{k,i} = [\boldsymbol{x}_{k,i}^{\mathrm{T}}, \boldsymbol{\theta}_{1,k}^{\mathrm{T}}, \ldots, \boldsymbol{\theta}_{S,k}^{\mathrm{T}}]^{\mathrm{T}}$ be the augmented state, then according to (9.4.1) and (9.4.4), the system dynamic model with the augmented state can be established as

$$\bar{\boldsymbol{x}}_{k,i} = \bar{\boldsymbol{F}}_{k-1}\bar{\boldsymbol{x}}_{k-1,i} + \bar{\boldsymbol{v}}_{k-1} \tag{9.4.5}$$

where the augmented transition matrix is $\bar{\boldsymbol{F}}_{k-1} = \text{blkdiag}(\boldsymbol{F}_{k-1}, \boldsymbol{I}_{n_1 \times n_1}, \ldots, \boldsymbol{I}_{n_S \times n_S})$, the augmented process noise is $\bar{\boldsymbol{v}}_{k-1} = [\boldsymbol{v}_{k-1}^{\mathrm{T}}, \boldsymbol{w}_{1,k-1}^{\mathrm{T}}, \ldots, \boldsymbol{w}_{S,k-1}^{\mathrm{T}}]^{\mathrm{T}}$, and its covariance is $\bar{\boldsymbol{Q}}_{k-1} = \text{blkdiag}(\boldsymbol{Q}_{k-1}, \boldsymbol{B}_{1,k-1}, \ldots, \boldsymbol{B}_{S,k-1})$.

The sensor's polar coordinate measurement contains two types of errors: systematic error (or bias) and zero-mean additive random noise. For the polar coordinate measurement m originating from target i, the non-linear measurement equation of sensor s can be expressed as

$$\boldsymbol{y}_{k,m}^{s} = h(\boldsymbol{x}_{k,i}, \boldsymbol{x}_{s,k}) + \boldsymbol{\theta}_{s,k} + \boldsymbol{n}_{s,k} \tag{9.4.6}$$

where $\boldsymbol{x}_{s,k} = [x_{s,k}, y_{s,k}]^{\mathrm{T}}$ is the known position of the fixed sensor s, $\boldsymbol{n}_{s,k} \sim \mathcal{N}(\boldsymbol{n}_{s,k}; \boldsymbol{0}, \boldsymbol{R}_{s,p})$ is the white Gaussian measurement noise with zero mean and covariance $\boldsymbol{R}_{s,p} = \text{blkdiag}(\sigma_{s,r}^2, \sigma_{s,a}^2)$, and the non-linear measurement function is

$$h(\boldsymbol{x}_{k,i}, \boldsymbol{x}_{s,k}) = \begin{bmatrix} r_{k,i} \\ a_{k,i} \end{bmatrix} = \begin{bmatrix} \sqrt{(x_{k,i} - x_{s,k})^2 + (y_{k,i} - y_{s,k})^2} \\ \arctan\big((y_{k,i} - y_{s,k})/(x_{k,i} - x_{s,k})\big) \end{bmatrix} \tag{9.4.7}$$

In order to obtain a linear measurement equation, a first-order Taylor expansion of $h(\cdot)$ at the predicted state $\hat{\boldsymbol{x}}_{k|k-1,i}$ can lead to

$$\begin{aligned} \boldsymbol{y}_{k,m}^{s} &\approx h\big(\hat{\boldsymbol{x}}_{k|k-1,i}, \boldsymbol{x}_{s,k}\big) + \boldsymbol{H}_{s,k}\big(\boldsymbol{x}_{k,i} - \hat{\boldsymbol{x}}_{k|k-1,i}\big) + \boldsymbol{\theta}_{s,k} + \boldsymbol{n}_{s,k} \\ &= \boldsymbol{H}_{s,k}\boldsymbol{x}_{k,i} + h\big(\hat{\boldsymbol{x}}_{k|k-1,i}, \boldsymbol{x}_{s,k}\big) - \boldsymbol{H}_{s,k}\hat{\boldsymbol{x}}_{k|k-1,i} + \boldsymbol{\theta}_{s,k} + \boldsymbol{n}_{s,k} \end{aligned} \tag{9.4.8}$$

where $\boldsymbol{H}_{s,k}$ is the Jacobian matrix of non-linear measurement function $h(\cdot)$ at $\hat{\boldsymbol{x}}_{k|k-1,i}$.

According to (9.4.8), the transformed measurement equation can be obtained as

$$\begin{aligned} \boldsymbol{y}_{c,k,m}^{s} &= \boldsymbol{y}_{k,m}^{s} + \boldsymbol{H}_{s,k}\hat{\boldsymbol{x}}_{k|k-1,i} - h\left(\hat{\boldsymbol{x}}_{k|k-1,i}, \boldsymbol{x}_{s,k}\right) \\ &= \boldsymbol{H}_{s,k}\boldsymbol{x}_{k,i} + \boldsymbol{\theta}_{s,k} + \boldsymbol{n}_{s,k} \end{aligned} \tag{9.4.9}$$

which can be further rewritten as

$$\begin{aligned} \boldsymbol{y}_{c,k,m}^{s} &= \left[\boldsymbol{H}_{s,k}, \boldsymbol{\Psi}_{1}, \ldots, \boldsymbol{\Psi}_{S}\right]\left[\boldsymbol{x}_{k,i}^{\mathrm{T}}, \boldsymbol{\theta}_{1,k}^{\mathrm{T}}, \ldots, \boldsymbol{\theta}_{S,k}^{\mathrm{T}}\right]^{\mathrm{T}} + \boldsymbol{n}_{s,k} \\ &= \bar{\boldsymbol{H}}_{s,k}\bar{\boldsymbol{x}}_{k,i} + \boldsymbol{n}_{s,k} \end{aligned} \tag{9.4.10}$$

where

$$\boldsymbol{\Psi}_{r} = \begin{cases} \boldsymbol{I}_{n_r \times n_r}; & r = s \\ \boldsymbol{0}_{n_r \times n_r}; & r \neq s \end{cases} \tag{9.4.11}$$

For a Doppler radar, the Doppler measurement equation is defined as

$$y_{d,k,m}^{s} = \dot{r}_{k,i} + n_{d,k}^{s} \tag{9.4.12}$$

where $n_{d,k}^{s} \sim \mathcal{N}(n_{d,k}^{s}; 0, \sigma_{s,d}^{2})$ is the Doppler measurement noise, which is independent of $\boldsymbol{n}_{s,k}$, and

$$\dot{r}_{k,i} = h_{d,k}\left(\boldsymbol{x}_{k,i}, \boldsymbol{x}_{s,k}\right) = \frac{\dot{x}_{k,i}\left(x_{k,i} - x_{s,k}\right) + \dot{y}_{k,i}\left(y_{k,i} - y_{s,k}\right)}{\sqrt{\left(x_{k,i} - x_{s,k}\right)^{2} + \left(y_{k,i} - y_{s,k}\right)^{2}}} \tag{9.4.13}$$

Note that Eq. (9.4.12) assumes that there is no bias in the Doppler measurement. Let $\boldsymbol{z}_{k,m}^{s} = [\boldsymbol{y}_{c,k,m}^{s}; y_{d,k,m}^{s}]$, then

$$\boldsymbol{z}_{k,m}^{s} = \begin{bmatrix} \bar{\boldsymbol{H}}_{s,k}\bar{\boldsymbol{x}}_{k,i} \\ \dot{r}_{k,i} \end{bmatrix} + \begin{bmatrix} \boldsymbol{n}_{s,k} \\ n_{d,k}^{s} \end{bmatrix} \tag{9.4.14}$$

Based on the RFS representation, the augmented dynamic model of multiple moving targets and the augmented measurement model of multiple sensors can be given, the latter including the slant range, azimuth and Doppler originated from targets and clutters. Let N_{k-1} be the number of targets and their state at time $k-1$ be $\bar{\boldsymbol{x}}_{k-1,1}, \ldots, \bar{\boldsymbol{x}}_{k-1,N_{k-1}}$. At the next moment, existing targets may either disappear or continue to survive, and new targets may either appear or be spawned from existing targets. This results in N_k new states $\bar{\boldsymbol{x}}_{k,1}, \ldots, \bar{\boldsymbol{x}}_{k,N_k}$. Assume that sensor s receives M_k^s measurements $\boldsymbol{z}_{k,1}^{s}, \ldots, \boldsymbol{z}_{k,M_k^s}^{s}$ at time k, and some of the measurements come from the targets and the rest from clutters. Note that the corresponding sets of target states and measurements have no order, and can be naturally represented as random sets, namely

$$\bar{X}_k = \left\{\bar{\boldsymbol{x}}_{k,1}, \ldots, \bar{\boldsymbol{x}}_{k,N_k}\right\} \tag{9.4.15}$$

$$Z_k^s = \left\{\boldsymbol{z}_{k,1}^s, \ldots, \boldsymbol{z}_{k,M_k^s}^s\right\} \tag{9.4.16}$$

Given the multi-target state $\bar{X}_{k-1}$ at time $k-1$, the multi-target state $\bar{X}_k$ at time k is the union of existing targets and birth targets (for simplicity, the spawning targets are ignored here)

$$\bar{X}_k = \bar{\Gamma}_k \cup \left[\cup_{\bar{\boldsymbol{x}}_{k-1,i} \in \bar{X}_{k-1}} S_{k|k-1}(\bar{\boldsymbol{x}}_{k-1,i})\right] \tag{9.4.17}$$

The above equation assumes that each RFS in the union is independent of each other, where $\bar{\Gamma}_k$ represents the RFS of the birth targets at time k; $S_{k|k-1}(\bar{\boldsymbol{x}}_{k-1,i})$ is the RFS of the surviving targets at time k based on $\bar{\boldsymbol{x}}_{k-1,i}$. It is $\{\bar{\boldsymbol{x}}_{k,j}\}$ when the target survives or $\emptyset$ when the target disappears. Each $\bar{\boldsymbol{x}}_{k-1,i} \in \bar{X}_{k-1}$ either continues to survive with probability $p_{S,k}(\bar{\boldsymbol{x}}_{k-1,i})$ or disappears with probability $1 - p_{S,k}(\bar{\boldsymbol{x}}_{k-1,i})$. Conditional on the existence of the target, the probability density function (PDF) for the state transition from $\bar{\boldsymbol{x}}_{k-1,i}$ to $\bar{\boldsymbol{x}}_{k,j}$ is

$$\phi_{k|k-1}\left(\bar{\boldsymbol{x}}_{k,j}|\bar{\boldsymbol{x}}_{k-1,i}\right) = \mathcal{N}\left(\bar{\boldsymbol{x}}_{k,j}; \bar{\boldsymbol{F}}_{k-1}\bar{\boldsymbol{x}}_{k-1,i}, \bar{\boldsymbol{Q}}_{k-1}\right) \tag{9.4.18}$$

Assume that the survival probability is independent of the state, i.e.,

$$p_{S,k}\left(\bar{\boldsymbol{x}}_{k-1,i}\right) = p_{S,k} \tag{9.4.19}$$

Given a multi-target state $\bar{X}_k$, the multi-target measurement Z_k^s received by the sensor is formed by the union of target measurements and clutters, i.e.,

$$Z_k^s = K_k^s \cup \left[\cup_{\bar{\boldsymbol{x}}_{k,i} \in \bar{X}_k} \Theta_k^s\left(\bar{\boldsymbol{x}}_{k,i}\right)\right] \tag{9.4.20}$$

In the above equation, it is assumed that each RFS in the union is also independent of each other, where K_k^s is the RFS of the clutter, $\Theta_k^s(\bar{\boldsymbol{x}}_{k,i})$ is the RFS of detected measurements from the targets, and it is $\{\boldsymbol{z}_{k,m}^s\}$ when the target is detected with detection probability $p_{D,k}^s(\bar{\boldsymbol{x}}_{k,i})$ or $\emptyset$ when the target is subject to missed detection with probability $1 - p_{D,k}^s(\bar{\boldsymbol{x}}_{k,i})$. Conditioned that the target is detected, the PDF of measurement $\boldsymbol{z}_{k,m}^s$ obtained from $\bar{\boldsymbol{x}}_{k,i}$ is

$$g_k\left(\boldsymbol{z}_{k,m}^s|\bar{\boldsymbol{x}}_{k,i}\right) = \mathcal{N}\left(\boldsymbol{y}_{c,k,m}^s; \bar{\boldsymbol{H}}_{s,k}\bar{\boldsymbol{x}}_{k,i}, \boldsymbol{R}_{s,p}\right)\mathcal{N}\left(y_{d,k,m}^s; \dot{r}_{k,i}, \sigma_{s,d}^2\right) \tag{9.4.21}$$

Assume that the detection probability is independent of the state, i.e.,

$$p_{D,k}^s(\bar{\boldsymbol{x}}_{k,i}) = p_{D,k}^s \tag{9.4.22}$$

9.4.2 Augmented State GM-PHD Filter with Registration Error

Assume that the evolution of each target and the measurement generation process are independent of each other, the predicted multi-target RFS is dominated by the Poisson distribution, and the clutter also obeys the Poisson distribution and is independent of the target measurements. It is assumed that the target state is unrelated to the sensor bias and the intensity of the birth RFS $\bar{\Gamma}_k$ is in Gaussian mixture form, i.e.,

$$\begin{aligned} v_{\gamma,k}(\bar{\boldsymbol{x}}) &= \sum_{j=1}^{J_{\gamma,k}} w_{\gamma,k}^{(j)} \mathcal{N}\left(\bar{\boldsymbol{x}}; \bar{\boldsymbol{m}}_{\gamma,k}^{(j)}, \bar{\boldsymbol{P}}_{\gamma,k}^{(j)}\right) \\ &= \sum_{j=1}^{J_{\gamma,k}} w_{\gamma,k}^{(j)} \mathcal{N}\left(\boldsymbol{x}; \boldsymbol{m}_{\gamma,k}^{(j)}, \boldsymbol{P}_{\gamma,k}^{(j)}\right) \prod_{s=1}^{S} \mathcal{N}\left(\boldsymbol{\theta}_s; \boldsymbol{u}_{\gamma,k}^{(s,j)}, \boldsymbol{\Xi}_{\gamma,k}^{(s,j)}\right) \end{aligned} \tag{9.4.23}$$

where $\bar{\boldsymbol{m}}_{\gamma,k}^{(j)} = [\boldsymbol{m}_{\gamma,k}^{(j)}; \boldsymbol{u}_{\gamma,k}^{(1,j)}; \ldots; \boldsymbol{u}_{\gamma,k}^{(S,j)}]^{\mathrm{T}}$, $\bar{\boldsymbol{P}}_{\gamma,k}^{(j)} = \mathrm{blkdiag}(\boldsymbol{P}_{\gamma,k}^{(j)}, \boldsymbol{\Xi}_{\gamma,k}^{(1,j)}, \ldots, \boldsymbol{\Xi}_{\gamma,k}^{(S,j)})$, the number of components $J_{\gamma,k}$, the component weight $w_{\gamma,k}^{(j)}$, the mean $\boldsymbol{m}_{\gamma,k}^{(j)}$ and covariance $\boldsymbol{P}_{\gamma,k}^{(j)}$ of the state component, the mean $\boldsymbol{u}_{\gamma,k}^{(s,j)}$ and covariance $\boldsymbol{\Xi}_{\gamma,k}^{(s,j)}$ $(j = 1, \ldots, J_{\gamma,k}, s = 1, \ldots, S)$ of the sensor bias component, are the given model parameters, which determine the shape of the birth intensity.

In deriving the augmented state GM-PHD filter, for the sake of simplicity, the same assumptions as in Sect. 4.4.1 are ignored for simplicity. Based on the standard GM-PHD and GM-PHD with Doppler (GM-PHD-D) [223] filters, the augmented state GM-PHD filter with the registration error (GM-PHD-R-D) consists of the prediction and the update. The specific process is as follows.

9.4.2.1 Prediction

Assume that prior intensity v_{k-1} at time $k-1$ is in the form of Gaussian mixture, i.e.,

$$\begin{aligned} v_{k-1}(\bar{\boldsymbol{x}}) &= \sum_{j=1}^{J_{k-1}} w_{k-1}^{(j)} \mathcal{N}\left(\bar{\boldsymbol{x}}; \bar{\boldsymbol{m}}_{k-1}^{(j)}, \bar{\boldsymbol{P}}_{k-1}^{(j)}\right) \\ &= \sum_{j=1}^{J_{k-1}} w_{k-1}^{(j)} \mathcal{N}\left(\boldsymbol{x}; \boldsymbol{m}_{k-1}^{(j)}, \boldsymbol{P}_{k-1}^{(j)}\right) \prod_{s=1}^{S} \mathcal{N}\left(\boldsymbol{\theta}_s; \boldsymbol{u}_{k-1}^{(s,j)}, \boldsymbol{\Xi}_{k-1}^{(s,j)}\right) \end{aligned} \tag{9.4.24}$$

where $\bar{\boldsymbol{m}}_{k-1}^{(j)} = [\boldsymbol{m}_{k-1}^{(j)}; \boldsymbol{u}_{k-1}^{(1,j)}; \ldots; \boldsymbol{u}_{k-1}^{(S,j)}]^{\mathrm{T}}$ and $\bar{\boldsymbol{P}}_{k-1}^{(j)} = \mathrm{blkdiag}$ $(\boldsymbol{P}_{k-1}^{(j)}, \boldsymbol{\Xi}_{k-1}^{(1,j)}, \ldots, \boldsymbol{\Xi}_{k-1}^{(S,j)})$. Equation (9.4.24) is based on the assumption that the state component is independent of the sensor bias component, then the predicted intensity $v_{k|k-1}$ at time k is also in Gaussian mixture form, i.e., [63]

$$v_{k|k-1}(\bar{\boldsymbol{x}}) = v_{\gamma,k}(\bar{\boldsymbol{x}}) + v_{S,k|k-1}(\bar{\boldsymbol{x}}) = \sum_{j=1}^{J_{k|k-1}} w_{k|k-1}^{(j)} \mathcal{N}\left(\bar{\boldsymbol{x}}; \bar{\boldsymbol{m}}_{k|k-1}^{(j)}, \bar{\boldsymbol{P}}_{k|k-1}^{(j)}\right) \tag{9.4.25}$$

where $v_{\gamma,k}(\bar{\boldsymbol{x}})$ is given by (9.4.23), and

$$v_{S,k|k-1}(\bar{\boldsymbol{x}}) = p_{S,k} \sum_{j=1}^{J_{k-1}} w_{k-1}^{(j)} \mathcal{N}\left(\bar{\boldsymbol{x}}; \bar{\boldsymbol{m}}_{S,k|k-1}^{(j)}, \bar{\boldsymbol{P}}_{S,k|k-1}^{(j)}\right) \tag{9.4.26}$$

$$\bar{\boldsymbol{m}}_{S,k|k-1}^{(j)} = \left[\boldsymbol{m}_{S,k|k-1}^{(j)}; \boldsymbol{u}_{k|k-1}^{(1,j)}; \ldots; \boldsymbol{u}_{k|k-1}^{(S,j)}\right] = \bar{\boldsymbol{F}}_{k-1} \bar{\boldsymbol{m}}_{k-1}^{(j)} \tag{9.4.27}$$

$$\bar{\boldsymbol{P}}_{S,k|k-1}^{(j)} = \text{blkdiag}\left(\boldsymbol{P}_{S,k|k-1}^{(j)}, \boldsymbol{\varXi}_{k|k-1}^{(1,j)}, \ldots, \boldsymbol{\varXi}_{k|k-1}^{(S,j)}\right) = \bar{\boldsymbol{F}}_{k-1} \bar{\boldsymbol{P}}_{k-1}^{(j)} \bar{\boldsymbol{F}}_{k-1}^{\mathrm{T}} + \bar{\boldsymbol{Q}}_{k-1} \tag{9.4.28}$$

Expanding (9.4.27) and (9.4.28) leads to

$$\boldsymbol{m}_{S,k|k-1}^{(j)} = \boldsymbol{F}_{k-1} \boldsymbol{m}_{k-1}^{(j)}, \ \boldsymbol{P}_{S,k|k-1}^{(j)} = \boldsymbol{F}_{k-1} \boldsymbol{P}_{k-1}^{(j)} \boldsymbol{F}_{k-1}^{\mathrm{T}} + \boldsymbol{Q}_{k-1} \tag{9.4.29}$$

$$\boldsymbol{u}_{k|k-1}^{(s,j)} = \boldsymbol{u}_{k-1}^{(s,j)}, \ \boldsymbol{\varXi}_{k|k-1}^{(s,j)} = \boldsymbol{\varXi}_{k-1}^{(s,j)} + \boldsymbol{B}_{s,k-1}, \ s = 1, \ldots, S \tag{9.4.30}$$

Therefore, Eq. (9.4.26) can be rewritten as

$$v_{S,k|k-1}(\bar{\boldsymbol{x}}) = p_{S,k} \sum_{j=1}^{J_{k-1}} w_{k-1}^{(j)} \mathcal{N}\left(\boldsymbol{x}; \boldsymbol{m}_{S,k|k-1}^{(j)}, \boldsymbol{P}_{S,k|k-1}^{(j)}\right) \prod_{s=1}^{S} \mathcal{N}\left(\boldsymbol{\theta}_s; \boldsymbol{u}_{k|k-1}^{(s,j)}, \boldsymbol{\varXi}_{k|k-1}^{(s,j)}\right) \tag{9.4.31}$$

Substituting (9.4.23) and (9.4.31) into (9.4.25) yields

$$\begin{aligned} v_{k|k-1}(\bar{\boldsymbol{x}}) = & \sum_{j=1}^{J_{\gamma,k}} w_{\gamma,k}^{(j)} \mathcal{N}\left(\boldsymbol{x}; \boldsymbol{m}_{\gamma,k}^{(j)}, \boldsymbol{P}_{\gamma,k}^{(j)}\right) \prod_{s=1}^{S} \mathcal{N}\left(\boldsymbol{\theta}_s; \boldsymbol{u}_{\gamma,k}^{(s,j)}, \boldsymbol{\varXi}_{\gamma,k}^{(s,j)}\right) \\ & + p_{S,k} \sum_{j=1}^{J_{k-1}} w_{k-1}^{(j)} \mathcal{N}\left(\boldsymbol{x}; \boldsymbol{m}_{S,k|k-1}^{(j)}, \boldsymbol{P}_{S,k|k-1}^{(j)}\right) \prod_{s=1}^{S} \mathcal{N}\left(\boldsymbol{\theta}_s; \boldsymbol{u}_{k|k-1}^{(s,j)}, \boldsymbol{\varXi}_{k|k-1}^{(s,j)}\right) \end{aligned} \tag{9.4.32}$$

9.4.2.2 Update

Assume that there are R ($R \subseteq S$) sensors working at time k, and the sequential sensor update method is adopted here, i.e., the update equation of the GM-PHD filter is sequentially applied to each sensor. For sensor r ($r = 1, \ldots, R$), the posterior intensity $v_{k|k,r}$ at time k is also a Gaussian mixture, given by [63]

$$v_{k|k,r}(\bar{\boldsymbol{x}}) = (1 - p_{D,k}^r)v_{k|k,r-1}(\bar{\boldsymbol{x}}) + \sum_{z \in Z_k^r} v_{D,k,r}(\bar{\boldsymbol{x}}; z) \tag{9.4.33}$$

where $v_{k|k,0}(\bar{\boldsymbol{x}}) = v_{k|k-1}(\bar{\boldsymbol{x}})$ and

$$\begin{aligned} v_{D,k,r}(\bar{\boldsymbol{x}}; z) &= \sum_{j=1}^{J_{k|k,r-1}} w_{k|k,r}^{(j)}(z)\mathcal{N}\left(\bar{\boldsymbol{x}}; \bar{\boldsymbol{m}}_{k|k,r}^{(j)}(z), \bar{\boldsymbol{P}}_{k|k,r}^{(j)}\right) \\ &\approx \sum_{j=1}^{J_{k|k,r-1}} w_{k|k,r}^{(j)}(z)\mathcal{N}\left(\boldsymbol{x}; \boldsymbol{m}_{k|k,r}^{(j)}(z), \boldsymbol{P}_{k|k,r}^{(j)}\right) \prod_{s=1}^{S} \mathcal{N}\left(\boldsymbol{\theta}_s; \boldsymbol{u}_{k|k,r}^{(s,j)}, \boldsymbol{\Xi}_{k|k,r}^{(s,j)}\right) \end{aligned} \tag{9.4.34}$$

with $J_{k|k,0} = J_{k|k-1}$. The calculation of $w_{k|k,r}^{(j)}(z)$, $\boldsymbol{m}_{k|k,r}^{(j)}(z)$, $\boldsymbol{P}_{k|k,r}^{(j)}$, $\boldsymbol{u}_{k|k,r}^{(s,j)}$, and $\boldsymbol{\Xi}_{k|k,r}^{(s,j)}$ and the reason for taking "$\approx$" are as follows.

In order to effectively utilize the Doppler measurement in the augmented state GM-PHD filter, a sequential processing method is adopted. That is, first, the target state and sensor bias are updated using the slant range and azimuth measurements; then, the target state is sequentially updated using the Doppler measurement; finally, the weight is calculated using the slant range, azimuth, and Doppler measurements. The detailed execution steps are as follows.

- Step 1: update the target state and sensor bias with measurement $\boldsymbol{y}_{c,k,m}^r$ ($m = 1, \ldots, M_k^r$) of sensor r

$$\bar{\boldsymbol{m}}_{k|k,r}^{(j)}\left(\boldsymbol{y}_{c,k,m}^r\right) = \bar{\boldsymbol{m}}_{k|k,r-1}^{(j)} + \bar{\boldsymbol{G}}_{c,k,r}^{(j)}\tilde{\boldsymbol{y}}_{r,k}^{(m,j)} \tag{9.4.35}$$

$$\bar{\boldsymbol{P}}_{k|k,r}^{(j)}\left(\boldsymbol{y}_{c,k,m}^r\right) = \left(\boldsymbol{I} - \bar{\boldsymbol{G}}_{c,k,r}^{(j)}\bar{\boldsymbol{H}}_{r,k}^{(j)}\right)\bar{\boldsymbol{P}}_{k|k,r-1}^{(j)} \tag{9.4.36}$$

$$\bar{\boldsymbol{G}}_{c,k,r}^{(j)} = \left[\boldsymbol{G}_{c,k,r}^{(j)}; \boldsymbol{K}_{1,k,r}^{(j)}; \ldots; \boldsymbol{K}_{S,k,r}^{(j)}\right] = \bar{\boldsymbol{P}}_{k|k,r-1}^{(j)}\bar{\boldsymbol{H}}_{r,k}^{(j)}\left[\boldsymbol{S}_{c,k,r}^{(j)}\right]^{-1} \tag{9.4.37}$$

$$\bar{\boldsymbol{H}}_{r,k}^{(j)} = \bar{\boldsymbol{H}}_{r,k}\Big|_{\bar{\boldsymbol{m}}_{k|k,r-1}^{(j)}} \tag{9.4.38}$$

where $\boldsymbol{m}_{k|k,0}^{(j)} = \boldsymbol{m}_{k|k-1}^{(j)}$, $\boldsymbol{u}_{k|k,0}^{(s,j)} = \boldsymbol{u}_{k|k-1}^{(s,j)}$, $\boldsymbol{P}_{k|k,0}^{(j)} = \boldsymbol{P}_{k|k-1}^{(j)}$, $\boldsymbol{\Xi}_{k|k,0}^{(s,j)} = \boldsymbol{\Xi}_{k|k-1}^{(s,j)}$, $\bar{\boldsymbol{m}}_{k|k,r-1}^{(j)} = [\boldsymbol{m}_{k|k,r-1}^{(j)}; \boldsymbol{u}_{k|k,r-1}^{(1,j)}; \ldots; \boldsymbol{u}_{k|k,r-1}^{(S,j)}]$,

$\bar{\boldsymbol{m}}_{k|k,r}^{(j)}(\boldsymbol{y}_{c,k,m}^{r}) = [\boldsymbol{m}_{k|k,r}^{(j)}; \boldsymbol{u}_{k|k,r}^{(1,j)}; \ldots; \boldsymbol{u}_{k|k,r}^{(S,j)}](\boldsymbol{y}_{c,k,m}^{r})$, $\bar{\boldsymbol{P}}_{k|k,r-1}^{(j)} = \text{blkdiag}(\boldsymbol{P}_{k|k,r-1}^{(j)}, \boldsymbol{\Xi}_{k|k,r-1}^{(1,j)}, \ldots, \boldsymbol{\Xi}_{k|k,r-1}^{(S,j)})$. Moreover, according to (9.4.36), we have

$$\bar{\boldsymbol{P}}_{k|k,r}^{(j)}\left(\boldsymbol{y}_{c,k,m}^{r}\right) = \left(\boldsymbol{I} - \bar{\boldsymbol{G}}_{c,k,r}^{(j)}\bar{\boldsymbol{H}}_{r,k}^{(j)}\right)\bar{\boldsymbol{P}}_{k|k,r-1}^{(j)} = \bar{\boldsymbol{P}}_{k|k,r-1}^{(j)} - \bar{\boldsymbol{G}}_{c,k,r}^{(j)}\bar{\boldsymbol{H}}_{r,k}^{(j)}\bar{\boldsymbol{P}}_{k|k,r-1}^{(j)} \tag{9.4.39}$$

Expanding the above equation leads to

$$\bar{\boldsymbol{P}}_{k|k,r}^{(j)}\left(\boldsymbol{y}_{c,k,m}^{r}\right) = \begin{bmatrix} \boldsymbol{P}_{k|k,r-1}^{(j)} - \boldsymbol{G}_{c,k,r}^{(j)}\boldsymbol{H}_{r,k}^{(j)}\boldsymbol{P}_{k|k,r-1}^{(j)} & -\boldsymbol{G}_{c,k,r}^{(j)}\boldsymbol{\Psi}_1\boldsymbol{\Xi}_{k|k,r-1}^{(1,j)} & \cdots & -\boldsymbol{G}_{c,k,r}^{(j)}\boldsymbol{\Psi}_S\boldsymbol{\Xi}_{k|k,r-1}^{(S,j)} \\ -\boldsymbol{K}_{1,k,r}^{(j)}\boldsymbol{H}_{r,k}^{(j)}\boldsymbol{P}_{k|k,r-1}^{(j)} & \boldsymbol{\Xi}_{k|k,r-1}^{(1,j)} - \boldsymbol{K}_{1,k,r}^{(j)}\boldsymbol{\Psi}_1\boldsymbol{\Xi}_{k|k,r-1}^{(1,j)} & \cdots & -\boldsymbol{K}_{1,k,r}^{(j)}\boldsymbol{\Psi}_S\boldsymbol{\Xi}_{k|k,r-1}^{(S,j)} \\ \vdots & \vdots & \ddots & \vdots \\ -\boldsymbol{K}_{S,k,r}^{(j)}\boldsymbol{H}_{r,k}^{(j)}\boldsymbol{P}_{k|k,r-1}^{(j)} & -\boldsymbol{K}_{S,k,r}^{(j)}\boldsymbol{\Psi}_1\boldsymbol{\Xi}_{k|k,r-1}^{(1,j)} & \cdots & \boldsymbol{\Xi}_{k|k,r-1}^{(S,j)} - \boldsymbol{K}_{S,k,r}^{(j)}\boldsymbol{\Psi}_S\boldsymbol{\Xi}_{k|k,r-1}^{(S,j)} \end{bmatrix} \tag{9.4.40}$$

In order to decouple the errors of the target state and the sensor bias, ignore the cross-covariance terms and set them as 0, then we have

$$\bar{\boldsymbol{P}}_{k|k,r}^{(j)}\left(\boldsymbol{y}_{c,k,m}^{r}\right) \approx \text{blkdiag}\left(\boldsymbol{P}_{k|k,r}^{(j)}, \boldsymbol{\Xi}_{k|k,r}^{(1,j)}, \ldots, \boldsymbol{\Xi}_{k|k,r}^{(S,j)}\right) \tag{9.4.41}$$

Expanding (9.4.37) yields

$$\boldsymbol{G}_{c,k,r}^{(j)} = \boldsymbol{P}_{k|k,r-1}^{(j)}(\boldsymbol{H}_{r,k}^{(j)})^{\mathrm{T}}(\boldsymbol{S}_{c,k,r}^{(j)})^{-1} \tag{9.4.42}$$

$$\boldsymbol{K}_{s,k,r}^{(j)} = \boldsymbol{\Xi}_{k|k,r-1}^{(s,j)}\boldsymbol{\Psi}_s^{\mathrm{T}}\left(\boldsymbol{S}_{c,k,r}^{(j)}\right)^{-1} = \begin{cases} \boldsymbol{\Xi}_{k|k,r-1}^{(s,j)}(\boldsymbol{S}_{c,k,r}^{(j)})^{-1}, & s = r \\ \boldsymbol{0}, & s \neq r \end{cases} \tag{9.4.43}$$

where

$$\boldsymbol{H}_{r,k}^{(j)} = \boldsymbol{H}_{r,k}\Big|_{\boldsymbol{m}_{k|k,r-1}^{(j)}} \tag{9.4.44}$$

$$\begin{aligned} \boldsymbol{S}_{c,k,r}^{(j)} &= \bar{\boldsymbol{H}}_{r,k}^{(j)}\bar{\boldsymbol{P}}_{k|k,r-1}^{(j)}\left(\bar{\boldsymbol{H}}_{r,k}^{(j)}\right)^{\mathrm{T}} + \boldsymbol{R}_{r,p} \\ &= \boldsymbol{H}_{r,k}^{(j)}\boldsymbol{P}_{k|k,r-1}^{(j)}\left(\boldsymbol{H}_{r,k}^{(j)}\right)^{\mathrm{T}} + \sum_{s=1}^{S}\boldsymbol{\Psi}_s\boldsymbol{\Xi}_{k|k,r-1}^{(s,j)}\boldsymbol{\Psi}_s^{\mathrm{T}} + \boldsymbol{R}_{r,p} \end{aligned} \tag{9.4.45}$$

Substituting (9.4.11) into the above equation can lead to

$$S_{c,k,r}^{(j)} = \boldsymbol{H}_{r,k}^{(j)} \boldsymbol{P}_{k|k,r-1}^{(j)} \left(\boldsymbol{H}_{r,k}^{(j)}\right)^{\mathrm{T}} + \boldsymbol{\Xi}_{k|k,r-1}^{(r,j)} + \boldsymbol{R}_{r,p} \tag{9.4.46}$$

Finally, according to (9.4.35) and (9.4.36), the update equation for decoupling the target state and sensor bias can be obtained as

$$\boldsymbol{m}_{k|k,r}^{(j)}\left(\boldsymbol{y}_{c,k,m}^{r}\right) = \boldsymbol{m}_{k|k,r-1}^{(j)} + \boldsymbol{G}_{c,k,r}^{(j)} \tilde{\boldsymbol{y}}_{r,k}^{(m,j)} \tag{9.4.47}$$

$$\boldsymbol{P}_{k|k,r}^{(j)}\left(\boldsymbol{y}_{c,k,m}^{r}\right) = \left(\boldsymbol{I} - \boldsymbol{G}_{c,k,r}^{(j)} \boldsymbol{H}_{r,k}^{(j)}\right) \boldsymbol{P}_{k|k,r-1}^{(j)} \tag{9.4.48}$$

$$\boldsymbol{u}_{k|k,r}^{(s,j)} = \boldsymbol{u}_{k|k,r-1}^{(s,j)} + \boldsymbol{K}_{s,k,r}^{(j)} \tilde{\boldsymbol{y}}_{r,k}^{(m,j)} \tag{9.4.49}$$

$$\boldsymbol{\Xi}_{k|k,r}^{(s,j)} = \boldsymbol{\Xi}_{k|k,r-1}^{(s,j)} - \boldsymbol{K}_{s,k,r}^{(j)} \boldsymbol{\Psi}_{s} \boldsymbol{\Xi}_{k|k,r-1}^{(s,j)}, \quad s = 1, \ldots, S \tag{9.4.50}$$

where

$$\begin{aligned} \tilde{\boldsymbol{y}}_{r,k}^{(m,j)} &= \boldsymbol{y}_{c,k,m}^{r} - \bar{\boldsymbol{H}}_{r,k}^{(j)} \boldsymbol{m}_{k|k,r-1}^{(j)} \\ &= \boldsymbol{y}_{c,k,m}^{r} - \boldsymbol{H}_{r,k}^{(j)} \boldsymbol{m}_{k|k,r-1}^{(j)} - \sum\nolimits_{s=1}^{S} \boldsymbol{\Psi}_{s} \boldsymbol{u}_{k|k,r-1}^{(s,j)} \\ &= \boldsymbol{y}_{c,k,m}^{r} - \boldsymbol{H}_{r,k}^{(j)} \boldsymbol{m}_{k|k,r-1}^{(j)} - \boldsymbol{u}_{k|k,r-1}^{(r,j)} \end{aligned} \tag{9.4.51}$$

- Step 2: sequentially update the target state using the Doppler measurement $y_{d,k,m}^{r}$. The obtained target state component $(\boldsymbol{m}_{k|k,r}^{(j)}(\boldsymbol{y}_{c,k,m}^{r}), \boldsymbol{P}_{k|k,r}^{(j)}(\boldsymbol{y}_{c,k,m}^{r}))$ is further sequentially updated with the Doppler measurement, i.e.,

$$\boldsymbol{m}_{k|k,r}^{(j)}\left(\boldsymbol{z}_{k,m}^{r}\right) = \boldsymbol{m}_{k|k,r}^{(j)}\left(\boldsymbol{y}_{c,k,m}^{r}\right) + \boldsymbol{G}_{d,k,r}^{(j)}\left(\boldsymbol{y}_{c,k,m}^{r}\right) \tilde{y}_{d,k,r}^{(m,j)} \tag{9.4.52}$$

$$\boldsymbol{P}_{k|k,r}^{(j)}\left(\boldsymbol{z}_{k,m}^{r}\right) = \left[\boldsymbol{I} - \boldsymbol{G}_{d,k,r}^{(j)}\left(\boldsymbol{y}_{c,k,m}^{r}\right) \boldsymbol{H}_{d,k,r}^{(j)}\left(\boldsymbol{y}_{c,k,m}^{r}\right)\right] \boldsymbol{P}_{k|k,r}^{(j)}\left(\boldsymbol{y}_{c,k,m}^{r}\right) \tag{9.4.53}$$

where

$$\tilde{y}_{d,k,r}^{(m,j)} = y_{d,k,m}^{r} - \hat{y}_{d,k,r}^{(m,j)} \tag{9.4.54}$$

$$\hat{y}_{d,k,r}^{(m,j)} = \dot{r}_{k}\left(\boldsymbol{m}_{k|k,r}^{(j)}(\boldsymbol{y}_{c,k,m}^{r})\right) \tag{9.4.55}$$

$$\boldsymbol{G}_{d,k,r}^{(j)}\left(\boldsymbol{y}_{c,k,m}^{r}\right) = \boldsymbol{P}_{k|k,r}^{(j)}\left(\boldsymbol{y}_{c,k,m}^{r}\right)\left[\boldsymbol{H}_{d,k,r}^{(j)}(\boldsymbol{y}_{c,k,m}^{r})\right]^{\mathrm{T}}\left[S_{d,k,r}^{(m,j)}\right]^{-1} \tag{9.4.56}$$

$$S_{d,k,r}^{(m,j)} = \boldsymbol{H}_{d,k,r}^{(j)}\left(\boldsymbol{y}_{c,k,m}^{r}\right) \boldsymbol{P}_{k|k,r}^{(j)}\left(\boldsymbol{y}_{c,k,m}^{r}\right)\left[\boldsymbol{H}_{d,k,r}^{(j)}(\boldsymbol{y}_{c,k,m}^{r})\right]^{\mathrm{T}} + \sigma_{r,d}^{2} \tag{9.4.57}$$

$$\boldsymbol{H}_{d,k,r}^{(j)}\left(\boldsymbol{y}_{c,k,m}^{r}\right) = \nabla \dot{r}_{k}\Big|_{\boldsymbol{m}_{k|k,r}^{(j)}(\boldsymbol{y}_{c,k,m}^{r})} = \left[h_{1}\ h_{2}\ h_{3}\ h_{4}\right] \tag{9.4.58}$$

$$h_1 = \left(\dot{x}_{k|k}^{(m,j)} - \hat{y}_{d,k,r}^{(m,j)} \cos \hat{a}_{k,m,j}\right) / \hat{r}_{k,m,j} \tag{9.4.59}$$

$$h_2 = \left(\dot{y}_{k|k}^{(m,j)} - \hat{y}_{d,k,r}^{(m,j)} \sin \hat{a}_{k,m,j}\right) / \hat{r}_{k,m,j} \tag{9.4.60}$$

$$h_3 = \left(x_{k|k}^{(m,j)} - x_{r,k}\right) / \hat{r}_{k,m,j} \triangleq \cos \hat{a}_{k,m,j} \tag{9.4.61}$$

$$h_4 = \left(y_{k|k}^{(m,j)} - y_{r,k}\right) / \hat{r}_{k,m,j} \triangleq \sin \hat{a}_{k,m,j} \tag{9.4.62}$$

In the above equations, $\hat{r}_{k,m,j} \triangleq \sqrt{(x_{k|k}^{(m,j)} - x_{r,k})^2 + (y_{k|k}^{(m,j)} - y_{r,k})^2}$, $\boldsymbol{m}_{k|k,r}^{(j)}(\boldsymbol{y}_{c,k,m}^r) = [x_{k|k}^{(m,j)}, y_{k|k}^{(m,j)}, \dot{x}_{k|k}^{(m,j)}, \dot{y}_{k|k}^{(m,j)}]^{\mathrm{T}}$, and $(x_{r,k}, y_{r,k})$ is the position of sensor r.

- Step 3: update weights with position and Doppler measurements

$$w_{k|k,r}^{(j)}(z) = \frac{p_{D,k}^r w_{k|k,r-1}^{(j)}(z) q_{c,k}^{(j)}(\boldsymbol{y}_{c,k,m}^r) q_{d,k}^{(j)}(y_{d,k,m}^r)}{\kappa_{c,k}^r \kappa_{d,k}^r + p_{D,k}^r \sum_{j=1}^{J_{k|k,r-1}} w_{k|k,r-1}^{(j)}(z) q_{c,k}^{(j)}(\boldsymbol{y}_{c,k,m}^r) q_{d,k}^{(j)}(y_{d,k,m}^r)} \tag{9.4.63}$$

where $w_{k|k,0}^{(j)} = w_{k|k-1}^{(j)}$, and

$$q_{c,k}^{(j)}(\boldsymbol{y}_{c,k,m}^r) = \mathcal{N}\left(\boldsymbol{y}_{c,k,m}^r; \boldsymbol{H}_{r,k}^{(j)} \boldsymbol{m}_{k|k,r-1}^{(j)} + \boldsymbol{u}_{k|k,r-1}^{(r,j)}, \boldsymbol{S}_{c,k,r}^{(j)}\right) \tag{9.4.64}$$

$$q_{d,k}^{(j)}(y_{d,k,m}^r) = \mathcal{N}\left(y_{d,k,m}^r; \hat{y}_{d,k,r}^{(m,j)}, S_{d,k,r}^{(m,j)}\right) \tag{9.4.65}$$

In (9.4.63), the clutter intensity in polar coordinates is $\kappa_{c,k}^r = \lambda_c^r \cdot V^r \cdot u^r$, λ_c^r represents the average number of clutters per unit volume of sensor r, V^r is the volume of the monitoring area of sensor r, and u^r represents the uniform distribution in the monitoring area. It is assumed that the clutter velocity is uniformly distributed within the range of $[-v_{\max}, v_{\max}]$, so that the Doppler clutter intensity is $\kappa_{d,k}^r = 1/(2v_{\max})$, where $v_{\max}$ is the maximum velocity that the sensor can detect.

For completeness, Table 9.3 summarizes the augmented state GM-PHD filter with the registration error. Lines 3–5 in Table 9.3 are the predictions of the target states and sensor biases for the birth targets. The "for" loops in lines 6–9 are the predictions of the target states and sensor biases for the surviving targets. Lines 11–38 are the process in which the update equation of the augmented state GM-PHD filter at time k is applied to R sensors in turn, where lines 12–17 correspond to the construction process of the PHD update component, lines 18–20 deal with missed detections, and lines 22–36 cope with the measurement update. Line 25 calculates the transformed polar coordinate measurements according to (9.4.9), lines 26–28 use these transformed measurements to update the target state and sensor bias, lines

29–33 use the Doppler measurement to sequentially update the target state, line 34 computes the likelihood function for each component using polar coordinates and Doppler measurements, and line 36 computes the weights.

It is worth pointing out that the GM-PHD filter with Doppler measurements proposed in [223] is a single-sensor version, which does not involve the sensor bias, so it is a special case of the proposed algorithm above. In addition, the proposed algorithm is a general version of the GM-PHD filter with registration errors (GM-PHD-R) proposed in [339]. If the Doppler measurement is not available, the proposed algorithm will degenerate to the GM-PHD-R filter.

Since the sensor bias is the same for all targets, it can be estimated by

$$\hat{\boldsymbol{\theta}}_{s,k} = \sum_{j=1}^{J_{k|k,R}} \frac{w_{k|k,R}^{(j)} \boldsymbol{u}_{k|k,R}^{(s,j)}}{\sum_{j=1}^{J_{k|k,R}} w_{k|k,R}^{(j)}}, \quad s = 1, \ldots, S \tag{9.4.66}$$

9.4.3 Simulation Analysis

In order to verify the effectiveness of the augmented state GM-PHD-R-D algorithm and analyze the ability of Doppler information to improve the accuracy of track estimation, it is compared with the GM-PHD-R filter without the Doppler information [339].

In the simulation, it is assumed that there are 2 sensors observing 4 targets. The first sensor (S1) and the second sensor (S2) are located at (0, − 50) km and (0, 50) km, respectively. The biases of both sensors are set to (1 km, 5°). The standard deviations of the measurement noise are set to $\sigma_{s,r} = 100\,\text{m}$, $\sigma_{s,a} = 0.5°$, and $\sigma_{s,d} = 0.5\,\text{m/s}$ $(s = 1, 2)$.

The total number of frames in the measurement set is $K = 24$, and the two sensors work asynchronously: Sensor 1 reports the measurements at $k = 1, 3, \ldots, 23$, while Sensor 2 reports its measurements at $k = 2, 4, \ldots, 24$. The time interval between frame $k - 1$ and frame k is a constant, equal to 3 s.

Each target starts at $k = 1$, with a survival probability of $p_{S,k} = 0.99$, and follows the linear Gaussian dynamics of (9.4.1), where the standard deviations of the process noise in the x–y plane are $\sigma_x = \sigma_y = 5\,\text{m/s}^2$. Table 9.4 gives the initial state of each target.

The sensor bias follows the first-order Gauss–Markov process of (9.4.4), and its process noise covariance is $\boldsymbol{B}_{s,k-1} = \text{blkdiag}(50\,\text{m}, 0.1°)^2$, $(s = 1, 2)$.

The clutters obey the uniform distribution in the observation areas of [60 km, 200 km] × [0°, 60°]×[− 350 m/s, 350 m/s] (for S1) and [60 km, 200 km] × [− 60°, 0°] × [− 350 m/s, 350 m/s] (for S2). The detection probability is $p_{D,k}^{s} = 0.98$, and the average number $\lambda_c^s \cdot V^s$ of clutters per frame of each sensor is abbreviated as $\lambda_c \cdot V$.

The RFS of birth targets obeys the Poisson distribution, and its intensity is

Table 9.3 Pseudo code for the augmented state GM-PHD-R-D filter

Input: $\{w_{k-1}^{(j)}, (\boldsymbol{m}_{k-1}^{(j)}, \boldsymbol{P}_{k-1}^{(j)}), \{\boldsymbol{u}_{k-1}^{(s,j)}, \boldsymbol{\Xi}_{k-1}^{(s,j)}\}_{s=1}^{S}\}_{j=1}^{J_{k-1}}$, a set of measurements $\{Z_k^r\}_{r=1}^{R}$
$i = 0$,
for $j = 1, \ldots, J_{\gamma,k}$
$i \leftarrow i+1,\ w_{k\mid k-1}^{(i)} = w_{\gamma,k}^{(j)},\ (\boldsymbol{m}_{k\mid k-1}^{(i)} = \boldsymbol{m}_{\gamma,k}^{(j)},\ \boldsymbol{P}_{k\mid k-1}^{(i)} = \boldsymbol{P}_{\gamma,k}^{(j)}),$ $\{\boldsymbol{u}_{k\mid k-1}^{(s,i)} = \boldsymbol{u}_{\gamma,k}^{(s,j)},\ \boldsymbol{\Xi}_{k\mid k-1}^{(s,i)} = \boldsymbol{\Xi}_{\gamma,k}^{(s,j)}\}_{s=1}^{S}$
end
for $j = 1, \ldots, J_{k-1}$
$i \leftarrow i+1,\ w_{k\mid k-1}^{(i)} = p_{S,k} w_{k-1}^{(j)},$
$(\boldsymbol{m}_{k\mid k-1}^{(i)} = \boldsymbol{F}_{k-1}\boldsymbol{m}_{k-1}^{(j)},\ \boldsymbol{P}_{k\mid k-1}^{(i)} = \boldsymbol{F}_{k-1}\boldsymbol{P}_{k-1}^{(j)}\boldsymbol{F}_{k-1}^{\mathrm{T}} + \boldsymbol{Q}_{k-1}),$ $\{\boldsymbol{u}_{k\mid k-1}^{(s,i)} = \boldsymbol{u}_{k-1}^{(s,j)},\ \boldsymbol{\Xi}_{k\mid k-1}^{(s,i)} = \boldsymbol{\Xi}_{k-1}^{(s,j)} + \boldsymbol{B}_{s,k-1}\}_{s=1}^{S},$
end
$J_{k\mid k,0} = J_{k\mid k-1} = i,$
for $r = 1, \ldots, R$
for $j = 1, \ldots, J_{k\mid k,r-1}$
Calculate $\boldsymbol{H}_{r,k}^{(j)}$ according to $\boldsymbol{m}_{k\mid k,r-1}^{(j)}$
$\boldsymbol{\eta}_{k\mid k,r-1}^{(j)} = \boldsymbol{H}_{r,k}^{(j)}\boldsymbol{m}_{k\mid k,r-1}^{(j)} + \boldsymbol{u}_{k\mid k,r-1}^{(r,j)},\ \boldsymbol{S}_{c,k,r}^{(j)} = \boldsymbol{H}_{r,k}^{(j)}\boldsymbol{P}_{k\mid k,r-1}^{(j)}(\boldsymbol{H}_{r,k}^{(j)})^{\mathrm{T}}$ $+\sum_{s=1}^{S}\boldsymbol{\Psi}_s\boldsymbol{\Xi}_{k\mid k,r-1}^{(s,j)}\boldsymbol{\Psi}_s^{\mathrm{T}} + \boldsymbol{R}_{r,p},$
$(\boldsymbol{G}_{c,k,r}^{(j)} = \boldsymbol{P}_{k\mid k,r-1}^{(j)}(\boldsymbol{H}_{r,k}^{(j)})^{\mathrm{T}}(\boldsymbol{S}_{c,k,r}^{(j)})^{-1},\ \boldsymbol{P}_{k\mid k,r}^{(j)}(\boldsymbol{y}_{c,k,m}^{r}) = (\boldsymbol{I} - \boldsymbol{G}_{c,k,r}^{(j)}\boldsymbol{H}_{r,k}^{(j)}) \times$ $\boldsymbol{P}_{k\mid k,r-1}^{(j)}),$
$\{\boldsymbol{K}_{s,k,r}^{(j)} = \boldsymbol{\Xi}_{k\mid k,r-1}^{(s,j)}\boldsymbol{\Psi}_s^{\mathrm{T}}(\boldsymbol{S}_{c,k,r}^{(j)})^{-1},\ \boldsymbol{\Xi}_{k\mid k,r}^{(s,j)} = \boldsymbol{\Xi}_{k\mid k,r-1}^{(s,j)} - \boldsymbol{K}_{s,k,r}^{(j)}\boldsymbol{\Psi}_s\boldsymbol{\Xi}_{k\mid k,r-1}^{(s,j)}\}_{s=1}^{S},$
end
for $j = 1, \ldots, J_{k\mid k,r-1}$
$w_{k\mid k,r}^{(j)} = (1 - p_{D,k}^{r})w_{k\mid k,r-1}^{(j)},\ (\boldsymbol{m}_{k\mid k,r}^{(j)} = \boldsymbol{m}_{k\mid k,r-1}^{(j)},\ \boldsymbol{P}_{k\mid k,r}^{(j)} = \boldsymbol{P}_{k\mid k,r-1}^{(j)}),$ $\{\boldsymbol{u}_{k\mid k,r}^{(s,j)} = \boldsymbol{u}_{k\mid k,r-1}^{(s,j)},\ \boldsymbol{\Xi}_{k\mid k,r}^{(s,j)} = \boldsymbol{\Xi}_{k\mid k,r-1}^{(s,j)}\}_{s=1}^{S},$
end
$l = 0$
for each $z \in Z_k^r$
$l \leftarrow l+1,$
for $j = 1, \ldots, J_{k\mid k,r-1}$
$\boldsymbol{y}_{c,k,m}^{r} = \boldsymbol{y}_{k,m}^{r} + \boldsymbol{H}_{r,k}^{(j)}\boldsymbol{m}_{k\mid k,r-1}^{(j)} - h(\boldsymbol{m}_{k\mid k,r-1}^{(j)}, \boldsymbol{x}_{s,k}),$
$\boldsymbol{m}_{k\mid k,r}^{(l\cdot J_{k\mid k,r-1}+j)}(\boldsymbol{y}_{c,k,m}^{r}) = \boldsymbol{m}_{k\mid k,r-1}^{(j)} + \boldsymbol{G}_{c,k,r}^{(j)}(\boldsymbol{y}_{c,k,m}^{r} - \boldsymbol{\eta}_{k\mid k,r-1}^{(j)}),$

(continued)

Table 9.3 (continued)

$$\boldsymbol{P}_{k|k,r}^{(l\cdot J_{k|k,r-1}+j)}(\boldsymbol{y}_{c,k,m}^{r}) = \boldsymbol{P}_{k|k,r}^{(j)}(\boldsymbol{y}_{c,k,m}^{r}),$$

$$\{\boldsymbol{u}_{k|k,r}^{(s,l\cdot J_{k|k,r-1}+j)} = \boldsymbol{u}_{k|k,r-1}^{(s,j)} + \boldsymbol{K}_{s,k,r}^{(j)}(\boldsymbol{y}_{c,k,m}^{r} - \boldsymbol{\eta}_{k|k,r-1}^{(j)}),\ \boldsymbol{\Xi}_{k|k,r}^{(s,l\cdot J_{k|k,r-1}+j)} = \boldsymbol{\Xi}_{k|k,r}^{(s,j)}\}_{s=1}^{S},$$

$$\boldsymbol{H}_{d,k,r}^{(j)}(\boldsymbol{y}_{c,k,m}^{r}) = \nabla \dot{r}_k \Big|_{\boldsymbol{m}_{k|k,r}^{(l\cdot J_{k|k,r-1}+j)}(\boldsymbol{y}_{c,k,m}^{r})},$$

$$\hat{y}_{d,k,r}^{(m,j)} = \dot{r}_k(\boldsymbol{m}_{k|k,r}^{(l\cdot J_{k|k,r-1}+j)}(\boldsymbol{y}_{c,k,m}^{r})),\ S_{d,k,r}^{(m,j)} = \boldsymbol{H}_{d,k,r}^{(j)}(\boldsymbol{y}_{c,k,m}^{r}) \times \boldsymbol{P}_{k|k,r}^{(l\cdot J_{k|k,r-1}+j)}(\boldsymbol{y}_{c,k,m}^{r})[\boldsymbol{H}_{d,k,r}^{(j)}(\boldsymbol{y}_{c,k,m}^{r})]^{\mathrm{T}} + \sigma_{r,d}^{2},$$

$$\boldsymbol{G}_{d,k,r}^{(j)}(\boldsymbol{y}_{c,k,m}^{r}) = \boldsymbol{P}_{k|k,r}^{(l\cdot J_{k|k,r-1}+j)}(\boldsymbol{y}_{c,k,m}^{r})[\boldsymbol{H}_{d,k,r}^{(j)}(\boldsymbol{y}_{c,k,m}^{r})]^{\mathrm{T}}[S_{d,k,r}^{(m,j)}]^{-1},$$

$$\boldsymbol{m}_{k|k,r}^{(l\cdot J_{k|k,r-1}+j)}(\boldsymbol{z}_{k,m}^{r}) = \boldsymbol{m}_{k|k,r}^{(l\cdot J_{k|k,r-1}+j)}(\boldsymbol{y}_{c,k,m}^{r}) + \boldsymbol{G}_{d,k,r}^{(j)}(\boldsymbol{y}_{c,k,m}^{r}) \times (y_{d,k,m}^{r} - \hat{y}_{d,k,r}^{(m,j)}),$$

$$\boldsymbol{P}_{k|k,r}^{(l\cdot J_{k|k,r-1}+j)}(\boldsymbol{z}_{k,m}^{r}) = [\boldsymbol{I} - \boldsymbol{G}_{d,k,r}^{(j)}(\boldsymbol{y}_{c,k,m}^{r})\boldsymbol{H}_{d,k,r}^{(j)}(\boldsymbol{y}_{c,k,m}^{r})] \times \boldsymbol{P}_{k|k,r}^{(l\cdot J_{k|k,r-1}+j)}(\boldsymbol{y}_{c,k,m}^{r}),$$

$$w_{k|k,r}^{(l\cdot J_{k|k,r-1}+j)}(\boldsymbol{z}) = p_{D,k}^{r} w_{k|k,r-1}^{(j)} \mathcal{N}(\boldsymbol{y}_{c,k,m}^{r}; \boldsymbol{\eta}_{k|k,r-1}^{(j)}, \boldsymbol{S}_{c,k,r}^{(j)}) \times \mathcal{N}(y_{d,k,m}^{r}; \hat{y}_{d,k,r}^{(m,j)}, S_{d,k,r}^{(m,j)}),$$

end

$$w_{k|k,r}^{(l\cdot J_{k|k,r-1}+j)}(\boldsymbol{z}) = \frac{w_{k|k,r}^{(l\cdot J_{k|k,r-1}+j)}(\boldsymbol{z})}{\kappa_{c,k}^{r}\kappa_{d,k}^{r} + \sum_{i=1}^{J_{k|k,r-1}} w_{k|k,r}^{(l\cdot J_{k|k,r-1}+i)}(\boldsymbol{z})},\ \text{for } j = 1, \ldots, J_{k|k,r-1}$$

end

$J_{k|k,r} = (l+1)J_{k|k,r-1}$,

end

$J_k = J_{k|k,R}$,

Output: $\{w_k^{(j)}, (\boldsymbol{m}_k^{(j)}, \boldsymbol{P}_k^{(j)}), \{\boldsymbol{u}_k^{(s,j)}, \boldsymbol{\Xi}_k^{(s,j)}\}_{s=1}^{S}\}_{j=1}^{J_k} = \{w_{k|k}^{(j)}, (\boldsymbol{m}_{k|k}^{(j)}, \boldsymbol{P}_{k|k}^{(j)}), \{\boldsymbol{u}_{k|k}^{(s,j)}, \boldsymbol{\Xi}_{k|k}^{(s,j)}\}_{s=1}^{S}\}_{j=1}^{J_k}$

Table 9.4 Initial target states

No.	Initial state $(x_0, y_0, \dot{x}_0, \dot{y}_0)$
1	(– 100 km, – 5 km, – 120 m/s, 150 m/s)
2	(– 100 km, 5 km, – 120 m/s, – 150 m/s)
3	(100 km, – 5 km, 150 m/s, 120 m/s)
4	(100 km, 5 km, 150 m/s, – 120 m/s)

$$v_{\gamma,k}(\bar{\boldsymbol{x}}) = 0.03 \sum_{j=1}^{2} \mathcal{N}\left(\boldsymbol{x}; \boldsymbol{m}_{\gamma,k}^{(j)}, \boldsymbol{P}_{\gamma,k}^{(j)}\right) \prod_{s=1}^{2} \mathcal{N}\left(\boldsymbol{\theta}_s; \boldsymbol{u}_{\gamma,k}^{(s,j)}, \boldsymbol{\Xi}_{\gamma,k}^{(s,j)}\right) \tag{9.4.67}$$

where $\boldsymbol{m}_{\gamma,k}^{(1)} = [100\,\text{km}, -5\,\text{km}, 0\,\text{m/s}, 0\,\text{m/s}]^{\mathrm{T}}$, $\boldsymbol{m}_{\gamma,k}^{(2)} = [100\,\text{km}, 5\,\text{km}\ 0\,\text{m/s}, 0\,\text{m/s}]^{\mathrm{T}}$, $\boldsymbol{u}_{\gamma,k}^{(1,j)} = [1\,\text{km}, 5°]^{\mathrm{T}}$, $\boldsymbol{u}_{\gamma,k}^{(2,j)} = [-1\,\text{km}, -5°]^{\mathrm{T}}$, $\boldsymbol{P}_{\gamma,k}^{(j)} = \text{blkdiag}(100\,\text{m}, 100\,\text{m}, 50\,\text{m/s}, 50\,\text{m/s})^2$, and $\boldsymbol{\Xi}_{\gamma,k}^{(s,j)} = \text{blkdiag}(500\,\text{m}, 1°)^2$, $s, j = 1, 2$. The pruning parameter is set as $T = 10^{-5}$, the merging threshold is $U = 4$, and the maximum number of Gaussian components is $J_{\max} = 100$.

Figure 9.14 shows the test scenario when $\lambda_c \cdot V = 10$. In this figure, the green diamonds and red rectangles represent the measurements from S1 and S2, respectively, and Figure (b) is an enlarged view of Figure (a). "Tn" indicates target n. Figure 9.15 compares the position estimates of the GM-PHD-R and GM-PHD-R-D filters. Although both methods successfully track the targets, it can be seen intuitively that the latter is better than the former in terms of the estimation accuracies of both the position and the number of targets.

In order to obtain the statistical performance, 1000 Monte Carlo (MC) experiments are performed to compare the optimal sub-pattern assignment (OSPA) error for multi-target state estimation and the root mean square (RMS) error for the sensor bias estimation.

Figure 9.16 shows the OSPA under the condition of the order $p = 2$ and the threshold $c = 2\,\text{km}$, and Fig. 9.17 shows the RMS error of the sensor bias estimation. As can be seen from Fig. 9.16, the additional Doppler information can significantly improve the performance of the localization (Loc.) and cardinality (Card.) components of the OSPA. Therefore, the total OSPA distance gap between the GM-PHD-R and GM-PHD-R-D filters is relatively large and the results in Fig. 9.17 also verify that the introduction of Doppler information can improve the estimation accuracy of the sensor bias.

Figures 9.18, 9.19, 9.20 and 9.21 show the results when $\lambda_c \cdot V = 20$, and a similar improvement trend can be seen. Moreover, by comparing the results in Figs. 9.14,

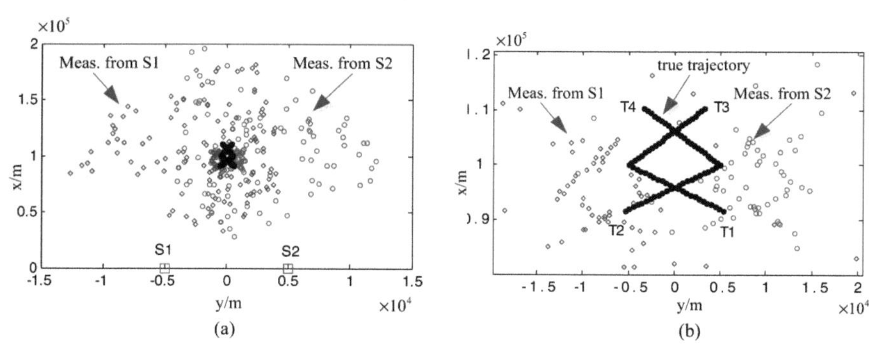

Fig. 9.14 Experimental scenario with 2 sensors and 4 targets ($\lambda_c \cdot V = 10$)

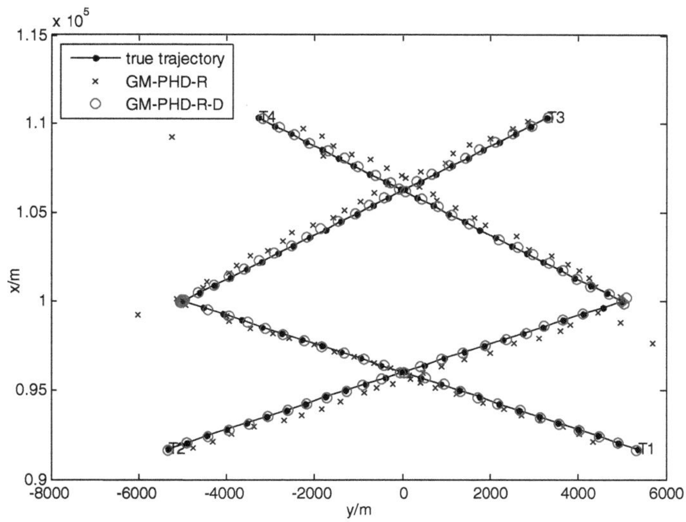

Fig. 9.15 Comparison of a typical position estimate at different times ($\lambda_c \cdot V = 10$)

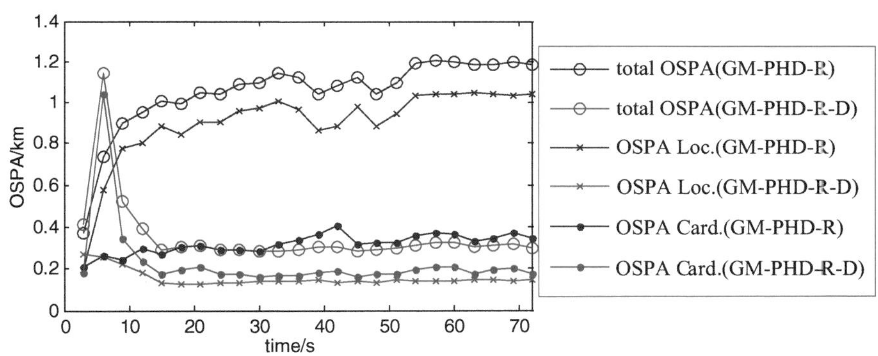

Fig. 9.16 Comparison of OSPA errors averaged on 1000 MC runs at different times ($\lambda_c \cdot V = 10$)

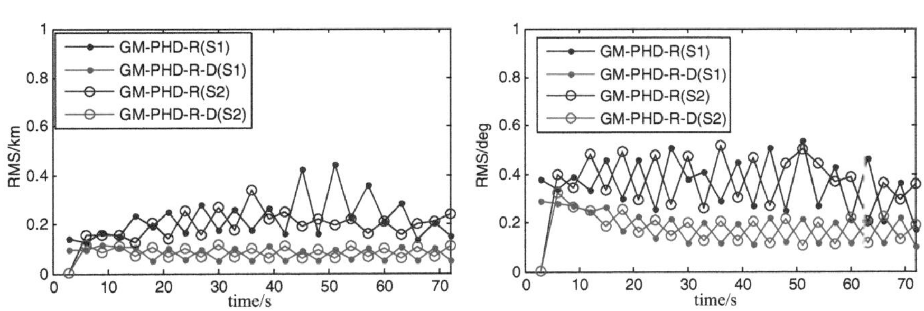

(a) RMS error of slant range bias

(b) RMS error of azimuth bias

Fig. 9.17 Comparison of sensor bias RMS errors averaged on 1000 MC runs at different times ($\lambda_c \cdot V = 10$)

9.15, 9.16 and 9.17 and Figs. 9.18, 9.19, 9.20 and 9.21, it can be seen that the OSPA localization performances of the GM-PHD-R and GM-PHD-R-D filters seem to maintain similar results as the clutter rate increases, and compared with the GM-PHD-R-D filter, the OSPA cardinality estimation performance and sensor bias estimation ability of the GM-PHD-R filter are more significantly decreased, which indicates that the introduction of Doppler information can effectively suppress the interference of clutter and improve the performance of multi-sensor multi-target tracking in dense clutter environment.

Figures 9.22 and 9.23 respectively show the time-averaged OSPA errors and sensor bias RMS errors at different clutter rates. It can be seen from these figures that for the GM-PHD-R and GM-PHD-R-D filters, the clutter mainly affects the OSPA

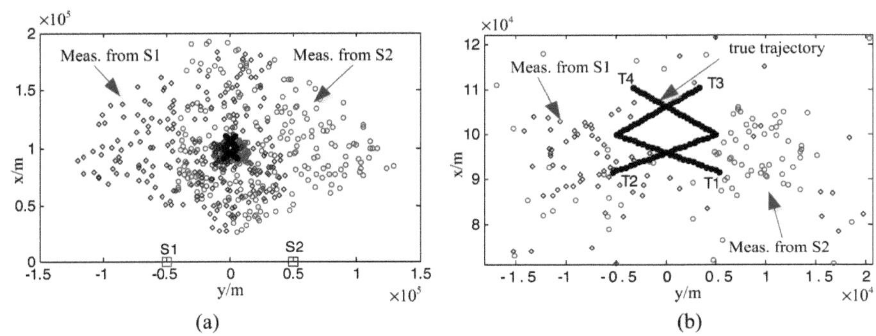

Fig. 9.18 Experimental scenario with 2 sensors and 4 targets ($\lambda_c \cdot V = 20$)

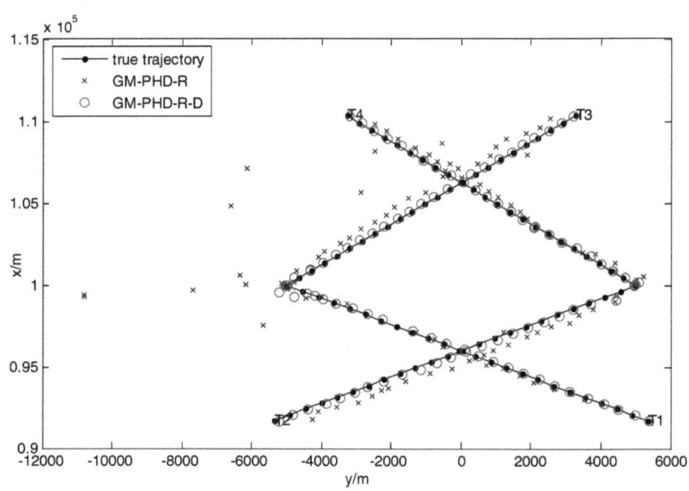

Fig. 9.19 Comparison of a typical position estimate at different times ($\lambda_c \cdot V = 20$)

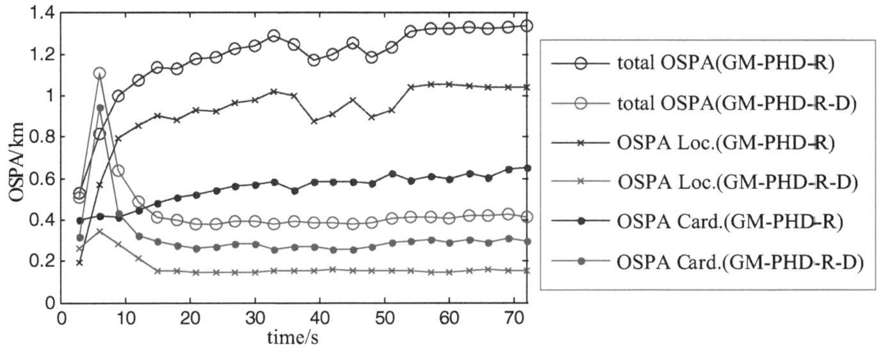

Fig. 9.20 Comparison of OSPA errors averaged on 1000 MC runs at different times ($\lambda_c \cdot V = 20$)

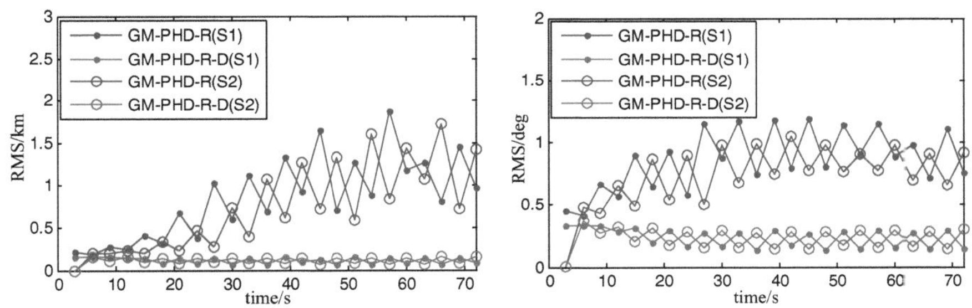

(a) RMS errors of slant range bias (b) RMS errors of azimuth bias

Fig. 9.21 Comparison of sensor bias RMS errors averaged on 1000 MC runs at different times ($\lambda_c \cdot V = 20$)

cardinality component and the estimation of the sensor bias. However, the GM-PHD-R filter is more sensitive than the GM-PHD-R-D filter, and in all cases, the GM-PHD-R-D filter is better than the GM-PHD-R filter due to the introduction of Doppler information.

The simulation results under the condition that the sensors are synchronous are similar to those under the asynchronous condition, and are not given due to space limitations. Therefore, the proposed algorithm is applicable to both synchronous and asynchronous sensors.

9.5 Summary

Aiming at the problem of target tracking with Doppler radars, this chapter, based on the RFS framework, presents the GM-CPHD filter with the Doppler measurement, GM-PHD filter with the MDV and Doppler measurement, and the augmented state

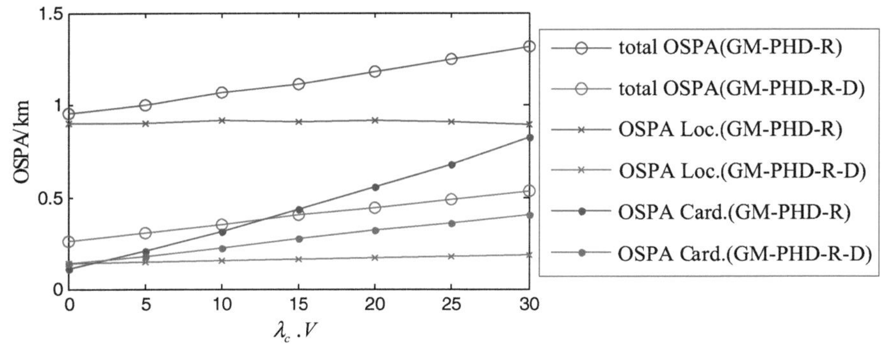

Fig. 9.22 Comparison of time-averaged OSPA errors at different clutter densities

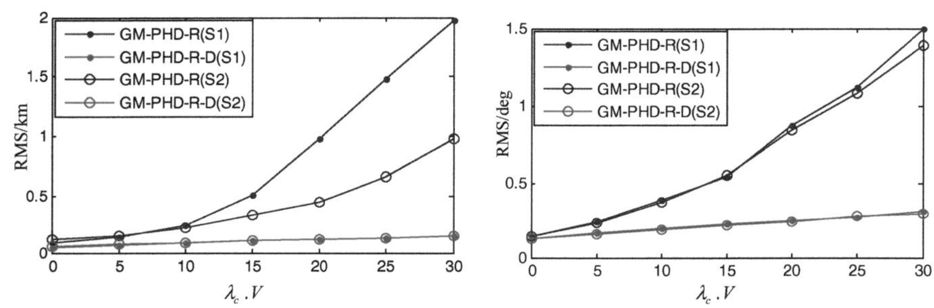

(a) RMS errors of slant range bias

(b) RMS errors of azimuth bias

Fig. 9.23 Comparison of time-averaged RMS errors of sensor bias at different clutter densities

GM-PHD filter with registration errors for networking Doppler radars. The simulation shows that the reasonable use of Doppler information and the DBZ-related MDV parameter can significantly improve the performance of Doppler radar target tracking in the presence of the DBZ. In addition, the augmented state GM-PHD filter with registration errors can effectively deal with the adverse effect of systematic errors on target tracking when Doppler radars are networked. It should be noted that although this chapter is mainly based on the PHD and CPHD filters, the method described can in principle be transplanted to other RFS filters.

Chapter 10
Track-Before-Detect for Dim Targets

10.1 Introduction

Track-before-detect (TBD) is an effective technique used for detection and tracking of dim targets. This technique does not claim the detection result regarding the presence or absence of a target based on the single-frame data. Instead, it firstly tracks a target according to hypothetical potential paths in the multi-frame data, filters clutters and constantly accumulates the target energy based on different characteristics of target echo, clutter and noise, and estimates the target trajectory at the time of target detection. Since the TBD sets no threshold or merely sets a lower threshold for the single-frame data, information of dim targets is retained as much as possible, thus avoiding the target loss problem faced by the traditional detect-before-track (DBT) method. From the perspective of the energy utilization, the TBD integrates detection and tracking. In this way, both single scanning pulse train coherent integration and inter-scanning non-coherent integration are used to improve the energy utilization efficiency. Therefore, the TBD is able to improve the capability of radar to detect dim and small targets.

There are many algorithms for the realization of this technique. Typical algorithms include time–space match filter [417], projection transformation [418], multiple hypothesis testing [419], particle filtering [420], and dynamic programming [421], and they can be divided into two categories: one is batch processing, which conducts joint treatment of multi-frame measurement data; another is recursive processing, which performs recursive iterative processing on multi-frame measurement data.

However, the above algorithms often aim at the single target case. In order to solve the multi-target TBD problem, most studies focus on the RFS algorithm. This chapter firstly presents multi-target TBD measurement model, then describes the analytical characteristics of multi-target posterior under four types of priors (e.g., Poisson prior, IIDC prior, MB prior and GLMB prior) based on the assumption of separable measurement likelihood. On this basis, the latter two are taken as examples. In other words, the MB filter-based TBD and labeled RFS filter-based TBD algorithms are introduced respectively.

W. Wu et al., *Target Tracking with Random Finite Sets*,
https://doi.org/10.1007/978-981-19-9815-7_10

10.2 Multi-target Track-Before-Detect (TBD) Measurement Model

10.2.1 TBD Measurement Likelihood and Its Separability

Let $\boldsymbol{x}_1, \ldots, \boldsymbol{x}_n \in \mathbb{X} \subseteq \mathbb{R}^d$ represent the state (or parameter) vector, and $Z = [z^{(1)}, \ldots, z^{(m)}]$ be the TBD measurement, which includes the values of m units (or pixels). Depending on the specific application, the ith pixel value $z^{(i)}$ may be a real number or a vector. For example, in a grey scale image, each pixel value is a real number, while in a colored image, each pixel value is a 3D vector representing the intensities of 3 color channels. Given the measurement Z, the TBD considers the joint estimation of target states and their number.

Let $T(\boldsymbol{x})$ denote the pixel set influenced by the target with state $\boldsymbol{x}$. For example, $T(\boldsymbol{x})$ may be the set of pixels centered on the target position within a certain distance. The value of pixel $i \in T(\boldsymbol{x})$ influenced by the target with state $\boldsymbol{x}$ obeys the distribution $p_T^{(i)}(\cdot|\boldsymbol{x})$, while the value of pixel $i \notin T(\boldsymbol{x})$ not influenced by any target follows the distribution $p_C^{(i)}(\cdot)$. More precisely, given state $\boldsymbol{x}$, the probability density of the value $z^{(i)}$ at the pixel i is

$$p(z^{(i)}|\boldsymbol{x}) = \begin{cases} p_T^{(i)}(z^{(i)}|\boldsymbol{x}), & i \in T(\boldsymbol{x}) \\ p_C^{(i)}(z^{(i)}), & i \notin T(\boldsymbol{x}) \end{cases} \tag{10.2.1}$$

For example, in the TBD, they are often as follows

$$p_C^{(i)}(z^{(i)}) = \mathcal{N}\big(z^{(i)}; 0, \sigma^2\big) \tag{10.2.2}$$

$$p_T^{(i)}(z^{(i)}|\boldsymbol{x}) = \mathcal{N}\big(z^{(i)}; h^{(i)}(\boldsymbol{x}), \sigma^2\big) \tag{10.2.3}$$

where $\mathcal{N}(\cdot; \mu, \sigma^2)$ indicates the Gaussian density of mean μ and variance σ^2, and $h^{(i)}(\boldsymbol{x})$ is the contribution of state $\boldsymbol{x}$ to pixel i and depends on the point spread function, target's position and reflected energy. It shall be noted that, Eq. (10.2.1) is also applicable to non-additive model.

Based on the following assumptions that:

- conditioned on the multi-target state, the pixel values are distributed independently;
- and the image regions influenced by targets don't overlap each other (whether the targets overlap is shown in Fig. 10.1), i.e., if $\boldsymbol{x} \neq \boldsymbol{x}'$, then $T(\boldsymbol{x}) \cap T(\boldsymbol{x}') = \emptyset$;

the multi-target likelihood function conditioned on multi-target state X is given by

$$g(Z|X) = \left(\prod_{\boldsymbol{x}\in X} \prod_{i\in T(\boldsymbol{x})} p_T^{(i)}(z^{(i)}|\boldsymbol{x})\right) \prod_{i\notin \cup_{\boldsymbol{x}\in X} T(\boldsymbol{x})} p_C^{(i)}(z^{(i)})$$

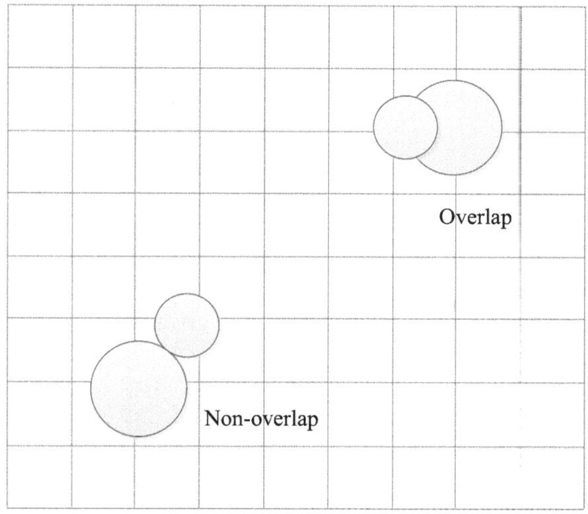

Fig. 10.1 Schematic diagram of target overlapping [249]

$$= g_C(Z) \prod_{x \in X} g_Z(\boldsymbol{x}) = g_C(Z) g_Z^X \tag{10.2.4}$$

where

$$g_Z(\boldsymbol{x}) = \prod_{i \in T(\boldsymbol{x})} p_T^{(i)}(z^{(i)}|\boldsymbol{x}) / p_C^{(i)}(z^{(i)}) \tag{10.2.5}$$

$$g_C(Z) = \prod_{i=1}^{m} p_C^{(i)}(z^{(i)}) \tag{10.2.6}$$

It is said that the multi-target likelihood function of the form (10.2.4) is separable. Generally, the true multi-target likelihood function is not separable; but, if targets don't overlap each other in the measurement space, separable likelihood assumption is a reasonable approximation.

10.2.2 Typical TBD Measurement Models

Taking the radar as an example and considering the fluctuation of the target echo, the Swerling fluctuation model is incorporated into the likelihood function and the target echo measured by the radar is assumed to follow the Swerling echo amplitude fluctuation model [422]. The radar located at the Cartesian origin collects m distance-Doppler-azimuth unit measurement $Z = [z^{(1)}, \ldots, z^{(m)}]$, which is composed of signal energy

$$z^{(i)} = |z_A^{(i)}|^2 \tag{10.2.7}$$

with $z_A^{(i)}$ being the complex signal at unit i, i.e.,

$$z_A^{(i)} = \sum_{x \in X} 1_{T(x)}(i) A(\boldsymbol{x}) h_A^{(i)}(\boldsymbol{x}) + \boldsymbol{n}^{(i)} \tag{10.2.8}$$

where $\boldsymbol{n}^{(i)}$ is the circularly symmetric complex Gaussian (CSCG) noise with zero mean and variance $2\sigma^2_{\boldsymbol{n}^{(i)}}$, and $h_A^{(i)}(\boldsymbol{x})$ represents the point spread function (PSF) over unit i influenced by target state $\boldsymbol{x}$, given by

$$h_A^{(i)}(\boldsymbol{x}) = \exp\left(-\frac{(r_i - r(\boldsymbol{x}))^2}{2R} - \frac{(d_i - d(\boldsymbol{x}))^2}{2D} - \frac{(b_i - b(\boldsymbol{x}))^2}{2B}\right) \tag{10.2.9}$$

in which, $r(\boldsymbol{x}) = \sqrt{p_x^2 + p_y^2}$, $d(\boldsymbol{x}) = -(\dot{p}_x p_x + \dot{p}_y p_y)/r(\boldsymbol{x})$, and $b(\boldsymbol{x}) = \operatorname{a\tan 2}(p_y, p_x)$ are respectively the distance, Doppler, and azimuth at a given target state $\boldsymbol{x}$, $[\, p_x \;\, p_y \;\, \dot{p}_x \;\, \dot{p}_y \,]$ represents the position and velocity components in respective dimensions, R, D and B are the distance, Doppler and azimuth resolutions, and r_i, d_i and b_i are the center of the corresponding unit.

$A(\boldsymbol{x})$ is the complex echo of target $\boldsymbol{x}$, given by

$$A(\boldsymbol{x}) = \overline{A}(\boldsymbol{x}) \exp(\iota\theta) + a(\boldsymbol{x}) \tag{10.2.10}$$

where ι stands for the imaginary unit, $\overline{A}(\boldsymbol{x})$ is a known amplitude, θ is an unknown phase uniformly distributed over $[0, 2\pi)$, and $a(\boldsymbol{x})$ is the complex Gaussian variable with zero mean and variance $\sigma^2_{a(\boldsymbol{x})}$. For the non-fluctuating (Swerling 0) target model, $A(\boldsymbol{x})$ degenerates to

$$A(\boldsymbol{x}) = \overline{A}(\boldsymbol{x}) \exp(\iota\theta) \tag{10.2.11}$$

Let $\hat{z}^{(i)} = \left|\hat{z}_A^{(i)}\right|^2$ be the power echo generated merely by the target at unit i, where

$$\hat{z}_A^{(i)} = \sum_{x \in X} 1_{T(x)}(i) \overline{A}(\boldsymbol{x}) h_A^{(i)}(\boldsymbol{x}) \tag{10.2.12}$$

Then, the power measurement over unit i is

$$\begin{aligned} z^{(i)} &= \left|\hat{z}_A^{(i)} \exp(\iota\theta) + \boldsymbol{n}^{(i)}\right|^2 \\ &= \left[\hat{z}_A^{(i)} \cos\theta + \mathcal{R}(\boldsymbol{n}^{(i)})\right]^2 + \left[\hat{z}_A^{(i)} \sin\theta + \mathcal{I}(\boldsymbol{n}^{(i)})\right]^2 \\ &= U_R^2 + U_I^2 \end{aligned} \tag{10.2.13}$$

where $U_R \sim \mathcal{N}(U_R; \hat{z}_A^{(i)} \cos\theta, \sigma^2_{n^{(i)}})$ and $U_I \sim \mathcal{N}(U_I; \hat{z}_A^{(i)} \sin\theta, \sigma^2_{n^{(i)}})$ are statistically independent normal random variables, and $\mathcal{R}(\cdot)$ and $\mathcal{I}(\cdot)$ represent the operations of taking the real part and the imaginary part, respectively. As a result, $\left|z_A^{(i)}\right| = \sqrt{U_R^2 + U_I^2}$ obeys the Ricean distribution, and when $\hat{z}_A^{(i)} = 0$, it degenerates into the Rayleigh distribution. Therefore, the measurement likelihood becomes

$$p_T^{(i)}\left(z^{(i)}|\boldsymbol{x}\right) = \frac{\sqrt{z^{(i)}}}{\sigma^2_{n^{(i)}}} \exp\left(-\frac{z^{(i)} + \hat{z}^{(i)}}{2\sigma^2_{n^{(i)}}}\right) I_0\left(\frac{\sqrt{z^{(i)}\hat{z}^{(i)}}}{\sigma^2_{n^{(i)}}}\right) \tag{10.2.14}$$

or

$$p_C^{(i)}\left(z^{(i)}\right) = \frac{\sqrt{z^{(i)}}}{\sigma^2_{n^{(i)}}} \exp\left(-\frac{z^{(i)}}{2\sigma^2_{n^{(i)}}}\right) \tag{10.2.15}$$

where $p_T^{(i)}(z^{(i)}|\boldsymbol{x})$ is the measurement likelihood when target $\boldsymbol{x}$ exists in unit i, $p_C^{(i)}(z^{(i)})$ is the likelihood under the assumption of no target, and $I_0(\cdot)$ is the (zero-order) modified Bessel function, given by [42]

$$I_0(x) = \frac{1}{2\pi}\int_0^{2\pi} \exp(x\cos\theta)d\theta = \sum_{j=0}^{\infty} \frac{(x^2/4)^j}{j!\Gamma(j+1)} \tag{10.2.16}$$

where $\Gamma(\cdot)$ is the Gamma function, $\Gamma(x) = \int_0^{\infty} t^{x-1}\exp(-t)dt$; for the integer j, $\Gamma(j+1) = j!$

Let SNR be the signal-to-noise ratio defined in dB form

$$\text{SNR} = 10\log\left[\overline{A}^2(\boldsymbol{x})/\left(2\sigma^2_{n^{(i)}}\right)\right] \tag{10.2.17}$$

When $\sigma^2_{n^{(i)}} = 1$, then $\overline{A}(\boldsymbol{x}) = \sqrt{2 \times 10^{\text{SNR}/10}}$. Since $\sqrt{U_R^2 + U_I^2} \sim \text{Rice}(\hat{z}_A^{(i)}, 1)$, the measurement $z^{(i)}$ of each unit follows the noncentral chi-square distribution with 2 degrees of freedom and non-central parameter $\hat{z}_A^{(i)}$, which degenerates into the central chi-square distribution with 2 degrees of freedom in case of $\hat{z}_A^{(i)} = 0$. Consequently, the likelihood ratio at unit i is

$$p_T^{(i)}\left(z^{(i)}|\boldsymbol{x}\right)/p_C^{(i)}\left(z^{(i)}\right) = \exp\left(-0.5\hat{z}^{(i)}\right) I_0\left(\sqrt{z^{(i)}\hat{z}^{(i)}}\right) \tag{10.2.18}$$

Assume that the measurement value at each unit is independently distributed, and given measurement Z, the likelihood function for multi-target state X is the product of the measurement likelihoods from all units, i.e.,

$$g(Z|\mathrm{X}) = g_C(Z)g_Z^X = g_C(Z)\prod_{x\in X} g_Z(\boldsymbol{x})$$
$$\propto \prod_{i\in\cup_{x\in X}T(\boldsymbol{x})} p_T^{(i)}(z^{(i)}|\boldsymbol{x})/p_C^{(i)}(z^{(i)}) \tag{10.2.19}$$

where $g_C(Z)$ and $g_Z(\boldsymbol{x})$ are given by (10.2.5) and (10.2.6), respectively.

10.3 Analytic Characteristics of Multi-target Posteriors

Since the measurement model doesn't exert any impact on the prediction step of a filter and it merely influences the update step, the following part only focuses on multi-target Bayesian update. With the concepts of the FISST integral and the FISST density, posterior probability density $\pi(\cdot|Z)$ of multi-target state can be obtained based on the prior π by using the following Bayes rule

$$\pi(X|Z) = \frac{g(Z|X)\pi(X)}{\int g(Z|X)\pi(X)\delta X} \tag{10.3.1}$$

where $g(Z|X)$ is the probability density of measurement Z given multi-target state X, i.e., multi-target likelihood function, whose specific expression is shown in (10.2.4).

According to (10.2.4), $g_C(Z)$ can be offset when calculating the posterior density $\pi(\cdot|Z)$, so that the posterior probability density of the multi-target state becomes

$$\pi(X|Z) = \frac{g_Z^X\pi(X)}{\int g_Z^{X'}\pi(X')\delta X'} \tag{10.3.2}$$

In the following part, analytical characteristics of the multi-target posterior distribution are provided for the aforementioned measurement model (10.2.4) and 4 types of multi-target priors (e.g., Poisson, IIDC, MB and GLMB distributions). First, the conclusion on the posterior PGFl under the condition of separable measurement likelihood is drawn, which allows an analytical description of posterior distributions of Poisson, IIDC, MB and GLMB RFSs.

By applying Bayes rule (10.3.1) and the PGFl definition, posterior probability density $\pi(\cdot|Z)$ can be obtained from the prior $\pi(\cdot)$ and its corresponding PGFl is

$$\begin{aligned} G[h|Z] &= \int h^X\pi(Z|X)\delta X \\ &= \frac{\int h^X g(Z|X)\pi(X)\delta X}{\int g(Z|X')\pi(X')\delta X'} = \frac{g_C(Z)\int [hg_Z]^X\pi(X)\delta X}{g_C(Z)\int g_Z^{X'}\pi(X')\delta X'} \\ &= \frac{G[hg_Z]}{G[g_Z]} \end{aligned} \tag{10.3.3}$$

Therefore, the following proposition can be obtained.

Proposition 28 *Assuming that X is a random finite set with prior PGFl G over $\mathbb{X}$ and the measurement Z is in the form of separable likelihood* (10.2.4), *the posterior PGFl $G[\cdot|Z]$ of X given Z is*

$$G[h|Z] = \frac{G[hg_Z]}{G[g_Z]} \tag{10.3.4}$$

10.3.1 Closed Form Measurement Update Under Poisson Prior

For a Poisson prior fully described by the PHD, let X be subject to a Poisson distribution with PHD of v and PGFl of $G[h] = \exp(\langle v, h-1\rangle)$, and according to Proposition 28, then we have

$$\begin{aligned} G[h|Z] &= \frac{G[hg_Z]}{G[g_Z]} = \frac{\exp(\langle v, hg_Z - 1\rangle)}{\exp(\langle v, g_Z - 1\rangle)} = \exp(\langle v, hg_Z - g_Z\rangle) \\ &= \exp(\langle vg_Z, h - 1\rangle) \end{aligned} \tag{10.3.5}$$

Therefore, the posterior is in Poisson form with PHD of vg_Z. As a consequence, the following corollary can be obtained.

Corollary 5 *Based on* Proposition 28, *if the prior distribution of RFS X is in Poisson form with PHD of v, the posterior distribution is also in Poisson form with PHD of $v(\cdot|Z)$ and the corresponding intensity is*

$$v(\boldsymbol{x}|Z) = v(\boldsymbol{x})g_Z(\boldsymbol{x}) \tag{10.3.6}$$

Corollary 5 shows how to update the PHD with measurement Z, i.e., how to calculate the posterior PHD based on the prior and the measurement. It shows that, the posterior PHD is equal to vg_Z and the posterior RFS follows a Poisson distribution. This conclusion can be applicable to the IIDC distributed RFS, which is fully characterized by the PHD and the cardinality distribution.

10.3.2 Closed Form Measurement Update Under IIDC Prior

For the IIDC distributed RFS, there is the following corollary.

Corollary 6 (see Appendix K for Proof) *Based on* Proposition 28, *if the prior distribution of X is IIDC with PHD of v and cardinality distribution of ρ, the posterior* *distribution is also IIDC with PHD of* $v(\cdot|Z)$ *and cardinality distribution of* $\rho(\cdot|Z)$, *and the corresponding PHD and cardinality distribution are respectively*

$$v(\boldsymbol{x}|Z) = v(\boldsymbol{x})g_Z(\boldsymbol{x})\frac{\sum_{i=0}^{\infty}(i+1)\rho(i+1)\langle v, g_Z\rangle^i/\langle v, 1\rangle^{i+1}}{\sum_{j=0}^{\infty}\rho(j)(\langle v, g_Z\rangle/\langle v, 1\rangle)^j} \tag{10.3.7}$$

$$\rho(n|Z) = \frac{\rho(n)(\langle v, g_Z\rangle/\langle v, 1\rangle)^n}{\sum_{j=0}^{\infty}\rho(j)(\langle v, g_Z\rangle/\langle v, 1\rangle)^j} \tag{10.3.8}$$

Corollary 5 is a special case of Corollary 6 when the cardinality is Poisson distributed. Corollaries 5 and 6 describe the posterior distribution through the PHD and cardinality distribution, while the subsequent Corollary 7 describes the posterior distribution through the set of existence probabilities and probability densities.

10.3.3 Closed Form Measurement Update Under Multi-Bernoulli Prior

For the MB RFS, let X be MB with the parameter set $\{(\varepsilon^{(i)}, p^{(i)})\}_{i=1}^N$ and PGFl $G[h] = \prod_{i=1}^N (1 - \varepsilon^{(i)} + \varepsilon^{(i)}\langle p^{(i)}, h\rangle)$. Based on Proposition 28, then we have

$$\begin{aligned} G[h|Z] &= \frac{G[hg_Z]}{G[g_Z]} = \frac{\prod_{i=1}^N \left(1 - \varepsilon^{(i)} + \varepsilon^{(i)}\langle p^{(i)}, hg_Z\rangle\right)}{\prod_{i=1}^N \left(1 - \varepsilon^{(i)} + \varepsilon^{(i)}\langle p^{(i)}, g_Z\rangle\right)} \\ &= \prod_{i=1}^N \frac{\left(1 - \varepsilon^{(i)} + \varepsilon^{(i)}\langle p^{(i)}g_Z, h\rangle\right)}{\left(1 - \varepsilon^{(i)} + \varepsilon^{(i)}\langle p^{(i)}, g_Z\rangle\right)} \\ &= \prod_{i=1}^N \left(1 - \frac{\varepsilon^{(i)}\langle p^{(i)}, g_Z\rangle}{1 - \varepsilon^{(i)} + \varepsilon^{(i)}\langle p^{(i)}, g_Z\rangle}\right. \\ &\quad \left. + \frac{\varepsilon^{(i)}\langle p^{(i)}, g_Z\rangle}{1 - \varepsilon^{(i)} + \varepsilon^{(i)}\langle p^{(i)}, g_Z\rangle}\left\langle \frac{p^{(i)}g_Z}{\langle p^{(i)}, g_Z\rangle}, h\right\rangle\right) \end{aligned} \tag{10.3.9}$$

The ith term in the above product is the PGFl of a Bernoulli RFS. Therefore, the posterior is an MB with the parameter set given by Eq. (10.3.10) and the following corollary is obtained.

Corollary 7 *Based on* Proposition 28, *if the prior distribution of* X *is an MB with the parameter set* $\{(\varepsilon^{(i)}, p^{(i)})\}_{i=1}^N$, *the posterior is also an MB and the corresponding parameter set is*

$$\left\{\left(\frac{\varepsilon^{(i)}\langle p^{(i)}, g_Z\rangle}{1 - \varepsilon^{(i)} + \varepsilon^{(i)}\langle p^{(i)}, g_Z\rangle}, \frac{p^{(i)}g_Z}{\langle p^{(i)}, g_Z\rangle}\right)\right\}_{i=1}^N \tag{10.3.10}$$

As long as the sensor likelihood function is separable and sensor measurements conditioned on the multi-target state are independent of each other, Corollaries 5–7 can be easily extended to the multi-sensor case. Assuming that the measurements of two independent sensors are $Z^{(1)}$ and $Z^{(2)}$, the multi-sensor multi-target posterior density is

$$\begin{aligned}\pi\left(X|Z^{(1)},Z^{(2)}\right) &\propto g\left(Z^{(2)}|X\right)g\left(Z^{(1)}|X\right)\pi(X)\\ &\propto g\left(Z^{(1)}|X\right)g\left(Z^{(2)}|X\right)\pi(X)\end{aligned} \tag{10.3.11}$$

Unlike the PHD or MB update approximation aiming for point measurements, the posterior parameter update in Corollaries 5–7 accurately capture the necessary and sufficient statistics of the posterior multi-target density. Similarly, for the multi-sensor update, sensor 1 is used to update the prior parameter, then the updated parameter is used as the new prior parameter and updated by sensor 2, so as to conduct iterative computation of the accurate posterior parameter. This step is repeated until the list of sensors are exhausted. Since each update is made in an accurate way, the final result is also accurate, regardless of the order in which the update is performed.

10.3.4 Closed Form Measurement Update Under GLMB Prior

It can be seen from above that, the Poisson, IIDC and MB densities are all conjugate with respect to the separable multi-target likelihood function [249]. Since the GLMB family is the extension of the (label-free) MB family, this conjugacy is also applicable to the GLMB family. In other words, the GLMB family is also the conjugate prior with respect to the separable likelihood function.

Given measurement Z, let the multi-target likelihood of state X be similar to that of Eq. (10.2.4) and have the following separable form

$$g(Z|\mathrm{X}) \propto g_Z^{\mathrm{X}} = \prod_{\mathbf{x}\in\mathrm{X}} g_Z(\mathbf{x}) \tag{10.3.12}$$

If the multi-target prior is in the GLMB form given by (3.3.38), according to the multi-target prior and multi-target likelihood (10.3.12), we can obtain

$$\begin{aligned}\pi(\mathrm{X}|Z) \propto \pi(\mathrm{X})g_Z^{\mathrm{X}} &= \Delta(\mathrm{X})\sum_{c\in\mathbb{C}} w^{(c)}(\mathcal{L}(\mathrm{X}))g_Z^{\mathrm{X}}\left[p^{(c)}\right]^{\mathrm{X}}\\ &= \Delta(\mathrm{X})\sum_{c\in\mathbb{C}} w^{(c)}(\mathcal{L}(\mathrm{X}))[\eta_Z]^{\mathcal{L}(\mathrm{X})}\frac{\left[g_Z p^{(c)}\right]^{\mathrm{X}}}{[\eta_Z]^{\mathcal{L}(\mathrm{X})}}\\ &= \Delta(\mathrm{X})\sum_{c\in\mathbb{C}} w_Z^{(c)}(\mathcal{L}(\mathrm{X}))\left[p^{(c)}(\cdot|Z)\right]^{\mathrm{X}}\end{aligned} \tag{10.3.13}$$

where

$$w_Z^{(c)}(L) = [\eta_Z]^L w^{(c)}(L) \tag{10.3.14}$$

$$p^{(c)}(\boldsymbol{x}, \ell|Z) = p^{(c)}(\boldsymbol{x}, \ell) g_Z(\boldsymbol{x}, \ell)/\eta_Z(\ell) \tag{10.3.15}$$

$$\eta_Z(\ell) = \langle p^{(c)}(\cdot, \ell), g_Z(\cdot, \ell)\rangle \tag{10.3.16}$$

Therefore, the following corollary can be obtained.

Corollary 8 *If the multi-target prior is in the GLMB form* (3.3.38) *and the multi-target likelihood is in separable form as shown in* (10.3.12)*, the multi-target posterior density is also in the GLMB form, given by*

$$\pi(\mathrm{X}|Z) \propto \Delta(\mathrm{X}) \sum_{c \in \mathbb{C}} w_Z^{(c)}(\mathcal{L}(\mathrm{X}))\big[p^{(c)}(\cdot|Z)\big]^{\mathrm{X}} \tag{10.3.17}$$

10.4 Multi-Bernoulli Filter-Based Track-Before-Detect

By modeling the state set as a random finite set, this section describes the problem of jointly detecting multiple targets and estimating their states from the TBD measurement under the Bayesian framework. Under the assumption that the measurement areas influenced by independent targets don't overlap each other and based on the analytic expression of the posterior distribution under the MB prior proposed in Sect. 10.3.3, a multi-target filter applicable to the TBD measurement at low signal-to-noise ratio is introduced. This track-before-detect filter, based on the MB filter, is able to conduct joint estimation of target states and their number according to the TBD measurement. Finally, the particle implementation of this multi-target filter is given.

In addition to using the above MB update (Corollary 7) to develop the multi-target filtering algorithm for TBD measurements, we also can use the PHD update (Corollary 5) or the IIDC update (Corollary 6) to develop similar algorithms. However, since the measurement model is highly non-linear, when calling the particle implementation to approximate the PHD, the clustering technique needs to be used to extract the states to be estimated from the particles. This clustering technique brings in the additional error sources and increases the computational burden, while the MB method avoids the above problems. Therefore, this section merely gives a description of MB-based TBD method, while Sect. 10.5 will introduce a more general labeled RFS-based TBD approach.

10.4.1 Multi-Bernoulli Filter for TBD Measurement Model

Assume that the target is small enough and the prior π_{k-1} is MB distributed, then the multi-target predicted density $\pi_{k|k-1}$ is also MB distributed. Furthermore, if targets don't overlap each other, according to Corollary 7, it can be obtained that the updated multi-target density $\pi_k(\cdot|Z)$ is also MB distributed. Therefore, the prediction and update steps are as follows.

MB prediction: since the TBD measurement model only influences the measurement update step, this step is exactly the same as that of the standard CBMeMBer filter, see Proposition 11 for details.

MB update: given predicted MB parameter $\pi_{k|k-1} = \{(\varepsilon_{k|k-1}^{(i)}, p_{k|k-1}^{(i)})\}_{i=1}^{M_{k|k-1}}$, the updated MB parameter is given by

$$\pi_k = \left\{(\varepsilon_k^{(i)}, p_k^{(i)})\right\}_{i=1}^{M_{k|k-1}} \tag{10.4.1}$$

where

$$\varepsilon_k^{(i)} = \frac{\varepsilon_{k|k-1}^{(i)}\left\langle p_{k|k-1}^{(i)}, g_Z\right\rangle}{1-\varepsilon_{k|k-1}^{(i)}+\varepsilon_{k|k-1}^{(i)}\left\langle p_{k|k-1}^{(i)}, g_Z\right\rangle} \tag{10.4.2}$$

$$p_k^{(i)} = p_{k|k-1}^{(i)} g_Z \Big/ \left\langle p_{k|k-1}^{(i)}, g_Z\right\rangle \tag{10.4.3}$$

Considering the assumption of targets no overlapping, the overlapping estimates should be merged.

10.4.2 SMC Implementation

Due to the strong nonlinearity of TBD measurement likelihood (10.2.4), the above MB prediction step and update step (10.4.1) generally adopt the sequential Monte Carlo (SMC) implementation method. The specific process is as follows.

SMC prediction: this step is exactly the same as that of the SMC implementation of the standard CBMeMBer filter (see Sect. 6.4.1 for details).

SMC update: it is assumed that the MB multi-target predicted density at time k is $\pi_{k|k-1} = \{(\varepsilon_{k|k-1}^{(i)}, p_{k|k-1}^{(i)})\}_{i=1}^{M_{k|k-1}}$ and each of $p_{k|k-1}^{(i)}, i = 1, \ldots, M_{k|k-1}$ is composed of a set of weighted samples $\{(w_{k|k-1}^{(i,j)}, \boldsymbol{x}_{k|k-1}^{(i,j)})\}_{j=1}^{L_{k|k-1}^{(i)}}$, i.e.,

$$p_{k|k-1}^{(i)}(\boldsymbol{x}) = \sum_{j=1}^{L_{k|k-1}^{(i)}} w_{k|k-1}^{(i,j)} \delta_{\boldsymbol{x}_{k|k-1}^{(i,j)}}(\boldsymbol{x}) \tag{10.4.4}$$

Then, the MB multi-target updated density (10.4.1) can be calculated as follows

$$\varepsilon_k^{(i)} = \frac{\varepsilon_{k|k-1}^{(i)} \eta_k^{(i)}}{1 - \varepsilon_{k|k-1}^{(i)} + \varepsilon_{k|k-1}^{(i)} \eta_k^{(i)}} \tag{10.4.5}$$

$$p_k^{(i)} = \frac{1}{\eta_k^{(i)}} \sum_{j=1}^{L_{k|k-1}^{(i)}} w_{k|k-1}^{(i,j)} g_{Z_k}\left(\boldsymbol{x}_{k|k-1}^{(i,j)}\right) \delta_{\boldsymbol{x}_{k|k-1}^{(i,j)}}(\boldsymbol{x}) \tag{10.4.6}$$

where

$$\eta_k^{(i)} = \sum_{j=1}^{L_{k|k-1}^{(i)}} w_{k|k-1}^{(i,j)} g_{Z_k}\left(\boldsymbol{x}_{k|k-1}^{(i,j)}\right) \tag{10.4.7}$$

Like the SMC implementation of the standard CBMeMBer filter, the particle resampling should be performed after the update step for each hypothetical target. The number of particles is re-assigned in a way proportional to the existence probability and is limited between the maximum $L_{\max}$ and minimum $L_{\min}$. In order to reduce the increasing number of tracks (or particles), it is necessary to discard particles with existence probabilities below a certain threshold T_h.

10.5 Mδ-GLMB Filter-Based Track-Before-Detect

In the following, the marginal δ-GLMB (Mδ-GLMB) filter is applied to multi-target tracking for the TBD measurement model and it doesn't assume that the multi-target likelihood function $g(\cdot|\cdot)$ has any special structure. Since the TBD measurement model only influences the measurement update step, the prediction step is exactly the same as that of the standard Mδ-GLMB filter, see the Mδ-GLMB prediction in Sect. 7.5.2 for details.

For the multi-target predicted density in the following Mδ-GLMB form

$$\pi_{k|k-1}(\mathrm{X}) = \Delta(\mathrm{X}) \sum_{I \in \mathcal{F}(\mathbb{L}_{0:k})} \delta_I(\mathcal{L}(\mathrm{X})) w_{k|k-1}^{(I)} \left[p_{k|k-1}^{(I)}\right]^{\mathrm{X}} \tag{10.5.1}$$

according to TBD measurement likelihood $g(Z|\mathrm{X})$, the multi-target posterior density can be obtained as follows

$$\pi_k(\mathrm{X}|\mathrm{Z}) = \Delta(\mathrm{X}) \sum_{I \in \mathcal{F}(\mathbb{L}_{0:k})} \delta_I(\mathcal{L}(\mathrm{X})) w_k^{(I)}(\mathrm{Z}) p_k^{(I)}(\mathrm{X}|\mathrm{Z}) \tag{10.5.2}$$

where

$$w_k^{(I)}(\mathrm{Z}) \propto w_{k|k-1}^{(I)} \eta_{\mathrm{Z}}(I) \tag{10.5.3}$$

$$p_k^{(I)}(\mathrm{X}|\mathrm{Z}) = \left[p_{k|k-1}^{(I)}\right]^{\mathrm{X}} g(\mathrm{Z}|\mathrm{X})/\eta_{\mathrm{Z}}(I) \tag{10.5.4}$$

$$\eta_{\mathrm{Z}}(\{\ell_1, \ldots, \ell_n\}) = \int g(\mathrm{Z}|\{(\boldsymbol{x}_1, \ell_1), \ldots, (\boldsymbol{x}_n, \ell_n)\}) \prod_{i=1}^{n} p_{k|k-1}^{(\{\ell_1, \ldots, \ell_n\})}(\boldsymbol{x}_i, \ell_i) d(\boldsymbol{x}_1, \ldots, \boldsymbol{x}_n) \tag{10.5.5}$$

It shall be noted that, according to Eq. (10.5.4), each multi-target exponential $[p_{k|k-1}^{(I)}]^{\mathrm{X}}$ from the prior δ-GLMB turns into $p_k^{(I)}(\mathrm{X}|\mathrm{Z})$ after the measurement update and it is not necessarily a multi-target exponential. As a result, in general, Eq. (10.5.2) will no longer provide the GLMB density.

(1) Separable likelihood case

If targets are well separable in the measurement space, the likelihood function can be approximated as a separable likelihood, i.e., $g(\mathrm{Z}|\mathrm{X}) \propto g_{\mathrm{Z}}^{\mathrm{X}}$. Thus, according to Corollary 8, the following approximate GLMB posterior can be obtained

$$\hat{\pi}_k(\mathrm{X}|\mathrm{Z}) = \Delta(\mathrm{X}) \sum_{I \in \mathcal{F}(\mathbb{L}_{0:k})} \delta_I(\mathcal{L}(\mathrm{X})) \hat{w}_k^{(I)}(\mathrm{Z}) \left[\hat{p}_k^{(I)}(\cdot|\mathrm{Z})\right]^{\mathrm{X}} \tag{10.5.6}$$

where

$$\hat{w}_k^{(I)}(\mathrm{Z}) \propto w_{k|k-1}^{(I)} [\eta_{\mathrm{Z}}]^I \tag{10.5.7}$$

$$\hat{p}_k^{(I)}(\boldsymbol{x}, \ell|\mathrm{Z}) = p_{k|k-1}^{(I)}(\boldsymbol{x}, \ell) g_{\mathrm{Z}}(\boldsymbol{x}, \ell)/\eta_{\mathrm{Z}}(\ell) \tag{10.5.8}$$

$$\eta_{\mathrm{Z}}(\ell) = \langle p_{k|k-1}^{(I)}(\cdot, \ell), g_{\mathrm{Z}}(\cdot, \ell)\rangle \tag{10.5.9}$$

(2) General case

On the contrary, if targets are close to each other and the separable likelihood assumption is no longer valid, the multi-target posterior of Eq. (10.5.2) shall be directly approximated, which can be written as follows

$$\pi_k(\mathrm{X}|\mathrm{Z}) = \Delta(\mathrm{X}) w_k^{(\mathcal{L}(\mathrm{X}))}(\mathrm{Z}) p_k^{(\mathcal{L}(\mathrm{X}))}(\mathrm{X}|\mathrm{Z}) \tag{10.5.10}$$

According to Proposition 21, in addition to minimizing the Kullback–Leibler divergence (KLD), the multi-target posterior that matches the cardinality and PHD of the above-mentioned multi-target posterior is approximated to

$$\hat{\pi}_k(\mathrm{X}|\mathrm{Z}) = \Delta(\mathrm{X}) \sum_{I \in \mathcal{F}(\mathbb{L}_{0:k})} \delta_I(\mathcal{L}(\mathrm{X})) w_k^{(I)}(\mathrm{Z}) \left[\hat{p}_k^{(I)}(\cdot|\mathrm{Z})\right]^{\mathrm{X}} \tag{10.5.11}$$

where, for each label set $I = \{\ell_1, \ldots, \ell_n\}$, $\hat{p}_k^{(\{\ell_1,\ldots,\ell_n\})}(\cdot, \ell_i|Z)$, $i = 1, \ldots, n$ is the marginal density of $p_k^{(\{\ell_1,\ldots,\ell_n\})}(\{(\cdot, \ell_1), \ldots, (\cdot, \ell_n)\}|Z)$ given by (10.5.4). We can notice that, the weight $w_k^{(I)}(Z)$ given by (10.5.3) is retained in (10.5.11) and it comes from the true posterior (10.5.2).

10.6 Summary

Based on the introduction of the multi-target TBD measurement model, this chapter presents the analytic-form measurement update expressions with respect to four types of priors (e.g., Poisson, IIDC, MB, and GLMB priors) under the condition that the multi-target likelihood function is separable, and describes two representative TBD algorithms under the RFS framework, ie., MB filter-based TBD and Mδ-GLMB filter-based TBD. Different from the typical TDB algorithms based on dynamic programming or particle filtering (which are mainly for the single target case), these algorithms here have a natural ability of multi-target tracking and can effectively track multiple dim targets.

Chapter 11
Target Tracking with Non-standard Measurements

11.1 Introduction

Multi-target tracking is the process of estimating the number of targets and their states based on imperfect sensor measurements contaminated by noises, missed detections and false alarms. The key challenges lie in the uncertainty of the number of targets and the uncertainty of the correlation between the measurement set and targets. The multi-target tracking is expected to filter out the influence of these three aspects in order to obtain an accurate estimation of true target states. A Bayesian solution to this problem requires a model describing the association between measurements and the underlying (unknown) target states. The vast majority of traditional trackers (such as the JPDA or MHT tracker) use the so-called standard measurement model: each target generates at most one measurement and each measurement originates from at most one target at a given time, which is the well-known point target model. In fact, in addition to traditional trackers, the RFS-based multi-target tracking algorithms introduced in the previous chapters are also based on the standard measurement model. This model assumption greatly simplifies the development of multi-target trackers and is also consistent with most real-world situations. However, in some special cases, it is an unrealistic representation of the real measurement process, typically manifested in extended target tracking and target tracking with merged measurement.

Since the measurement models under these conditions violate the standard measurement model, the aforementioned target tracking algorithms based on the standard measurement model are no longer valid, and therefore the target tracking algorithms with non-standard measurements need to be adopted in these cases. This chapter introduces two main types of target tracking algorithms with non-standard measurements under the RFS framework, namely, extended target tracking and target tracking with merged measurements. Considering that the extended target tracking algorithm is roughly divided into two categories: ignoring the estimation of the target extension and estimating the target extension, the GM-PHD-based extended target

W. Wu et al., *Target Tracking with Random Finite Sets*,
https://doi.org/10.1007/978-981-19-9815-7_11

tracking algorithm and GGIW-based extended target tracking algorithm are introduced respectively. The former is relatively simple and it only focuses on the target's reference point, ignoring the estimation of the target extension, while the latter is more complicated and it can estimate the target extension by using the Gamma Gaussian inverse Wishart (GGIW) distribution. Extended target measurements mean that a single target may produce multiple measurements. Conversely, merged measurements mean that multiple targets may generate a single measurement. As a result, after the introduction of extended target tracking, the target tracking algorithm with merged measurements is finally introduced.

11.2 GM-PHD Filter-Based Extended Target Tracking

This section simplifies the extended target tracking problem by assuming that measurements from one target are approximately distributed around the target's reference point [261], which may be the center of mass or any other point related to the target extension. Although all targets obviously have a certain spatial extension, the estimation of the target extension is temporarily ignored here, and only the target's reference point is concerned. Therefore, there is no information about target's size (dimension) and shape in the state vector. However, it must be emphasized that this does not undermine the generality. For example, similar to [423], the extended target GM-PHD (ET-GM-PHD) filter can be used to deal with the joint estimation of the sizes (dimensions), shapes and motion variables of square or elliptical extended targets. Moreover, for the sake of simplicity, the label information is not used here, however, similar to [424], it is allowed to be included to provide track continuity.

11.2.1 Extended Target Tracking Problem

The target-related characteristics to be estimated constitute the target state vector $\boldsymbol{x}$. Generally, in addition to such motion variables as position, velocity, and heading, the state vector may also include information about target's spatial extension. As mentioned earlier, when the target state does not contain any variables related to the target extension, although the target is estimated as a point target, the tracking algorithm still needs to consider the problem of multiple measurements originated from one target. As a result, the generalized definition of the extended target is adopted here (which is independent of whether or not the target extension needs to be estimated): an extended target may generate more than one measurement at each time.

Denote the set of target states to be estimated at time k as $X_k = \{\boldsymbol{x}_k^{(i)}\}_{i=1}^{|X_k|}$, and the set of measurements obtained at this time as $Z_k = \{\boldsymbol{z}_k^{(i)}\}_{i=1}^{|Z_k|}$, where $|X_k|$ is the number of unknown targets and $|Z_k|$ is the number of measurements. In RFS X_k, the dynamic evolution of each target state $\boldsymbol{x}_k^{(i)}$ can be modeled using the linear Gaussian dynamic model as follows

$$\boldsymbol{x}_{k+1}^{(i)} = \boldsymbol{F}_k \boldsymbol{x}_k^{(i)} + \boldsymbol{\Gamma}_k \boldsymbol{v}_k^{(i)}, \quad i = 1, \ldots, |X_k| \tag{11.2.1}$$

where $\boldsymbol{v}_k^{(i)}$ is the zero-mean Gaussian white noise with covariance $\boldsymbol{Q}_k^{(i)}$. The above equation assumes that each target state evolves according to the same dynamic model.

At each instant, the number of measurements produced by the ith target is a Poisson distributed random variable with a ratio of $\gamma(\boldsymbol{x}_k^{(i)})$ measurements per frame, where $\gamma(\cdot)$ is a known non-negative function defined over the target state space, then the probability that the ith target generates at least one measurement is

$$1 - \exp\left(-\gamma(\boldsymbol{x}_k^{(i)})\right) \tag{11.2.2}$$

Let the probability of the ith target being detected be $p_D(\boldsymbol{x}_k^{(i)})$, where $p_D(\cdot)$ is a known non-negative function defined over the target state space, then the effective detection probability can be obtained as

$$\left(1 - \exp\left(-\gamma(\boldsymbol{x}_k^{(i)})\right)\right) p_D\left(\boldsymbol{x}_k^{(i)}\right) \tag{11.2.3}$$

Assume that the measurements generated by each target are independent of other targets and the measurements of the ith target are related to the target state, each of which is modeled as the following linear Gaussian model

$$\boldsymbol{z}_k^{(j)} = \boldsymbol{H}_k \boldsymbol{x}_k^{(i)} + \boldsymbol{n}_k^{(j)} \tag{11.2.4}$$

where $\boldsymbol{n}_k^{(j)}$ is the zero-mean Gaussian white noise with covariance $\boldsymbol{R}_k$. It should be emphasized that under the RFS framework, both the measurement set Z_k and the target state set X_k are not labeled, so the relationship between targets and measurements is not known.

It is assumed that the clutter measurements are uniformly distributed in the surveillance area, and the number of clutter measurements generated at each moment is a Poisson distributed random variable with a ratio of κ_k clutter measurements per unit monitoring volume. Therefore, if the volume of the monitoring area is V_s, the expected number of clutter measurements is $\kappa_k V_s$ per frame.

The goal now is to obtain the estimation of the set of target states $X^K = \{X_k\}_{k=1}^K$ given a set of measurements $Z^K = \{Z_k\}_{k=1}^K$. This can be accomplished by using the PHD filter to propagate the predicted intensity $v_{k|k-1}$ and the updated intensity v_k of the set X_k of target states.

11.2.2 GM-PHD Filter for Extended Target Tracking

According to the derivation of the GM-PHD filter for the standard measurement model, the GM-PHD recursion in the extended target case can be obtained. Since the prediction equation of the extended target PHD filter is the same as that of the standard PHD filter, the Gaussian mixture (GM) prediction equation of the extended target PHD filter is also the same as that of the standard GM-PHD filter. As a result, only the measurement update equation of the extended target GM-PHD (ET-GM-PHD) filter is considered here.

The predicted intensity (PHD) is assumed to be in the following GM form

$$v_{k|k-1}(\boldsymbol{x}) = \sum_{i=1}^{J_{k|k-1}} w_{k|k-1}^{(i)} \mathcal{N}\left(\boldsymbol{x}; \boldsymbol{m}_{k|k-1}^{(i)}, \boldsymbol{P}_{k|k-1}^{(i)}\right) \tag{11.2.5}$$

where $J_{k|k-1}$ is the number of predicted components, $w_{k|k-1}^{(i)}$ is the weight of the ith component, and $\boldsymbol{m}_{k|k-1}^{(i)}$ and $\boldsymbol{P}_{k|k-1}^{(i)}$ are the predicted mean and covariance of the ith component, respectively. For the extended target Poisson model [263], the updated PHD is given by the product of the predicted PHD and the measurement pseudo-likelihood function L_{Z_k}, i.e., [273]

$$v_k(\boldsymbol{x}) = L_{Z_k}(\boldsymbol{x}) v_{k|k-1}(\boldsymbol{x}) \tag{11.2.6}$$

with the measurement pseudo-likelihood function L_{Z_k} being [95]

$$\begin{aligned} L_{Z_k}(\boldsymbol{x}) = 1 - \left[1 - \exp(-\gamma(\boldsymbol{x}))\right] p_D(\boldsymbol{x}) + \exp(-\gamma(\boldsymbol{x})) p_D(\boldsymbol{x}) \\ \times \sum_{\mathcal{P} \angle Z_k} w_{\mathcal{P}} \sum_{W \in \mathcal{P}} \frac{\gamma(\boldsymbol{x})^{|W|}}{\varpi_W} \prod_{z \in W} \frac{g(z|\boldsymbol{x})}{\lambda_k p_{C,k}(z)} \end{aligned} \tag{11.2.7}$$

where $\lambda_k = \kappa_k V_s$ is the expected number of clutter measurements, $p_{C,k}(z) = 1/V_s$ is the spatial distribution of clutters over the monitoring area, the symbol $\mathcal{P} \angle Z_k$ represents the partition $\mathcal{P}$ of measurement set Z_k (see Sect. 11.2.3 for the explanation of the partition), $W \in \mathcal{P}$ is the non-empty cell W constituting the partition $\mathcal{P}$, $w_{\mathcal{P}}$ and ϖ_W are non-negative coefficients for each partition $\mathcal{P}$ and each cell W, and $g(z|\boldsymbol{x})$ is the likelihood function of single-target measurement, which is assumed to be a Gaussian density here. The first summation operation on the right side of (11.2.7) traverses all partitions $\mathcal{P}$ of measurement set Z_k, while the second summation traverses all cells W in the current partition $\mathcal{P}$.

In order to derive the measurement update equation of the ET-GM-PHD filter, the same six assumptions as those for the GM-PHD filter in Sect. 4.4 are followed. Nevertheless, the fifth assumption (A.5) about the detection probability is relaxed as follows.

A.11: For all $\boldsymbol{x}$ and $i = 1, \ldots, J_{k|k-1}$, the following approximation of detection probability function $p_D(\cdot)$ holds [95]

$$p_D(\boldsymbol{x})\mathcal{N}\left(\boldsymbol{x}; \boldsymbol{m}_{k|k-1}^{(i)}, \boldsymbol{P}_{k|k-1}^{(i)}\right) \approx p_D\left(\boldsymbol{m}_{k|k-1}^{(i)}\right)\mathcal{N}\left(\boldsymbol{x}; \boldsymbol{m}_{k|k-1}^{(i)}, \boldsymbol{P}_{k|k-1}^{(i)}\right) \quad (11.2.8)$$

A.11 is easier to be satisfied than A.5. When $p_D(\cdot)$ is a constant, i.e., $p_D(\cdot) = p_D$, (11.2.8) often holds. Generally, A.11 approximately holds when the function $p_D(\cdot)$ doesn't change significantly in the uncertainty area determined by covariance $\boldsymbol{P}_{k|k-1}^{(i)}$. Moreover, A.11 also approximately holds when $p_D(\cdot)$ is a sufficiently smooth function or when the signal-to-noise ratio (SNR) is high enough so that $\boldsymbol{P}_{k|k-1}^{(i)}$ is sufficiently small. It shall be noted that, A.11 is just for the sake of simplicity rather than approximation, because $p_D(\cdot)$ can always be approximated by an exponential mixture of quadratic function (or equivalently approximated as a Gaussian form), which causes no damage to the GM structure of the updated PHD. However, this will lead to a multiplied increase in the Gaussian components in the updated PHD, which in turn makes the algorithm require more complex pruning and merging operations. In the GMTI target tracking, a similar method with variable detection probability has also been adopted in order to model the clutter notch [425].

In addition, the following assumption is made about the expected number $\gamma(\cdot)$ of measurements originating from a target.

A.12: For all $\boldsymbol{x}$, $n = 1, 2, \ldots$ and $i = 1, \ldots, J_{k|k-1}$, the following approximation of $\gamma(\cdot)$ holds [95]

$$\begin{aligned} &\exp(-\gamma(\boldsymbol{x}))\gamma^n(\boldsymbol{x})\mathcal{N}\left(\boldsymbol{x}; \boldsymbol{m}_{k|k-1}^{(i)}, \boldsymbol{P}_{k|k-1}^{(i)}\right) \\ &\approx \exp\left(-\gamma(\boldsymbol{m}_{k|k-1}^{(i)})\right)\gamma^n(\boldsymbol{m}_{k|k-1}^{(i)})\mathcal{N}\left(\boldsymbol{x}; \boldsymbol{m}_{k|k-1}^{(i)}, \boldsymbol{P}_{k|k-1}^{(i)}\right) \end{aligned} \quad (11.2.9)$$

Obviously, when $\gamma(\cdot)$ is a constant, i.e., $\gamma(\cdot) = \gamma$, it is a special case satisfying A.12. Generally, due to the presence of the exponential function, the Gaussian mixture assumption of $\gamma(\cdot)$ is difficult to hold, and it is more difficult to satisfy A.12 than to satisfy A.11. However, when $\gamma(\cdot)$ is smooth enough or when SNR is high enough to make $\boldsymbol{P}_{k|k-1}^{(i)}$ small enough, A.12 is expected to approximately hold.

Under the above assumptions, the posterior intensity at time k is in the following Gaussian mixture (GM) form [95]

$$v_k(\boldsymbol{x}) = v_{M,k}(\boldsymbol{x}) + \sum_{\mathcal{P}\angle Z_k} \sum_{W\in\mathcal{P}} v_{D,k}(\boldsymbol{x}; \boldsymbol{z}_W) \quad (11.2.10)$$

where $v_{M,k}(\boldsymbol{x})$ corresponds to the result of processing missed detections, given by

$$v_{M,k}(\boldsymbol{x}) = \sum_{i=1}^{J_{k|k-1}} w_{k|k}^{(i)}\mathcal{N}\left(\boldsymbol{x}; \boldsymbol{m}_{k|k}^{(i)}, \boldsymbol{P}_{k|k}^{(i)}\right) \quad (11.2.11)$$

$$w_{k|k}^{(i)} = \left\{1 - \left[1 - \exp(-\gamma^{(i)})\right] p_D^{(i)}\right\} w_{k|k-1}^{(i)} \tag{11.2.12}$$

$$\boldsymbol{m}_{k|k}^{(i)} = \boldsymbol{m}_{k|k-1}^{(i)}, \ \boldsymbol{P}_{k|k}^{(i)} = \boldsymbol{P}_{k|k-1}^{(i)} \tag{11.2.13}$$

with $\gamma^{(i)}$ and $p_D^{(i)}$ being the abbreviations of $\gamma(\boldsymbol{m}_{k|k-1}^{(i)})$ and $p_D(\boldsymbol{m}_{k|k-1}^{(i)})$, respectively.

$v_{D,k}(\boldsymbol{x}; \boldsymbol{z}_W)$ corresponds to the result of processing target detections and it is

$$v_{D,k}(\boldsymbol{x}; \boldsymbol{z}_W) = \sum_{i=1}^{J_{k|k-1}} w_{k|k}^{(i)} \mathcal{N}\left(\boldsymbol{x}; \boldsymbol{m}_{k|k}^{(i)}(\boldsymbol{z}_W), \boldsymbol{P}_{k|k}^{(i)}\right) \tag{11.2.14}$$

with the formula for calculating the weight $w_{k|k}^{(i)}$ being

$$w_{k|k}^{(i)} = w_{\mathcal{P}} \frac{\Gamma^{(i)} p_D^{(i)}}{\varpi_W} \varphi_W^{(i)} w_{k|k-1}^{(i)} \tag{11.2.15}$$

where

$$w_{\mathcal{P}} = \frac{\prod_{W \in \mathcal{P}} \varpi_W}{\sum_{\mathcal{P}' \angle Z_k} \prod_{W' \in \mathcal{P}'} \varpi_{W'}} \tag{11.2.16}$$

$$\varpi_W = \delta_{|W|}(1) + \sum_{i=1}^{J_{k|k-1}} \Gamma^{(i)} p_D^{(i)} \varphi_W^{(i)} w_{k|k-1}^{(i)} \tag{11.2.17}$$

$$\Gamma^{(i)} = \exp\left(-\gamma^{(i)}\right)\left(\gamma^{(i)}\right)^{|W|} \tag{11.2.18}$$

$$\varphi_W^{(i)} = g_W^{(i)} \prod_{\boldsymbol{z} \in W} \left[\lambda_k p_{C,k}(\boldsymbol{z})\right]^{-1} \tag{11.2.19}$$

In the above equations, the partition weight $w_{\mathcal{P}}$ can be interpreted as the probability that the partition is true, $|W|$ is the number of elements in W, δ is the Kronecker delta function, and the coefficient $g_W^{(i)}$ is

$$g_W^{(i)} = \mathcal{N}\left(\boldsymbol{z}_W; \boldsymbol{H}_W \boldsymbol{m}_{k|k-1}^{(i)}, \boldsymbol{H}_W \boldsymbol{P}_{k|k-1}^{(i)} \boldsymbol{H}_W^{\mathrm{T}} + \boldsymbol{R}_W\right) \tag{11.2.20}$$

where $\boldsymbol{z}_W$, $\boldsymbol{H}_W$ and $\boldsymbol{R}_W$ are respectively defined as

$$\boldsymbol{z}_W \triangleq \bigoplus_{\boldsymbol{z} \in W} \boldsymbol{z} \tag{11.2.21}$$

$$\boldsymbol{H}_W \triangleq \underbrace{\left[\boldsymbol{H}_k^{\mathrm{T}}, \boldsymbol{H}_k^{\mathrm{T}}, \ldots, \boldsymbol{H}_k^{\mathrm{T}}\right]^{\mathrm{T}}}_{|W|\, times} \tag{11.2.22}$$

$$\boldsymbol{R}_W \triangleq \text{blkdiag}\,\underbrace{(\boldsymbol{R}_k, \boldsymbol{R}_k, \ldots, \boldsymbol{R}_k)}_{|W|\,times} \tag{11.2.23}$$

with $\oplus$ being the column vector concatenation operator.

The mean $\boldsymbol{m}_{k|k}^{(i)}(z_W)$ and covariance $\boldsymbol{P}_{k|k}^{(i)}$ of the Gaussian component are updated using the standard Kalman measurement update equation, namely

$$\boldsymbol{m}_{k|k}^{(i)}(z_W) = \boldsymbol{m}_{k|k-1}^{(i)} + \boldsymbol{G}_k^{(i)}\left(z_W - \boldsymbol{\eta}_{k|k-1}^{(i)}\right) \tag{11.2.24}$$

$$\boldsymbol{P}_{k|k}^{(i)} = \left(\boldsymbol{I} - \boldsymbol{G}_k^{(j)}\boldsymbol{H}_W\right)\boldsymbol{P}_{k|k-1}^{(j)} \tag{11.2.25}$$

$$\boldsymbol{G}_k^{(j)} = \boldsymbol{P}_{k|k-1}^{(j)}\boldsymbol{H}_W^{\mathrm{T}}\left(\boldsymbol{S}_{k|k-1}^{(j)}\right)^{-1} \tag{11.2.26}$$

$$\boldsymbol{\eta}_{k|k-1}^{(j)} = \boldsymbol{H}_W\boldsymbol{m}_{k|k-1}^{(j)} \tag{11.2.27}$$

$$\boldsymbol{S}_{k|k-1}^{(j)} = \boldsymbol{H}_W\boldsymbol{P}_{k|k-1}^{(j)}\boldsymbol{H}_W^{\mathrm{T}} + \boldsymbol{R}_W \tag{11.2.28}$$

Remark In order to keep the number of Gaussian components at a computationally manageable level, pruning and merging are also required.

11.2.3 Partitioning of Measurement Set

As described earlier, a necessary step of the ET-GM-PHD filter is to partition the measurement set. Partitioning is very important because the same target can generate more than one measurement. A partition refers to partitioning the measurement set into non-empty subsets called cells, each of which can be interpreted as containing all measurements from the same source (or a target, or a clutter source). Strictly speaking in the mathematical sense, a partition of a set A is defined as a set of pairwise disjoint mutually exclusive non-empty subsets of A whose union is equal to A. The number of partitions of a set with cardinality n (with n elements) is given by the nth Bell number B_n [426]. The Bell number comes from the combinatorial mathematics and it is named after mathematician Eric Temple Bell, with a recursion equation of $B_{n+1} = \sum_{k=0}^{n} C_k^n B_k$. Starting from $B_0 = B_1 = 1$, the first several Bell numbers are: 1, 1, 2, 5, 15, 52, 203, 877, 4140, 21147, 115975, …. It can be seen that the Bell number B_n will increase rapidly with the increase of n. However, in the context of multi-target tracking, many less likely partitions can often be eliminated.

For ease of understanding, the partitioning process is illustrated by taking a measurement set $Z_k = \{z_k^{(1)}, z_k^{(2)}, z_k^{(3)}\}$ containing 3 independent measurements as an example. This 3-element set can be partitioned in $B_3 = 5$ different ways, namely,

$$
\begin{aligned}
&\mathcal{P}_1 : W_1^1 = \{z_k^{(1)}, z_k^{(2)}, z_k^{(3)}\};\\
&\mathcal{P}_2 : W_1^2 = \{z_k^{(1)}, z_k^{(2)}\},\ W_2^2 = \{z_k^{(3)}\};\\
&\mathcal{P}_3 : W_1^3 = \{z_k^{(1)}, z_k^{(3)}\},\ W_2^3 = \{z_k^{(2)}\};\\
&\mathcal{P}_4 : W_1^4 = \{z_k^{(2)}, z_k^{(3)}\},\ W_2^4 = \{z_k^{(1)}\};\\
&\mathcal{P}_5 : W_1^5 = \{z_k^{(1)}\},\ W_2^5 = \{z_k^{(2)}\},\ W_3^5 = \{z_k^{(3)}\}.
\end{aligned}
$$

where $\mathcal{P}_i$ is the ith partition, W_j^i is the jth cell of partition i, $|\mathcal{P}_i|$ is denoted as the number of cells in partition i, and $|W_j^i|$ denotes the number of measurements in cell j of partition i.

Note that the aforementioned set symbol implies that neither the order of partitions nor the order of elements within each partition is considered. This means that the following partitions are the same: $\{\{z_k^{(2)}\}, \{z_k^{(1)}, z_k^{(3)}\}\}$, $\{\{z_k^{(1)}, z_k^{(3)}\}, \{z_k^{(2)}\}\}$, $\{\{z_k^{(2)}\}, \{z_k^{(3)}, z_k^{(1)}\}\}$, $\{\{z_k^{(3)}, z_k^{(1)}\}, \{z_k^{(2)}\}\}$.

As mentioned earlier, as the dimension of the measurement set increases, the number of possible partitions becomes large. In order to obtain a computationally tractable target tracking method, only a subset of all possible partitions can be considered, which in turn must represent the most probable ones of all possible partitions in order to achieve good extended target tracking results. A heuristic method based on the distance between measurements (i.e., distance-based partitioning method) is introduced as follows to find out the most likely subset of partitions.

Consider a set of measurements $Z = \{z^{(i)}\}_{i=1}^{|Z|}$, where $z^{(i)} \in \mathbb{Z}$, $\forall i$, $\mathbb{Z}$ is the measurement space. Let $d(\cdot, \cdot) : \mathbb{Z} \times \mathbb{Z} \to [0, \infty)$ be the distance measure. For all $z^{(i)}, z^{(j)} \in \mathbb{Z}$, the measure satisfies the following two conditions

$$\text{Non-negativity:}\ d\big(z^{(i)}, z^{(j)}\big) \geq 0 \tag{11.2.29a}$$

$$\text{Symmetry:}\ d\big(z^{(i)}, z^{(j)}\big) = d\big(z^{(j)}, z^{(i)}\big) \tag{11.2.29b}$$

The distance-based partitioning method depends on the following proposition [274] (see Appendix L for its proof).

Proposition 29 *Let $T_l \geq 0$ be an arbitrary distance threshold, for each unique measurement set and each unique distance threshold, if and only if one of the following two conditions holds, there is a unique partition $\mathcal{P}^*$ that makes any pair of measurements lie in the same cell.*

① Condition 1

$$d(z^{(i)}, z^{(j)}) \leq T_l \tag{11.2.30}$$

② Condition 2

There exists a non-empty subset $\{z^{(r_1)}, z^{(r_2)}, \ldots, z^{(r_R)}\} \subset W - \{z^{(i)}, z^{(j)}\}$ such that the following condition hold:

When $R = 1$, then

$$d(z^{(i)}, z^{(r_1)}) \leq T_l \tag{11.2.31a}$$

$$d(z^{(r_1)}, z^{(j)}) \leq T_l \tag{11.2.31b}$$

Otherwise, when $R > 1$, then

$$d(z^{(i)}, z^{(r_1)}) \leq T_l \tag{11.2.31c}$$

$$d\left(z^{(r_s)}, z^{(r_{s+1})}\right) \leq T_l, \quad s = 1, 2, \ldots, R-1 \tag{11.2.31d}$$

$$d(z^{(r_R)}, z^{(j)}) \leq T_l \tag{11.2.31e}$$

It can be understood from the following two aspects that two measurements should be in the same cell if and only if one of the following two conditions is true [95]:

(1) The distance between them is less than this threshold (11.2.30);
(2) Through bridging of other measurements, stepping can be made between the two measurements, and the length of the step is less than this threshold (11.2.31).

Proposition 29 shows that, the partition is unique for each unique measurement set and each unique distance threshold. That is, there is a unique partition such that all measurement pairs (i, j) in the same cell satisfy $d(z^{(i)}, z^{(j)}) \leq T_l$.

This algorithm can be used to generate N_T alternative partitions of measurement set Z by choosing N_T different thresholds as follows

$$\{T_l\}_{l=1}^{N_T}, \; T_l < T_{l+1}, \; l = 1, \ldots, N_T - 1 \tag{11.2.32}$$

As T_l increases, there are fewer cells in alternative partitions and therefore there are usually more measurements in a cell.

Threshold $\{T_l\}_{l=1}^{N_T}$ can be selected from the following set

$$\mathcal{T} \triangleq \{0\} \cup \left\{d(z^{(i)}, z^{(j)}) | 1 \leq i < j \leq |Z|\right\} \tag{11.2.33}$$

where elements of $\mathcal{T}$ are arranged in ascending order. If all the elements in $\mathcal{T}$ are used to form alternative partitions, $|\mathcal{T}| = |Z|(|Z| - 1)/2 + 1$ partitions will be obtained. Since some partitions resulting from this selection may still be the same, it is necessary to discard them so that each final partition is unique. Among the unique partitions obtained, the first partition (corresponding to the threshold $T_1 = 0$) will contain $|Z|$ cells, in which case there is only one measurement per cell. The final partition will have one cell containing all $|Z|$ measurements.

The above partitioning strategy can significantly reduce the number of partitions. In order to further reduce the computational burden, the partitions can be calculated only for a subset of the threshold set $\mathcal{T}$. This subset is determined based on the statistical properties of the distance between measurements belonging to the same target. Assuming that the selected distance measure $d(\cdot, \cdot)$ is the Mahalanobis distance, i.e.,

$$d_M\left(z^{(i)}, z^{(j)}\right) = \sqrt{\left(z^{(i)} - z^{(j)}\right)^{\mathrm{T}} \boldsymbol{R}^{-1}\left(z^{(i)} - z^{(j)}\right)} \tag{11.2.34}$$

then, for two measurements ($z^{(i)}$ and $z^{(j)}$) generated from the same target, $d_M(z^{(i)}, z^{(j)})$ is a chi-square distribution χ^2 with degrees of freedom (DOF) equal to the dimension of the measurement vector. For a given probability p_G, using the inverse cumulative χ^2 distribution function invchi2$(\cdot)$, the dimensionless distance threshold can be calculated as

$$T_{p_G} = \text{invchi2}(p_G) \tag{11.2.35}$$

It has been shown that [95], according to the subset of distance thresholds in $\mathcal{T}$ that satisfy $T_{p_{\mathrm{L}}} < T_l < T_{p_{\mathrm{U}}}$, the calculated partitions can achieve a better target tracking result, with a lower probability $p_{\mathrm{L}} \leq 0.3$ and a higher probability $p_{\mathrm{U}} \geq 0.8$.

Take a simple example to illustrate. If there are 4 targets, the expected number of measurements is 20 for each target, and the mean number of clutter measurements is $\lambda_k = \kappa_k V_s = 50$, then the expected number of measurements collected at each moment will be 130. For 130 measurements, the number of all possible partitions is given by the Bell number $B_{130} \propto 10^{161}$. By using all thresholds in the set $\mathcal{T}$, 130 different partitions will be calculated on average, while by using the lower probability and the higher probability (i.e., $p_{\mathrm{L}} = 0.3$ and $p_{\mathrm{U}} = 0.8$), Monte Carlo simulation shows that, only 27 partitions will be computed on average [95], which implies that the computational complexity is reduced by several orders of magnitude.

When multiple (including two) extended targets are close to each other in space, the ET-GM-PHD filter will exhibit the cardinality underestimation problem of the target set. The reason for this is that, when targets are close to each other, the measurements they produce are also close. As a consequence, when the above distance-based partitioning method is used, the same cell W in all partitions $\mathcal{P}$ will contain measurements from more than one measurement source, and therefore the ET-GM-PHD filter will interpret the measurements from multiple targets as measurements from only one target. In an ideal scenario where all possible partitions of the measurement set are taken into consideration, there will be alternative partitions containing a subset of erroneously merged cells, which will dominate the output of the ET-GM-PHD filter and make it tend to the correct estimated number of targets. However, since these partitions are eliminated by using the distance-based partitioning method, the ET-GM-PHD filter lacks the means to correct the estimated number of targets.

One of the remedies for this problem is to perform additional subpartitioning after performing the distance-based partitioning and append it to the list of partitions considered by the ET-GM-PHD filter. Obviously, this is only done when there is a risk of merging measurements belonging to more than one target and whether this risk exists or not will be determined based on the expected number of measurements derived from one target.

It is assumed that a series of partitions have been computed using the distance-based partitioning method. For each obtained partition $\mathcal{P}_i$, the maximum likelihood estimation (MLE) $\hat{N}_{\boldsymbol{x}}^j$ of the number of targets is calculated for each cell W_j^i. If this estimation is greater than 1, the cell W_j^i will be divided into $\hat{N}_{\boldsymbol{x}}^j$ smaller cells, which are denoted as $\{W_s^+\}_{s=1}^{\hat{N}_{\boldsymbol{x}}^j}$. Then, new partitions (composed of new cells and other cells in $\mathcal{P}_i$) are added to the list of partitions obtained from the distance-based partitioning.

Table 11.1 presents the subpartitioning algorithm in which the split operation on a cell is done by the function split$(W_j^i, \hat{N}_{\boldsymbol{x}}^j)$.

In Table 11.1, the key is the calculation of the maximum likelihood estimation $\hat{N}_{\boldsymbol{x}}^j$ and the choice of the function split$(\cdot, \cdot)$.

(1) Calculate $\hat{N}_{\boldsymbol{x}}^j$. For this step, assume that the expected number $\gamma(\cdot)$ of measurements generated by a target is constant, i.e., $\gamma(\cdot) = \gamma$, the measurements produced by each target are independent of those produced by other targets, and the number of measurements produced by each target follows the Poisson distribution $\mathcal{PS}(\cdot, \gamma)$. Then, the likelihood function of the number of targets corresponding to cell W_j^i is

$$p\left(|W_j^i| \middle| N_{\boldsymbol{x}}^j = n\right) = \mathcal{PS}\left(|W_j^i|, \gamma n\right) \tag{11.2.36}$$

Table 11.1 Subpartitioning algorithm

No.	Input: the set $\{\mathcal{P}_1, \ldots, \mathcal{P}_{N_\mathcal{P}}\}$ of measurement partitions, where $N_\mathcal{P}$ is the number of partitions	
1:	Initialization: initialize the counter $c = N_\mathcal{P}$ of partitions	
2:	for $i = 1 : N_\mathcal{P}$	
3:	for $j = 1 : \lvert\mathcal{P}_i\rvert$	
4:	$\hat{N}_{\boldsymbol{x}}^j = \arg\max_n p(\lvert W_j^i\rvert \mid N_{\boldsymbol{x}}^j = n)$	
5:	if $\hat{N}_{\boldsymbol{x}}^j > 1$	
6:	$c = c + 1$	% increase the counter of partitions
7:	$\mathcal{P}_c = \mathcal{P}_i - W_j^i$	% eliminate current partition of current cell
8:	$\{W_k^+\}_{k=1}^{\hat{N}_{\boldsymbol{x}}^j} = \text{split}(W_j^i, \hat{N}_{\boldsymbol{x}}^j)$	% split the current cell
9:	$\mathcal{P}_c = \mathcal{P}_c \cup \{W_k^+\}_{k=1}^{\hat{N}_{\boldsymbol{x}}^j}$	% append to the current partition
10:	end	
11:	end	
12:		end

In addition, it is assumed that the volume covered by a cell is small enough that the number of false alarms in the cell is negligible, i.e., there are no false alarms in W_j^i. Then, the maximum likelihood estimation $\hat{N}_x^j$ can be calculated as

$$\hat{N}_x^j = \arg\max_n \mathcal{PS}(|W_j^i|, \gamma n) \tag{11.2.37}$$

(2) split(·, ·) function. Another important part of the subpartitioning function in Table 11.1 is the split(·, ·) function, which is used to split the measurements in one cell into smaller cells. The k-means clustering can be used to split the measurements in the cell. Of course, other methods of splitting the measurements can also be adopted.

Limitations of subpartitioning: the subpartitioning algorithm can be interpreted as only a first-order remedy for the above problem and, therefore, has only the limited correction capability. This is because the combination of cells is not considered when adding new partitions.

11.3 GGIW Distribution-Based Extended Target Tracking

As mentioned earlier, the ET-GM-PHD filter doesn't consider the estimation of the target extension. In order to estimate the extension (such as the ellipse) of the extended target, the GGIW model can be used. In the following part, the GGIW model for extended target is presented, and on this basis, the Bayesian filtering for single extended target as well as the CPHD and LMB filters for multiple extended targets are introduced respectively based on the GGIW distribution.

11.3.1 GGIW Model for a Extended Target

Let $\mathbb{R}^+$ represent the space of positive real numbers, $\mathbb{R}^n$ be the space of n-dimensional real vectors, $\mathbb{S}_+^n$ denote the space of symmetric positive semi-definite matrices of size $n \times n$, and $\mathbb{S}_{++}^n$ be the space of symmetric positive definite matrices of size $n \times n$. For a single extended target, it can be modeled as a GGIW probability distribution. Before introducing this distribution, we first introduce the following probability distributions.

- $\mathcal{GAM}(x_{\mathcal{R}}; \alpha, \beta)$: a Gamma probability density function (PDF) defined on $x_{\mathcal{R}} \in \mathbb{R}^+$ with shape parameter $\alpha > 0$ and inverse scale parameter $\beta > 0$

$$\mathcal{GAM}(x_{\mathcal{R}}; \alpha, \beta) = \frac{\beta^\alpha}{\Gamma(\alpha)} x_{\mathcal{R}}^{\alpha-1} \exp(-\beta x_{\mathcal{R}}) \tag{11.3.1}$$

where $\Gamma(\cdot)$ is the Gamma function.

- $\mathcal{N}(\boldsymbol{x}_\mathcal{K}; \boldsymbol{m}, \boldsymbol{P})$: a multi-dimensional Gaussian PDF defined on $\boldsymbol{x}_\mathcal{K} \in \mathbb{R}^n$ with mean $\boldsymbol{m} \in \mathbb{R}^n$ and covariance $\boldsymbol{P} \in \mathbb{S}^n_+$

$$\mathcal{N}(\boldsymbol{x}_\mathcal{K}; \boldsymbol{m}, \boldsymbol{P}) = \left(\sqrt{(2\pi)^n|\boldsymbol{P}|}\right)^{-1} \exp\left[-(\boldsymbol{x}_\mathcal{K} - \boldsymbol{m})^\mathrm{T}\boldsymbol{P}^{-1}(\boldsymbol{x}_\mathcal{K} - \boldsymbol{m})/2\right] \tag{11.3.2}$$

where $|\boldsymbol{P}|$ is the determinant of matrix $\boldsymbol{P}$.

- $\mathcal{IW}_d(\boldsymbol{P}_\mathcal{E}; u, \boldsymbol{U})$: an inverse Wishart (IW) distribution defined over matrix $\boldsymbol{P}_\mathcal{E} \in \mathbb{S}^d_{++}$ with degree of freedom $u > 2d$ and scale matrix $\boldsymbol{U} \in \mathbb{S}^d_{++}$ [279, 427]

$$\mathcal{IW}_d(\boldsymbol{P}_\mathcal{E}; u, \boldsymbol{U}) = \frac{2^{-d(u-d-1)/2}|\boldsymbol{U}|^{(u-d-1)/2}}{\Gamma_d((u-d-1)/2)|\boldsymbol{P}_\mathcal{E}|^{u/2}} \exp\left(-\mathrm{tr}(\boldsymbol{U}\boldsymbol{P}_\mathcal{E}^{-1})/2\right) \tag{11.3.3}$$

where $\Gamma_d(\cdot)$ is a multi-dimensional Gamma function with $\Gamma_d(a) = \pi^{d(d-1)/4}\prod_{i=1}^{d}\Gamma(a + (1-i)/2)$, and $\mathrm{tr}(\cdot)$ represents the trace of a matrix.

The extended target state is modeled as the following triple

$$\boldsymbol{x} = (x_\mathcal{R}, \boldsymbol{x}_\mathcal{K}, \boldsymbol{P}_\mathcal{E}) \in \mathbb{R}^+ \times \mathbb{R}^n \times \mathbb{S}^d_{++} \tag{11.3.4}$$

where $x_\mathcal{R} \in \mathbb{R}^+$ is the parameter of the measurement ratio of the Poisson distribution, which is used to model the number of measurements produced by the target, $\boldsymbol{x}_\mathcal{K} \in \mathbb{R}^n$ is the vector describing the kinetic state of the center of mass of the target, and $\boldsymbol{P}_\mathcal{E} \in \mathbb{S}^d_{++}$ is the covariance matrix describing the extension of the target around the center of mass. Thus, the purpose of extended target tracking is to estimate these 3 kinds of information about each target: the average number of measurements it produces, the kinetic state (dynamic for short), and the extension state. In order to construct the probability distribution of the extended target state, the following assumptions are made.

A.13: An extended target is characterized by the number of scattering points diffused over its extension [276], and the number of measurements produced by each extended target follows a Poisson distribution, while its ratio parameter (or mean) $x_\mathcal{R}$ follows a Gamma distribution.

The following assumptions about the dynamic and extension of the extended target were firstly proposed by Koch and have been widely adopted.

A.14: The dynamic $\boldsymbol{x}_\mathcal{K}$ (e.g., position, velocity and acceleration) and extension $\boldsymbol{P}_\mathcal{E}$ (e.g., shape and size) of the extended target can be decomposed into a random vector and a random matrix [264, 276, 428].

According to this assumption, the extension is modeled as a random matrix, which means that the extended target is restricted to have the shape of an ellipse. For comments on the rationality of this model, see [264, 265, 276].

A.15: The measurement ratio $x_{\mathcal{R}}$ is conditionally independent of $\boldsymbol{x}_{\mathcal{K}}$ and $\boldsymbol{P}_{\mathcal{E}}$.

According to the above assumptions, by modeling densities of the ratio parameter, target dynamic and extension covariance as the Gamma, Gaussian and inverse Wishart distributions respectively, it can be obtained that the density of the extended target state is the product of these three distributions, and the result obtained is the gamma Gaussian inverse Wishart (GGIW) distribution on space $\mathbb{R}^{+} \times \mathbb{R}^{n} \times \mathbb{S}_{++}^{d}$ [264, 279]

$$
\begin{aligned}
p(\boldsymbol{x}) &\triangleq \mathcal{GGIW}(\boldsymbol{x}; \boldsymbol{\zeta}) = p(x_{\mathcal{R}}) p(\boldsymbol{x}_{\mathcal{K}} | \boldsymbol{P}_{\mathcal{E}}) p(\boldsymbol{P}_{\mathcal{E}}) \\
&= \mathcal{GAM}(x_{\mathcal{R}}; \alpha, \beta) \times \mathcal{N}(\boldsymbol{x}_{\mathcal{K}}; \boldsymbol{m}, \boldsymbol{P} \otimes \boldsymbol{P}_{\mathcal{E}}) \times \mathcal{IW}_{d}(\boldsymbol{P}_{\mathcal{E}}; u, \boldsymbol{U}) \qquad (11.3.5)
\end{aligned}
$$

where $\boldsymbol{\zeta} = (\alpha, \beta, \boldsymbol{m}, \boldsymbol{P}, u, \boldsymbol{U})$ is an array that encapsulates the parameters of the GGIW density, $\boldsymbol{A} \otimes \boldsymbol{B}$ represents the Kronecker product of matrix $\boldsymbol{A}$ and matrix $\boldsymbol{B}$, and the Gaussian covariance is $(\boldsymbol{P} \otimes \boldsymbol{P}_{\mathcal{E}}) \in \mathbb{S}_{+}^{n}$, where $\boldsymbol{P} \in \mathbb{S}_{+}^{n/d}$.

11.3.2 *GGIW Distribution-Based Bayesian Filter for Single Extended Target*

For the GGIW distribution of a single extended target, the Bayesian prediction and update steps are as follows.

11.3.2.1 GGIW Prediction

In the Bayesian state estimation, the predicted density $p_{+}(\cdot)$ of a single extended target is given by the following Chapman–Kolmogorov (C–K) equation

$$
p_{+}(\boldsymbol{x}) = \int \phi(\boldsymbol{x} | \boldsymbol{x}') p(\boldsymbol{x}') d\boldsymbol{x}' \qquad (11.3.6)
$$

where $p(\boldsymbol{x}') = \mathcal{GGIW}(\boldsymbol{x}'; \boldsymbol{\zeta}')$ is the previous posterior density with parameter $\boldsymbol{\zeta}' = (\alpha', \beta', \boldsymbol{m}', \boldsymbol{P}', u', \boldsymbol{U}')$, and $\phi(\cdot|\cdot)$ is the transition density from the previous time to the current time. Since there is generally no closed form solution to the above equation, a GGIW approximation is required for $p_{+}(\boldsymbol{x})$. Assuming that the extended target state transition density can be written as the following product form [264, 279]

$$
\begin{aligned}
\phi(\boldsymbol{x} | \boldsymbol{x}') &\approx \phi_{\mathcal{R}}(x_{\mathcal{R}} | x'_{\mathcal{R}}) \phi_{\mathcal{K},\mathcal{E}}(\boldsymbol{x}_{\mathcal{K}}, \boldsymbol{P}_{\mathcal{E}} | \boldsymbol{x}'_{\mathcal{K}}, \boldsymbol{P}'_{\mathcal{E}}) \\
&\approx \phi_{\mathcal{R}}(x_{\mathcal{R}} | x'_{\mathcal{R}}) \phi_{\mathcal{K}}(\boldsymbol{x}_{\mathcal{K}} | \boldsymbol{x}'_{\mathcal{K}}, \boldsymbol{P}_{\mathcal{E}}) \phi_{\mathcal{E}}(\boldsymbol{P}_{\mathcal{E}} | \boldsymbol{P}'_{\mathcal{E}}) \qquad (11.3.7)
\end{aligned}
$$

then the predicted density can be obtained as

$$
\begin{aligned}
p_{+}(\boldsymbol{x}) = & \int \mathcal{GAM}(x'_{\mathcal{R}}; \alpha', \beta') \phi_{\mathcal{R}}(x_{\mathcal{R}} | x'_{\mathcal{R}}) dx'_{\mathcal{R}} \\
& \times \int \mathcal{N}(\boldsymbol{x}'_{\mathcal{K}}; \boldsymbol{m}', \boldsymbol{P}' \otimes \boldsymbol{P}_{\mathcal{E}}) \phi_{\mathcal{K}}(\boldsymbol{x}_{\mathcal{K}} | \boldsymbol{x}'_{\mathcal{K}}, \boldsymbol{P}_{\mathcal{E}}) d\boldsymbol{x}'_{\mathcal{K}} \\
& \times \int \mathcal{IW}_d(\boldsymbol{P}'_{\mathcal{E}}; u', \boldsymbol{U}') \phi_{\mathcal{E}}(\boldsymbol{P}_{\mathcal{E}} | \boldsymbol{P}'_{\mathcal{E}}) d\boldsymbol{P}'_{\mathcal{E}}
\end{aligned} \tag{11.3.8}
$$

If the dynamic model is in the following linear Gaussian form

$$
\phi_{\mathcal{K}}(\boldsymbol{x}_{\mathcal{K}} | \boldsymbol{x}'_{\mathcal{K}}, \boldsymbol{P}_{\mathcal{E}}) = \mathcal{N}(\boldsymbol{x}_{\mathcal{K}}; (\boldsymbol{F} \otimes \boldsymbol{I}_d)\boldsymbol{x}'_{\mathcal{K}}, \boldsymbol{Q} \otimes \boldsymbol{P}_{\mathcal{E}}) \tag{11.3.9}
$$

the following closed-form solution can be obtained for the motion component (i.e., the second line in (11.3.8))

$$
\int \mathcal{N}(\boldsymbol{x}'_{\mathcal{K}}; \boldsymbol{m}', \boldsymbol{P}' \otimes \boldsymbol{P}_{\mathcal{E}}) \phi_{\mathcal{K}}(\boldsymbol{x}_{\mathcal{K}} | \boldsymbol{x}'_{\mathcal{K}}, \boldsymbol{P}_{\mathcal{E}}) d\boldsymbol{x}'_{\mathcal{K}} = \mathcal{N}(\boldsymbol{x}_{\mathcal{K}}; \boldsymbol{m}, \boldsymbol{P} \otimes \boldsymbol{P}_{\mathcal{E}}) \tag{11.3.10}
$$

where

$$
\boldsymbol{m} = (\boldsymbol{F} \otimes \boldsymbol{I}_d)\boldsymbol{m}', \quad \boldsymbol{P} = \boldsymbol{F}\boldsymbol{P}'\boldsymbol{F}^{\mathrm{T}} + \boldsymbol{Q} \tag{11.3.11}
$$

$$
\boldsymbol{F} = \begin{bmatrix} 1 & T_s & T_s^2/2 \\ 0 & 1 & T_s \\ 0 & 0 & \exp(-T_s/\theta) \end{bmatrix} \tag{11.3.12}
$$

$$
\boldsymbol{Q} = \sigma_a^2 \cdot \left[1 - \exp(-2T_s/\theta)\right] \cdot \mathrm{blkdiag}\left(\left[\,0\ 0\ 1\,\right]\right) \tag{11.3.13}
$$

In the above, $\boldsymbol{I}_d$ is the identity matrix with dimension d, T_s is the sampling time, σ_a is the standard deviation of the scalar acceleration, and θ is the maneuvering correlation time.

However, for the components of the measurement ratio and the target extension, the closed form solution cannot be directly obtained, and some approximations are required.

For the component of the measurement ratio, the following approximation can be used

$$
\int \mathcal{GAM}(x'_{\mathcal{R}}; \alpha', \beta') \phi_{\mathcal{R}}(x_{\mathcal{R}} | x'_{\mathcal{R}}) dx'_{\mathcal{R}} \approx \mathcal{GAM}(x_{\mathcal{R}}; \alpha, \beta) \tag{11.3.14}
$$

where

$$
\alpha = \alpha'/\mu, \quad \beta = \beta'/\mu \tag{11.3.15}
$$

In the above equation, $\mu = 1/(1 - 1/w) = w/(w - 1)$ is an exponential forgetting factor with a window length of $w > 1$. The above approximation is based on heuristic assumptions $\mathrm{E}[x_{\mathcal{R}}] = \mathrm{E}[x'_{\mathcal{R}}]$ and $\mathrm{Var}[x_{\mathcal{R}}] = \mathrm{Var}[x'_{\mathcal{R}}] \times \mu$, i.e., the prediction operator preserves the expected value of the density and scales the variance by a factor μ.

For the extension component, the following approximation is used [264]

$$\int \mathcal{IW}_d(\boldsymbol{P}'_{\mathcal{E}}; u', \boldsymbol{U}')\phi_{\mathcal{E}}(\boldsymbol{P}_{\mathcal{E}}|\boldsymbol{P}'_{\mathcal{E}})d\boldsymbol{P}'_{\mathcal{E}} \approx \mathcal{IW}_d(\boldsymbol{P}_{\mathcal{E}}; u, \boldsymbol{U}) \tag{11.3.16}$$

where

$$u = \exp(-T_s/\tau)u', \quad \boldsymbol{U} = \frac{u-d-1}{u'-d-1}\boldsymbol{U}' \tag{11.3.17}$$

Similar to the measurement ratio, the above approximation assumes that the prediction preserves the expected value and reduces the precision of the density. For the inverse Wishart distribution, the degree of freedom (DOF) parameter u is related to the accuracy, with a lower value leading to a less accurate density. As a result, the time decay constant τ is used in (11.3.17) in order to control the reduction of the DOF. Based on the calculated value of u, the expression of $\boldsymbol{U}$ preserves the expected value of the inverse Wishart distribution through the prediction.

In summary, the above equation is the approximate expression for the predicted GGIW density $p_+(\boldsymbol{x}) \approx \mathcal{GGIW}(\boldsymbol{x}; \boldsymbol{\zeta})$, where $\boldsymbol{\zeta} = (\alpha, \beta, \boldsymbol{m}, \boldsymbol{P}, u, \boldsymbol{U})$ is the array of predicted parameters defined by (11.3.10), (11.3.14) and (11.3.16).

11.3.2.2 GGIW Update

The different subsets of measurements received each frame are used to perform the measurement update for each extended target. For a given set of measurements W generated by an extended target, a single target measurement update with the predicted density $p(\cdot) = \mathcal{GGIW}(\cdot; \boldsymbol{\zeta})$ is described as follows. It should be noted that, different from the GGIW prediction, the GGIW update has an exact closed-form and does not require any approximations.

Assume that each independent detection in W is generated according to the following model

$$z_k = (\boldsymbol{H} \otimes \boldsymbol{I}_d)\boldsymbol{x}_{\mathcal{K},k} + \boldsymbol{n}_k \tag{11.3.18}$$

where $\boldsymbol{H} = [1\ \ 0\ \ 0]$, $\boldsymbol{n}_k \sim \mathcal{N}(\boldsymbol{n}_k; \boldsymbol{0}, \boldsymbol{P}_{\mathcal{E},k})$ is the IID Gaussian measurement noise whose covariance is given by the target extension matrix $\boldsymbol{P}_{\mathcal{E},k}$. If a target is detected, the target will generate a set of measurements $W = \{z_1, \ldots, z_{|W|}\}$ according to the model (11.3.18), and its cardinality $|W|$ follows a Poisson distribution. Based on this model, the single-target likelihood function is [276]

$$\tilde{g}(W|\boldsymbol{x}) = \mathcal{PS}(|W|; x_{\mathcal{R}}) \prod_{j=1}^{|W|} \mathcal{N}\big(\boldsymbol{z}_j; (\boldsymbol{H} \otimes \boldsymbol{I}_d)\boldsymbol{x}_{\mathcal{K}}, \boldsymbol{P}_{\mathcal{E}}\big) \tag{11.3.19}$$

where $\mathcal{PS}(\cdot\,; \lambda)$ represents the Poisson PDF with mean λ. Note that for the multi-extended target likelihood, (11.3.19) needs to be substituted into (11.3.85).

Assume that the predicted density $p(\boldsymbol{x})$ is the GGIW given by (11.3.8), according to the Bayesian rule, the posterior density can be calculated as follows

$$p(\boldsymbol{x}|W) = \frac{\tilde{g}(W|\boldsymbol{x})p(\boldsymbol{x})}{\int \tilde{g}(W|\boldsymbol{x}')p(\boldsymbol{x}')d\boldsymbol{x}'} \tag{11.3.20}$$

where the numerator is

$$\begin{aligned}\tilde{g}(W|\boldsymbol{x})p(\boldsymbol{x}) = {} & \mathcal{PS}(|W|; x_{\mathcal{R}}) \prod_{j=1}^{|W|} \mathcal{N}\big(\boldsymbol{z}_j; (\boldsymbol{H} \otimes \boldsymbol{I}_d)\boldsymbol{x}_{\mathcal{K}}, \boldsymbol{P}_{\mathcal{E}}\big) \\ & \times \mathcal{GAM}(x_{\mathcal{R}}; \alpha, \beta)\mathcal{N}(\boldsymbol{x}_{\mathcal{K}}; \boldsymbol{m}, \boldsymbol{P} \otimes \boldsymbol{P}_{\mathcal{E}})\mathcal{IW}_d(\boldsymbol{P}_{\mathcal{E}}; u, \boldsymbol{U})\end{aligned} \tag{11.3.21}$$

Since the measurement ratio component is independent of the dynamic-extension component, it can be handled separately, so that the measurement ratio component in (11.3.21) can be obtained as follows

$$\mathcal{PS}(|W|; x_{\mathcal{R}})\mathcal{GAM}(x_{\mathcal{R}}; \alpha, \beta) = \eta_{\mathcal{R}}(W; \boldsymbol{\zeta})\mathcal{GAM}(x_{\mathcal{R}}; \alpha_W, \beta_W) \tag{11.3.22}$$

where

$$\alpha_W = \alpha + |W| \tag{11.3.23}$$

$$\beta_W = \beta + 1 \tag{11.3.24}$$

$$\eta_{\mathcal{R}}(W; \boldsymbol{\zeta}) = \frac{1}{|W|!} \frac{\Gamma(\alpha_W)\beta^{\alpha}}{\Gamma(\alpha)(\beta_W)^{\alpha_W}} \tag{11.3.25}$$

As a result, the integral of the denominator in (11.3.20) with respect to $x_{\mathcal{R}}$ is

$$\int \mathcal{PS}(|W|; x_{\mathcal{R}})\mathcal{GAM}(x_{\mathcal{R}}; \alpha, \beta)dx_{\mathcal{R}} = \eta_{\mathcal{R}}(W; \boldsymbol{\zeta}) \tag{11.3.26}$$

On the other hand, the joint dynamic-extension component in (11.3.21) is [276]

$$\prod_{j=1}^{|W|} \mathcal{N}\big(\boldsymbol{z}_j; (\boldsymbol{H} \otimes \boldsymbol{I}_d)\boldsymbol{x}_{\mathcal{K}}, \boldsymbol{P}_{\mathcal{E}}\big)\mathcal{N}(\boldsymbol{x}_{\mathcal{K}}; \boldsymbol{m}, \boldsymbol{P} \otimes \boldsymbol{P}_{\mathcal{E}})\mathcal{IW}_d(\boldsymbol{P}_{\mathcal{E}}; u, \boldsymbol{U})$$

$$= \eta_{\mathcal{K},\mathcal{E}}(W;\boldsymbol{\zeta})\mathcal{N}(\boldsymbol{x}_{\mathcal{K}};\boldsymbol{m}_W,\boldsymbol{P}_W\otimes\boldsymbol{P}_{\mathcal{E}})\mathcal{IW}_d(\boldsymbol{P}_{\mathcal{E}};u_W,\boldsymbol{U}_W) \tag{11.3.27}$$

where

$$\eta_{\mathcal{K},\mathcal{E}}(W;\boldsymbol{\zeta}) = \frac{(\pi^{|W|}|W|)^{-d/2}|\boldsymbol{U}|^{u/2}\Gamma_d(u_W/2)}{S^{d/2}|\boldsymbol{U}_W|^{u_W/2}\Gamma_d(u/2)} \tag{11.3.28}$$

$$\boldsymbol{m}_W = \boldsymbol{m} + (\boldsymbol{G}\otimes\boldsymbol{I}_d)\tilde{z} \tag{11.3.29}$$

$$\boldsymbol{P}_W = \boldsymbol{P} - \boldsymbol{G}S\boldsymbol{G}^{\mathrm{T}} \tag{11.3.30}$$

$$u_W = u + |W| \tag{11.3.31}$$

$$\boldsymbol{U}_W = \boldsymbol{U} + S^{-1}\tilde{z}\tilde{z}^{\mathrm{T}} + \boldsymbol{P}_U \tag{11.3.32}$$

and

$$\tilde{z} = \bar{z} - (\boldsymbol{H}\otimes\boldsymbol{I}_d)\boldsymbol{m} \tag{11.3.33}$$

$$\boldsymbol{G} = \boldsymbol{P}\boldsymbol{H}^{\mathrm{T}}S^{-1} \tag{11.3.34}$$

$$S = \boldsymbol{H}\boldsymbol{P}\boldsymbol{H}^{\mathrm{T}} + 1/|W| \tag{11.3.35}$$

$$\boldsymbol{P}_U = \sum_{z\in W}(z-\bar{z})(z-\bar{z})^{\mathrm{T}} \tag{11.3.36}$$

$$\bar{z} = \frac{1}{|W|}\sum_{z\in W} z \tag{11.3.37}$$

Therefore, the integral of (11.3.27) with respect to $\boldsymbol{x}_{\mathcal{K}}$ and $\boldsymbol{P}_{\mathcal{E}}$ is

$$\iint \prod_{j=1}^{|W|}\mathcal{N}\big(z_j;(\boldsymbol{H}\otimes\boldsymbol{I}_d)\boldsymbol{x}_{\mathcal{K}},\boldsymbol{P}_{\mathcal{E}}\big)\mathcal{N}(\boldsymbol{x}_{\mathcal{K}};\boldsymbol{m},\boldsymbol{P}\otimes\boldsymbol{P}_{\mathcal{E}})\mathcal{IW}_d(\boldsymbol{P}_{\mathcal{E}};u,\boldsymbol{U})d\boldsymbol{x}_{\mathcal{K}}d\boldsymbol{P}_{\mathcal{E}}$$
$$= \eta_{\mathcal{K},\mathcal{E}}(W;\boldsymbol{\zeta}) \tag{11.3.38}$$

Note $\eta_{\mathcal{R}}(W;\boldsymbol{\zeta})$ in (11.3.25) is a function of the measurement set and the prior GGIW parameter $\boldsymbol{\zeta}$, which actually corresponds to the negative binomial PDF $p(|W|;\alpha,1-(\beta+1)^{-1}) = C^{\alpha-1}_{|W|+\alpha-1}(1-(\beta+1)^{-1})^{\alpha}((\beta+1)^{-1})^{|W|}$, while $\eta_{\mathcal{K},\mathcal{E}}(W;\boldsymbol{\zeta})$ in (11.3.28) is proportional to the matrix variable generalized beta II PDF. Thus, the denominator in (11.3.20) (i.e., the normalization constant of the full GGIW Bayesian update) is given by the product $\eta_{\mathcal{R}}(W; \boldsymbol{\zeta}) \times \eta_{\mathcal{K},\mathcal{E}}(W; \boldsymbol{\zeta})$, which will be used to calculate the weight of the posterior density component in the update step of the filter.

11.3.3 GGIW Distribution-Based CPHD Filter for Multiple Extended Targets

In combination with the GGIW extended target model, the GGIW distribution-based CPHD (GGIW-CPHD) filter is introduced as follows, which can estimate the number and states of multiple extended targets in clutter. These states include the kinetic state (dynamic), measurement ratio, and extension.

11.3.3.1 Model Assumptions

In order to derive the prediction and update equations for the GGIW-CPHD filter, additional assumptions are also required in addition to the GGIW model assumptions for the extended target as already described in Sect. 11.3.1.

A.16: The kinetic state of each target follows a linear Gaussian dynamic model (11.3.9).
A.17: The sensor has a linear Gaussian measurement model (11.3.18).
A.18: The estimated PHD $v_{k-1}(\cdot)$ at the previous moment is a mixture of unnormalized GGIW distributions, i.e.,

$$v_{k-1}(\boldsymbol{x}) \approx \sum_{i=1}^{J_{k-1}} w_{k-1}^{(i)} \mathcal{GGIW}(\boldsymbol{x}; \boldsymbol{\zeta}_{k-1}^{(i)}) \tag{11.3.39}$$

where J_{k-1} is the number of components, and $w_{k-1}^{(i)}$ and $\boldsymbol{\zeta}_{k-1}^{(i)}$ are the weight and density parameter of the ith component.
A.19: The density of the birth RFS is also a mixture of unnormalized GGIW distributions.
A.20: The survival probability is independent of the state, i.e., $p_{S,k}(\boldsymbol{x}) = p_{S,k}$.
A.21: The following approximation holds for detection probability $p_D(\cdot)$

$$p_D(\boldsymbol{x})\mathcal{GGIW}\left(\boldsymbol{x}; \boldsymbol{\zeta}_{k|k-1}^{(i)}\right) \approx p_D\left(\boldsymbol{\zeta}_{k|k-1}^{(i)}\right)\mathcal{GGIW}\left(\boldsymbol{x}; \boldsymbol{\zeta}_{k|k-1}^{(i)}\right) \tag{11.3.40}$$

Assumption A.21, similar to Assumption A.11, is reasonable when $p_D(\cdot)$ is a constant, i.e., $p_D(\cdot) = p_D$. Generally, when the function $p_D(\cdot)$ does not change much in the uncertainty area in augmented target state space (see (11.3.4)), for example, when $p_D(\cdot)$ is a sufficiently smooth function, or when the signal-to-noise ratio (SNR) is high enough such that the uncertainty area is smaller, A.21 approximately holds.

A.22: The independent measurement likelihood is assumed to be

$$g(z|\boldsymbol{x}) = g(z|\boldsymbol{x}_{\mathcal{K}}, \boldsymbol{P}_{\mathcal{E}}) = \mathcal{N}(z; (\boldsymbol{H} \otimes \boldsymbol{I}_d)\boldsymbol{x}_{\mathcal{K}}, \boldsymbol{P}_{\mathcal{E}}) \tag{11.3.41}$$

The independent measurement likelihood $g(z|\boldsymbol{x})$ describes the relationship between the measurement z generated by a target and the corresponding target state $\boldsymbol{x}$. Note that the above likelihood doesn't depend on measurement ratio $x_{\mathcal{R}}$, because the CPHD update Eq. (11.3.50) already provides the likelihood for the update of the measurement ratio parameter by using the product term $G_z^{(|W|)}(0|\cdot)$ with respect to $x_{\mathcal{R}}$.

11.3.3.2 GGIW-CPHD Prediction

The GGIW-CPHD filter, like the CPHD filter, also propagates the posterior intensity v_k and the cardinality distribution ρ_k. Moreover, in the prediction step, the prediction of the cardinality distribution of the GGIW-CPHD filter is exactly the same as that of the standard GM-CPHD filter, see (5.4.3), which is not repeated here. Hence, only the predicted PHD $v_{k|k-1}$ is considered. For simplicity, the target spawning case is ignored here, and only the PHDs corresponding to the surviving targets and the birth targets in (5.4.2) are considered. The birth PHD corresponding to the birth targets at time k is

$$v_{\gamma,k}(\boldsymbol{x}) = \sum_{i=1}^{J_{\gamma,k}} w_{\gamma,k}^{(i)} \mathcal{GGIW}\left(\boldsymbol{x}; \zeta_{\gamma,k}^{(i)}\right) \tag{11.3.42}$$

and the predicted PHD corresponding to the survival targets is

$$v_{S,k|k-1}(\boldsymbol{x}) = \int p_{S,k}(\boldsymbol{x}')\phi_{k|k-1}(\boldsymbol{x}|\boldsymbol{x}')v_{k-1}(\boldsymbol{x}')d\boldsymbol{x}' \tag{11.3.43}$$

Using (11.3.7), (11.3.39) and A.20, the above integral simplifies to

$$\begin{aligned} v_{S,k|k-1}(\boldsymbol{x}) = p_{S,k} \sum_{j=1}^{J_{k-1}} w_{k-1}^{(j)} & \int \phi_{\mathcal{R}}(x_{\mathcal{R}}|x'_{\mathcal{R}}) \mathcal{GAM}\left(x'_{\mathcal{R}}; \alpha_{k-1}^{(j)}, \beta_{k-1}^{(j)}\right) dx'_{\mathcal{R}} \\ & \times \int \phi_{\mathcal{K}}(\boldsymbol{x}_{\mathcal{K}}|\boldsymbol{x}'_{\mathcal{K}}, \boldsymbol{P}_{\mathcal{E}}) \mathcal{N}\left(\boldsymbol{x}_{\mathcal{K}}; \boldsymbol{m}_{k-1}^{(j)}, \boldsymbol{P}_{k-1}^{(j)} \otimes \boldsymbol{P}_{\mathcal{E}}\right) d\boldsymbol{x}'_{\mathcal{K}} \\ & \times \int \phi_{\mathcal{E}}(\boldsymbol{P}_{\mathcal{E}}|\boldsymbol{P}'_{\mathcal{E}}) \mathcal{IW}_d\left(\boldsymbol{P}_{\mathcal{E}}; u_{k-1}^{(j)}, \boldsymbol{U}_{k-1}^{(j)}\right) d\boldsymbol{P}'_{\mathcal{E}} \end{aligned} \tag{11.3.44}$$

Using the linear Gaussian model given by (11.3.9), the prediction of the kinetic state part becomes [264]

$$\int \phi_{\mathcal{K}}(\boldsymbol{x}_{\mathcal{K}}|\boldsymbol{x}'_{\mathcal{K}}, \boldsymbol{P}_{\mathcal{E}})\mathcal{N}\Big(\boldsymbol{x}_{\mathcal{K}}; \boldsymbol{m}_{k-1}^{(j)}, \boldsymbol{P}_{k-1}^{(j)} \otimes \boldsymbol{P}_{\mathcal{E}}\Big)d\boldsymbol{x}'_{\mathcal{K}}$$
$$= \mathcal{N}\Big(\boldsymbol{x}_{\mathcal{K}}; \boldsymbol{m}_{S,k|k-1}^{(j)}, \boldsymbol{P}_{S,k|k-1}^{(j)} \otimes \boldsymbol{P}_{\mathcal{E}}\Big) \tag{11.3.45}$$

where

$$\boldsymbol{m}_{S,k|k-1}^{(j)} = (\boldsymbol{F} \otimes \boldsymbol{I}_d)\boldsymbol{m}_{k-1}^{(j)} \tag{11.3.46a}$$

$$\boldsymbol{P}_{S,k|k-1}^{(j)} = \boldsymbol{F}\boldsymbol{P}_{k-1}^{(j)}\boldsymbol{F}^{\mathrm{T}} + \boldsymbol{Q} \tag{11.3.46b}$$

For the prediction of the measurement ratio, we have

$$\alpha_{k|k-1}^{(j)} = \alpha_{k-1}^{(j)}/\mu_k \tag{11.3.47a}$$

$$\beta_{k|k-1}^{(j)} = \beta_{k-1}^{(j)}/\mu_k \tag{11.3.47b}$$

where the exponential forgetting factor $\mu_k > 1$ corresponds to an effective window length $w_e = 1/(1 - 1/\mu_k) = \mu_k/(\mu_k - 1)$.

For the prediction of the extension, the predicted DOF and inverse scale matrix can be approximated as [264]

$$u_{k|k-1}^{(j)} = \exp(-T_s/\tau)u_{k-1}^{(j)} \tag{11.3.48a}$$

$$\boldsymbol{U}_{k|k-1}^{(j)} = \frac{u_{k|k-1}^{(j)} - d - 1}{u_{k-1}^{(j)} - d - 1}\boldsymbol{U}_{k-1}^{(j)} \tag{11.3.48b}$$

where T_s is the sampling time and τ is the time decay constant.

Therefore, the PHD (11.3.43) corresponding to the survival targets is

$$v_{S,k|k-1}(\boldsymbol{x}) = \sum_{j=1}^{J_{k-1}} w_{k|k-1}^{(j)}\mathcal{GGIW}\Big(\boldsymbol{x}; \boldsymbol{\zeta}_{k|k-1}^{(j)}\Big) \tag{11.3.49}$$

where $w_{k|k-1}^{(j)} = p_{S,k}w_{k-1}^{(j)}$ and the predicted parameter $\boldsymbol{\zeta}_{k|k-1}^{(j)}$ is given by (11.3.46), (11.3.47) and (11.3.48).

The full predicted PHD is the sum of the birth PHD (11.3.42) and the predicted survival PHD (11.3.49), thus, there are a total of $J_{k|k-1} = J_{\gamma,k} + J_{k-1}$ GGIW components.

11.3.3.3 GGIW-CPHD Update

The updated PHD $v_k(\boldsymbol{x})$ and cardinality distribution $\rho_k(n)$ are calculated as follows [278, 279]

$$v_k(\boldsymbol{x}) = \begin{cases} \left(\begin{array}{l} \varpi(1 - p_D(\boldsymbol{x}) + p_D(\boldsymbol{x})G_z(0|\boldsymbol{x})) \\ + \dfrac{p_D(\boldsymbol{x}) \sum_{\mathcal{P}\angle Z_k} \sum_{W\in\mathcal{P}} \sigma_{\mathcal{P},W} G_z^{(|W|)}(0|\boldsymbol{x}) \prod_{z'\in W} \frac{g(z'|\boldsymbol{x})}{p_{C,k}(z')}}{\sum_{\mathcal{P}\angle Z_k} \sum_{W\in\mathcal{P}} \iota_{\mathcal{P},W} \psi_{\mathcal{P},W}} \end{array} \right) \times \\ p_{k|k-1}(\boldsymbol{x}), & |Z_k| \neq 0 \\ \varpi(1 - p_D(\boldsymbol{x}) + p_D(\boldsymbol{x})G_z(0|\boldsymbol{x}))p_{k|k-1}(\boldsymbol{x}), & |Z_k| = 0 \end{cases} \tag{11.3.50}$$

$$\rho_k(n) = \begin{cases} \dfrac{\sum_{\mathcal{P}\angle Z_k} \sum_{W\in\mathcal{P}} \psi_{\mathcal{P},W} G_{k|k-1}^{(n)}(0) \left(\begin{array}{l} G_{\mathrm{FA}}(0) \dfrac{\eta_W}{|\mathcal{P}|} \dfrac{\varsigma^{n-|\mathcal{P}|}}{(n-|\mathcal{P}|)!} \delta_{n\geq|\mathcal{P}|} \\ +G_{\mathrm{FA}}^{(|W|)}(0) \dfrac{\varsigma^{n-|\mathcal{P}|+1}}{(n-|\mathcal{P}|+1)!} \delta_{n\geq|\mathcal{P}|-1} \end{array} \right)}{\sum_{\mathcal{P}\angle Z_k} \sum_{W\in\mathcal{P}} \iota_{\mathcal{P},W} \psi_{\mathcal{P},W}}, & |Z_k| \neq 0 \\ \dfrac{\varsigma^n G_{k|k-1}^{(n)}(0)}{G_{k|k-1}(\varsigma)}, & |Z_k| = 0 \end{cases} \tag{11.3.51}$$

where $g(\cdot|\boldsymbol{x})$ represents the likelihood of a single measurement conditioned on the target state $\boldsymbol{x}$, $p_{C,k}(\cdot)$ is the likelihood of a single false alarm, $p_{k|k-1}(\boldsymbol{x})$ is the predicted single target density (11.3.59), functions $G_{k|k-1}(\cdot)$, $G_{\mathrm{FA}}(\cdot)$, and $G_z(\cdot|\boldsymbol{x})$ respectively represent the probability generating function (PGF) of the predicted state, false alarm, and the measurement conditioned on state $\boldsymbol{x}$, the superscript (n) in the PGF indicates the n-order derivative of the PGF, and $\mathcal{P}\angle Z$ means that $\mathcal{P}$ partitions the measurement set Z into non-empty subsets and when it is used under a summation notation, it implies that the summation traverses all possible partitions $\mathcal{P}$. $|\mathcal{P}|$ represents the number of non-empty subsets in partition $\mathcal{P}$, the non-empty subset is denoted by W and is called the cell. When $W \in \mathcal{P}$ is used under a summation notation, the summation traverses all cells in the partition, and $|W|$ represents the number of measurements in the cell. Moreover, other coefficients used in (11.3.50) and (11.3.51) are defined as follows

$$\varsigma = \langle p_{k|k-1}(\cdot), 1 - p_D(\cdot) + p_D(\cdot)G_z(0|\cdot) \rangle \tag{11.3.52}$$

$$\eta_W = \left\langle p_{k|k-1}(\cdot), p_D(\cdot)G_z^{(|W|)}(0|\cdot) \prod_{z'\in W} \frac{g(z'|\boldsymbol{x})}{p_{C,k}(z')} \right\rangle \tag{11.3.53}$$

$$\iota_{\mathcal{P},W} = G_{\mathrm{FA}}(0)G_{k|k-1}^{(|\mathcal{P}|)}(\varsigma)\frac{\eta_W}{|\mathcal{P}|} + G_{\mathrm{FA}}^{(|W|)}(0)G_{k|k-1}^{(|\mathcal{P}|-1)}(\varsigma) \tag{11.3.54}$$

$$\chi_{\mathcal{P},W} = G_{\mathrm{FA}}(0)G_{k|k-1}^{(|\mathcal{P}|+1)}(\varsigma)\frac{\eta_W}{|\mathcal{P}|} + G_{\mathrm{FA}}^{(|W|)}(0)G_{k|k-1}^{(|\mathcal{P}|)}(\varsigma) \tag{11.3.55}$$

$$\psi_{\mathcal{P},W} = \prod_{W' \in \mathcal{P}-W} \eta_{W'} \tag{11.3.56}$$

$$\varpi = \begin{cases} \frac{\sum_{\mathcal{P}\angle Z_k} \sum_{W\in\mathcal{P}} \chi_{\mathcal{P},W}\psi_{\mathcal{P},W}}{\sum_{\mathcal{P}\angle Z_k} \sum_{W\in\mathcal{P}} \iota_{\mathcal{P},W}\psi_{\mathcal{P},W}}, & |Z_k| \neq 0 \\ N_{k|k-1}, & |Z_k| = 0 \end{cases} \tag{11.3.57}$$

$$\sigma_{\mathcal{P},W} = \frac{\psi_{\mathcal{P},W}}{|\mathcal{P}|} G_{\text{FA}}(0) G_{k|k-1}^{(|\mathcal{P}|)}(\varsigma) + \frac{\prod_{W'\in\mathcal{P}-W} \psi_{\mathcal{P},W'}\iota_{\mathcal{P},W'}}{\eta_W} \tag{11.3.58}$$

$$p_{k|k-1}(\boldsymbol{x}) = N_{k|k-1}^{-1} v_{k|k-1}(\boldsymbol{x}) \tag{11.3.59}$$

$$N_{k|k-1}(\boldsymbol{x}) = \int v_{k|k-1}(\boldsymbol{x}) d\boldsymbol{x} \tag{11.3.60}$$

When actually computing the updated PHD $v_k(\boldsymbol{x})$, the following steps are implemented.

(a) For all components j and all sets W in all partitions $\mathcal{P}$ of Z_k, first calculate the following centroid measurement, scatter matrix, innovation factor, gain vector, innovation vector, and innovation covariance

$$\bar{z}_k(W) = \frac{1}{|W|} \sum_{z_k \in W} z_k \tag{11.3.61}$$

$$\boldsymbol{P}_U(W) = \sum_{z_k \in W} [z_k - \bar{z}_k(W)][z_k - \bar{z}_k(W)]^{\mathrm{T}} \tag{11.3.62}$$

$$S_{k|k-1}^{(j)}(W) = \boldsymbol{H}\boldsymbol{P}_{k|k-1}^{(j)}\boldsymbol{H}^{\mathrm{T}} + \frac{1}{|W|} \tag{11.3.63}$$

$$\boldsymbol{G}_{k|k-1}^{(j)}(W) = \boldsymbol{P}_{k|k-1}^{(j)}\boldsymbol{H}^{\mathrm{T}}\left[S_{k|k-1}^{(j)}(W)\right]^{-1} \tag{11.3.64}$$

$$\tilde{z}_{k|k-1}^{(j)}(W) = \bar{z}_k(W) - (\boldsymbol{H} \otimes \boldsymbol{I}_d)\boldsymbol{m}_{k|k-1}^{(j)} \tag{11.3.65}$$

$$\boldsymbol{S}_{k|k-1}^{(j)}(W) = \left[S_{k|k-1}^{(j)}(W)\right]^{-1} \tilde{z}_{k|k-1}^{(j)}(W)\left[\tilde{z}_{k|k-1}^{(j)}(W)\right]^{\mathrm{T}} \tag{11.3.66}$$

Then, calculate the components of the posterior GGIW parameter $\boldsymbol{\zeta}_{k|k}^{(j)}$

$$\alpha_{k|k}^{(j)}(W) = \alpha_{k|k-1}^{(j)} + |W| \tag{11.3.67}$$

$$\beta_{k|k}^{(j)}(W) = \beta_{k|k-1}^{(j)} + 1 \tag{11.3.68}$$

$$\boldsymbol{m}_{k|k}^{(j)}(W) = \boldsymbol{m}_{k|k-1}^{(j)} + \left(\boldsymbol{G}_{k|k-1}^{(j)}(W) \otimes \boldsymbol{I}_d\right)\tilde{z}_{k|k-1}^{(j)}(W) \tag{11.3.69}$$

$$\boldsymbol{P}_{k|k}^{(j)}(W) = \boldsymbol{P}_{k|k-1}^{(j)} - \boldsymbol{G}_{k|k-1}^{(j)}(W)S_{k|k-1}^{(j)}(W)\left[\boldsymbol{G}_{k|k-1}^{(j)}(W)\right]^{\mathrm{T}} \tag{11.3.70}$$

$$u_{k|k}^{(j)}(W) = u_{k|k-1}^{(j)} + |W| \tag{11.3.71}$$

$$\boldsymbol{U}_{k|k}^{(j)}(W) = \boldsymbol{U}_{k|k-1}^{(j)} + \boldsymbol{S}_{k|k-1}^{(j)}(W) + \boldsymbol{P}_U(W) \tag{11.3.72}$$

(b) According to (11.3.52), calculate

$$\varsigma = \sum_{j=1}^{J_{k|k-1}} \overline{w}_{k|k-1}^{(j)}\left(1 - p_D^{(j)} + p_D^{(j)} G_z(0, j)\right) \tag{11.3.73}$$

where $G_z(0, j)$ is the expected probability that the jth component will not lead to any detection, and $\overline{w}_{k|k-1}^{(j)}$ is the normalized prior PHD weight, calculated respectively as follows

$$G_z(0, j) = \left[\beta_{k|k-1}^{(j)}/(\beta_{k|k-1}^{(j)} + 1)\right]^{\alpha_{k|k-1}^{(j)}} \tag{11.3.74}$$

$$\overline{w}_{k|k-1}^{(j)} = w_{k|k-1}^{(j)} / \sum_{i=1}^{J_{k|k-1}} w_{k|k-1}^{(i)}, \quad j = 1, \ldots, J_{k|k-1} \tag{11.3.75}$$

(c) According to (11.3.53), for all sets W in all partitions $\mathcal{P}$ of Z_k, calculate the following equation

$$\eta_W = \sum_{j=1}^{J_{k|k-1}} \overline{w}_{k|k-1}^{(j)} p_D^{(j)} \mathcal{L}_k^{(j)}(W)/\mathcal{L}_{\text{FA}}(W) \tag{11.3.76}$$

where

$$\mathcal{L}_{\text{FA}}(W) = \prod_{z \in W} p_{C,k}(z) \tag{11.3.77}$$

$$\mathcal{L}_k^{(j)}(W) = \eta_{k,\mathcal{R}}^{(j)} \eta_{k,\mathcal{K},\mathcal{E}}^{(j)} \tag{11.3.78}$$

$$\eta_{k,\mathcal{R}}^{(j)} = \frac{1}{|W|!} \frac{\Gamma\left(\alpha_{k|k}^{(j)}(W)\right)\left(\beta_{k|k-1}^{(j)}\right)^{\alpha_{k|k-1}^{(j)}}}{\Gamma\left(\alpha_{k|k-1}^{(j)}\right)\left(\beta_{k|k}^{(j)}(W)\right)^{\alpha_{k|k}^{(j)}(W)}} \tag{11.3.79}$$

$$\eta_{k,\mathcal{K},\mathcal{E}}^{(j)} = \frac{\left(\pi^{|W|}|W|\right)^{-d/2}\left|\boldsymbol{U}_{k|k-1}^{(j)}\right|^{u_{k|k-1}^{(j)}/2}}{\left(s_{k|k-1}^{(j)}(W)\right)^{d/2}\left|\boldsymbol{U}_{k|k}^{(j)}(W)\right|^{u_{k|k}^{(j)}(W)/2}} \frac{\Gamma_d\left(u_{k|k}^{(j)}(W)/2\right)}{\Gamma_d\left(u_{k|k-1}^{(j)}/2\right)} \tag{11.3.80}$$

In the Gaussian inverse Wishart (GIW) likelihood (11.3.80), $|\boldsymbol{U}|$ is the determinant of the matrix $\boldsymbol{U}$, and $|W|$ denotes the number of measurements in cell W. Note that the meaning of $|\cdot|$ is distinguished according to the corresponding parameters. The GGIW likelihood (11.3.78) is the product of measurement ratio likelihood (11.3.79) and the GIW likelihood (11.3.80).

(d) For all sets W and all partitions $\mathcal{P}$, using (11.3.54), (11.3.55), (11.3.56), (11.3.57), and (11.3.58), calculate the coefficients $\iota_{\mathcal{P},W}$, $\chi_{\mathcal{P},W}$, $\psi_{\mathcal{P},W}$, ϖ, and $\sigma_{\mathcal{P},W}$, respectively.

(e) Calculate the posterior weight

$$w_{k|k,\mathcal{P},W}^{(j)} = \frac{\overline{w}_{k|k-1}^{(j)} p_D^{(j)} \sigma_{\mathcal{P},W} \mathcal{L}_k^{(j)}(W)/\mathcal{L}_{\text{FA}}(W)}{\sum_{\mathcal{P}\angle Z_k}\sum_{W\in\mathcal{P}} \iota_{\mathcal{P},W}\psi_{\mathcal{P},W}} \tag{11.3.81}$$

Finally, the obtained posterior PHD is the GGIW mixture, which is

$$v_k(\boldsymbol{x}) = v_{M,k}(\boldsymbol{x}) + \sum_{\mathcal{P}\angle Z_k}\sum_{W\in\mathcal{P}}\sum_{j=1}^{J_{k|k-1}} w_{k|k,\mathcal{P},W}^{(j)} \mathcal{GGIW}(\boldsymbol{x}; \boldsymbol{\zeta}_{k|k}^{(j)}) \tag{11.3.82}$$

where

$$\begin{aligned} v_{M,k}(\boldsymbol{x}) = {} & \varpi \sum_{j=1}^{J_{k|k-1}} \overline{w}_{k|k-1}^{(j)}(1 - p_D^{(j)})\mathcal{GGIW}(\boldsymbol{x}; \boldsymbol{\zeta}_{k|k-1}^{(j)}) \\ & + \varpi \sum_{j=1}^{J_{k|k-1}} \overline{w}_{k|k-1}^{(j)} p_D^{(j)} \left(\beta_{k|k-1}^{(j)}/(\beta_{k|k-1}^{(j)} + 1)\right)^{\alpha_{k|k-1}^{(j)}} \mathcal{GGIW}\left(\boldsymbol{x}; \overline{\boldsymbol{\zeta}}_{k|k-1}^{(j)}\right) \end{aligned} \tag{11.3.83}$$

The first summation on the right-hand side of the above equation corresponds to the case in which targets are not detected, and the second summation deals with the case in which targets are detected but no measurement is generated. In the second summation, all correction parameters of $\overline{\boldsymbol{\zeta}}_{k|k-1}^{(j)}$ are the same as those of $\boldsymbol{\zeta}_{k|k-1}^{(j)}$, except that $\overline{\beta}_{k|k-1}^{(j)} = \beta_{k|k-1}^{(j)} + 1$.

The updated cardinality distribution can be calculated by substituting the obtained results into (11.3.51).

In order to reduce the number of exponentially growing GGIW components, merging and pruning techniques are also required. For the merging method for the

GGIW mixture, refer to [279, 428]. Due to space limitations, no more detail will be presented here.

11.3.3.4 Improvement of Measurement Set Partitioning

Note that both the GGIW-CPHD filter here and the ET-GM-PHD filter as described in Sect. 11.2 and [273] require all partitions of the current measurement set when updated. As mentioned earlier, as the total number of measurements increases, the number of possible partitions grows dramatically. Even in the case of as few as 5 measurements, there are still 120 possible partitions. Therefore, it is computationally infeasible to consider all partitions and some approximation is required. It has been shown previously that the set of all partitions can be effectively approximated by using the subset of partitions containing the most probable partitions. The distance-based partitioning method described in Sect. 11.2.3 imposes an upper bound on the distance between measurements in the same cell W in partition $\mathcal{P}$, which reduces the number of partitions to be considered by several orders of magnitude, while losing as little tracking performance as possible [95, 276, 423].

Careful examination of the update equation of the GGIW-CPHD filter shows that, it needs to include the partitions of false alarms in a single cell. As an example, assume that the current measurement set consists of two clusters of close-distance measurements generated by two targets and 10 well-separated false alarms, as shown in Fig. 11.1a. In this case, it is very possible to obtain a single partition composed of 12 cells by using the distance partitioning method: two cells contain measurements generated by respective targets, and the other 10 cells contain false alarms. These 12 cells are represented by different colors as shown in Fig. 11.1b. In the update of the cardinality distribution (11.3.51), the probability of n less than $|\mathcal{P}| - 1$ will become 0. As a consequence, this will result in severely inaccurate cardinality estimation, thus leading to a large number of false tracks.

The main reason for this problem is to approximate all partition sets with a limited number of distance-based partitions. Note that if a complete set of partitions is used, there will always be a partition that includes the measurements produced by each

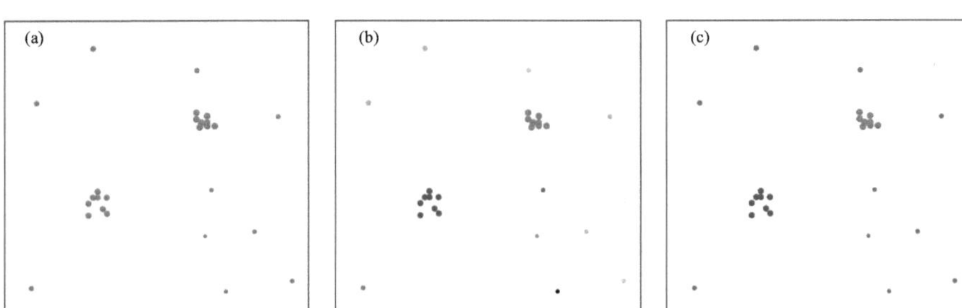

Fig. 11.1 An example of an improved measurement partitioning [279]

target into their respective cells and all clutter measurements into a single cell. Therefore, this problem can be solved using the following method, which is an improvement of the distance-based partitioning method in Sect. 11.2.3. Here, only the necessary modifications are discussed. As shown in Fig. 11.1b, the improved distance-based partitioning method places false alarms into an independent cell, which is unchanged. The reason is that they are isolated measurements and are clusters of measurements far away from targets. Therefore, the method needs to be remedied from the perspective of modifying the partitions. The remedied method is to merge all cells of n measurements containing $|W| \leq N_{\text{low}}$ into a single cell for the partitions computed by the distance-based partitioning method. Figure 11.1c illustrates the case of $N_{\text{low}} = 1$, where the clutter measurements are indicated by the same color. In this way, the set of partitions obtained by the distance-based partitioning method is further processed to form new partitions to solve the problem of cardinality overestimation in the case of a higher number of false alarms.

11.3.4 GGIW Distribution-Based Labeled RFS Filters for Multiple Extended Targets

Based on the GGIW distribution, the labeled RFS filter can be generalized to be suitable for extended target tracking. The following first introduces the measurement model of multiple extended targets. On this basis, the GGIW-GLMB filter and the GGIW-LMB filter are described respectively.

11.3.4.1 Measurement Model for Multiple Extended Targets

The dynamical model of extended targets is modeled using the GGIW distribution in Sect. 11.3.1, while the measurement model of extended targets is based on the following three assumptions.

A.23: The detection probability of a extended target with state $(\boldsymbol{x}, \ell)$ is $p_D(\boldsymbol{x}, \ell)$, and the probability of missed detection is $q_D(\boldsymbol{x}, \ell) = 1 - p_D(\boldsymbol{x}, \ell)$.
A.24: If a target is detected, the extended target with state $(\boldsymbol{x}, \ell)$ generates a set of detections W with likelihood $\tilde{g}(W|\boldsymbol{x}, \ell)$ given by (11.3.19), independent of all other targets.
A.25: The false alarm measurements produced by the sensor is a Poisson RFS K with an intensity function $\kappa(\cdot)$, i.e., K follows the distribution $g_C(K) = \exp(-\langle\kappa, 1\rangle)\kappa^K$, and the false alarms are independent of target-generated measurements.

Let $\mathcal{G}_i(Z)$ denote the set of all partitions containing exactly i cells according to the finite measurement set Z, and $\mathcal{P}(Z) \in \mathcal{G}_i(Z)$ represent a particular partition of Z. Denote the labeled RFS of multiple extended targets as $\mathrm{X} =$

$\{(\boldsymbol{x}_1, \ell_1), \ldots, (\boldsymbol{x}_{|\mathrm{X}|}, \ell_{|\mathrm{X}|})\}$. For a given multi-target state X, let $\Theta(\mathcal{P}(Z))$ denote the space of association map $\theta : \mathcal{L}(\mathrm{X}) \to \{0, 1, \ldots, |\mathcal{P}(Z)|\}$ satisfying $\theta(\ell) = \theta(\ell') > 0 \Rightarrow \ell = \ell'$. In addition, denote by $\mathcal{P}_{\theta(\ell)}(Z)$ the elements corresponding to the label ℓ in the partition $\mathcal{P}(Z)$ under the map θ.

Proposition 30 (see Appendix M for proof) *Under Assumptions A.23, A.24 and A.25, the measurement likelihood function is*

$$\begin{aligned} g(Z|\mathrm{X}) &= g_C(Z) \sum_{i=1}^{|\mathrm{X}|+1} \sum_{\substack{\mathcal{P}(Z)\in\mathcal{G}_i(Z)\\ \theta\in\Theta(\mathcal{P}(Z))}} \left[\varphi_{\mathcal{P}(Z)}(\cdot;\theta)\right]^{\mathrm{X}} \\ &= \exp(-\langle\kappa, 1\rangle)\kappa^Z \sum_{i=1}^{|\mathrm{X}|+1} \sum_{\substack{\mathcal{P}(Z)\in\mathcal{G}_i(Z)\\ \theta\in\Theta(\mathcal{P}(Z))}} \left[\varphi_{\mathcal{P}(Z)}(\cdot;\theta)\right]^{\mathrm{X}} \end{aligned} \tag{11.3.84}$$

where

$$\varphi_{\mathcal{P}(Z)}(\boldsymbol{x}, \ell; \theta) = \begin{cases} p_D(\boldsymbol{x}, \ell)\tilde{g}\big(\mathcal{P}_{\theta(\ell)}(Z)|\boldsymbol{x}, \ell\big)/[\kappa]^{\mathcal{P}_{\theta(\ell)}(Z)}, & \theta(\ell) > 0 \\ q_D(\boldsymbol{x}, \ell), & \theta(\ell) = 0 \end{cases} \tag{11.3.85}$$

with $\tilde{g}(W|\boldsymbol{x})$ *being given by* (11.3.19).

Since the number of measurement partitions and the number of sets of cell-target maps are extremely large, the exact calculation of the likelihood (11.3.84) is generally difficult to numerically process. However, in many practical conditions, it is only necessary to consider a smaller subset of these partitions to achieve a good performance [95, 276]. In addition, utilizing the ranked assignment algorithm can significantly reduce the number of sets of cell-target maps, thereby further reducing the number of secondary terms in the likelihood.

11.3.4.2 GGIW-GLMB Filter

Based on the above measurement likelihood and the dynamical model in Sect. 11.3.1, a GLMB filter suitable for extended targets is obtained, which consists of two steps of prediction and update. Sine the standard birth/death model is used for multi-target dynamics, the prediction step is the same as that of the standard GLMB filter, as detailed in Proposition 15. However, it is necessary to replace the single-target transition kernel function $\phi(\cdot|\cdot, \ell)$ in Proposition 15 with the GGIW transition kernel for extended targets defined in (11.3.7), and then carry out the similar derivation as in Sects. 11.3.2.1 and 11.3.3.2, which will not be repeated due to space limitations. Since the measurement likelihood functions have different forms, the difference between the extended target GLMB filter and the standard GLMB filter mainly depends on the measurement update step.

According to the prior distribution (3.3.38) and the multi-target likelihood function (11.3.84), we can obtain

$$\begin{aligned}
\pi(\mathrm{X})g(Z|\mathrm{X}) &= \Delta(\mathrm{X})g_C(Z)\sum_{c\in\mathbb{C}}\sum_{i=1}^{|\mathrm{X}|+1}\sum_{\substack{\mathcal{P}(Z)\in\mathcal{G}_i(Z)\\ \theta\in\Theta(\mathcal{P}(Z))}} w^{(c)}(\mathcal{L}(\mathrm{X}))\left[p^{(c)}(\cdot)\varphi_{\mathcal{P}(Z)}(\cdot;\theta)\right]^{\mathrm{X}}\\
&= \Delta(\mathrm{X})g_C(Z)\sum_{c\in\mathbb{C}}\sum_{i=1}^{|\mathrm{X}|+1}\sum_{\substack{\mathcal{P}(Z)\in\mathcal{G}_i(Z)\\ \theta\in\Theta(\mathcal{P}(Z))}} w^{(c)}(\mathcal{L}(\mathrm{X}))\times\\
&\quad\left[p^{(c,\theta)}(\cdot|\mathcal{P}(Z))\eta^{(c,\theta)}_{\mathcal{P}(Z)}(\cdot)\right]^{\mathrm{X}}\\
&= \Delta(\mathrm{X})g_C(Z)\sum_{c\in\mathbb{C}}\sum_{i=1}^{|\mathcal{L}(\mathrm{X})|+1}\sum_{\substack{\mathcal{P}(Z)\in\mathcal{G}_i(Z)\\ \theta\in\Theta(\mathcal{P}(Z))}} w^{(c)}(\mathcal{L}(\mathrm{X}))\left[\eta^{(c,\theta)}_{\mathcal{P}(Z)}(\cdot)\right]^{\mathcal{L}(\mathrm{X})}\times\\
&\quad\left[p^{(c,\theta)}(\cdot|\mathcal{P}(Z))\right]^{\mathrm{X}}
\end{aligned}\tag{11.3.86}$$

where $\varphi_{\mathcal{P}(Z)}(\boldsymbol{x},\ell;\theta)$ is given by (11.3.85), and

$$p^{(c,\theta)}(\boldsymbol{x},\ell|\mathcal{P}(Z)) = p^{(c)}(\boldsymbol{x},\ell)\varphi_{\mathcal{P}(Z)}(\boldsymbol{x},\ell;\theta)/\eta^{(c,\theta)}_{\mathcal{P}(Z)}(\ell) \tag{11.3.87}$$

$$\eta^{(c,\theta)}_{\mathcal{P}(Z)}(\ell) = \left\langle p^{(c)}(\cdot,\ell),\varphi_{\mathcal{P}(Z)}(\cdot,\ell;\theta)\right\rangle \tag{11.3.88}$$

Using the abbreviated notations $(\boldsymbol{x},\ell)_{1:j} \equiv ((\boldsymbol{x}_1,\ell_1),\ldots,(\boldsymbol{x}_j,\ell_j))$, $\ell_{1:j} \equiv (\ell_1,\ldots,\ell_j)$ and $\boldsymbol{x}_{1:j}\equiv(\boldsymbol{x}_1,\ldots,\boldsymbol{x}_j)$ to represent vectors, and using $\{(\boldsymbol{x},\ell)_{1:j}\}$ and $\{\ell_{1:j}\}$ represent the corresponding sets, then, the set integral of (11.3.86) with respect to X is

$$\begin{aligned}
&\int\pi(\mathrm{X})g(Z|\mathrm{X})\delta\mathrm{X}\\
&= g_C(Z)\sum_{c\in\mathbb{C}}\int\sum_{i=1}^{|\mathcal{L}(\mathrm{X})|+1}\sum_{\substack{\mathcal{P}(Z)\in\mathcal{G}_i(Z)\\ \theta\in\Theta(\mathcal{P}(Z))}}\Delta(\mathrm{X})w^{(c)}(\mathcal{L}(\mathrm{X}))\left[\eta^{(c,\theta)}_{\mathcal{P}(Z)}(\cdot)\right]^{\mathcal{L}(\mathrm{X})}\times\\
&\quad\left[p^{(c,\theta)}(\cdot|\mathcal{P}(Z))\right]^{\mathrm{X}}\delta\mathrm{X}\\
&= g_C(Z)\sum_{c\in\mathbb{C}}\sum_{j=0}^{\infty}\frac{1}{j!}\sum_{\ell_{1:j}\in\mathbb{L}^j}\sum_{i=1}^{j+1}\sum_{\substack{\mathcal{P}(Z)\in\mathcal{G}_i(Z)\\ \theta\in\Theta(\mathcal{P}(Z))}}\Delta\big(\{(\boldsymbol{x},\ell)_{1:j}\}\big)w^{(c)}\big(\{\ell_{1:j}\}\big)\times\\
&\quad\left[\eta^{(c,\theta)}_{\mathcal{P}(Z)}(\cdot)\right]^{\{\ell_{1:j}\}}\times\int\left[p^{(c,\theta)}(\cdot|\mathcal{P}(Z))\right]^{\{(\boldsymbol{x},\ell)_{1:j}\}}d\boldsymbol{x}_{1:j}
\end{aligned}$$

$$= g_C(Z) \sum_{c \in \mathbb{C}} \sum_{L \subseteq \mathbb{L}} \sum_{i=1}^{|L|+1} \sum_{\substack{\mathcal{P}(Z) \in \mathcal{G}_i(Z) \\ \theta \in \Theta(\mathcal{P}(Z))}} w^{(c)}(L) \left[\eta_{\mathcal{P}(Z)}^{(c,\theta)} \right]^L \tag{11.3.89}$$

where the second line is obtained by applying (3.3.18) and taking the outer part of the resulting integral only related to labels, while the last line is obtained by using the following result: the distinct label indicator (DLI) limits the sum on $j : 0 \to \infty$ and $\ell_{1:j} \in \mathbb{L}^j$ to the sum on a subset of $\mathbb{L}$.

Substituting (11.3.86) and (11.3.89) into (3.5.5), the posterior density can be obtained as follows

$$\pi(\mathbf{X}|Z) = \Delta(\mathbf{X}) \sum_{c \in \mathbb{C}} \sum_{i=1}^{|\mathbf{X}|+1} \sum_{\substack{\mathcal{P}(Z) \in \mathcal{G}_i(Z) \\ \theta \in \Theta(\mathcal{P}(Z))}} w_{\mathcal{P}(Z)}^{(c,\theta)}(\mathcal{L}(\mathbf{X})) \left[p^{(c,\theta)}(\cdot|\mathcal{P}(Z)) \right]^{\mathbf{X}} \tag{11.3.90}$$

where $p^{(c,\theta)}(\cdot|\mathcal{P}(Z))$ is given by (11.3.87), and

$$w_{\mathcal{P}(Z)}^{(c,\theta)}(L) = \frac{w^{(c)}(L) \left[\eta_{\mathcal{P}(Z)}^{(c,\theta)} \right]^L}{\sum_{c \in \mathbb{C}} \sum_{J \subseteq \mathbb{L}} \sum_{i=1}^{|J|+1} \sum_{\substack{\mathcal{P}(Z) \in \mathcal{G}_i(Z) \\ \theta \in \Theta(\mathcal{P}(Z))}} w^{(c)}(J) \left[\eta_{\mathcal{P}(Z)}^{(c,\theta)} \right]^J} \tag{11.3.91}$$

As a result, if the extended target prior is a GLMB, then the extended multi-target posterior is also a GLMB (11.3.90) after the extended multi-target likelihood function (11.3.84) is applied. This conclusion shows that, the GLMB is a conjugate prior with respect to the extended multi-target measurement likelihood function.

11.3.4.3 GGIW-LMB Filter

In order to reduce the computational complexity of the GGIW-GLMB tracking algorithm, a GGIW-LMB tracking algorithm suitable for extended targets can also be obtained based on the LMB filter. An important principle of the LMB filter is to simplify the representation of the multi-target density after each iteration, thereby reducing the computational complexity of the algorithm. Specifically, it does not maintain a full GLMB posterior distribution from one cycle to the next, but it is approximated to the LMB distribution after the measurement update step. In subsequent iterations, this LMB approximation is used for the prediction, and the predicted LMB is converted back into the GLMB in preparation for the next measurement update. Therefore, transforming the GGIW-GLMB filter into the GGIW-LMB filter requires 3 modifications: (1) replace the GLMB prediction with the LMB prediction; (2) convert the LMB expression into the GLMB expression; (3) approximate the updated GLMB distribution as the LMB distribution.

(1) LMB prediction: this step is the same as that of the standard LMB filter, as detailed in Proposition 19. However, it is also necessary to replace the single-target transition kernel function $\phi(\cdot|\cdot, \ell)$ in Proposition 19 with the GGIW transition kernel defined by (11.3.7) for extended targets.
(2) Conversion of the LMB to the GLMB: in the update step, it is necessary to convert the LMB expression of the predicted multi-target density into the GLMB expression. In principle, converting the predicted LMB $\pi_+ = \{(\varepsilon_+^{(\ell)}, p_+^{(\ell)})\}_{\ell\in\mathbb{L}_+}$ into the GLMB involves computing the GLMB weights of all subsets of $\mathbb{L}_+$, but in practice it is possible to reduce the number of components and improve the efficiency of the conversion by approximation.
(3) Approximation of the GLMB as the LMB: after the measurement update step, the posterior GLMB given by (11.3.90) can be approximated by the LMB matching the PHD, and the corresponding parameters are

$$\varepsilon^{(\ell)} = \sum_{\substack{c\in\mathbb{C}\\ L\in\mathbb{L}_+}} \sum_{i=1}^{|L|+1} \sum_{\substack{\mathcal{P}(Z)\in\mathcal{G}_i(Z)\\ \theta\in\Theta(\mathcal{P}(Z))}} 1_L(\ell) w_{\mathcal{P}(Z)}^{(c,\theta)}(L) \tag{11.3.92}$$

$$p^{(\ell)}(\boldsymbol{x}) = \frac{1}{\varepsilon^{(\ell)}} \sum_{\substack{c\in\mathbb{C}\\ L\in\mathbb{L}_+}} \sum_{i=1}^{|L|+1} \sum_{\substack{\mathcal{P}(Z)\in\mathcal{G}_i(Z)\\ \theta\in\Theta(\mathcal{P}(Z))}} 1_L(\ell) w_{\mathcal{P}(Z)}^{(c,\theta)}(L) p^{(c,\theta)}(\boldsymbol{x}, \ell|\mathcal{P}(Z)) \tag{11.3.93}$$

The existence probability corresponding to each label is the sum of weights of GLMB components containing that label, and the corresponding PDF turns into the weighted sum of the corresponding PDF from the GLMB. Therefore, the PDF of each track in the LMB becomes a mixture of GGIW densities, where each mixture component corresponds to a different measurement association process. In order to prevent the number of components from becoming too large, it is necessary to reduce the number of mixture components through pruning and merging. The concrete implementations of the GGIW-GLMB and GGIW-LMB algorithms introduced above can be found in [148].

11.4 Target Tracking with Merged Measurements

Target tracking with merged measurements is also referred to as unresolved target tracking. Mahler modeled the multi-target state as a set of point groups and firstly proposed the PHD filter applicable to unresolved targets. The likelihood function for this point group model involves the ergodic summation of the partitions of the measurement set, so a direct implementation is computationally infeasible. Moreover, this point group model is intuitionistic and attractive, when targets are close to each other in the state space such that the measurements become merged. However, this implicit assumed merging only depends on the target state, and thus, this model is limited to merged measurements that are produced when targets are only close to each other in the state space rather than the measurement space. In some important cases, such as azimuth-only tracking, video tracking, etc., this assumption is too strict. In these cases, targets that are unresolved along the line of sight may be separated by a considerable distance in the state space. This usually occurs when the measurement space is a lower-dimensional projection of the state space (e.g., azimuth-only tracking), where well-separated points in the state space may become close to each other (unresolved) after being projected into the measurement space, thus, the well-separated states may potentially yield close-range measurements. In addition, the proposed PHD filter can't provide labeled track estimates, i.e., it conducts multi-target filtering instead of multi-target tracking.

In order to solve these problems, a likelihood function applicable to merged measurements is required, not only for the case where the targets are close in the state space, but also for the more general case where the targets are close in the measurement space. To achieve this goal, partitioning the set of targets must be considered. At this time, each group in one partition produces at most one (merged) measurement. In order to provide the target track information, the GLMB filter is further extended to the sensor model including merged measurements, so that it can be applied to a wider range of practical problems. By considering feasible partitions of the set of targets and the feasible assignments between measurements and target groups within these partitions, a multi-target likelihood function is introduced, which considers possible merging of measurements originating from the targets. One advantage of this method is that it can be parallelized, thereby facilitating the potential real-time implementation. The exact form of the Bayesian optimal multi-target tracker for merged measurements is intractable, so a computationally more economical approximation algorithm is introduced.

11.4.1 Multi-target Measurement Likelihood Model for Merged Measurements

The likelihood model for merged measurements depends on the partition of the target set. Based on the concept of partitioning, the multi-target measurement likelihood model is defined as the summation over the traversal of the partitions of the target set, i.e.,

$$g(Z|\mathrm{X}) = \sum_{\mathcal{P}(\mathrm{X}) \in \mathcal{G}(\mathrm{X})} \tilde{g}(Z|\mathcal{P}(\mathrm{X})) \tag{11.4.1}$$

where $\mathcal{G}(\mathrm{X})$ is the set of all partitions of X (note that the partitioning object here is the target set X, while the partitioning object in the extended target tracking is the measurement set Z), and $\tilde{g}(Z|\mathcal{P}(\mathrm{X}))$ is the measurement likelihood obtained according to the partition $\mathcal{P}(\mathrm{X})$ conditioned on targets being detected, see (11.4.2). This is similar to the method used in [291], both of which consider the target partitioning to form resolved events, the main difference is that the likelihood is defined here as a function of a set of target states instead of a fixed-dimensional vector. Thus, the expression for the posterior density can be derived using the FISST principle.

In order to obtain the expression of $\tilde{g}(Z|\mathcal{P}(\mathrm{X}))$, the standard likelihood function in (3.4.20) is extended to the target group, so that the following equation can be obtained

$$\tilde{g}(Z|\mathcal{P}(\mathrm{X})) = \exp(-\langle\kappa, 1\rangle)\kappa^{Z} \sum_{\theta\in\Theta(\mathcal{P}(\mathcal{L}(\mathrm{X})))} \left[\tilde{\varphi}_{Z}(\cdot;\theta)\right]^{\mathcal{P}(\mathrm{X})} \tag{11.4.2}$$

where for convenience of description, $\mathcal{P}(\mathrm{X})$ is used to represent a partition of the state-label in the labeled set X, $\mathcal{P}(\mathcal{L}(\mathrm{X}))$ represents the relevant partition that only contains labels, $\Theta(\mathcal{P}(\mathcal{L}(\mathrm{X})))$ is defined as the set of all one-to-one maps from target label groups in $\mathcal{P}(\mathcal{L}(\mathrm{X}))$ to measurement indices in Z, i.e., $\theta : \mathcal{P}(\mathcal{L}(\mathrm{X})) \rightarrow \{0, 1, \ldots, |Z|\}$, where $\theta(L) = \theta(J) > 0 \Rightarrow L = J$, and $\tilde{\varphi}_Z(\mathrm{Y};\theta)$ is the group likelihood defined as

$$\tilde{\varphi}_Z(\mathrm{Y};\theta) = \begin{cases} \frac{\tilde{p}_D(\mathrm{Y})g\left(z_{\theta(\mathcal{L}(\mathrm{Y}))}|\mathrm{Y}\right)}{\kappa\left(z_{\theta(\mathcal{L}(\mathrm{Y}))}\right)}, & \theta(\mathcal{L}(\mathrm{Y})) > 0 \\ \tilde{q}_D(\mathrm{Y}), & \theta(\mathcal{L}(\mathrm{Y})) = 0 \end{cases} \tag{11.4.3}$$

where $\tilde{p}_D(\mathrm{Y})$ is the detection probability of target group Y, $\tilde{q}_D(\mathrm{Y}) = 1 - \tilde{p}_D(\mathrm{Y})$ is the missed detection probability of group Y, and $g(z_{\theta(\mathcal{L}(\mathrm{Y}))}|\mathrm{Y})$ is the likelihood of measurement $z_{\theta(\mathcal{L}(\mathrm{Y}))}$ given target group Y, whose form generally depends on the sensor characteristics. For simplicity, $g(\cdot|\mathrm{Y})$ can be modeled as a Gaussian distribution with its center located at the group mean, and the corresponding standard deviation is related to the group size. Note that in (11.4.2), the exponent $\mathcal{P}(\mathrm{X})$ is a set of targets, while the base $\tilde{\varphi}_Z(\cdot;\theta)$ is the real-valued function whose parameter is the target set.

Substituting (11.4.2) into (11.4.1) leads to the likelihood function as

$$g(Z|\mathrm{X}) = \exp(-\langle\kappa, 1\rangle)\kappa^{Z} \sum_{\substack{\mathcal{P}(\mathrm{X})\in\mathcal{G}(\mathrm{X}) \\ \theta\in\Theta(\mathcal{P}(\mathcal{L}(\mathrm{X})))}} \left[\tilde{\varphi}_{Z}(\cdot;\theta)\right]^{\mathcal{P}(\mathrm{X})} \tag{11.4.4}$$

11.4.2 General Form of Target Tracker with Merged Measurements

The labeled RFS mixture density over the state-space $\mathbb{X}$ and the discrete label space $\mathbb{L}$ is given by (7.5.9). Note that the labeled RFS mixture density (7.5.9) is different from the GLMB density given by (3.3.38), where each $p^{(c)}(\mathrm{X})$ is the joint density of all targets in X, while in (3.3.38) it is limited to the product of single-target densities. In fact, any labeled RFS density can be written as the mixture form of (7.5.9), so the mixture form in (7.5.9) is more general. The mixture form is used here, because it can conveniently provide the update step of the tracker with merged measurements.

Proposition 31 *If the multi-target prior is a labeled RFS mixture density of the form* (7.5.9), *after the transition kernel* (3.4.13) *is applied, the predicted multi-target density is also the labeled RFS mixture density, which is*

$$\pi_+(\mathrm{X}_+) = \Delta(\mathrm{X}_+) \sum_{c\in\mathbb{C}} \sum_{L\subseteq\mathbb{L}} w_{+,L}^{(c)}(\mathcal{L}(\mathrm{X}_+)) p_{+,L}^{(c)}(\mathrm{X}_+) \tag{11.4.5}$$

where

$$w_{+,L}^{(c)}(J) = 1_L(J \cap \mathbb{L}) w_\gamma(J - \mathbb{L}) w^{(c)}(L) \eta_S^{(c)}(L) \tag{11.4.6}$$

$$p_{+,L}^{(c)}(\mathrm{X}_+) = \left[p_\gamma(\cdot)\right]^{\mathrm{X}_+ - \mathbb{X}\times\mathbb{L}} p_{S,L}^{(c)}(\mathrm{X}_+ \cap \mathbb{X}\times\mathbb{L}) \tag{11.4.7}$$

$$p_{S,L}^{(c)}(\mathrm{S}) = \int p_{\ell_{1:|L|}}^{(c)}(\boldsymbol{x}_{1:|L|}) \prod_{i=1}^{|L|} \Phi(\mathrm{S}; \boldsymbol{x}_i, \ell_i) d\boldsymbol{x}_{1:|L|} / \eta_S^{(c)}(L) \tag{11.4.8}$$

$$\eta_S^{(c)}(L) = \iint p_{\ell_{1:|L|}}^{(c)}(\boldsymbol{x}_{1:|L|}) \prod_{i=1}^{|L|} \Phi(\mathrm{S}; \boldsymbol{x}_i, \ell_i) d\boldsymbol{x}_{1:|L|} \delta\mathrm{S} \tag{11.4.9}$$

$$p_{\ell_{1:|L|}}^{(c)}(\boldsymbol{x}_{1:|L|}) \triangleq p^{(c)}\left(\{(\boldsymbol{x}_1, \ell_1), \ldots, (\boldsymbol{x}_{|L|}, \ell_{|L|})\}\right) \tag{11.4.10}$$

In the above equations, parameters w_γ and p_γ are the weight and the density of the birth density π_γ (3.4.11), *$\Phi(\mathrm{S}; \boldsymbol{x}, \ell)$ is defined in* (3.4.10), *and for any function $f : \mathcal{F}(\mathbb{X} \times \mathbb{L}) \to \mathbb{R}$, the abbreviation $f_{\ell_{1:n}}(\boldsymbol{x}_{1:n}) = f(\{(\boldsymbol{x}_1, \ell_1), \ldots, (\boldsymbol{x}_n, \ell_n)\})$ is used to represent the function defined over $\mathbb{X}^n$ given label $\ell_{1:n}$.*

Proposition 32 *If the prior is the labeled RFS mixture density of the form* (7.5.9), *after the likelihood function* (11.4.4) *is applied, the updated (posterior) multi-target density is also the labeled RFS mixture density, which is*

$$\pi(\mathrm{X}|\mathrm{Z}) = \Delta(\mathrm{X}) \sum_{c\in\mathbb{C}} \sum_{\substack{\mathcal{P}(\mathrm{X})\in\mathcal{G}(\mathrm{X}) \\ \theta\in\Theta(\mathcal{P}(\mathcal{L}(\mathrm{X})))}} w_Z^{(c,\theta)}(\mathcal{P}(\mathcal{L}(\mathrm{X}))) p^{(c,\theta)}(\mathcal{P}(\mathrm{X})|\mathrm{Z}) \tag{11.4.11}$$

where

$$w_Z^{(c,\theta)}(\mathcal{P}(L)) = \frac{w^{(c)}(L) \eta_Z^{(c,\theta)}(\mathcal{P}(L))}{\sum_{c\in\mathbb{C}} \sum_{J\subseteq\mathbb{L}} \sum_{\substack{\mathcal{P}(J)\in\mathcal{G}(J) \\ \theta\in\Theta(\mathcal{P}(J))}} w^{(c)}(J) \eta_Z^{(c,\theta)}(\mathcal{P}(J))} \tag{11.4.12}$$

$$p^{(c,\theta)}(\mathcal{P}(\mathrm{X})|Z) = \left[\tilde{\varphi}_Z(\cdot;\theta)\right]^{\mathcal{P}(\mathrm{X})} p^{(c)}(\mathrm{X})/\eta_Z^{(c,\theta)}(\mathcal{P}(\mathcal{L}(\mathrm{X}))) \tag{11.4.13}$$

$$\eta_Z^{(c,\theta)}(\mathcal{P}(L)) = \int \left[\tilde{\varphi}_Z(\cdot;\theta)\right]^{\mathcal{P}(L)} p^{(c)}_{\ell_{1:|L|}}(\boldsymbol{x}_{1:|L|})d\boldsymbol{x}_{1:|L|} \tag{11.4.14}$$

Appendix N gives the proofs of Propositions 31 and 32. Technically, since the labeled RFS mixture density family is closed under the Chapman–Kolmogorov (C–K) prediction and Bayesian update, it can be regarded as a conjugate prior with respect to the standard measurement likelihood and merged measurement likelihood. However, this is a very general class of priors that are usually difficult to deal with in practice.

Although the recursion equation of this filter is obtained based on Proposition 31 or Proposition 32, the recursion will quickly become intractable if there isn't an explicit method to prune the summation over the target group-measurement association space $\Theta(\mathcal{P}(\mathcal{L}(\mathrm{X})))$. The reason is that, each component in the density (7.5.9) is composed of a single joint density containing all targets, which means that the standard ranked assignment technique cannot be applied. In order to obtain a tractable version of the GLMB filter under the merged measurement likelihood, an approximation method is introduced as follows.

11.4.3 Tractable Approximation

Due to the existence of the joint density $p^{(c)}(\mathrm{X})$ in the prior (7.5.9), the exact form of the filter given by Propositions 31 and 32 is intractable. In order to obtain a practical and feasible algorithm, the labeled RFS mixture prior is approximated with the GLMB. However, applying the merged measurement likelihood function to the GLMB prior will lead to a labeled RFS mixture posterior rather than the GLMB posterior. However, by marginalizing the updated joint density, the posterior can be approximated as a GLMB posterior, thus realizing a recursive filter. According to Proposition 22, the GLMB approximation here preserves the first-order moment and the cardinality distribution.

For the prior density of the GLMB form (3.3.38), applying the likelihood function (11.4.4) will give the following labeled RFS mixture form posterior

$$\pi(\mathrm{X}|Z) = \Delta(\mathrm{X}) \sum_{c\in\mathbb{C}} \sum_{\substack{\mathcal{P}(\mathrm{X})\in\mathcal{G}(\mathrm{X}) \\ \theta\in\Theta(\mathcal{P}(\mathcal{L}(\mathrm{X})))}} w_Z^{(c,\theta)}(\mathcal{P}(\mathcal{L}(\mathrm{X})))\left[p^{(c,\theta)}(\cdot|Z)\right]^{\mathcal{P}(\mathrm{X})} \tag{11.4.15}$$

where

$$w_Z^{(c,\theta)}(\mathcal{P}(L)) = \frac{w^{(c)}(L)\left[\eta_Z^{(c,\theta)}\right]^{\mathcal{P}(L)}}{\sum_{c\in\mathbb{C}}\sum_{J\subseteq\mathbb{L}}\sum_{\substack{\mathcal{P}(J)\in\mathcal{G}(J)\\ \theta\in\Theta(\mathcal{P}(J))}} w^{(c)}(J)\left[\eta_Z^{(c,\theta)}\right]^{\mathcal{P}(J)}} \tag{11.4.16}$$

$$p^{(c,\theta)}(\mathrm{Y}|\mathrm{Z}) = \tilde{\varphi}_{\mathrm{Z}}(\mathrm{Y};\theta)\tilde{p}^{(c)}(\mathrm{Y})/\eta_Z^{(c,\theta)}(\mathcal{L}(\mathrm{Y})) \tag{11.4.17}$$

$$\tilde{p}^{(c)}(\mathrm{Y}) = \left[p^{(c)}(\cdot)\right]^{\mathrm{Y}} \tag{11.4.18}$$

$$\eta_Z^{(c,\theta)}(L) = \left\langle \tilde{p}^{(c)}_{\ell_{1:|L|}}(\cdot), (\tilde{\varphi}_{\mathrm{Z}})_{\ell_{1:|L|}}(\cdot;\theta)\right\rangle \tag{11.4.19}$$

with $\tilde{\varphi}_{\mathrm{Z}}(\mathrm{Y};\theta)$ being given by (11.4.3). The above derivation process is given in Appendix N.

In order to convert the posterior density back to the desired GLMB form, marginalization is performed on the joint density in each component of the labeled RFS mixture, i.e.,

$$p^{(c,\theta)}(\mathrm{Y}|\mathrm{Z}) \approx \left[\hat{p}^{(c,\theta)}(\cdot|\mathrm{Z})\right]^{\mathrm{Y}} \tag{11.4.20}$$

$$\hat{p}^{(c,\theta)}(\boldsymbol{x}_i,\ell_i|\mathrm{Z}) \triangleq \int p^{(c,\theta)}_{\ell_{1:|L|}}\left(\boldsymbol{x}_{1:|L|}|\mathrm{Z}\right)d\left(\boldsymbol{x}_{1:i-1},\boldsymbol{x}_{i+1:|L|}\right) \tag{11.4.21}$$

After the marginalization operation, the product of joint densities in partition $\mathcal{P}(\mathrm{X})$ degenerates into the product of independent densities in X, thus, the following approximation can be obtained

$$\left[p^{(c,\theta)}(\cdot|\mathrm{Z})\right]^{\mathcal{P}(\mathrm{X})} \approx \left[\hat{p}^{(c,\theta)}(\cdot|\mathrm{Z})\right]^{\mathrm{X}} \tag{11.4.22}$$

This leads to the following GLMB approximation of the multi-target posterior

$$\pi(\mathrm{X}|\mathrm{Z}) \approx \Delta(\mathrm{X})\sum_{c\in\mathbb{C}}\sum_{\substack{\mathcal{P}(\mathrm{X})\in\mathcal{G}(\mathrm{X})\\ \theta\in\Theta(\mathcal{P}(\mathcal{L}(\mathrm{X})))}} w_Z^{(c,\theta)}(\mathcal{P}(\mathcal{L}(\mathrm{X})))\left[\hat{p}^{(c,\theta)}(\cdot|\mathrm{Z})\right]^{\mathrm{X}} \tag{11.4.23}$$

Obviously, computing this posterior is still very computationally expensive, as it traverses the triple-nested summation of the prior component, the target set partition, and the target group-measurement association. Therefore, a corresponding pruning method is required to deal with the triple-nested summation so as to achieve an efficient implementation of the filter.

The most natural way to calculate the terms in (11.4.23) is enumerate all possible partitions of the target set, and then generate measurement-target group assignment within each partition. Obviously, in the presence of a large number of targets, this step is not feasible due to the exponential growth of the number of partitions. However, by using such two types of information as the group structure of the target set and the sensor resolution model, it is possible to delete many impossible partitions. By constructing mutually exclusive subsets of well separated targets which are unlikely to share measurements, the targets are clustered, and then an independent filter is implemented for each cluster. Hence, for each filter, it is only required to partition a smaller set of targets, thus significantly reducing the total number of partitions to be calculated [54].

One way to further reduce the number of partitions for each cluster is to use a sensor resolution model to remove less likely partitions. There are already many resolution models that can be used to approximate the probability of a resolution event (or equivalently, a partition of targets), and some partitions can also be eliminated by retaining only partitions whose weights exceed some minimum probability threshold. In order to delete impossible partitions, a simple model is introduced, which requires two parameters: the maximum divergence $d_{\max}^{t}$ of targets in the unresolved group in the measurement space, and the minimum distance $d_{\min}^{g}$ between the centroids of the unresolved group in the measurement space.

A partition $\mathcal{P}(\mathrm{X})$ is said to be infeasible if any of the following conditions is true

$$D_{\min}\left(\{\overline{\mathrm{Y}};\ \mathrm{Y} \in \mathcal{P}(\mathrm{X})\}\right) < d_{\min}^{g} \tag{11.4.24}$$

$$\max(\{D_{\max}(\mathrm{Y});\ \mathrm{Y} \in \mathcal{P}(\mathrm{X})\}) > d_{\max}^{t} \tag{11.4.25}$$

where, $\overline{\mathrm{Y}}$ is the mean of the set Y, $D_{\min}(\mathrm{Y})$ is the minimum distance between two points in Y, and $D_{\max}(\mathrm{Y})$ is the maximum distance between two points in Y. The above model can be applied to each measurement dimension separately. If the condition (11.4.24) holds for all dimensions or the condition (11.4.25) holds for any dimension, the partition will be rejected. Intuitively, this model imposes the constraint that any unresolved target group cannot be spread too far, and the resolved target group cannot be too close to each other. By applying this constraint, impossible partitions are removed, thereby allowing the filter to devote the limited computational resources to processing more likely partitions.

After feasible partitions are obtained, the posterior GLMB density can be calculated. Similar to the standard GLMB filter, the Murty algorithm is used to generate a set of alternative measurement assignment rankings. The concrete implementation of this algorithm can be found in [94]. Nevertheless, the difference is that measurements are assigned to target groups within each partition, instead of being assigned to independent targets.

11.5 Summary

This chapter introduces the corresponding target tracking algorithms with non-standard measurements for such two special targets as the extended target and the unresolved (merged measurement) target. For extended target tracking, a GM-PHD-based extended target tracking algorithm is first given. This algorithm is relatively simple and ignores the extension of the extended target. Nevertheless, the measurement set partitioning step in this algorithm is the key to understand the complete target tracking with non-standard measurements and even multi-sensor target tracking in the next chapter. Then, in order to further estimate the extension of the extended targets, the GGIW distribution-based extended target tracking algorithm is introduced. On the basis of the GGIW model for extended targets, the Bayesian filter based on the GGIW distribution for a single extended target as well as the CPHD and labeled RFS filters for multiple extended targets are introduced from easy to difficult. It is worth emphasizing that, unlike the Doppler measurement model in Chap. 9 and the TBD measurement model in Chap. 10, which only influence the update step of the filters, in the extended target model, the GGIW model modifies the state transition function. Hence, for RFS filters suitable for extended target tracking, not only the update step is changed by the extended target measurement model, but also the prediction step is accordingly changed by the GGIW model. Finally, (unresolved) target tracking with merged measurements is provided, and then the generalized form of target tracker with merged measurements and its tractable approximation are given based on the multi-target likelihood model of merged measurements. It should be noted that, these algorithms utilize more complex non-standard measurement models and relax the assumptions on standard multi-target measurement models, thus usually at the expense of increased computational complexity.

Chapter 12
Distributed Multi-sensor Target Tracking

12.1 Introduction

The RFS-based multi-target tracking algorithms introduced in the previous chapters are mainly aimed at a single sensor. The recent advances in sensor networking technology have given rise to the development of large-scale sensor networks composed of interconnected nodes (or agents) with sensing, communication and processing capabilities. Given all sensor data, one of the ways of establishing the posterior density of the multi-target state is to send all measurements from all sensing systems to a central station for fusion. Although this centralized scheme is optimal, it is required to send all measurements to a single station, which may result in a heavy communication burden. Moreover, since the entire network will stop working once the central station fails, this method makes the sensor network vulnerable. Another method is to treat each sensing system as a node in the distributed system. These nodes collect and process measurements locally to obtain local estimates, and then these local estimates (rather than original measurements) are periodically broadcast to other nodes for data fusion across the entire network. This method is referred to as the distributed fusion.

Distributed fusion has many potential advantages. In distributed data fusion, fusion is performed on the entire network rather than a single central site, which has the advantage of reliability. If a single node fails, other nodes can continue to work and information can be communicated through other nodes in the network. Moreover, the distributed fusion system has better flexibility and scalability (relative to the number of nodes), and additional nodes with different processing algorithms and sensing systems can be added or removed as needed at any time.

In a word, distributed fusion technology can establish a more comprehensive environmental situation by combining the information of each node (usually with limited observation capabilities), and has the characteristics of scalability, flexibility

W. Wu et al., *Target Tracking with Random Finite Sets*,
https://doi.org/10.1007/978-981-19-9815-7_12

and reliability. However, in order to give full play to the above advantages, it is necessary to re-design structures and algorithms to solve the following problems:

- Lack of central fusion node;
- They should be scalable relative to the scale of the network (that is, the computational burden of each node is independent of the scale of the network), and should have scalable processing capabilities;
- Each node is able to operate under unknown network topology (e.g., the number of nodes and the connection relationship between nodes are unknown);
- Each node can operate without knowing the correlation between its own information and information received from other nodes.

In order to combine the limited information of all nodes, an appropriate information fusion step is required to reconstruct target states existing in the surrounding environment. The need for scalability, the lack of fusion centers, and the lack of knowledge of network topology require a consensus approach to achieve overall fusion on the network through iterative local fusion between adjacent nodes. Furthermore, a robust (but sub-optimal) fusion rule is often required, due to the data incest problem that may lead to double computation of information, especially when there is a network loop.

Under the RFS framework, this chapter focuses on the distributed multi-target tracking on a network of heterogeneous and geographically separated nodes. As mentioned in Chap. 1, in order to fuse the multi-target densities calculated by different nodes of the network, a generalized covariance intersection (GCI) is used. The GCI fusion is also referred to as the Chernoff fusion, KLA fusion, exponential mixture density (EMD) fusion, etc. First, the distributed multi-target tracking problem is described from the perspective of the system model and the solving goal. Then, based on the single-target KLA and consensus algorithm, a relatively simple distributed single-target filtering and fusion is presented. Next, based on the multi-target KLA, the fusion formulas for the CPHD, Mδ-GLMB and LMB densities are provided. Finally, the distributed fusion for both the SMC-CPHD filter and the Gaussian mixture RFS filters are presented, and the latter includes consensus the GM-CPHD filter, the consensus GM-M δ-GLMB and the consensus GM-LMB filter.

12.2 Formulation of Distributed Multi-target Tracking Problem

12.2.1 System Model

In a distributed network, each node (agent) locally updates the multi-target information using multi-target dynamics and obtained local measurements, exchanges the information with adjacent nodes through communication, and combines the information of all adjacent nodes for fusion. Consider the network of tracking nodes as

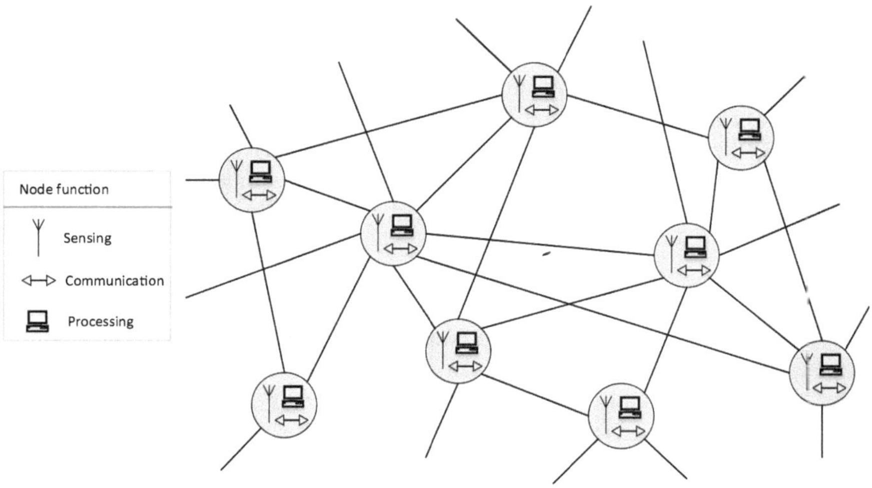

Fig. 12.1 Multi-sensor network model [92, 329]

shown in Fig. 12.1, which is composed of geographically dispersed heterogeneous nodes with sensing, processing, and communication capabilities. Specifically, each node is able to acquire the measurement information (such as distance, angle, Doppler frequency shift, etc.) of motion variables related to moving targets in the surrounding environment (monitoring area), process local data, and exchange data with adjacent nodes.

From a mathematical point of view, the above network can be described as a directed graph $\mathcal{G} = (\mathcal{S}, \mathcal{A})$, where $\mathcal{S}$ is the set of nodes, the total number of nodes in the network is denoted as $|\mathcal{S}|$ (i.e., the cardinality of $\mathcal{S}$), and $\mathcal{A} \subseteq \mathcal{S} \times \mathcal{S}$ is the set of arcs representing connections (or links). Specifically, if node j can receive the data of node i, then $(i, j) \in \mathcal{A}$. For any node $j \in \mathcal{S}$, $\mathcal{I}_j \triangleq \{i \in \mathcal{S} : (i, j) \in \mathcal{A}\}$ indicates the set of in-neighbour nodes of j (including node j itself), that is, the set of all nodes that node j can receive data from. By definition, for any node j, we have $(j, j) \in \mathcal{A}$. Therefore, for all j, we have $j \in \mathcal{I}_j$.

12.2.2 *Solving Goal*

In a multi-sensor multi-target scenario, the multi-target motion model and measurement model are the same as those in Sect. 3.4 and the main difference is that the number of sensors is no longer 1 but greater than 1. The measurements received by sensor i at time k are denoted as $Z_{k,i}$, and the measurements up to time k are denoted as $Z_{1:k,i} = \{Z_{1,i}, \ldots, Z_{k,i}\}$. Given all sensor data, the goal here is to construct the posterior density of the multi-target state, i.e.,

$$\pi_k\big(X_k|\{Z_{1:k,i}\}_{i\in\mathcal{S}}\big) \tag{12.2.1}$$

For the convenience of exposition, consider any two nodes (i and j) in the sensor fusion network. At any given time, the update of a node only uses the information from a subset of other nodes in the network, and each node receives the sensor information and has a posterior distribution $\pi(X_k|Z_{1:k,i})$. Node i periodically sends its posterior to node j, and node j fuses this posterior and calculates the joint posterior

$$\pi\big(X_k|Z_{1:k,i}, Z_{1:k,j}\big) = \pi\big(X_k|Z_{1:k,i} \cup Z_{1:k,j}\big) \tag{12.2.2}$$

When the information of different nodes is optimally fused, it cannot be simply assumed that $\pi(X_k|Z_{1:k,i})$ and $\pi(X_k|Z_{1:k,j})$ are conditionally independent of each other, i.e., there is actually some correlation between $\pi(X_k|Z_{1:k,i})$ and $\pi(X_k|Z_{1:k,j})$. This correlation comes from two reasons. First, when two nodes track the same target, the common process noise will be generated. Second, when nodes exchange their local estimates with each other, the common measurement noise will also be generated. To accurately model the above correlation, we have

$$\pi\big(X_k|Z_{1:k,i}, Z_{1:k,j}\big) \propto \frac{\pi\big(X_k|Z_{1:k,i}\big)\pi\big(X_k|Z_{1:k,j}\big)}{\pi\big(X_k|Z_{1:k,i} \cap Z_{1:k,j}\big)} \tag{12.2.3}$$

This update formula indicates that the common information between nodes must be "removed". For different network topologies, there are different corresponding formulas [429]. However, in almost all cases, only when some global managers continuously monitor the state of the entire network, $\pi(X_k|Z_{1:k,i} \cap Z_{1:k,j})$ can be solved. The only exception is the tree topology where a single path exists between any two nodes. By using the so-called channel filter for monitoring the information flowing over the arcs, the common information can be solved. However, since the failure of a single node will split the network, the tree topology is inherently vulnerable, and thus, the optimal distributed data fusion algorithm can only be implemented under harsh conditions.

Assume that each node i in the sensor network can calculate the multi-target density $\pi_i(X)$ based on the information it obtains (collected locally or from other nodes). The goal now is to design a suitable distributed algorithm to ensure that all nodes in the network reach consensus on the multi-target density of an unknown multi-target set. An important feature that the designed algorithm should have is scalability, and for this reason, the optimal (Bayesian) fusion method described above [301] can be excluded, because the optimal method requires that each pair of adjacent nodes (i, j) needs to know the multi-target density $\pi(X_k|Z_{1:k,i} \cap Z_{1:k,j})$ conditioned on common information $Z_{1:k,i} \cap Z_{1:k,j}$, while in real networks it is impossible to track such common information in a scalable manner. Therefore, most of the advantages of distributed fusion can only be achieved under strictly local conditions (i.e., each node only knows the properties of its immediate neighbors). In this case, the nodes don't need to know the overall topology of the network. To achieve these goals, some robust multi-target tracking and sub-optimal distributed fusion techniques must be employed.

12.3 Distributed Single-Target Filtering and Fusion

For ease of understanding, a simple distributed single target filtering and fusion problem is described first. It is assumed that the dynamic model has the following Markov transition density

$$\phi_{k|k-1}(\boldsymbol{x}_k|\boldsymbol{x}_{k-1}) \tag{12.3.1}$$

For each node $i \in \mathcal{S}$, the likelihood function is

$$g_{k,i}\left(z_{k,i}|\boldsymbol{x}_k\right) \tag{12.3.2}$$

Let $p_{k|k-1}$ denote the conditional PDF of $\boldsymbol{x}_k$ given $z_{1:k-1} \triangleq \{z_1, \ldots, z_{k-1}\}$. Similarly, $p_k(\boldsymbol{x}_k|z_k)$ represents the conditional PDF of $\boldsymbol{x}_k$ given $z_{1:k} \triangleq \{z_1, \ldots, z_k\}$. Strictly speaking, $p_{k|k-1}$ and p_k shall be written as $p_{k|k-1}(\cdot|z_1, \ldots, z_{k-1})$ and $p_k(\cdot|z_1, \ldots, z_{k-1}, z_k)$, respectively, but their dependence on the measurement history is ignored for the sake of simplicity.

In a centralized fusion structure, each node can obtain the information of all measurements $z_k = (z_{k,1}, \ldots, z_{k,|\mathcal{S}|})$. Given a suitable initial distribution p_0, the solution to the state estimation problem is recursively calculated by the following Bayesian filter

$$p_{k|k-1}(\boldsymbol{x}_k) = \left\langle \phi_{k|k-1}(\boldsymbol{x}_k|\cdot), p_{k-1}(\cdot|z_{k-1}) \right\rangle \tag{12.3.3}$$

$$p_k(\boldsymbol{x}_k|z_k) = \left(g_k(z_k|\boldsymbol{x}_k) \oplus p_{k|k-1}(\boldsymbol{x}_k)\right) \tag{12.3.4}$$

where information fusion operator $\oplus$ is defined in Appendix O. The measurement of each sensor is assumed to be independent of the target state, hence, $g_k(\cdot|\cdot)$ is

$$g_k(z_k|\boldsymbol{x}_k) = \prod_{i \in \mathcal{S}} g_{k,i}\left(z_{k,i}|\boldsymbol{x}_k\right) \tag{12.3.5}$$

According to the above, the multi-sensor measurement likelihood has no effect on the prediction step, and only exerts the impact on the update step. Therefore, in the multi-sensor target tracking in this chapter, only the update step is concerned. In fact, as we will see later, the main object of distributed fusion is the posterior density after the update step.

On the other hand, in a distributed configuration, each node (agent) $i \in \mathcal{S}$ updates the local posterior density $p_{k,i}$ by appropriately fusing the information provided by sub-networks $\mathcal{I}_i$ (including node i). As a result, the key to distributed estimation is to fuse the posterior densities of targets of interest provided by different nodes in a mathematically consistent manner. The KLD-based Kullback–Leibler average (KLA), a concept from the information theory [305], offers a consistent way for the PDF fusion.

12.3.1 Single Target KLA

Since the main focus is on the fusion of posterior densities, the subscript representing the timestamp k will be omitted in the following as long as it does not cause confusion. Given the set $\{(\omega_i, p_i)\}_{i\in\mathcal{I}}$ of single-target PDFs and the related weights, where $\omega_i \geq 0$, $\sum_{i\in\mathcal{I}} \omega_i = 1$, then the weighted KLA $\overline{p}$ is defined as [430]

$$\overline{p} = \arg_p \inf \sum_{i\in\mathcal{I}} \omega_i \mathcal{D}_{\text{KL}}(p||p_i) \tag{12.3.6}$$

where inf indicates the infimum, and $\mathcal{D}_{\text{KL}}$ is the KLD between the single target densities p and p_i, which is calculated by

$$\mathcal{D}_{\text{KL}}(p||p_i) = \int p(\boldsymbol{x}) \log \frac{p(\boldsymbol{x})}{p_i(\boldsymbol{x})} d\boldsymbol{x} \tag{12.3.7}$$

In fact, the weighted KLA in (12.3.6) is equivalent to the normalized weighted geometric mean (NWGM) of PDFs [305], i.e.,

$$\overline{p}(\boldsymbol{x}) \triangleq \bigoplus_{i\in\mathcal{I}} (\omega_i \odot p_i(\boldsymbol{x})) = \frac{\prod_{i\in\mathcal{I}} p_i^{\omega_i}(\boldsymbol{x})}{\int \prod_{i\in\mathcal{I}} p_i^{\omega_i}(\boldsymbol{x}) d\boldsymbol{x}} \tag{12.3.8}$$

where the weighting operator $\odot$ is defined in Appendix O and the above formula is known as the Chernoff fusion rule [304]. For unweighted KLA, i.e., $\omega_i = 1/|\mathcal{I}|$, we have

$$\overline{p}(\boldsymbol{x}) = \oplus_{i\in\mathcal{I}}((1/|\mathcal{I}|) \odot p_i(\boldsymbol{x})) \tag{12.3.9}$$

Remark The weighted KLA of Gaussian distributions is still a Gaussian [305]. More specifically, let $(\boldsymbol{u}, \boldsymbol{I}) \triangleq (\boldsymbol{P}^{-1}\boldsymbol{m}, \boldsymbol{P}^{-1})$ represent the vector–matrix information tuple associated with Gaussian $\mathcal{N}(\cdot; \boldsymbol{m}, \boldsymbol{P})$, then the information tuple $(\overline{\boldsymbol{u}}, \overline{\boldsymbol{I}})$ of KLA $\overline{p}(\cdot) = \mathcal{N}(\cdot; \overline{\boldsymbol{m}}, \overline{\boldsymbol{P}})$ is the weighted arithmetic mean of the information tuple $(\boldsymbol{u}_i, \boldsymbol{I}_i)$ of $p_i(\cdot) = \mathcal{N}(\cdot; \boldsymbol{m}_i, \boldsymbol{P}_i)$. This actually corresponds to the well-known covariance intersection (CI) fusion rule.

After establishing the KLA concept of the PDF fusion, the following describes the distributed and scalable consensus algorithm for computing the KLA for each node in the network.

12.3.2 Consensus Algorithm

The consensus algorithm [299, 431] is a powerful tool commonly used for distributed computing (e.g., minimization, maximization, averaging, etc.) over the entire network, and has been widely applied to distributed parameter/state estimation [432–435]. In its basic form, the consensus algorithm can be viewed as a distributed averaging technique over the entire network. The purpose of each node is to calculate the global mean of a given quantity from iterative local means, in which the terms “global” and “local” mean for all network nodes and only neighboring nodes, respectively. The core idea of the consensus algorithm is to iteratively update and transmit its local information to neighboring nodes for each node, thus reaching a collective agreement across the entire network. These repeated iterative local operations provide a mechanism for transferring information across the entire network. In order to obtain a reliable and scalable distributed multi-target tracking algorithm, at each moment k, the consensus is used for distributed computation of the global unweighted KLA of the posterior densities $p_{k,i}$ of nodes $i \in \mathcal{S}$.

It starts with a brief introduction to the basic average consensus problem. Let node i provide estimation $\hat{\theta}_i$ of a given variable θ. The goal now is to obtain an algorithm, which can calculate the following mean in a distributed fashion on each node

$$\overline{\theta} = \sum_{i \in \mathcal{S}} \hat{\theta}_i / |\mathcal{S}| \tag{12.3.10}$$

Let $\hat{\theta}_{i,0} = \hat{\theta}_i$, then a simple consensus algorithm takes the following iterative form

$$\hat{\theta}_{i,c} = \sum_{j \in \mathcal{I}_i} \omega_{i,j} \hat{\theta}_{j,c-1}, \quad \forall i \in \mathcal{S} \tag{12.3.11}$$

where the subscript c indicates the cth iteration step and the consensus weight satisfies the following condition

$$\omega_{i,j} \geq 0 \quad \forall i, j \in \mathcal{S}; \qquad \sum_{j \in \mathcal{I}_i} \omega_{i,j} = 1 \quad \forall i \in \mathcal{S} \tag{12.3.12}$$

Note that Eq. (12.3.11) indicates that the consensus estimate given by any node is computed by a convex combination of the previous estimates in the previous consensus step of neighboring nodes. In other words, the iteration (12.3.11) is only the local mean computed by node i, and the goal of the consensus step is to make the local mean converge to the global mean (12.3.10). It shall be emphasized that, the consensus weight has an important impact on the convergence property [299, 431]. Denote $\boldsymbol{\Omega}$ as the consensus matrix, and its (i, j) element is consensus weight $\omega_{i,j}$ (if $j \notin \mathcal{I}_i$, then $\omega_{i,j}$ is 0). Therefore, if the consensus matrix $\boldsymbol{\Omega}$ is not only a primitive matrix but also a doubly stochastic matrix,[1] the consensus algorithm (12.3.11) gradually tends to be the mean (12.3.10), i.e.,

$$\lim_{c\to\infty} \hat{\theta}_{i,c} = \overline{\theta}, \quad \forall i \in \mathcal{S} \tag{12.3.13}$$

A necessary condition for matrix $\boldsymbol{\Omega}$ to be a primitive matrix is that, the network $\mathcal{G}$ is strongly connected [432], that is, for any pair of nodes $i, j \in \mathcal{S}$, there is a directed path from i to j, and vice versa. For $\forall i \in \mathcal{S}$, $j \in \mathcal{I}_i$, when $\omega_{i,j} > 0$, this condition is satisfied. Moreover, when network $\mathcal{G}$ is undirected (i.e., node i can receive information from node j, and j is also able to receive information from i), to guarantee that $\boldsymbol{\Omega}$ is not only a primitive matrix but also a doubly stochastic matrix, and thus ensuring that the local mean converges to the global mean, the following Metropolis weights can be chosen [431, 432]

$$\omega_{i,j} = \frac{1}{1 + \max\{|\mathcal{I}_i|, |\mathcal{I}_j|\}}, \quad i \in \mathcal{S}, j \in \mathcal{I}_i, i \neq j \tag{12.3.14}$$

$$\omega_{i,i} = 1 - \sum_{j\in\mathcal{I}_i, j\neq i} \omega_{i,j} \tag{12.3.15}$$

12.3.3 Consensus-Based Suboptimal Distributed Single Target Fusion

It is assumed that, at time k, each node i starts from posterior p_i, takes it as the initial iteration $p_{i,0}$, and then calculates the cth consensus iteration with the following equation

$$p_{i,c}(\boldsymbol{x}) = \bigoplus_{j\in\mathcal{I}_i} \left(\omega_{i,j} \odot p_{j,c-1}(\boldsymbol{x})\right) \tag{12.3.16}$$

where $\omega_{i,j} \geq 0$ is the consensus weight associated with node i and node $j \in \mathcal{I}_i$, satisfying $\sum_{j\in\mathcal{I}_i} \omega_{i,j} = 1$. By using the properties of operators $\oplus$ and $\odot$ (see Appendix O), we can obtain [305]

[1] If the sum of all rows and columns of a non-negative square matrix $\boldsymbol{\Omega}$ is 1, then $\boldsymbol{\Omega}$ is said to be a doubly stochastic matrix. Further, if there is an integer m such that all elements of $\boldsymbol{\Omega}^m$ strictly positive, then $\boldsymbol{\Omega}$ is said to be a primitive matrix.

Table 12.1 Consensus single-object filtering (CSOF)

CSOF step (node i, time k)	
Local prediction	▷ See (12.3.3)
Local update	▷ See (12.3.4)
for $c = 1, \ldots, C$	
Information exchange	
Fusion	▷ See (12.3.16)
end	

$$p_{i,c}(\boldsymbol{x}) = \bigoplus_{j \in \mathcal{S}} \left(\omega_{i,j,c} \odot p_j(\boldsymbol{x})\right) \tag{12.3.17}$$

where $\omega_{i,j,c}$ is the (i, j)th term of $\boldsymbol{\Omega}^c$, $\boldsymbol{\Omega}$ is the consensus matrix with equal number of rows and columns, and its (i, j)th term is $\omega_{i,j} 1_{\mathcal{I}_i}(j)$, which can be interpreted that, as long as $\omega_{i,j,c} = 0$, p_j can be negligible in the fusion process. More importantly, if the consensus matrix $\boldsymbol{\Omega}$ is not only a primitive matrix but also a doubly stochastic matrix, then for any $i, j \in \mathcal{S}$, we have [299]

$$\lim_{c \to \infty} \omega_{i,j,c} = 1/|\mathcal{S}| \tag{12.3.18}$$

In other words, if the consensus matrix is not only a primitive matrix but also a doubly stochastic matrix, the consensus iteration of each node in the network tends to be the overall unweighted KLA of the posterior densities across the entire network [305].

In summary, in order to calculate the unweighted KLA over the entire network at a given time, each node $i \in \mathcal{S}$ independently performs the steps as listed in Table 12.1.

The above method assumes that there is only one target in the monitoring area. However, in most of practical applications, the number of targets is unknown and time-varying, and the measurements also faces such problems as missed detection, clutter and association uncertainty. It is not easy to apply the consensus method under such a general condition. In order to tackle the multi-target problem, the concept of the multi-target density is required, and the RFS framework provides a powerful tool for dealing with the multi-target density.

12.4 Fusion of Multi-target Densities

The RFS provides a rigorous notion for the probability density of the multi-target state. Therefore, the concept of single-target KLA (12.3.8) devised in Sect. 12.3.1 can be directly extended to the multi-target density (including labeled multi-target density. For the labeled RFS, it is only necessary to replace the labeled version with the corresponding label-free version) in a mathematical principle manner. Here, the concept of multi-target density in the measure theory proposed in [62] is adopted and it doesn't face the dimensional compatibility problem in the integration involving the exponential product of multi-target densities.

12.4.1 Multi-target KLA

Obviously, the first problem to be solved is how to define the mean of local multi-target densities $\pi_i(X)$. Drawing on the KLA (12.3.7) that has been used for the single-target PDFs, we can introduce the concept of the KLA of two multi-target densities. First, the concept of the KLD is extended to multi-target densities $\pi_i(X)$ and $\pi_j(X)$, i.e., the KLD of two multi-target densities π_i and π_j is defined as

$$\mathcal{D}_{\text{KL}}\big(\pi_i||\pi_j\big) = \int \pi_i(X) \log \frac{\pi_i(X)}{\pi_j(X)} \delta X \tag{12.4.1}$$

where the integral is a set integral. Therefore, the weighted KLA $\overline{\pi}(X)$ of multi-target densities $\pi_i(X), i \in \mathcal{I}$ is defined as

$$\overline{\pi}(X) \triangleq \arg_\pi \inf \sum_{i\in\mathcal{I}} \omega_i \mathcal{D}_{\text{KL}}(\pi||\pi_i) \tag{12.4.2}$$

where the parameter $\omega_i \in [0, 1]$ determines the relative weight assigned to each distribution, satisfying

$$0 \leq \omega_i \leq 1, i \in \mathcal{I}, \quad \sum_{i\in\mathcal{I}} \omega_i = 1 \tag{12.4.3}$$

It shall be noted that, Eq. (12.4.2) indicates that the weighted KLA of densities is the density that minimizes weighted sum of the KLD of these densities [436]. If $|\mathcal{I}|$ is the number of nodes, and $\omega_i = 1/|\mathcal{I}|$ for $i = 1, \ldots, |\mathcal{I}|$, then Eq. (12.4.2) represents the (unweighted) KLA. In the Bayesian statistics, KLD (12.4.1) is considered to be the information gain that is achieved when moving from prior $\pi_j(X)$ to posterior $\pi_i(X)$. Thus, according to (12.4.2), the average PDF is the PDF that minimizes the sum of information gains of the initial multi-target densities. Therefore, the optimal probability density representing the current information state will yield the least information gain, which is consistent with the principle of minimum discrimination information (PMDI) (for discussion about this principle and its relation with Gauss principle and maximum likelihood estimation, see [437]). It is important to follow the PMDI in order to deal with the so-called data incest phenomenon, because it is difficult to be aware of repeated use of the same block of information due to the existence of loops within the network. In other words, this will help avoid repeated computation or double counting in arbitrary network topologies [303].

Proposition 33 *The weighted KLA defined by* (12.4.2) *is the normalized weighted geometric mean of multi-target densities* $\pi_i, i \in \mathcal{I}$ *and it is (see* Appendix P *for proof)*

$$\overline{\pi}(X) \triangleq \bigoplus_{i\in\mathcal{I}} (\omega_i \odot \pi_i(X)) = \frac{\prod_{i\in\mathcal{I}} \pi_i^{\omega_i}(X)}{\int \prod_{i\in\mathcal{I}} \pi_i^{\omega_i}(X)\delta X} \tag{12.4.4}$$

In the case of labeled RFSs, following the proof in Appendix P, similar results can be obtained, which are formally equivalent to replacing the label-free X in the above equation with the corresponding labeled X.

When $\omega_i = 1/|\mathcal{I}|$, Eq. (12.4.4) becomes

$$\overline{\pi}(X) = \oplus_{i\in\mathcal{I}}((1/|\mathcal{I}|) \odot \pi_i(X)) \tag{12.4.5}$$

The KLA of local multi-target densities can be obtained according to fusion rule (12.4.4). In essence, fusion rule (12.4.4) is the multi-target version of Chernoff fusion (12.3.8) and it is consistent with the GCI proposed first by Mahler for multi-target fusion [301]. Specifically, $\overline{\pi}(X)$ in (12.4.4) is obviously the normalized weighted geometric mean of multi-target densities $\pi_i(X), i \in \mathcal{I}$, which is also referred to as the GCI or exponential mixture density (EMD) [301, 325]. The reason why it is named the GCI is that: the covariance intersection (CI) is a (single-target) PDF fusion rule established for Gaussian PDFs [301, 303], while Eq. (12.4.4) is a multi-target version of the (single-target) PDF fusion rule and therefore it is a generalization of the CI method. For the CI method, given the estimation $\hat{\boldsymbol{x}}_i$, related covariance $\boldsymbol{P}_i$ and unknown correlation from multiple estimators for the same variable $\boldsymbol{x}$, the CI fusion is expressed as

$$\hat{\boldsymbol{x}} = \boldsymbol{P} \sum_i \omega_i \boldsymbol{P}_i^{-1} \hat{\boldsymbol{x}}_i, \quad \boldsymbol{P} = \left(\sum_i \omega_i \boldsymbol{P}_i^{-1} \right)^{-1} \tag{12.4.6}$$

From the perspective of reducing the determinant of the posterior covariance matrix (or making the peak of the fusion distribution larger than that of the prior distribution), the CI fusion can obtain the information gain. Equation (12.4.6) is unique in that: for any weight ω_i satisfying (12.4.3), all estimations are assumed to be consensus, i.e.,

$$\mathrm{E}\left[(\boldsymbol{x} - \hat{\boldsymbol{x}}_i)(\boldsymbol{x} - \hat{\boldsymbol{x}}_i)^{\mathrm{T}}\right] \le \boldsymbol{P}_i, \quad \forall i \tag{12.4.7}$$

then the fused estimation is also consensus, i.e.,

$$\mathrm{E}\left[(\boldsymbol{x} - \hat{\boldsymbol{x}})(\boldsymbol{x} - \hat{\boldsymbol{x}})^{\mathrm{T}}\right] \le \boldsymbol{P} \tag{12.4.8}$$

For estimates that obey a normal distribution, it is easy to know that (12.4.6) is equivalent to

$$\overline{p}(\boldsymbol{x}) = \frac{\prod_i p_i^{\omega_i}(\boldsymbol{x})}{\int \prod_i p_i^{\omega_i}(\boldsymbol{x})d\boldsymbol{x}} \tag{12.4.9}$$

where $p_i(\cdot) \triangleq \mathcal{N}(\cdot; \hat{\boldsymbol{x}}_i, \boldsymbol{P}_i)$ is a Gaussian PDF with mean $\hat{\boldsymbol{x}}_i$ and covariance $\boldsymbol{P}_i$. In conclusion, it shall be noted that, the consensus property is the major motivation for developing the CI fusion rule and it is actually consistent with the PMDI considered in the multi-target fusion.

12.4.2 Weighted KLA of CPHD Densities

First, the EMD of the IIDC distributions is first computed, which can be used for the fusion of posteriors from the CPHD filter.

The set of targets is modeled as an IIDC process, that is, the multi-target density of node i to be fused has the following form

$$\pi_i(X) = |X|!\rho_i(|X|) \prod_{\boldsymbol{x}\in X} p_i(\boldsymbol{x}) \tag{12.4.10}$$

where $(\rho_i(n), p_i(\boldsymbol{x}))$ is the posterior output from the CPHD filter at node i.

Without loss of generality, let's take two multi-target distributions as an example. Consider the following two IIDC distributions π_i and π_j

$$\pi_i(X) = |X|!\rho_i(|X|) \prod_{\boldsymbol{x}\in X} p_i(\boldsymbol{x}) \tag{12.4.11}$$

$$\pi_j(X) = |X|!\rho_j(|X|) \prod_{\boldsymbol{x}\in X} p_j(\boldsymbol{x}) \tag{12.4.12}$$

according to (12.4.4), their fused density is

$$\overline{\pi}(X) = \frac{\pi_i^{1-\omega}(X)\pi_j^{\omega}(X)}{\int \pi_i^{1-\omega}(X')\pi_j^{\omega}(X')\delta X'} \tag{12.4.13}$$

Substituting (12.4.11) and (12.4.12) into (12.4.13) leads to

$$\overline{\pi}(X) = \frac{1}{K}|X|!\rho_i^{(1-\omega)}(|X|)\rho_j^{\omega}(|X|) \prod_{\boldsymbol{x}\in X} p_i^{(1-\omega)}(\boldsymbol{x})p_j^{\omega}(\boldsymbol{x}) \tag{12.4.14}$$

with the normalization factor K being the set integral of the numerator, i.e.,

$$K = \sum_{|X|=0}^{\infty} \rho_i^{(1-\omega)}(|X|)\rho_j^{\omega}(|X|)\left(\int p_i^{(1-\omega)}(\boldsymbol{x}')p_j^{\omega}(\boldsymbol{x}')d\boldsymbol{x}'\right)^{|X|}$$

$$= \sum_{|X|=0}^{\infty} \rho_i^{(1-\omega)}(|X|)\rho_j^{\omega}(|X|)\eta_{i,j}^{|X|}(\omega) \tag{12.4.15}$$

where

$$\eta_{i,j}(\omega) = \left\langle p_i^{(1-\omega)}, p_j^{\omega} \right\rangle = \int p_i^{(1-\omega)}(\boldsymbol{x})p_j^{\omega}(\boldsymbol{x})d\boldsymbol{x} \tag{12.4.16}$$

Multiplying the numerator and denominator of (12.4.14) by $\eta_{i,j}^{|X|}(\omega)$ yields

$$\overline{\pi}(X) = |X|!\frac{\rho_i^{(1-\omega)}(|X|)\rho_j^{\omega}(|X|)\eta_{i,j}^{|X|}(\omega)}{K}\prod_{\boldsymbol{x}\in X}\overline{p}(\boldsymbol{x}) \tag{12.4.17}$$

where $\overline{p}(\boldsymbol{x})$ is the probability density over the single-target state space, i.e.,

$$\overline{p}(\boldsymbol{x}) = p_i^{(1-\omega)}(\boldsymbol{x})p_j^{\omega}(\boldsymbol{x})/\eta_{i,j}(\omega) \tag{12.4.18}$$

Therefore, the positional density $\overline{p}(\boldsymbol{x})$ of the fusion distribution is exactly the EMD of input positional densities p_i and p_j.

Further, by substituting (12.4.15) into (12.4.17), (12.4.17) can be written in the following IIDC form

$$\overline{\pi}(X) = |X|!\overline{\rho}(|X|)\prod_{\boldsymbol{x}\in X}\overline{p}(\boldsymbol{x}) \tag{12.4.19}$$

where

$$\overline{\rho}(|X|) = \frac{\rho_i^{(1-\omega)}(|X|)\rho_j^{\omega}(|X|)\eta_{i,j}^{|X|}(\omega)}{\sum_{m=0}^{\infty}\rho_i^{(1-\omega)}(m)\rho_j^{\omega}(m)\eta_{i,j}^{m}(\omega)} \tag{12.4.20}$$

The above equation shows that, the fused cardinality density is the scaling product of the fractional power of the input cardinality and the $|X|$th power of the scale factor (12.4.16) obtained from the fused positional density.

Using induction, the fusion method applicable to two sensors can be generalized to a more general case, resulting in the following proposition.

Proposition 34 *The EMD of multi-target densities* $\pi_i, i \in \mathcal{I}$ *subject to the IIDC distribution defined by* (12.4.10) *is also an IIDC distribution of the form*

$$\overline{\pi}(X) = |X|!\overline{\rho}(|X|)\prod_{\boldsymbol{x}\in X}\overline{p}(\boldsymbol{x}) \tag{12.4.21}$$

where

$$\overline{p}(\boldsymbol{x})=\frac{\prod_i p_i^{\omega_i}(\boldsymbol{x})}{\int\prod_i p_i^{\omega_i}(\boldsymbol{x})d\boldsymbol{x}} \tag{12.4.22}$$

$$\overline{\rho}(n)=\frac{\prod_i \rho_i^{\omega_i}(n)\left[\int\prod_i p_i^{\omega_i}(\boldsymbol{x})d\boldsymbol{x}\right]^n}{\sum_{j=0}^{\infty}\prod_i \rho_i^{\omega_i}(j)\left[\int\prod_i p_i^{\omega_i}(\boldsymbol{x})d\boldsymbol{x}\right]^j} \tag{12.4.23}$$

Proposition 34 actually corresponds to the GCI fusion (12.4.4). According to Proposition 34, the posterior distributions propagated by the CPHD filters can be fused. Equations (12.4.21)–(12.4.23) show: the fusion result of IIDC distributions is still an IIDC distribution, its positional density $\overline{p}(\cdot)$ is the weighted geometric mean of positional densities $p_i(\cdot)$ of each node, and the fused cardinality $\overline{\rho}(\cdot)$ has a more complicated expression (12.4.23), which involves the positional PDF and the cardinality PMF.

For the distributed fusion of PHD filters and the distributed fusion of Bernoulli filters, following similar derivations, the corresponding conclusions about multi-target Poisson distribution and (single-target) Bernoulli distribution can be drawn, respectively. For details, please refer to Appendix Q. In addition, for Mδ-GLMB densities and LMB densities, their KLAs are also Mδ-GLMB density and LMB density, respectively. In the following, the closed form solutions to normalized weighted geometric mean are presented, and these results are necessary to perform the fusion of Mδ-GLMB and LMB tracking filters.

12.4.3 Weighted KLA of Mδ-GLMB Densities

The following proposition gives a summary of the fusion rule of Mδ-GLMB densities. In other words, applying this proposition can obtain the KLA (12.4.4) of δ-GLMB densities $\pi_i, i\in\mathcal{I}$.

Proposition 35 (see Appendix R for proof) *Let* $\pi_i(\mathrm{X}) = \Delta(\mathrm{X})\sum_{I\in\mathcal{F}(\mathbb{L})}\delta_I(\mathcal{L}(\mathrm{X}))w_i^{(I)}[p_i^{(I)}]^{\mathrm{X}}$, $i\in\mathcal{I}$ *be the Mδ-GLMB density defined by* (7.5.16), *and the weight* ω_i *satisfy* (12.4.3), *then the weighted KLA (normalized weighted geometric mean) of Mδ-GLMB densities is also in the Mδ-GLMB form, i.e.,*

$$\overline{\pi}(\mathrm{X})=\oplus_{i\in\mathcal{I}}(\omega_i\odot\pi_i(\mathrm{X}))=\Delta(\mathrm{X})\sum_{I\in\mathcal{F}(\mathbb{L})}\delta_I(\mathcal{L}(\mathrm{X}))\overline{w}^{(I)}\left[\overline{p}^{(I)}\right]^{\mathrm{X}} \tag{12.4.24}$$

where

$$\overline{w}^{(I)}=\frac{\prod_{i\in\mathcal{I}}\left(w_i^{(I)}\right)^{\omega_i}\left[\int\prod_{i\in\mathcal{I}}\left(p_i^{(I)}(\boldsymbol{x},\cdot)\right)^{\omega_i}d\boldsymbol{x}\right]^I}{\sum_{J\subseteq\mathbb{L}}\prod_{i\in\mathcal{I}}\left(w_i^{(J)}\right)^{\omega_i}\left[\int\prod_{i\in\mathcal{I}}\left(p_i^{(J)}(\boldsymbol{x},\cdot)\right)^{\omega_i}d\boldsymbol{x}\right]^J} \tag{12.4.25}$$

$$\overline{p}^{(I)} = \oplus_{i\in\mathcal{I}}\left(\omega_i \odot p_i^{(I)}\right) = \frac{\prod_{i\in\mathcal{I}}\left(p_i^{(I)}\right)^{\omega_i}}{\int\prod_{i\in\mathcal{I}}\left(p_i^{(I)}\right)^{\omega_i}dx} \tag{12.4.26}$$

Remark Variables $\overline{w}^{(I)}$ and $\overline{p}^{(I)}$ can be determined by (12.4.25) and (12.4.26), respectively, hence, the entire fusion step is performed in a completely parallel way. Additionally, it shall be noted that (12.4.26) is actually the Chernoff fusion rule for single-target PDFs.

12.4.4 Weighted KLA of LMB Densities

The following proposition summarizes the fusion rule for LMB densities. In other words, applying this proposition yields the KLA (12.4.4) of LMB densities $\{(\varepsilon_i^{(\ell)}, p_i^{(\ell)})\}_{\ell\in\mathbb{L}}, i \in \mathcal{S}$.

Proposition 36 (see Appendix S for proof) *Let* $\pi_i = \{(\varepsilon_i^{(\ell)}, p_i^{(\ell)})\}_{\ell\in\mathbb{L}}, i \in \mathcal{I}$ *be the LMB densities defined by* (3.3.35), *and weight* ω_i *satisfy* (12.4.3), *then the weighted KLA (normalized weighted geometric mean) of LMB densities is also in the LMB form, i.e.,*

$$\overline{\pi}(\mathrm{X}) = \oplus_{i\in\mathcal{I}}(\omega_i \odot \pi_i(\mathrm{X})) = \left\{(\overline{\varepsilon}^{(\ell)}, \overline{p}^{(\ell)})\right\}_{\ell\in\mathbb{L}} \tag{12.4.27}$$

where

$$\overline{\varepsilon}^{(\ell)} = \frac{\int\prod_{i\in\mathcal{I}}\left(\varepsilon_i^{(\ell)}p_i^{(\ell)}(\boldsymbol{x})\right)^{\omega_i}dx}{\prod_{i\in\mathcal{I}}\left(1-\varepsilon_i^{(\ell)}\right)^{\omega_i} + \int\prod_{i\in\mathcal{I}}\left(\varepsilon_i^{(\ell)}p_i^{(\ell)}(\boldsymbol{x})\right)^{\omega_i}dx} \tag{12.4.28}$$

$$\overline{p}^{(\ell)} = \oplus_{i\in\mathcal{I}}\left(\omega_i \odot p_i^{(\ell)}\right) \tag{12.4.29}$$

Remark Since each Bernoulli component $(\overline{\varepsilon}^{(\ell)}, \overline{p}^{(\ell)})$ can be determined separately according to (12.4.28) and (12.4.29), the entire fusion is completely parallel. In addition, it shall also be noted that, (12.4.29) is actually the Chernoff fusion rule of single-target PDFs.

12.5 Distributed Fusion of SMC-CPHD Filters

Based on the conclusion about the weighted KLA of CPHD densities (see Sect. 12.4.2), the CPHD filter is used for distributed multi-target tracking in this section. At this point, each node will conduct local update and fusion of cardinality PMFs $\rho(\cdot)$ and positional PDFs $p(\cdot)$. For the sake of simplicity, the CPHD filter and the CPHD distribution are collectively referred to as the CPHD. Strictly speaking, the CPHD-based distributed multi-target tracking problem over the network $\mathcal{G} = (\mathcal{S}, \mathcal{A})$ can be described as follows: given local measurement $Z_{k,i}$ of node i and data received from its all neighboring nodes $j \in \mathcal{I}_i \backslash i$, at each time $k \in \{1, 2, \ldots\}$, each node $i \in \mathcal{S}$ should estimate the CPHD of an unknown multi-target set X_k such that the two-tuple estimate $(\rho_{k|k,i}(n), p_{k|k,i}(\boldsymbol{x}))$ are as close as possible to the estimate provided when the centralized CPHD fusion processes all node information simultaneously.

The implementation of distributed fusion based on the EMD faces two major challenges. The first challenge is that there is almost no closed form solution for the EMD. For example, if $p_i(\boldsymbol{x})$ in (12.4.10) is in the GM form, the weighted geometric mean calculated by (12.4.22) is generally no longer in the GM form. Although it can be approximated to the GM form using the Newton series expansion, the series expansion suffers from the numerical instability unless an extremely large number of components are used, thus requiring robust computational methods. The second challenge is that, it shall take into consideration the influence of the strategy of choosing ω in the update step. This is an optimization process, and the updated distribution and divergence values must be repeatedly calculated for different ω values. Therefore, an effective calculation scheme is required.

Considering that there is generally no closed form solution to the EMD, the SMC implementation of the CPHD filter is used here. Moreover, to fuse other sensor data, it must be able to calculate (12.4.19), (12.4.18), and (12.4.16) for a wide range of ω values. However, sine each node has its own SMC-CPHD filter, this is difficult to implement directly. Therefore, in the following, the clustering technique is used to obtain the continuous approximation of distributions, and then sampling is conducted based on these distributions to calculate the EMD.

12.5.1 Representation of Local Information Fusion

Each node has a local IIDC distribution. For node $i \in \mathcal{S}$, its local IIDC distribution is denoted as $\{\rho_{k|k,i}(n), \{w_{k|k,i}^{(m)}, \boldsymbol{x}_{k|k,i}^{(m)}, \ell_i^{(m)}\}_{m=1:M_i}\}$, where $\rho_{k|k,i}(n)$ is the cardinality distribution, the positional distribution is expressed by a total of M_i particles, the remaining 3 terms store information about each particle, $\boldsymbol{x}_{k|k,i}^{(m)}$ is the particle generated by the positional distribution, component $w_{k|k,i}^{(m)}$ is the weight related to the mth particle $\boldsymbol{x}_{k|k,i}^{(m)}$, and $\ell_i^{(m)}$ is the label of the particle. Since the labels can identify the measurements that produced the particles, the association of measurements to particles is maintained by particle labeling, and this information can be used when fusing distributions from different nodes.

Given two components $w_{k|k,i}^{(m)}$ and $\boldsymbol{x}_{k|k,i}^{(m)}$, the positional distribution can be expressed as

$$\hat{p}_{k|k,i}(\boldsymbol{x}) = \sum_{m=1}^{M_i} w_{k|k,i}^{(m)} \delta_{\boldsymbol{x}_{k|k,i}^{(m)}}(\boldsymbol{x}) \tag{12.5.1}$$

and the average number of targets is

$$\mu_{k|k,i} = \sum_{n=0}^{n_{\max}} n\rho_{k|k,i}(n) \tag{12.5.2}$$

where, $n_{\max}$ is the storage length of $\rho_{k|k,i}(n)$, and the PHD is

$$\hat{v}_{k|k,i}(\boldsymbol{x}) = \mu_{k|k,i}\hat{p}_{k|k,i}(\boldsymbol{x}) \tag{12.5.3}$$

12.5.2 Continuous Approximation of SMC-CPHD Distribution

Since each node is expressed by its own SMC representation and has its own unique particle set, it cannot be guaranteed that each node has the same support domain and number of particles. Hence, the intensities (PHD) of node i and node j can't be directly fused. Therefore, the continuous approximation is used here. This approximation problem can be solved by the kernel density estimation (KDE) method [438]. Fraley has reviewed model-based clustering methods for density estimation [439]. However, these methods are not robust to the outliers and may lead to high uncertainty in the cardinality distribution, resulting in many mixture components. On the contrary, the labeling technique is invoked to generate clusters here. Each cluster $\ell \in \mathcal{L} \triangleq \{\ell_1, \ldots, \ell_L\}$ is associated to parameter set $\boldsymbol{C}_\ell$ and the following density estimation is used (for the ease of description, the subscripts k, i of the variables are omitted here)

$$\hat{p}(\boldsymbol{x}) = \frac{1}{M}\sum_{m=1}^{M} \mathcal{K}\big(\boldsymbol{x}, \boldsymbol{x}^{(m)}; \boldsymbol{C}_{\ell^{(m)}}\big) \tag{12.5.4}$$

where $\mathcal{K}(\boldsymbol{x}, \boldsymbol{x}^{(m)}; \boldsymbol{C}_{\ell^{(m)}})$ represents a Gaussian density with mean $\boldsymbol{x}^{(m)}$ and covariance $\boldsymbol{C}_{\ell^{(m)}}$.

In order to obtain the kernel parameter $\boldsymbol{C}_\ell$ of cluster ℓ (i.e., $\{\boldsymbol{x}^{(m')}|\ell^{(m')} = \ell\}$), we first need a transformation, which can diagonalize the empirical covariance $\boldsymbol{\Sigma}_\ell$ of cluster ℓ in the transform domain, so that the multi-dimensional kernel parameter problem is reduced to multiple independent single-dimensional problems. The transformation uses the negative square root of the empirical covariance matrix $\boldsymbol{\Sigma}_\ell$, i.e., the following equation is used to transform all $\boldsymbol{x}^{(m')} \in \{\boldsymbol{x}^{(m')}|\ell^{(m')} = \ell\}$

$$\boldsymbol{y}^{(m')} = \boldsymbol{W}_{\ell}\boldsymbol{x}^{(m')} \tag{12.5.5}$$

where

$$\boldsymbol{W}_{\ell} = \boldsymbol{\Sigma}_{\ell}^{-1/2} \tag{12.5.6}$$

Assuming that the covariance of $\boldsymbol{y}^{(m')}$ is a diagonal matrix, the D-dimensional kernel in the transform domain can be simplified as

$$\mathcal{K}\left(\boldsymbol{y}, \boldsymbol{y}^{(m')}\right) = \prod_{d=1}^{D} \frac{1}{\sqrt{2\pi} B_d} \exp\left(-\frac{\left(y_d - y_d^{(m')}\right)^2}{2B_d^2}\right) \tag{12.5.7}$$

where $\boldsymbol{y}_d$ and $\boldsymbol{y}_d^{(m')}$ indicate the dth dimensional components of $\boldsymbol{y}$ and $\boldsymbol{y}^{(m')}$, respectively, D is the dimension of the state space, and B_d is bandwidth parameter of the one-dimensional Gaussian kernel [438].

Bandwidth B_d of each dimension can be obtained through the following rule-of-thumb (RUT) [438, 440]

$$B_d = \sigma_d \cdot (4/(3N))^{1/5} \tag{12.5.8}$$

where σ_d is the empirical standard deviation of $\boldsymbol{y}_d^{(m')}$, and N is the number of these points. Compared with other methods, this method is simple with lower computational complexity.

Therefore, for cluster ℓ, the covariance matrix in (12.5.4) is determined by

$$\boldsymbol{C}_{\ell} = \boldsymbol{T}_{\ell}\boldsymbol{\Lambda}_{\ell}\boldsymbol{T}_{\ell}^{\mathrm{T}} \tag{12.5.9}$$

where

$$\boldsymbol{T}_{\ell} = \boldsymbol{W}_{\ell}^{-1} \tag{12.5.10}$$

$$\boldsymbol{\Lambda}_{\ell} = \mathrm{blkdiag}\left(B_1^2, B_2^2, \ldots, B_D^2\right) \tag{12.5.11}$$

12.5.3 Construction of Exponential Mixture Densities (EMD)

In the following, the EMD in Proposition 34 is considered, and the Monte Carlo method is used to construct the multi-target EMD for any $\omega \in [0, 1]$. First, the sampling step is used to generate particles representing the fused positional density

$\overline{p}(\boldsymbol{x})$ given by (12.4.18). Then, the fused cardinality distribution (12.4.20) is obtained after estimation of the scale factor $\eta_{i,j}(\omega)$ given by (12.4.16).

(1) Sampling according to the EMD positional distribution: samples are drawn according to the fused positional density (12.4.18) using equal weight particle sets $\{\boldsymbol{x}_i^{(m_i)}\}_{m_i=1:M_i}$ and $\{\boldsymbol{x}_j^{(m_j)}\}_{m_j=1:M_j}$ representing $p_i(\boldsymbol{x})$ and $p_j(\boldsymbol{x})$, and KDE parameters $\{\boldsymbol{C}_{\ell_i}\}_{\ell_i \in \mathcal{L}_i}$ and $\{\boldsymbol{C}_{\ell_j}\}_{\ell_j \in \mathcal{L}_j}$, respectively.

The consensus nature of the EMD allows a mixture of $p_i(\boldsymbol{x})$ and $p_j(\boldsymbol{x})$ to be used as a proposed density for importance sampling. Compared with $\overline{p}(\boldsymbol{x})$, this mixture density has more severe trailing [441], and its expression is as follows

$$p_q(\boldsymbol{x}) = \frac{M_i p_i(\boldsymbol{x}) + M_j p_j(\boldsymbol{x})}{M_i + M_j} \tag{12.5.12}$$

Sampling is performed according to the proposed density (12.5.12) to obtain the input particle set composed of $\overline{M} = M_i + M_j$ samples, i.e.,

$$P_U = \left\{\boldsymbol{x}_i^{(m_i)}\right\}_{m_i=1:M_i} \cup \left\{\boldsymbol{x}_j^{(m_j)}\right\}_{m_j=1:M_j} \tag{12.5.13}$$

Therefore, P_U represents the particle set of $\overline{p}(\boldsymbol{x})$, and then for $\boldsymbol{x}^{(m')} \in P_U$, the weight of importance sampling (IS) is

$$w^{(m')} \propto \frac{p_i^{(1-\omega)}\left(\boldsymbol{x}^{(m')}\right) p_j^{\omega}\left(\boldsymbol{x}^{(m')}\right)}{M_i p_i\left(\boldsymbol{x}^{(m')}\right) + M_j p_j\left(\boldsymbol{x}^{(m')}\right)} \tag{12.5.14}$$

After resampling $\{w^{(m')}, \boldsymbol{x}^{(m')}\}_{m'=1:M_i+M_j}$ M_i+M_j times, particles approximately generated from $\overline{p}(\boldsymbol{x})$ are obtained. However, in order to calculate the IS weight (12.5.14), it is necessary to calculate the values of $p_i(\boldsymbol{x})$ and $p_j(\boldsymbol{x})$ at all points of P_U. In order to estimate these values, KDE parameters $\{\boldsymbol{C}_{\ell_i}\}_{\ell_i \in \mathcal{L}_i}$ and $\{\boldsymbol{C}_{\ell_j}\}_{\ell_j \in \mathcal{L}_j}$ in (12.5.4) are used and then KDEs $\hat{p}_i(\boldsymbol{x})$ and $\hat{p}_j(\boldsymbol{x})$ are obtained. Then, $\hat{p}_i(\boldsymbol{x})$ and $\hat{p}_j(\boldsymbol{x})$ at P_U are calculated. Therefore, a feasible estimation of $w^{(m')}$ is obtained by substituting these quantities into (12.5.14)

$$\hat{w}^{(m')} \propto \frac{\hat{p}_i^{(1-\omega)}\left(\boldsymbol{x}^{(m')}\right) \hat{p}_j^{\omega}\left(\boldsymbol{x}^{(m')}\right)}{M_i \hat{p}_i\left(\boldsymbol{x}^{(m')}\right) + M_j \hat{p}_j\left(\boldsymbol{x}^{(m')}\right)} \tag{12.5.15}$$

After resampling $\{\hat{w}^{(m)}, \boldsymbol{x}^{(m)}\}$, the equal-weighted samples representing $\overline{p}(\boldsymbol{x})$ are obtained.

(2) Construction of the cardinality distribution of the EMD. In order to calculate the fused cardinality distribution (12.4.20), it is necessary to estimate $\eta_{i,j}(\omega)$ given by (12.4.16). With the proposed density $p_q(\boldsymbol{x})$ given by (12.5.12), the IS estimation of $\eta_{i,j}(\omega)$ is [441]

$$\tilde{\eta}_{i,j}(\omega) \triangleq \sum_{x \in P_U} \frac{p_i^{(1-\omega)}(x) p_j^{\omega}(x)}{M_i p_i(x) + M_j p_j(x)} \tag{12.5.16}$$

Substituting KDEs $\hat{p}_i(x)$ and $\hat{p}_j(x)$ into the above equation leads to

$$\hat{\eta}_{i,j}(\omega) \triangleq \sum_{x \in P_U} \frac{\hat{p}_i^{(1-\omega)}(x) \hat{p}_j^{\omega}(x)}{M_i \hat{p}_i(x) + M_j \hat{p}_j(x)} \tag{12.5.17}$$

After estimating the scale factor, $\overline{\rho}(n)$, $n = 0, 1, \ldots, n_{\max}$ can be constructed by substituting $\rho_i(n)$, $\rho_j(n)$, and $\hat{\eta}_{i,j}(\omega)$ into (12.4.20).

12.5.4 Determination of Weighting Parameter

In order to determine the fused density, it is also necessary to select the EMD weights. From the perspective that the mixing parameter ω must be specified, the EMD fusion rule is somewhat subjective, which is different from the Bayesian rule. This parameter dominates the relative weights of π_i and π_j. Assuming that $J(\omega)$ is a cost function, the goal now is to select ω to minimize the cost, i.e.,

$$\omega^* = \arg_{\omega \in [0,1]} \min J(\omega) \tag{12.5.18}$$

By reference to the derivation of the CI method, available options for $J(\omega)$ include the determinant or trace of the covariance of $\overline{\pi}$. However, when $\overline{\pi}$ is a multiple model distribution, the covariance is not necessarily a good representation of uncertainty. Another possible option is the Shannon entropy of $\overline{\pi}$ [302]. Nevertheless, this entropy may contain the local minima, making the optimization problem of (12.5.18) difficult to solve [325]. Moreover, Herley has proposed the following criteria in the probabilistic analysis of the CI: the Kullback–Leibler divergence (KLD) of $\overline{\pi}$ and π_i and the KLD of $\overline{\pi}$ and π_j should be the same [302]. Specifically, $J(\omega)$ can be chosen as

$$J(\omega) = \left(\mathcal{D}_{\mathrm{KL}}(\overline{\pi}||\pi_i) - \mathcal{D}_{\mathrm{KL}}(\overline{\pi}||\pi_j)\right)^2 \tag{12.5.19}$$

where KLD $\mathcal{D}_{\mathrm{KL}}(\cdot||\cdot)$ is defined in (12.4.1).

Though Hurley's conclusion is only strictly applicable to discrete distributions, Dabak has extended it to continuous distributions [442]. Moreover, it has been proven that $\mathcal{D}(\overline{\pi}||\pi_i)$ is a non-decreasing function of ω. Therefore, (12.5.19) has a unique minimum, thus significantly simplifying the optimization problem. However, the information-theoretic proof using this divergence metric as a cost function is not clear. Hurley argues that, this option is based on the correlation of the resulting distribution with the Chernoff information. However, though it is relevant to the binary classification problem, its relevance to information fusion is not clear. As a result, another measure, Renyi divergence (RD), is employed here, which is easy to solve but still conveys the potentially useful information.

The RD is defined as

$$\mathcal{R}_\alpha(\pi_i||\pi_j) = \frac{1}{\alpha - 1}\int \log \pi_i^{1-\alpha}(X)\pi_j^{\alpha}(X)\delta X \tag{12.5.20}$$

Compared with the KLD, the RD is more useful in the sensor management problem. The RD generalizes the KLD by introducing a free parameter α, which can be used to highlight differences in specific aspects (such as tails) of distributions of interest. When $\alpha \rightarrow 1$, the RD converges to the KLD, and when $\alpha = 0.5$, it is equal to the Hellinger affinity.

12.5.5 Calculation of Renyi Divergence

Using the RD formula (12.5.21) between two IIDC processes, we can obtain

$$\begin{aligned}\mathcal{R}_\alpha(\overline{\pi}||\pi_i) &= \frac{1}{\alpha-1}\log\sum_{n=0}^{\infty}\overline{\rho}^{(1-\alpha)}(n)\rho_i^{\alpha}(n)\left[\int \overline{p}^{(1-\alpha)}(\boldsymbol{x})p_i^{\alpha}(\boldsymbol{x})d\boldsymbol{x}\right]^n \\ &= \frac{1}{\alpha-1}\log\sum_{n=0}^{\infty}\overline{\rho}^{\alpha}(n)\rho_i^{(1-\alpha)}(n)\left[\frac{\eta_{i,j}(\alpha\omega)}{\eta_{i,j}^{\alpha}(\omega)}\right]^n\end{aligned} \tag{12.5.21}$$

where $\eta_{i,j}(\alpha\omega)$ and $\eta_{i,j}(\omega)$ are obtained with (12.4.16).

Taking similar steps, the RD of the EMD with respect to $\pi_j(X)$ is

$$\mathcal{R}_\alpha(\overline{\pi}||\pi_j) = \frac{1}{\alpha-1}\log\sum_{n=0}^{\infty}\overline{\rho}^{\alpha}(n)\rho_j^{(1-\alpha)}(n)\left[\frac{\eta_{j,i}(\alpha(1-\omega))}{\eta_{i,j}^{\alpha}(\omega)}\right]^n \tag{12.5.22}$$

Finally, $J(\omega)$ in (12.5.19) becomes

$$J(\omega) = \left(\mathcal{R}_\alpha(\overline{\pi}||\pi_i) - \mathcal{R}_\alpha(\overline{\pi}||\pi_j)\right)^2 \tag{12.5.23}$$

The specific calculation of the above formula is as follows. Given α and ω, the cardinality distribution $\overline{\rho}(n)$ of the EMD is first constructed (see Sect. 12.5.3). To achieve this, it is only required to calculate the KDEs $\hat{p}_i(\boldsymbol{x})$ and $\hat{p}_j(\boldsymbol{x})$ once at P_U. Then, using these results, $\eta_{i,j}(\omega)$, $\eta_{i,j}(\alpha\omega)$ and $\eta_{j,i}(\alpha(1-\omega))$ are estimated according to (12.5.17). Finally, these quantities are substituted into (12.5.21), (12.5.22), and (12.5.23).

The discussion on different choices of $J(\omega)$ and different values of α can be found in [325]. Compared with other types of cost measures, divergence measures are easier to implement and has the least impact on the overall performance of the system.

12.5.6 Distributed Fusion Algorithm for SMC-CPHD Filters

The pseudo-code of the distributed fusion algorithm for SMC-CPHD filters is summarized in Table 12.2. The first input of the algorithm is the particle representation of IIDC posteriors from local sensor i and sensor j. The second input is RD parameter α, which is used to calculate the cost $J(\omega)$ of (12.5.19). The third input is an increment value Δ_ω, which is used to find out the optimal EMD weight ω^* in the exhaustive search.

Table 12.2 SMC implementation of multi-target EMD fusion algorithm at local sensor i

	Input: • IIDC posterior particles from local sensor i and sensor j: $\{\rho_i(n), \{\boldsymbol{x}_i^{(m_i)}, \ell_i^{(m_i)}\}_{m_i=1:M_i}\}$ and $\{\rho_j(n), \{\boldsymbol{x}_j^{(m_j)}, \ell_j^{(m_j)}\}_{m_j=1:M_j}\}$ • RD parameter: α • Increment of ω: Δ_ω
1:	Calculate $\{\boldsymbol{C}_{\ell_i}\}_{\ell_i \in \mathcal{L}_i}$ and $\{\boldsymbol{C}_{\ell_j}\}_{\ell_j \in \mathcal{L}_j}$ as described in Sect. 12.5.2
2:	Construct P_U given by (12.5.13)
3:	Calculate KDE densities $\hat{p}_i(\boldsymbol{x})$, $\boldsymbol{x} \in P_U$ and $\hat{p}_j(\boldsymbol{x})$, $\boldsymbol{x} \in P_U$ in (12.5.4) according to KDE parameters $\{\boldsymbol{C}_{\ell_i}\}_{\ell_i \in \mathcal{L}_i}$ and $\{\boldsymbol{C}_{\ell_j}\}_{\ell_j \in \mathcal{L}_j}$ % Note: $\hat{p}_i$ and $\hat{p}_j$ have been calculated before the "for" loop
4:	for $\omega = 0, \Delta_\omega, \ldots, 1$
5:	Estimate $\hat{\eta}_{i,j}(\omega)$, $\hat{\eta}_{i,j}(\alpha\omega)$ and $\hat{\eta}_{j,i}(\alpha(1-\omega))$ in (12.5.17) according to $\hat{p}_i$ and $\hat{p}_j$
6:	Calculate $\overline{\rho}$ in (12.4.20) based on the estimated $\hat{\eta}_{i,j}(\omega)$
7:	Calculate $\mathcal{R}_\alpha(\overline{\pi}\|\|\pi_i)$ in (12.5.21) according to $\overline{\rho}$ and the estimated $\hat{\eta}_{i,j}(\omega)$ and $\hat{\eta}_{i,j}(\alpha\omega)$
8:	Calculate $\mathcal{R}_\alpha(\overline{\pi}\|\|\pi_j)$ in (12.5.22) according to $\overline{\rho}$ and the estimated $\hat{\eta}_{i,j}(\omega)$ and $\hat{\eta}_{j,i}(\alpha(1-\omega))$
9:	Calculate $J(\omega)$ in (12.5.23) based on $\mathcal{R}_\alpha(\overline{\pi}\|\|\pi_i)$ and $\mathcal{R}_\alpha(\overline{\pi}\|\|\pi_j)$
10:	end
11:	Calculate $\omega^* = \arg_{\omega \in \{0, \Delta_\omega, \ldots, 1\}} \min J(\omega)$
12:	Calculate IS weights $\hat{w}^{(m')}$ in (12.5.15) for $\omega = \omega^*$ and each $\boldsymbol{x}^{(m')} \in P_U$ according to $\hat{p}_i$ and $\hat{p}_j$
13:	Save labels $\mathcal{L} = \mathcal{L}_i \cup \mathcal{L}_j$ and KDE parameters $\mathcal{C} \triangleq \{\boldsymbol{C}_{\ell_i}\}_{\ell_i \in \mathcal{L}_i} \cup \{\boldsymbol{C}_{\ell_j}\}_{\ell_j \in \mathcal{L}_j}$
14:	Output: $\{\overline{\rho}^*(n), \{\hat{w}^{(m')}, \boldsymbol{x}^{(m')}, \ell^{(m')}\}_{m'=1:M_i+M_j}\}$ and $\mathcal{C}$ % $\overline{\rho}^*$ has been calculated through the "for" loop

First, the KDE parameters of the particle set are solved. Then, the sample set P_U is constructed based on the proposed (suggested) density, and the KDE of the input positional density is calculated at the particles in this set. Once the KDE at P_U is calculated, the RD of the EMD with respect to the input and the cost (12.5.23) can be calculated. Meanwhile, starting from $\omega = 0$, ω varies as Δ_ω increases. The optimal EMD weight ω^* can be obtained after the cost is obtained on the grids determined by Δ_ω. In subsequent steps, the IS weight of the suggested sample can be calculated according to ω^*.

The output of the algorithm includes the fused cardinality distribution $\overline{\rho}^*(n)$ and the set of particles representing the fused positional density $\overline{p}^*(\boldsymbol{x})$. The most computationally expensive step in the whole algorithm is to calculate the KDE. However, since this step needs to be executed only once before the "for" loop, the computational cost of the exhaustive search is still affordable.

12.6 Distributed Fusion of Gaussian Mixture RFS Filters

The above section presents a distributed multi-target information fusion implementation for the SMC method, and since it involves a convex combination of Dirac delta functions, additional techniques shall be used, such as the kernel density estimation [91], least squares estimation [443] or parametric model method [444], resulting in an increase in the computational burden. Moreover, the SMC implementation also requires more resources than the GM implementation.

For distributed multi-target tracking, each node has the limited processing capacity and energy resource. Therefore, it is quite important to reduce the computational burden and inter-node data communication as much as possible. In this respect, the GM method is more economical and is preferred, because the number of Gaussian components required to achieve the similar tracking performance is usually several orders of magnitude lower than that of particles. To this end, a GM implementation of the fusion rule (12.4.4) is described here, which utilizes the consensus algorithm, enabling the fusion to be implemented in a fully distributed fashion.

12.6.1 Consensus Algorithm in Multi-target Context

Computing the global (overall) KLA (12.4.4) over the entire network requires obtaining all local multi-target densities. Through the iteration of the local mean, the global KLA can be computed in a distributed and scalable manner by using a consensus algorithm [92, 305]

$$\pi_{i,c+1}(X) = \bigoplus_{j\in\mathcal{I}_i} \left(\omega_{i,j} \odot \pi_{j,c}(X)\right), \quad \forall i \in \mathcal{S} \tag{12.6.1}$$

where $\pi_{i,0}(X) = \pi_i(X)$, $\omega_{i,j} \geq 0$ is the consensus weight related to node i and node $j \in \mathcal{I}_i$, satisfying $\sum_{j\in\mathcal{I}_i} \omega_{i,j} = 1$. The consensus iteration (12.6.1) is a multi-target version of (12.3.16).

According to the properties (O.3)–(O.8), we can obtain[2]

$$\pi_{i,c}(X) = \bigoplus_{j\in\mathcal{S}} \big(\omega_{i,j,c} \odot \pi_j(X)\big), \quad \forall i \in \mathcal{S} \tag{12.6.2}$$

where $\omega_{i,j,c}$ is defined as the (i, j) element of matrix $\boldsymbol{\Omega}^c$. As mentioned earlier, when selecting consensus weight $\omega_{i,j}$, it shall ensure that $\boldsymbol{\Omega}$ is a doubly stochastic matrix, then

$$\lim_{c\to\infty} \omega_{i,j,c} = 1/|\mathcal{S}|, \quad \forall i, j \in \mathcal{S} \tag{12.6.3}$$

As a result, similar to the single-target case, if the consensus matrix is not only a primitive matrix but also a doubly stochastic matrix, as c tends to infinity, the local multi-target density of each node in the network will converge to the global unweighted KLA (12.4.5) of multi-target posterior densities. For the convergence properties, see [305]. In fact, the iteration will generally stop at some finite c.

12.6.2 Gaussian Mixture Approximation of Fused Density

Suppose that each single-target density $p^{(*)}$ can be expressed in the form of Gaussian mixture (GM)

$$p^{(*)}(\boldsymbol{x}) = \sum_{j=1}^{J^{(*)}} w^{(j)} \mathcal{N}\big(\boldsymbol{x}; \boldsymbol{m}^{(j)}, \boldsymbol{P}^{(j)}\big) \tag{12.6.4}$$

where $(*)$ is a wildcard character, $(*)$ is (I) for the Mδ-GLMB filter, $(*)$ is (ℓ) for the LMB filter, and $(*)$ is negligible for the CPHD filter. In the following, for the ease of description, the upper-right wildcard $(*)$ is omitted.

For simplicity, consider the case of two nodes (labeled a and b) with the following GM positional densities

$$p_s(\boldsymbol{x}) = \sum_{j=1}^{J_s} w_s^{(j)} \mathcal{N}\big(\boldsymbol{x}; \boldsymbol{m}_s^{(j)}, \boldsymbol{P}_s^{(j)}\big), \quad s = a, b \tag{12.6.5}$$

---

[2] It shall be noted that, the weighting operator $\odot$ is only defined for strictly positive scalars. However, in (12.6.2), some scalar weights $\omega_{i,j,c}$ are allowed to be 0. It can be understood as follows: once $\omega_{i,j,c}$ is equal to 0, the corresponding multi-target density $\pi_j(X)$ will be ignored when using the information fusion operator $\oplus$. This is always feasible, because there is always a strictly positive weight $\omega_{i,j,c}$ for each $i \in \mathcal{S}$ and each c.

Naturally, the first question is whether the following fused positional PDF is still in the GM form

$$\overline{p}(\boldsymbol{x}) = \frac{p_a^{\omega}(\boldsymbol{x})p_b^{1-\omega}(\boldsymbol{x})}{\int p_a^{\omega}(\boldsymbol{x})p_b^{1-\omega}(\boldsymbol{x})d\boldsymbol{x}} \tag{12.6.6}$$

Note (12.6.6) involves the exponent and product operations of GMs. Therefore, the following conclusions regarding the basic operations of Gaussian components and Gaussian mixtures are first introduced.

(1) The exponent of a Gaussian component is also a Gaussian component, i.e.,

$$[w\mathcal{N}(\boldsymbol{x};\boldsymbol{m},\boldsymbol{P})]^{\omega} = w^{\omega}\alpha(\omega,\boldsymbol{P})\mathcal{N}(\boldsymbol{x};\boldsymbol{m},\boldsymbol{P}/\omega) \tag{12.6.7}$$

where

$$\alpha(\omega,\boldsymbol{P}) = \left[\det(2\pi\boldsymbol{P}\omega^{-1})\right]^{1/2}/[\det(2\pi\boldsymbol{P})]^{\omega/2} \tag{12.6.8}$$

(2) The product of Gaussian components is also a Gaussian component, i.e.,

$$w^{(i)}\mathcal{N}\left(\boldsymbol{x};\boldsymbol{m}^{(i)},\boldsymbol{P}^{(i)}\right)\cdot w^{(j)}\mathcal{N}\left(\boldsymbol{x};\boldsymbol{m}^{(j)},\boldsymbol{P}^{(j)}\right) = w^{(i,j)}\mathcal{N}\left(\boldsymbol{x};\boldsymbol{m}^{(i,j)},\boldsymbol{P}^{(i,j)}\right) \tag{12.6.9}$$

where

$$\boldsymbol{P}^{(i,j)} = \left[(\boldsymbol{P}^{(i)})^{-1} + (\boldsymbol{P}^{(j)})^{-1}\right]^{-1} \tag{12.6.10}$$

$$\boldsymbol{m}^{(i,j)} = \boldsymbol{P}^{(i,j)}\left[(\boldsymbol{P}^{(i)})^{-1}\boldsymbol{m}^{(i)} + (\boldsymbol{P}^{(j)})^{-1}\boldsymbol{m}^{(j)}\right] \tag{12.6.11}$$

$$w^{(i,j)} = w^{(i)}w^{(j)}\mathcal{N}\left(\boldsymbol{m}^{(i)} - \boldsymbol{m}^{(j)};\boldsymbol{0},\boldsymbol{P}^{(i)} + \boldsymbol{P}^{(j)}\right) \tag{12.6.12}$$

(3) According to (12.6.9) and the distribution property, the product of GMs is also a GM. More specifically, if $p_a(\cdot)$ and $p_b(\cdot)$ have J_a and J_b Gaussian components, respectively, $p_a(\cdot)p_b(\cdot)$ will have J_aJ_b components.
(4) However, the exponent of a GM is generally no longer a GM.

According to the final conclusion, it is noted that the fusion rules (12.4.22), (12.4.26), and (12.4.29) involve the exponent and product of Gaussian mixtures in (12.6.4), hence the result is generally no longer a GM. Therefore, in order to maintain the GM form of the positional PDF throughout the calculation, a suitable approximation of the exponent of a GM must be made, and the following approximation can be used

$$\left[\sum_{j=1}^{J} w^{(j)}\mathcal{N}\left(\boldsymbol{x}; \boldsymbol{m}^{(j)}, \boldsymbol{P}^{(j)}\right)\right]^{\omega} \cong \sum_{j=1}^{J}\left[w^{(j)}\mathcal{N}(\boldsymbol{x}; \boldsymbol{m}^{(j)}, \boldsymbol{P}^{(j)})\right]^{\omega}$$
$$= \sum_{j=1}^{J}\left(w^{(j)}\right)^{\omega}\alpha\left(\omega, \boldsymbol{P}^{(j)}\right)\mathcal{N}\left(\boldsymbol{x}; \boldsymbol{m}^{(j)}, \boldsymbol{P}^{(j)}/\omega\right) \tag{12.6.13}$$

In fact, the above approximation is reasonable, as long as the cross product of different terms in the GM can be ignored for all $\boldsymbol{x}$. Equivalently, the above approximation also holds, assuming that Gaussian components $\boldsymbol{m}^{(i)}$ and $\boldsymbol{m}^{(j)}$ $(i \neq j)$ are well separated, which can be measured by the associated covariances $\boldsymbol{P}^{(i)}$ and $\boldsymbol{P}^{(j)}$. From a geometrical point of view, the farther apart confidence ellipsoids of Gaussian components are, the smaller the approximation error involved in (12.6.13) will be. From a mathematical point of view, the condition for the validity of (12.6.13) can be expressed as the Mahalanobis distance inequality in the following form [445]

$$\left(\boldsymbol{m}^{(i)} - \boldsymbol{m}^{(j)}\right)^{\mathrm{T}}\left(\boldsymbol{P}^{(i)}\right)^{-1}\left(\boldsymbol{m}^{(i)} - \boldsymbol{m}^{(j)}\right) \gg 1 \tag{12.6.14}$$

$$\left(\boldsymbol{m}^{(i)} - \boldsymbol{m}^{(j)}\right)^{\mathrm{T}}\left(\boldsymbol{P}^{(j)}\right)^{-1}\left(\boldsymbol{m}^{(i)} - \boldsymbol{m}^{(j)}\right) \gg 1 \tag{12.6.15}$$

The approximation (12.6.13) is reasonable because the Gaussian mixture implementation of RFS filters requires a merging step to fuse Gaussian components whose Mahalanobis (or other types) distances are less than a given threshold.

In summary, according to (12.6.13), the fusions (12.4.22), (12.4.26) and (12.4.29) can be approximated as

$$\overline{p}^{(*)}(\boldsymbol{x}) = \frac{\sum_{i=1}^{J_a^{(*)}}\sum_{j=1}^{J_b^{(*)}} w_{ab}^{(i,j)}\mathcal{N}\left(\boldsymbol{x}; \boldsymbol{m}_{ab}^{(i,j)}, \boldsymbol{P}_{ab}^{(i,j)}\right)}{\left\langle\sum_{i=1}^{J_a^{(*)}}\sum_{j=1}^{J_b^{(*)}} w_{ab}^{(i,j)}\mathcal{N}\left(\boldsymbol{x}; \boldsymbol{m}_{ab}^{(i,j)}, \boldsymbol{P}_{ab}^{(i,j)}\right), 1\right\rangle}$$
$$= \frac{\sum_{i=1}^{J_a^{(*)}}\sum_{j=1}^{J_b^{(*)}} w_{ab}^{(i,j)}\mathcal{N}\left(\boldsymbol{x}; \boldsymbol{m}_{ab}^{(i,j)}, \boldsymbol{P}_{ab}^{(i,j)}\right)}{\sum_{i=1}^{J_a^{(*)}}\sum_{j=1}^{J_b^{(*)}} w_{ab}^{(i,j)}} \tag{12.6.16}$$

where

$$\boldsymbol{P}_{ab}^{(i,j)} = \left[\omega(\boldsymbol{P}_a^{(i)})^{-1} + (1-\omega)(\boldsymbol{P}_b^{(j)})^{-1}\right]^{-1} \tag{12.6.17}$$

$$\boldsymbol{m}_{ab}^{(i,j)} = \boldsymbol{P}_{ab}^{(i,j)}\left[\omega(\boldsymbol{P}_a^{(i)})^{-1}\boldsymbol{m}_a^{(i)} + (1-\omega)(\boldsymbol{P}_b^{(j)})^{-1}\boldsymbol{m}_b^{(j)}\right] \tag{12.6.18}$$

$$w_{ab}^{(i,j)} = \left(w_a^{(i)}\right)^{\omega}\left(w_b^{(j)}\right)^{1-\omega}\alpha\left(\omega, \boldsymbol{P}_a^{(i)}\right)\alpha\left(1-\omega, \boldsymbol{P}_b^{(j)}\right)$$

$$\mathcal{N}\left(\boldsymbol{m}_a^{(i)} - \boldsymbol{m}_b^{(j)}; \boldsymbol{0}, \omega^{-1}\boldsymbol{P}_a^{(i)} + (1-\omega)^{-1}\boldsymbol{P}_b^{(j)}\right) \tag{12.6.19}$$

Note that (12.6.16)–(12.6.19) are equivalent to performing the Chernoff fusion (CI fusion) on any possible 2-tuple formed by Gaussian components of node a and Gaussian components of node b. Moreover, the resulting coefficient $w_{ab}^{(i,j)}$ of the fused component includes a factor $\mathcal{N}(\boldsymbol{m}_a^{(i)} - \boldsymbol{m}_b^{(j)}; \boldsymbol{0}, \omega^{-1}\boldsymbol{P}_a^{(i)} + (1-\omega)^{-1}\boldsymbol{P}_b^{(j)})$, which measures the separation between two components $(\boldsymbol{m}_a^{(i)}, \boldsymbol{P}_a^{(i)})$ and $(\boldsymbol{m}_b^{(j)}, \boldsymbol{P}_b^{(j)})$ to be fused.

In (12.6.16), the Gaussian component with negligible coefficient $w_{ab}^{(i,j)}$ can be appropriately deleted, which can be accomplished by setting a threshold for the coefficient or checking whether the following Mahalanobis distance is lower than a given threshold

$$\sqrt{\left(\boldsymbol{m}_a^{(i)} - \boldsymbol{m}_b^{(j)}\right)^{\mathrm{T}}\left[\omega^{-1}\boldsymbol{P}_a^{(i)} + (1-\omega)^{-1}\boldsymbol{P}_b^{(j)}\right]^{-1}\left(\boldsymbol{m}_a^{(i)} - \boldsymbol{m}_b^{(j)}\right)} \tag{12.6.20}$$

By sequentially applying the pairwise fusion rules (12.6.17) and (12.6.18) $|\mathcal{S}|-1$ times, the fusion (12.6.6) can be generalized to $|\mathcal{S}| > 2$ nodes. Note that, according to the associative and commutative properties of the product operator, the final result is independent of the order of pairwise fusions.

12.6.3 Consensus GM-CPHD Filter

Table 12.3 presents the consensus GM-CPHD (CGM-CPHD) filtering algorithm, which lists the order of operations performed by each node $i \in \mathcal{S}$ in the network at each sampling time k. All nodes $i \in \mathcal{S}$ work in parallel in the same way at each sampling time k, and each node starts with the previous cardinality PMF and the positional PDF estimation with the following GM form

$$\left\{\rho_{k-1|k-1,i}(n)\right\}_{n=0}^{n_{\max}}, \quad \left\{(w_i^{(j)}, \boldsymbol{m}_i^{(j)}, \boldsymbol{P}_i^{(j)})_{k-1|k-1}\right\}_{j=1}^{(N_i)_{k-1|k-1}} \tag{12.6.21}$$

and get the new CPHD and the estimates of target set, i.e.,

$$\left\{\rho_{k|k,i}(n)\right\}_{n=0}^{n_{\max}}, \quad \left\{(w_i^{(j)}, \boldsymbol{m}_i^{(j)}, \boldsymbol{P}_i^{(j)})_{k|k}\right\}_{j=1}^{(N_i)_{k|k}}, \quad \hat{X}_{k|k}^i \tag{12.6.22}$$

The steps of the CGM-CPHD algorithm are described briefly as follows.

(1) First, each node i performs the local GM-CPHD update according to the multi-target dynamics and local measurement set $Z_{k,i}$. Details of the GM-CPHD prediction and update are given in Sect. 5.4. In order to reduce the number of Gaussian components to relieve communication and computational burdens,

Table 12.3 Pseudo-code of the CGM-CPHD filter

	Steps of the CGM-CPHD filter (node i, time k)	
1:	Local GM-CPHD prediction	⊳ See Proposition 9
2:	Local GM-CPHD update	⊳ See Proposition 10
3:	GM merging	⊳ See Table 4.2
4:	for $c = 1, \ldots, C$	
5:	Information exchange	
6:	GM-GCI fusion	⊳ See (12.4.23) and (12.6.16)
7:	GM merging	⊳ See Table 4.2
8:	end	
9:	Pruning	⊳ See Table 4.2
10:	Estimate extraction	⊳ See Table 4.3

a merging step shall be implemented after the local update and before the consensus stage.

(2) Then, the consensus algorithm is implemented for in-neighbour nodes $\mathcal{I}_i$ of each node i. Each node exchanges information (i.e., the cardinality PMF and GM expression of the positional PDF) with neighboring nodes. More precisely, node i sends its data to node j that satisfies the condition of $i \in \mathcal{I}_j$, and waits until data from node $j \in \mathcal{I}_i \backslash \{i\}$ is received. Next, node i performs the GM-GCI fusion as shown by (12.4.23) and (12.6.16)–(12.6.19) on the network $\mathcal{I}_i$. Finally, applying the merging step reduces the joint communication-computation burdens of the next consensus step. This step is applied repeatedly according to the chosen number $C \geq 1$ of consensus steps.

(3) After the consensus step, the resulting GM is further simplified through the pruning step (i.e., only the first $N_{\max}$ maximum acceptable Gaussian components are retained, while other GM components are deleted). Finally, according to the cardinality PMF and the pruned positional GM, an estimate of the target set can be obtained through the same estimate extraction step as that of the conventional GM-CPHD filter.

12.6.4 Consensus GM-Mδ-GLMB Filter

Based on the Mδ-GLMB and LMB filters and according to Propositions 35 and 36, and the consensus, two completely new fully distributed and scalable multi-target tracking algorithms can be obtained, namely the consensus Gaussian mixture Mδ-GLMB (CGM-Mδ-GLMB) and consensus Gaussian mixture LMB (CGM-LMB) filters. For the Mδ-GLMB multi-target densities, (12.4.4) can be calculated with (12.4.25) and (12.4.26), while for the LMB densities, (12.4.4) is calculated with (12.4.28) and (12.4.29).

Table 12.4
CGM-Mδ-GLMB filter

	CGM-Mδ-GLMB steps (node i, time k)
1:	Local prediction ▷ See Table 7.2
2:	Local update ▷ See Table 7.3
3:	Marginalization ▷ See (7.5.30) and (7.5.31)
4:	for $c = 1, \ldots, C$
5:	Information exchange
6:	GM-Mδ-GLMB fusion ▷ See (12.4.25) and (12.4.26)
7:	GM merging ▷ See Table 4.2
8:	end
9:	Estimate extraction ▷ See Table 7.7

For the CGM-Mδ-GLMB filter, Table 12.4 lists the local sequential implementation steps for each node $i \in \mathcal{S}$ in the network.

At each sampling interval k, each node i starts from the estimation of its own multi-target distribution π_i at the previous moment, which has a positional PDF $p_i^{(I)}(\boldsymbol{x}, \ell), \forall \ell \in I, I \in \mathcal{F}(\mathbb{L})$ in the GM form. A new multi-target distribution $\pi_i = \pi_{i,C}$ is obtained at the last moment of the sequential operation and taken as the result of the consensus step. The steps of the CGM-Mδ-GLMB algorithm are summarized as follows.

(1) Each node $i \in \mathcal{S}$ performs the GM-δ-GLMB prediction and update steps locally. The details of the two steps can be found in Sect. 7.3.2.
(2) At each consensus step, node i sends its data to its neighboring node $j \in \mathcal{I}_i \backslash \{i\}$ and waits until data from neighboring nodes is received. Next, the node executes the fusion rule of Proposition 35 over $\mathcal{I}_i$, that is, computes (12.4.4) using the local information and the information received from $\mathcal{I}_i$. Finally, the merging step is applied to each positional PDF to reduce the communication/computational burdens of the next consensus step. This step is repeated according to the chosen number $C \geq 1$ of the consensus steps.
(3) After the consensus step, an estimate of the target set can be obtained from the cardinality PMF and positional PDF through the estimate extraction step as described in Table 7.7.

12.6.5 Consensus GM-LMB Filter

Table 12.5 lists the CGM-LMB filtering algorithm, which is implemented locally by each node $i \in \mathcal{S}$ in the network in a sequential way. The CGM-LMB step is basically the same as that of the above CGM-Mδ-GLMB tracking filter, except that the Mδ-GLMB prediction and update are replaced with the LMB prediction and update.

Table 12.5 CGM-LMB filter

	CGM-LMB steps (node i, time k)
1:	Local prediction ▷ See Proposition 19
2:	Local update ▷ See Sect. 7.4.2
3:	for $c = 1, \ldots, C$
4:	Information exchange
5:	GM-LMB fusion ▷ See (12.4.28) and (12.4.29)
6:	GM merging ▷ See Table 4.2
7:	end
8:	Estimate extraction ▷ See Table 7.6

12.7 Summary

Aiming at the problem of distributed multi-sensor multi-target tracking, this chapter respectively introduces such distributed fusion algorithms as SMC-CPHD filter-based distributed fusion, consensus GM-CPHD, consensus GM-Mδ-GLMB and consensus GM-LMB filters under the RFS framework, based on conclusions about multi-target KLA and the consensus principle. Different from the classical non-RFS multi-sensor fusion methods, the RFS framework provides the concept of multi-target probability densities for multi-target states (note that this concept is not available in the MHT and JPDA methods), thus allowing the existing tools for the single-target probability density in the classical distributed estimation to be directly extended to the multi-target case, which greatly facilitates the development of RFS-based multi-sensor fusion algorithms.

It should be noted that, although these algorithms provide preliminary solutions for multi-sensor multi-target tracking, for practical multi-sensor multi-target tracking systems, a series of problems still need to be solved, such as spatial–temporal registration and out-of-sequence measurement processing, etc. Therefore, the corresponding improvement of the above algorithms is still worthy of further research.

Appendix A
Product Formulas of Gaussian Functions

Lemma 4 *Given* $\boldsymbol{F}$, $\boldsymbol{u}$, $\boldsymbol{Q}$, $\boldsymbol{m}$, *and* $\boldsymbol{P}$ *of appropriate dimensions, and assuming that* $\boldsymbol{Q}$ *and* $\boldsymbol{P}$ *are positive definite, then*

$$\int \mathcal{N}(\boldsymbol{x}; \boldsymbol{F}\boldsymbol{\xi} + \boldsymbol{u}, \boldsymbol{Q})\mathcal{N}(\boldsymbol{\xi}; \boldsymbol{m}, \boldsymbol{P})d\boldsymbol{\xi} = \mathcal{N}\big(\boldsymbol{x}; \boldsymbol{F}\boldsymbol{m} + \boldsymbol{u}, \boldsymbol{Q} + \boldsymbol{F}\boldsymbol{P}\boldsymbol{F}^{\mathrm{T}}\big) \tag{A.1}$$

Lemma 5 *Given* $\boldsymbol{H}$, $\boldsymbol{R}$, $\boldsymbol{d}$, $\boldsymbol{m}$, *and* $\boldsymbol{P}$ *of appropriate dimensions, and assuming that* $\boldsymbol{R}$ *and* $\boldsymbol{P}$ *are positive definite, then*

$$\mathcal{N}(z; \boldsymbol{H}\boldsymbol{x} + \boldsymbol{d}, \boldsymbol{R})\mathcal{N}(\boldsymbol{x}; \boldsymbol{m}, \boldsymbol{P}) = \mathcal{N}(z; \boldsymbol{H}\boldsymbol{m} + \boldsymbol{d}, S)\mathcal{N}\Big(\boldsymbol{x}; \tilde{\boldsymbol{m}}, \tilde{\boldsymbol{P}}\Big) \tag{A.2}$$

where $\tilde{\boldsymbol{m}} = \boldsymbol{m}+\boldsymbol{G}(z-\boldsymbol{H}\boldsymbol{m})$, $\tilde{\boldsymbol{P}} = \boldsymbol{P}-\boldsymbol{G}\boldsymbol{S}\boldsymbol{G}^{\mathrm{T}}$, $\boldsymbol{G} = \boldsymbol{P}\boldsymbol{H}^{\mathrm{T}}\boldsymbol{S}^{-1}$, *and* $\boldsymbol{S} = \boldsymbol{H}\boldsymbol{P}\boldsymbol{H}^{\mathrm{T}}+\boldsymbol{R}$.

Remark (A.1) can be derived from (A.2) [211].

W. Wu et al., *Target Tracking with Random Finite Sets*,
https://doi.org/10.1007/978-981-19-9815-7

Appendix B
Functional Derivatives and Set Derivatives

The functional $G[h]$ takes the function $h(\boldsymbol{x})$ as an independent variable, and its gradient derivative $\frac{\partial G}{\partial \zeta}[h]$ in the function direction ζ is defined as

$$\frac{\partial G}{\partial \zeta}[h] = \lim_{\lambda \to 0^+} \frac{G[h + \lambda\zeta] - G[h]}{\lambda} \tag{B.1}$$

where $\lambda \to 0^+$ represents the right limit of λ tending to 0, and the above equation is also referred to as the functional Frechet derivative. In particular, the gradient derivative of the functional in the function $\zeta = \delta_{\boldsymbol{x}}$ direction (i.e., at $\boldsymbol{x}$) is called the functional derivative, i.e.,

$$\frac{\partial G}{\partial \delta_{\boldsymbol{x}}}[h] = \lim_{\lambda \to 0^+} \frac{G[h + \lambda\delta_{\boldsymbol{x}}] - G[h]}{\lambda} \tag{B.2}$$

Further, the iterative functional derivative can be defined recursively as follows

$$\frac{\partial^n G}{\partial \delta_{\boldsymbol{x}_n} \cdots \partial \delta_{\boldsymbol{x}_1}}[h] = \frac{\partial}{\partial \delta_{\boldsymbol{x}_n}} \frac{\partial^n G}{\partial \delta_{\boldsymbol{x}_{n-1}} \cdots \partial \delta_{\boldsymbol{x}_1}}[h] \tag{B.3}$$

As a consequence, the functional derivative at the set $X = \{\boldsymbol{x}_1, \ldots, \boldsymbol{x}_n\}$ is

$$\frac{\partial G}{\partial X}[h] = \begin{cases} G[h], & X = \emptyset \\ \frac{\partial^n G}{\partial \delta_{\boldsymbol{x}_n} \cdots \partial \delta_{\boldsymbol{x}_1}}[h], & X = \{\boldsymbol{x}_1, \ldots, \boldsymbol{x}_n\}, |X| = n \end{cases} \tag{B.4}$$

Each functional can generate the corresponding set function as follows

$$\beta(S) = G[1_S] \tag{B.5}$$

W. Wu et al., *Target Tracking with Random Finite Sets*,
https://doi.org/10.1007/978-981-19-9815-7

In fact, the above set function is also called the belief mass function in the RFS theory. Given a set function $\beta(S)$, its set derivative in the $\delta_{\boldsymbol{x}}$ direction is defined as [14]

$$\begin{aligned}\frac{\partial\beta}{\partial\delta_{\boldsymbol{x}}}(S) = \frac{\partial G}{\partial\delta_{\boldsymbol{x}}}[1_S] &= \lim_{|\sigma_{\boldsymbol{x}}|\to 0^+}\frac{G[1_{S\cup\sigma_{\boldsymbol{x}}}] - G[1_S]}{|\sigma_{\boldsymbol{x}}|}\\ &= \lim_{|\sigma_{\boldsymbol{x}}|\to 0^+}\frac{\beta[S\cup\sigma_{\boldsymbol{x}}] - \beta[S]}{|\sigma_{\boldsymbol{x}}|}\end{aligned} \tag{B.6}$$

where $\sigma_{\boldsymbol{x}}$ denotes the minimal neighborhood at $\boldsymbol{x}$, whose volume is $|\sigma_{\boldsymbol{x}}|$. Further, the iterative set derivative can be defined recursively as follows

$$\frac{\partial^n\beta}{\partial\delta_{\boldsymbol{x}_n}\cdots\partial\delta_{\boldsymbol{x}_1}}(S) = \frac{\partial}{\partial\delta_{\boldsymbol{x}_n}}\frac{\partial^n\beta}{\partial\delta_{\boldsymbol{x}_{n-1}}\cdots\partial\delta_{\boldsymbol{x}_1}}(S) \tag{B.7}$$

As a consequence, the set derivative at the set $X = \{\boldsymbol{x}_1, \ldots, \boldsymbol{x}_n\}$ is

$$\frac{\partial\beta}{\partial X}(S) = \begin{cases}\beta(S), & X = \emptyset\\ \frac{\partial^n\beta}{\partial\delta_{\boldsymbol{x}_n}\cdots\partial\delta_{\boldsymbol{x}_1}}(S), & X = \{\boldsymbol{x}_1, \ldots, \boldsymbol{x}_n\}, |X| = n\end{cases} \tag{B.8}$$

Therefore, the set derivative and the functional derivative have the following relationship

$$\frac{\partial\beta}{\partial X}(S) = \frac{\partial G}{\partial X}[1_S] \tag{B.9}$$

It can be seen from the above formula that the set derivative is a special kind of functional derivative. More precisely, the set derivative of the belief mass function is a special case of the functional derivative of the probability generating functional (PGFl).

It should be noted that the set derivative introduced here and the Radon–Nikodym derivative are two different ways to define the multi-target density, but the set integral is the indefinite integral of the set derivative rather than the Radon–Nikodym derivative.

Appendix C
Probability Generating Function (PGF) and Probability Generating Functional (PGFl)

For the general functions, the integral transform methods (such as Fourier transform, Laplace transform, z transform, etc.) have the strong analysis capability, mainly because many mathematical operations are difficult to express in the time domain, but the processing in the transform domain is very simple. For example, by means of the integral transform method, the convolution of two signals in the time domain can be transformed into the product of the corresponding signals in the transform domain, or the derivation or integration operation in the time domain can be converted into a simple algebraic operation in the transform domain.

In the field of probability and statistics, the integral transforms are of equal importance. For example, if the density function of random variable $\boldsymbol{y}$ is $p(\boldsymbol{y})$, then its moment generating function $M(\boldsymbol{x})$ is

$$M(\boldsymbol{x}) = \int_{-\infty}^{\infty} \exp(\boldsymbol{x}\boldsymbol{y})p(\boldsymbol{y})d\boldsymbol{y} = \mathrm{E}(\exp(\boldsymbol{x}\boldsymbol{y})) \tag{C.1}$$

Conversely, the n-order statistical moment $\boldsymbol{m}_n = \int_{-\infty}^{\infty} \boldsymbol{y}^n p(\boldsymbol{y})d\boldsymbol{y}$ of $\boldsymbol{y}$ can be recovered from the moment generating function $M(\boldsymbol{x})$ by

$$\boldsymbol{m}_n = \frac{d^n M}{d\boldsymbol{x}^n}(0) \tag{C.2}$$

Therefore, the term "moment generating function" is called so.

In addition, let the probability distribution of random non-negative integer n be $\rho(n)$, then its probability generating function (PGF) is

$$G(\boldsymbol{x}) = \sum_{n=0}^{\infty} \boldsymbol{x}^n \rho(n) = \mathrm{E}(\boldsymbol{x}^n) \tag{C.3}$$

W. Wu et al., *Target Tracking with Random Finite Sets*,
https://doi.org/10.1007/978-981-19-9815-7

Similarly, the probability distribution of n can also be recovered from the probability generating function $G(\boldsymbol{x})$ by the following formula

$$\rho(n) = \frac{1}{n!} \frac{d^n G}{d\boldsymbol{x}^n}(0) \tag{C.4}$$

Therefore, the term "PGF" is called so.

The integral transformation is also of great significance in multi-target statistics. The probability generating functional (PGFl) $G[\cdot]$ of RFS X takes a non-negative real-valued function h as its independent variable, defined as

$$G[h] \triangleq \mathrm{E}\left[h^X\right] = \int h^X \pi(X) \delta X \tag{C.5}$$

where $\pi(X)$ is the density function of RFS X. Therefore, PGFl $G[\cdot]$ is the generalization of PGF $G(\cdot)$ in multi-target statistics, and the PGFl can be regarded as an integral transformation of the PGF. Moreover, due to the equivalence relationship between the PGFl and multi-target probability density, the PGFl can be used to obtain the multi-target probability density. Therefore, the term "probability generating functional (PGFl)" gets its name.

Appendix D
Proof of Related Labeled RFS Formulas

(1) Proof of Lemma 2

$$
\begin{aligned}
&\int \Delta(\mathrm{X}) h(\mathcal{L}(\mathrm{X})) g^{\mathrm{X}} \delta \mathrm{X} \\
&= \int \delta_{|\mathrm{X}|}(|\mathcal{L}(\mathrm{X})|) h(\mathcal{L}(\mathrm{X})) g^{\mathrm{X}} \delta \mathrm{X} \\
&= \sum_{n=0}^{\infty} \frac{1}{n!} \sum_{(\ell_1,\ldots,\ell_n)\in\mathbb{L}^n} \delta_n(|\{\ell_1,\ldots,\ell_n\}|) h(\{\ell_1,\ldots,\ell_n\}) \times \\
&\quad \int \left(\prod_{i=1}^{n} g(\boldsymbol{x}_i,\ell_i) \right) d\boldsymbol{x}_1 \ldots d\boldsymbol{x}_n \\
&= \sum_{n=0}^{\infty} \frac{1}{n!} \sum_{(\ell_1,\ldots,\ell_n)\in\mathbb{L}^n} \delta_n(|\{\ell_1,\ldots,\ell_n\}|) h(\{\ell_1,\ldots,\ell_n\}) \prod_{i=1}^{n} \left(\int g(\boldsymbol{x}_i,\ell_i) d\boldsymbol{x}_i \right) \\
&= \sum_{n=0}^{\infty} \sum_{\{\ell_1,\ldots,\ell_n\}\in\mathcal{F}_n(\mathbb{L})} h(\{\ell_1,\ldots,\ell_n\}) \prod_{i=1}^{n} \left(\int g(\boldsymbol{x}_i,\ell_i) d\boldsymbol{x}_i \right) \\
&= \sum_{L\subseteq\mathbb{L}} h(L) \left[\int g(\boldsymbol{x},\cdot) d\boldsymbol{x} \right]^L
\end{aligned}
\tag{D.1}
$$

where the penultimate line is derived from the symmetry of $h(\{\ell_1,\ldots,\ell_n\})\prod_{i=1}^{n}\left(\int g(\boldsymbol{x}_i,\ell_i)d\boldsymbol{x}_i\right)$ with respect to $(\ell_1,\ldots,\ell_n)$ and Lemma 1. Finally, the double summation can be combined into a single summation over the subsets of $\mathbb{L}$, thus leading to (3.3.20).

(2) Proof of Proposition 1

W. Wu et al., *Target Tracking with Random Finite Sets*,
https://doi.org/10.1007/978-981-19-9815-7

According to (3.2.13), the PHD of label-free RFS is [14]

$$
\begin{aligned}
v(\boldsymbol{x}) &= \int \pi(\{\boldsymbol{x}\} \cup X)\delta X \\
&= \sum_{n=0}^{\infty} \frac{1}{n!} \int \pi(\{\boldsymbol{x}\} \cup \{\boldsymbol{x}_1, \ldots, \boldsymbol{x}_n\}) d(\boldsymbol{x}_1, \ldots, \boldsymbol{x}_n)
\end{aligned} \tag{D.2}
$$

and according to (3.3.16), we have

$$
\pi(\{\boldsymbol{x}\} \cup \{\boldsymbol{x}_1, \ldots, \boldsymbol{x}_n\}) = \sum_{(\ell, \ell_1, \ldots, \ell_n) \in \mathbb{L}^{n+1}} \pi(\{(\boldsymbol{x}, \ell), (\boldsymbol{x}_1, \ell_1), \ldots, (\boldsymbol{x}_n, \ell_n)\}) \tag{D.3}
$$

Using the GLMB defined by (3.3.38) to replace π in (3.3.25), then moving the integral into the summation over $\mathbb{C}$, and using the property that the integral of each $p^{(c)}(\cdot, \ell_i)$ is 1, we can obtain

$$
\begin{aligned}
v(\boldsymbol{x}) &= \sum_{n=0}^{\infty} \frac{1}{n!} \sum_{c \in \mathbb{C}} \sum_{(\ell, \ell_1, \ldots, \ell_n) \in \mathbb{L}^{n+1}} \delta_{n+1}(|\{\ell, \ell_1, \ldots, \ell_n\}|) w^{(c)}(\{\ell, \ell_1, \ldots, \ell_n\}) p^{(c)}(\boldsymbol{x}, \ell) \\
&= \sum_{n=0}^{\infty} \frac{1}{n!} \sum_{c \in \mathbb{C}} \sum_{\ell \in \mathbb{L}} \sum_{(\ell_1, \ldots, \ell_n) \in \mathbb{L}^n} \delta_n(|\{\ell_1, \ldots, \ell_n\}|)(1 - 1_{\{\ell_1, \ldots, \ell_n\}}(\ell)) \times \\
&\quad w^{(c)}(\{\ell, \ell_1, \ldots, \ell_n\}) p^{(c)}(\boldsymbol{x}, \ell)
\end{aligned} \tag{D.4}
$$

Since $(1 - 1_{\{\ell_1, \ldots, \ell_n\}}(\ell)) w^{(c)}(\{\ell, \ell_1, \ldots, \ell_n\})$, as a function of $(\ell_1, \ldots, \ell_n)$, is of permutation invariance, applying Lemma 1 can lead to

$$
\begin{aligned}
v(\boldsymbol{x}) &= \sum_{n=0}^{\infty} \sum_{c \in \mathbb{C}} \sum_{\ell \in \mathbb{L}} p^{(c)}(\boldsymbol{x}, \ell) \sum_{L \in \mathcal{F}_n(\mathbb{L})} (1 - 1_L(\ell)) w^{(c)}(\{\ell\} \cup L) \\
&= \sum_{c \in \mathbb{C}} \sum_{\ell \in \mathbb{L}} p^{(c)}(\boldsymbol{x}, \ell) \sum_{L \subseteq \mathbb{L}} (1 - 1_L(\ell)) w^{(c)}(\{\ell\} \cup L) \\
&= \sum_{c \in \mathbb{C}} \sum_{\ell \in \mathbb{L}} p^{(c)}(\boldsymbol{x}, \ell) \sum_{L \subseteq \mathbb{L}} 1_L(\ell) w^{(c)}(L)
\end{aligned} \tag{D.5}
$$

(3) Proof of Proposition 2

Similar to the proof on the PHD (i.e., the proof of Proposition 1), using the GLMB defined by (3.3.38) to replace π in (3.3.25), then moving the integral into the sum over $\mathbb{C}$ and using Lemma 1 and the property that the integral of each $p^{(c)}(\cdot, \ell_i)$ is 1, we can obtain

$\rho(|\mathrm{X}| = n)$

$$
\begin{aligned}
&= \frac{1}{n!} \sum_{(\ell_1,\ldots,\ell_n)\in\mathbb{L}^n} \delta_n(|\{\ell_1,\ldots,\ell_n\}|) \sum_{c\in\mathbb{C}} w^{(c)}(\{\ell_1,\ldots,\ell_n\}) \times \\
&\quad \left[\int_{\mathbb{X}^n} \prod_{i=1}^{n} p^{(c)}(\boldsymbol{x}_i,\ell_i) d(\boldsymbol{x}_1,\ldots,\boldsymbol{x}_n) \right] \\
&= \sum_{c\in\mathbb{C}} \frac{1}{n!} \sum_{(\ell_1,\ldots,\ell_n)\in\mathbb{L}^n} \delta_n(|\{\ell_1,\ldots,\ell_n\}|) w^{(c)}(\{\ell_1,\ldots,\ell_n\}) \left(\prod_{i=1}^{n} \int_{\mathbb{X}} p^{(c)}(\boldsymbol{x}_i,\ell_i) d\boldsymbol{x}_i \right) \\
&= \sum_{c\in\mathbb{C}} \frac{1}{n!} \sum_{(\ell_1,\ldots,\ell_n)\in\mathbb{L}^n} \delta_n(|\{\ell_1,\ldots,\ell_n\}|) w^{(c)}(\{\ell_1,\ldots,\ell_n\}) \\
&= \sum_{c\in\mathbb{C}} \frac{1}{n!} n! \sum_{\{\ell_1,\ldots,\ell_n\}\in\mathcal{F}_n(\mathbb{L})} w^{(c)}(\{\ell_1,\ldots,\ell_n\}) \\
&= \sum_{L\in\mathcal{F}_n(\mathbb{L})} \sum_{c\in\mathbb{C}} w^{(c)}(L) \qquad \text{(D.6)}
\end{aligned}
$$

Appendix E
Derivation of CPHD Recursion

Let $v_{k|k-1}$ and v_k denote the predicted intensity and posterior intensity of the multi-target state, respectively, $\rho_{k|k-1}$ and ρ_k denote the predicted cardinality distribution and posterior cardinality distribution of the multi-target state, respectively, $G_{k|k-1}$ and G_k denote the PGFs of $\rho_{k|k-1}$ and ρ_k, respectively, and $G_{\gamma,k}$ and $G_{C,k}$ represent the PGFs of the birth target cardinality distribution $\rho_{\gamma,k}$ and the clutter cardinality distribution $\rho_{C,k}$, respectively; $G^{(i)}(\cdot)$ represents the i-order derivative of $G(\cdot)$, and $\hat{G}^{(i)}(\cdot) = G^{(i)}(\cdot)/[G^{(1)}(1)]^i$ [64]; $p_{C,k}(z) = \kappa_k(z)/\langle 1, \kappa_k \rangle$ is the clutter density and $q_{D,k}(\boldsymbol{x}) = 1 - p_{D,k}(\boldsymbol{x})$ is the probability of missed detection; and $\sigma_j(Z)$ represents $e_j(\alpha_k(v_{k|k-1}, Z))$. Moreover, for any non-normalized density v, let $\overline{v} = v/\langle 1, v \rangle$.

(1) Proof of Proposition 7

According to [3], the predicted CPHD is

$$v_{k|k-1}(\boldsymbol{x}) = \int p_{S,k}(\boldsymbol{\xi})\phi_{k|k-1}(\boldsymbol{x}|\boldsymbol{\xi})v_{k-1}(\boldsymbol{\xi})d\boldsymbol{\xi} + v_{\gamma,k}(\boldsymbol{x}) \tag{E.1}$$

$$\rho_{k|k-1}(n) = \sum_{j=0}^{n} \rho_{\gamma,k}(n-j)\Big[G_{k-1}^{(j)}\big(1 - \langle p_{S,k}, \overline{v}_{k-1} \rangle\big)\langle p_{S,k}, \overline{v}_{k-1} \rangle^j / j!\Big] \tag{E.2}$$

Note that the PHD prediction (E.1) is exactly the same as (5.2.1). To simplify the cardinality prediction (E.2), by using $\overline{v}_{k-1} = v_{k-1}/\langle 1, v_{k-1} \rangle$ and $G_{k-1}^{(j)}(y) = \sum_{i=j}^{\infty} P_j^i \rho_{k-1}(i) y^{i-j}$ [64], where P_j^i represents the permutation coefficient $i!/(i-j)!$, then the expression in the square bracket in (E.2) can be simplified as

$$\begin{aligned}
&G_{k-1}^{(j)}(1 - \langle p_{S,k}, \overline{v}_{k-1} \rangle)\langle p_{S,k}, \overline{v}_{k-1} \rangle^j / j! \\
&= \frac{1}{j!}\sum_{i=j}^{\infty} P_j^i \rho_{k-1}(i)\left[1 - \frac{\langle p_{S,k}, v_{k-1} \rangle}{\langle 1, v_{k-1} \rangle}\right]^{i-j}\left[\frac{\langle p_{S,k}, v_{k-1} \rangle}{\langle 1, v_{k-1} \rangle}\right]^j \\
&= \sum_{i=j}^{\infty} C_j^i \frac{\langle p_{S,k}, v_{k-1} \rangle^j (\langle 1, v_{k-1} \rangle - \langle p_{S,k}, v_{k-1} \rangle)^{i-j}}{\langle 1, v_{k-1} \rangle^i} \rho_{k-1}(i)
\end{aligned}$$

W. Wu et al., *Target Tracking with Random Finite Sets*,
https://doi.org/10.1007/978-981-19-9815-7

$$= \sum_{i=j}^{\infty} C_j^i \frac{\langle p_{S,k}, v_{k-1}\rangle^j \langle 1 - p_{S,k}, v_{k-1}\rangle^{i-j}}{\langle 1, v_{k-1}\rangle^i} \rho_{k-1}(i)$$
$$= \Psi_{k|k-1}[v_{k-1}, \rho_{k-1}](j) \tag{E.3}$$

Therefore, substituting (E.3) into (E.2) can lead to the CPHD cardinality prediction (5.2.2).

(2) Proof of Proposition 8

According to [3], the updated CPHD is

$$v_k(\boldsymbol{x}) = q_{D,k}(\boldsymbol{x}) \left[\frac{\sum_{j=0}^{|Z_k|} G_{C,k}^{(|Z_k|-j)}(0) \cdot G_{k|k-1}^{(j+1)}\left(\langle q_{D,k}, \overline{v}_{k|k-1}\rangle\right) \cdot \sigma_j(Z_k)}{\sum_{i=0}^{|Z_k|} G_{C,k}^{(|Z_k|-i)}(0) \cdot G_{k|k-1}^{(i)}\left(\langle q_{D,k}, \overline{v}_{k|k-1}\rangle\right) \cdot \sigma_i(Z_k)} \right] v_{k|k-1}(\boldsymbol{x})$$
$$+ p_{D,k}(\boldsymbol{x}) \sum_{z \in Z_k} \frac{g_k(z|\boldsymbol{x})}{p_{C,k}(z)} \times$$
$$\left[\frac{\sum_{j=0}^{|Z_k|-1} G_{C,k}^{(|Z_k|-j-1)}(0) \cdot G_{k|k-1}^{(j+1)}\left(\langle q_{D,k}, \overline{v}_{k|k-1}\rangle\right) \cdot \sigma_j(Z_k - \{z\})}{\sum_{i=0}^{|Z_k|} G_{C,k}^{(|Z_k|-i)}(0) \cdot G_{k|k-1}^{(i)}\left(\langle q_{D,k}, \overline{v}_{k|k-1}\rangle\right) \cdot \sigma_i(Z_k)} \right] v_{k|k-1}(\boldsymbol{x}) \tag{E.4}$$

$$\rho_k(n) = \frac{\sum_{j=0}^{|Z_k|} G_{C,k}^{(|Z_k|-j)}(0) \cdot \frac{1}{(n-j)!} \hat{G}_{k|k-1}^{(j)(n-j)}(0) \cdot \left\langle q_{D,k}, \overline{v}_{k|k-1}\right\rangle^{n-j} \cdot \sigma_j(Z_k)}{\sum_{i=0}^{|Z_k|} G_{C,k}^{(|Z_k|-i)}(0) \cdot G_{k|k-1}^{(i)}\left(\langle q_{D,k}, \overline{v}_{k|k-1}\rangle\right) \cdot \sigma_i(Z_k)} \tag{E.5}$$

First, simplify the intensity update (E.4). Note that the numerator and denominator in two square brackets in (E.4) can be written in the following general form

$$\sum_{j=0}^{|Z|} G_{C,k}^{(|Z|-j)}(0) \cdot G_{k|k-1}^{(j+u)}\left(\langle q_{D,k}, \overline{v}_{k|k-1}\rangle\right) \cdot \sigma_j(Z) \tag{E.6}$$

By using $\overline{v}_{k|k-1} = v_{k|k-1}/\langle 1, v_{k|k-1}\rangle$, $G_{C,k}^{(j)}(0) = i!\rho_{C,k}(i)$, $\hat{G}_{k|k-1}^{(i)}(y) = \langle 1, v_{k|k-1}\rangle^{-i} \sum_{n=i}^{\infty} P_i^n \cdot \rho_{k|k-1}(n) \cdot y^{n-i}$ [64] and $P_i^n = 0$ for any integer $n < i$, (E.6) can be simplified to

$$\sum_{j=0}^{|Z_k|} (|Z_k| - j)! \rho_{C,k}(|Z_k| - j) \cdot \frac{\sigma_j(Z_k)}{\left\langle 1, v_{k|k-1}\right\rangle^{j+u}} \times$$
$$\sum_{n=j+u}^{\infty} P_{j+u}^n \cdot \rho_{k|k-1}(n) \cdot \left(\langle q_{D,k}, \overline{v}_{k|k-1}\rangle\right)^{n-(j+u)}$$
$$= \sum_{j=0}^{|Z_k|} (|Z_k| - j)! \rho_{C,k}(|Z_k| - j) \cdot \frac{\sigma_j(Z_k)}{\left\langle 1, v_{k|k-1}\right\rangle^{j+u}} \times$$

$$\sum_{n=j+u}^{\infty} P_{j+u}^{n} \cdot \rho_{k|k-1}(n) \cdot \frac{\langle q_{D,k}, v_{k|k-1}\rangle^{n-(j+u)}}{\langle 1, v_{k|k-1}\rangle^{n-(j+u)}}$$

$$= \sum_{j=0}^{|Z_k|} (|Z_k| - j)!\rho_{C,k}(|Z_k| - j) \cdot \sigma_j(Z_k) \times$$

$$\sum_{n=j+u}^{\infty} P_{j+u}^{n} \cdot \rho_{k|k-1}(n) \cdot \frac{\langle q_{D,k}, v_{k|k-1}\rangle^{n-(j+u)}}{\langle 1, v_{k|k-1}\rangle^{n}}$$

$$= \sum_{j=0}^{\min(|Z_k|,n)} (|Z_k| - j)!\rho_{C,k}(|Z_k| - j) \cdot \sigma_j(Z_k) \times$$

$$\sum_{n=0}^{\infty} P_{j+u}^{n} \cdot \rho_{k|k-1}(n) \cdot \frac{\langle q_{D,k}, v_{k|k-1}\rangle^{n-(j+u)}}{\langle 1, v_{k|k-1}\rangle^{n}}$$

$$= \sum_{n=0}^{\infty} \rho_{k|k-1}(n)\Bigg[\sum_{j=0}^{\min(|Z_k|,n)} (|Z_k| - j)!\rho_{C,k}(|Z_k| - j)$$

$$\cdot \sigma_j(Z_k) P_{j+u}^{n} \frac{\langle q_{D,k}, v_{k|k-1}\rangle^{n-(j+u)}}{\langle 1, v_{k|k-1}\rangle^{n}}\Bigg] \tag{E.7}$$

Since the expression in the square bracket in the last line of the above equation is exactly the definition of $\Upsilon_k^{(u)}[v_{k|k-1}, Z](n)$ (5.2.6), (E.6) can be rewritten as

$$\sum_{n=0}^{\infty} \rho_{k|k-1}(n)\Upsilon_k^{(u)}[v_{k|k-1}, Z](n) = \left\langle \rho_{k|k-1}, \Upsilon_k^{(u)}[v_{k|k-1}, Z]\right\rangle \tag{E.8}$$

Substituting $Z = Z_k - \{z\}$ and $u = 1$ into (E.8) can get the numerator in the first square bracket in (E.4); substituting $Z = Z_k$ and $u = 1$ into (E.8) can obtain the numerator in the second square bracket in (E.4); and substituting $Z = Z_k$ and $u = 0$ into (E.8) can lead to the denominator in two square brackets in (E.4). Thus, the CPHD intensity update (5.2.4) can be obtained.

The cardinality update (E.5) is simplified as follows. Using $G_{C,k}^{(j)}(0) = i!\rho_{C,k}(i)$ and $\hat{G}_{k|k-1}^{(j)(n-j)}(0) = \langle 1, v_{k|k-1}\rangle^{-j} n!\rho_{k|k-1}(n)$ [64], the numerator in (E.5) can be simplified to

$$\sum_{j=0}^{|Z_k|} G_{C,k}^{(|Z_k|-j)}(0) \cdot \frac{1}{(n-j)!}\hat{G}_{k|k-1}^{(j)(n-j)}(0) \cdot \langle q_{D,k}, \overline{v}_{k|k-1}\rangle^{n-j} \cdot \sigma_j(Z_k)$$

$$= \sum_{j=0}^{|Z_k|} \frac{(|Z_k| - j)!\rho_{C,k}(|Z_k| - j)}{(n-j)!} \frac{n!\rho_{k|k-1}(n)}{\langle 1, v_{k|k-1}\rangle^{j}} \cdot \frac{\langle q_{D,k}, v_{k|k-1}\rangle^{n-j}}{\langle 1, v_{k|k-1}\rangle^{n-j}} \cdot \sigma_j(Z_k)$$

$$
\begin{aligned}
&= \sum_{j=0}^{|Z_k|} (|Z_k| - j)! \rho_{C,k}(|Z_k| - j) P_j^n \frac{\langle q_{D,k}, v_{k|k-1} \rangle^{n-j}}{\langle 1, v_{k|k-1} \rangle^n} \sigma_j(Z_k) \rho_{k|k-1}(n) \\
&= \Upsilon_k^{(0)}[v_{k|k-1}, Z_k](n) \rho_{k|k-1}(n)
\end{aligned} \tag{E.9}
$$

Furthermore, since the denominator in (E.5) is in the form of (E.8), the CPHD cardinality update (5.2.5) can be obtained.

Appendix F
Derivation of the Mean of MeMBer Posterior Cardinality

Proof of Proposition 13 (6.2.12) can be rewritten in the following Bernoulli form

$$G_{L,k}^{(i)}[h] = 1 - \varepsilon_{L,k}^{(i)} + \varepsilon_{L,k}^{(i)}\left\langle p_{L,k}^{(i)}, h\right\rangle \tag{F.1}$$

where $\varepsilon_{L,k}^{(i)}$ and $p_{L,k}^{(i)}$ are given by Proposition 12. Therefore, the product $\prod_{i=1}^{M_{k|k-1}} G_{L,k}^{(i)}[h]$ in (6.2.11) corresponds to the set of legacy tracks, which is the MB RFS, so the mean cardinality of legacy tracks is

$$\sum_{i=1}^{M_{k|k-1}} \varepsilon_{L,k}^{(i)} \tag{F.2}$$

The product $\prod_{z \in Z_k} G_{U,k}[h; z]$ in (6.2.11) corresponds to the set of measurement updated tracks, but is not the MB RFS. However, the corresponding cardinality mean can be calculated accurately. In fact, substituting $h(\boldsymbol{x}) = \boldsymbol{y}$ into (6.2.13)–(6.2.14), and then differentiating at $\boldsymbol{y} = 1$ can lead to $G'_{U,k}[1; z] = \varepsilon_{U,k}^{*}(z)$, where $\varepsilon_{U,k}^{*}(z)$ is given by (6.2.21), and the upper apostrophe represents the first-order derivative. Therefore, the cardinality mean of measurement updated tracks is

$$\sum_{z \in Z_k} G'_{U,k}[1; z] \tag{F.3}$$

According to (6.2.11), it can be found that, the posterior cardinality mean is the sum of the cardinality of legacy tracks (F.2) and the cardinality of measurement updated tracks (F.3). Therefore, the proposition is proved.

W. Wu et al., *Target Tracking with Random Finite Sets*,
https://doi.org/10.1007/978-981-19-9815-7

Appendix G
Derivation of GLMB Recursion

(1) Proof of Proposition 15

According to the Chapman–Kolmogorov (C–K) equation, the density of the survival multi-target state at the next moment is

$$
\begin{aligned}
\pi_S(\mathrm{S}) &= \int \pi_S(\mathrm{S}|\mathrm{X})\pi(\mathrm{X})\delta\mathrm{X} \\
&= \Delta(\mathrm{S}) \int 1_{\mathcal{L}(\mathrm{X})}(\mathcal{L}(\mathrm{S}))[\Phi(\mathrm{S};\cdot)]^{\mathrm{X}}\Delta(\mathrm{X}) \sum_{c\in\mathbb{C}} w^{(c)}(\mathcal{L}(\mathrm{X}))[p^{(c)}]^{\mathrm{X}}\delta\mathrm{X} \\
&= \Delta(\mathrm{S}) \sum_{c\in\mathbb{C}} \int \Delta(\mathrm{X}) 1_{\mathcal{L}(\mathrm{X})}(\mathcal{L}(\mathrm{S}))w^{(c)}(\mathcal{L}(\mathrm{X}))[\Phi(\mathrm{S};\cdot)p^{(c)}]^{\mathrm{X}}\delta\mathrm{X} \\
&= \Delta(\mathrm{S}) \sum_{c\in\mathbb{C}} \sum_{I\subseteq\mathbb{L}} 1_I(\mathcal{L}(\mathrm{S}))w^{(c)}(I) \prod_{\ell\in I} \langle \Phi(\mathrm{S};\cdot,\ell), p^{(c)}(\cdot,\ell)\rangle
\end{aligned} \tag{G.1}
$$

where $\pi_S(\mathrm{S}|\mathrm{X})$ is given by (3.4.9), and the last line makes use of Lemma 2. Due to the existence of term $1_I(\mathcal{L}(\mathrm{S}))$, only $I \supseteq \mathcal{L}(\mathrm{S})$ needs to be considered, and then we have

$$
\begin{aligned}
\prod_{\ell\in I} \langle \Phi(\mathrm{S};\cdot,\ell), p^{(c)}(\cdot,\ell)\rangle &= \prod_{\ell\in\mathcal{L}(\mathrm{S})} \langle \Phi(\mathrm{S};\cdot,\ell), p^{(c)}(\cdot,\ell)\rangle \times \\
&\quad \prod_{\ell\in I-\mathcal{L}(\mathrm{S})} \langle \Phi(\mathrm{S};\cdot,\ell), p^{(c)}(\cdot,\ell)\rangle \\
&= \prod_{\ell\in\mathcal{L}(\mathrm{S})} \sum_{(\boldsymbol{x}_+,\ell_+)\in\mathrm{S}} \delta_\ell(\ell_+)\langle p_S(\cdot,\ell)\phi(\boldsymbol{x}_+|\cdot,\ell), p^{(c)}(\cdot,\ell)\rangle \\
&\times \prod_{\ell\in I-\mathcal{L}(\mathrm{S})} \langle q_S(\cdot,\ell), p^{(c)}(\cdot,\ell)\rangle \\
&= \prod_{\ell\in\mathcal{L}(\mathrm{S})} \sum_{(\boldsymbol{x}_+,\ell_+)\in\mathrm{S}} \delta_\ell(\ell_+) p_{+,S}^{(c)}(\boldsymbol{x}_+,\ell)\eta_S^{(c)}(\ell) \times
\end{aligned}
$$

W. Wu et al., *Target Tracking with Random Finite Sets*,
https://doi.org/10.1007/978-981-19-9815-7

$$\prod_{\ell \in I-\mathcal{L}(\mathrm{S})} q_S^{(c)}(\ell)$$

$$= \prod_{(\boldsymbol{x}_+,\ell)\in \mathrm{S}} p_{+,S}^{(c)}(\boldsymbol{x}_+,\ell)\eta_S^{(c)}(\ell) \prod_{\ell \in I-\mathcal{L}(\mathrm{S})} q_S^{(c)}(\ell)$$

$$= [p_{+,S}^{(c)}]^{\mathrm{S}}[\eta_S^{(c)}]^{\mathcal{L}(\mathrm{S})}[q_S^{(c)}]^{I-\mathcal{L}(\mathrm{S})} \tag{G.2}$$

Substituting the above equation into (G.1) and using $w_S^{(c)}(L)$ in (7.2.3), we can obtain

$$\pi_S(\mathrm{S}) = \Delta(\mathrm{S})\sum_{c\in\mathbb{C}} w_S^{(c)}(\mathcal{L}(\mathrm{S}))\left[p_{+,S}^{(c)}\right]^{\mathrm{S}} \tag{G.3}$$

For the predicted multi-target density, using the birth density (3.4.11), letting $\mathrm{B} = \mathrm{X}_+ - \mathbb{X}\times\mathbb{L}$ and $\mathrm{S} = \mathrm{X}_+ \cap \mathbb{X}\times\mathbb{L}$, and substituting (3.4.13) into (3.5.4), we have

$$\begin{aligned}
\pi_+(\mathrm{X}_+) &= \int \phi(\mathrm{X}_+|\mathrm{X})\pi(\mathrm{X})\delta\mathrm{X} \\
&= \pi_\gamma(\mathrm{B})\int \pi_S(\mathrm{S}|\mathrm{X})\pi(\mathrm{X})\delta\mathrm{X} \\
&= \pi_\gamma(\mathrm{B})\pi_S(\mathrm{S}) \\
&= \Delta(\mathrm{B})\Delta(\mathrm{S})\sum_{c\in\mathbb{C}} w_\gamma(\mathcal{L}(\mathrm{B}))w_S^{(c)}(\mathcal{L}(\mathrm{S}))[p_\gamma]^{\mathrm{B}}[p_{+,S}^{(c)}]^{\mathrm{S}} \\
&= \Delta(\mathrm{X}_+)\sum_{c\in\mathbb{C}} w_\gamma(\mathcal{L}(\mathrm{X}_+)-\mathbb{L}))w_S^{(c)}(\mathcal{L}(\mathrm{X}_+)\cap\mathbb{L}))[p_+^{(c)}]^{\mathrm{X}_+} \\
&= \Delta(\mathrm{X}_+)\sum_{c\in\mathbb{C}} w_+^{(c)}(\mathcal{L}(\mathrm{X}_+))[p_+^{(c)}]^{\mathrm{X}_+}
\end{aligned} \tag{G.4}$$

(2) Proof of Proposition 16

For the multi-target likelihood (3.4.20), note that $\delta_{\theta^{-1}(\{0:|\mathrm{Z}|\})}(\mathcal{L}(\mathrm{X}))$ restricts the summation to θ with domain $\mathcal{L}(\mathrm{X})$, so (3.4.20) can be rewritten in the following form [14]

$$g(Z|\mathrm{X}) = \exp(-\langle\kappa,1\rangle)\kappa^Z\sum_{\theta\in\Theta}\delta_{\theta^{-1}(\{0:|\mathrm{Z}|\})}(\mathcal{L}(\mathrm{X}))[\varphi_Z(\cdot;\theta)]^{\mathrm{X}} \tag{G.5}$$

Then

$$g(Z|\mathrm{X})\pi(\mathrm{X}) = \Delta(\mathrm{X})\exp(-\langle\kappa,1\rangle)\kappa^Z \times$$
$$\sum_{c\in\mathbb{C}}\sum_{\theta\in\Theta}\delta_{\theta^{-1}(\{0:|\mathrm{Z}|\})}(\mathcal{L}(\mathrm{X}))w^{(c)}(\mathcal{L}(\mathrm{X}))\left[p^{(c)}\varphi_Z(\cdot;\theta)\right]^{\mathrm{X}}$$

$$= \Delta(\mathrm{X}) \exp(-\langle \kappa, 1\rangle)\kappa^{Z} \times$$
$$\sum_{c\in\mathbb{C}}\sum_{\theta\in\Theta} \delta_{\theta^{-1}(\{0:|Z|\})}(\mathcal{L}(\mathrm{X}))w^{(c)}(\mathcal{L}(\mathrm{X}))\left[\eta_Z^{(c,\theta)}\right]^{\mathcal{L}(\mathrm{X})}\left[p^{(c,\theta)}(\cdot|Z)\right]^{\mathrm{X}} \tag{G.6}$$

where the relationship $p^{(c)}(\boldsymbol{x}, \ell)\varphi_Z(\boldsymbol{x}, \ell; \theta) = \eta_Z^{(c,\theta)}(\ell)p^{(c,\theta)}(\boldsymbol{x}, \ell|Z)$ given by (7.2.10) is used. The integral of the above equation is

$$\int g(Z|\mathrm{X})\pi(\mathrm{X})\delta\mathrm{X} = \exp(-\langle \kappa, 1\rangle)\kappa^{Z} \int \Delta(\mathrm{X}) \sum_{c\in\mathbb{C}}\sum_{\theta\in\Theta} \delta_{\theta^{-1}(\{0:|Z|\})}(\mathcal{L}(\mathrm{X})) \times$$
$$w^{(c)}(\mathcal{L}(\mathrm{X}))\left[\eta_Z^{(c,\theta)}\right]^{\mathcal{L}(\mathrm{X})}\left[p^{(c,\theta)}(\cdot|Z)\right]^{\mathrm{X}}\delta\mathrm{X}$$
$$= \exp(-\langle \kappa, 1\rangle)\kappa^{Z} \sum_{c\in\mathbb{C}}\sum_{\theta\in\Theta} \int \Delta(\mathrm{X})\delta_{\theta^{-1}(\{0:|Z|\})}(\mathcal{L}(\mathrm{X})) \times$$
$$w^{(c)}(\mathcal{L}(\mathrm{X}))\left[\eta_Z^{(c,\theta)}\right]^{\mathcal{L}(\mathrm{X})}\left[p^{(c,\theta)}(\cdot|Z)\right]^{\mathrm{X}}\delta\mathrm{X}$$
$$= \exp(-\langle \kappa, 1\rangle)\kappa^{Z} \sum_{c\in\mathbb{C}}\sum_{\theta\in\Theta}\sum_{J\subseteq\mathbb{L}} \delta_{\theta^{-1}(\{0:|Z|\})}(J)w^{(c)}(J)\left[\eta_Z^{(c,\theta)}\right]^{J} \tag{G.7}$$

where the last line is derived by Lemma 2. Therefore, according to (3.5.5), we have

$$\pi(\mathrm{X}|Z) = \frac{g(Z|\mathrm{X})\pi(\mathrm{X})}{\int g(Z|\mathrm{X})\pi(\mathrm{X})\delta\mathrm{X}}$$
$$= \frac{\Delta(\mathrm{X}) \sum_{c\in\mathbb{C}}\sum_{\theta\in\Theta} \delta_{\theta^{-1}(\{0:|Z|\})}(\mathcal{L}(\mathrm{X}))w^{(c)}(\mathcal{L}(\mathrm{X}))\left[\eta_Z^{(c,\theta)}\right]^{\mathcal{L}(\mathrm{X})}\left[p^{(c,\theta)}(\cdot|Z)\right]^{\mathrm{X}}}{\sum_{c\in\mathbb{C}}\sum_{\theta\in\Theta}\sum_{J\subseteq\mathbb{L}} \delta_{\theta^{-1}(\{0:|Z|\})}(J)w^{(c)}(J)\left[\eta_Z^{(c,\theta)}\right]^{J}}$$
$$= \Delta(\mathrm{X}) \sum_{c\in\mathbb{C}}\sum_{\theta\in\Theta} w_Z^{(c,\theta)}(\mathcal{L}(\mathrm{X}))\left[p^{(c,\theta)}(\cdot|Z)\right]^{\mathrm{X}} \tag{G.8}$$

where (7.2.9) is used in the last row.

Appendix H
Derivation of δ-GLMB Recursion

(1) Proof of Proposition 17

According to Proposition 15 and Definition 3, we have

$$\pi_+(\mathbf{X}_+) = \Delta(\mathbf{X}_+) \sum_{(I,\vartheta)\in\mathcal{F}(\mathbb{L})\times\Xi} w_+^{(I,\vartheta)}(\mathcal{L}(\mathbf{X}_+))\left[p_+^{(\vartheta)}\right]^{\mathbf{X}_+} \tag{H.1}$$

where

$$w_+^{(I,\vartheta)}(L) = w_\gamma(L-\mathbb{L})w_S^{(I,\vartheta)}(L\cap\mathbb{L}) \tag{H.2}$$

$$w_S^{(I,\vartheta)}(J) = \left[\eta_S^{(\vartheta)}\right]^J \sum_{L\subseteq\mathbb{L}} 1_L(J)\delta_I(L)\left[q_S^{(\vartheta)}\right]^{L-J} w^{(I,\vartheta)} \tag{H.3}$$

Note that if $L = I$, then $1_L(J)\delta_I(L)[q_S^{(\vartheta)}]^L w^{(I,\vartheta)} = 1_I(J)[q_S^{(\vartheta)}]^I w^{(I,\vartheta)}$, otherwise it is 0. Therefore, we have

$$\begin{aligned} w_S^{(I,\vartheta)}(J) &= \left[\eta_S^{(\vartheta)}\right]^J 1_I(J)\left[q_S^{(\vartheta)}\right]^{I-J} w^{(I,\vartheta)} \\ &= \left[\eta_S^{(\vartheta)}\right]^J \sum_{Y\subseteq I}\delta_Y(J)\left[q_S^{(\vartheta)}\right]^{I-J} w^{(I,\vartheta)} \\ &= \sum_{Y\subseteq I}\delta_Y(J)\left[\eta_S^{(\vartheta)}\right]^J\left[q_S^{(\vartheta)}\right]^{I-J} w^{(I,\vartheta)} \\ &= \sum_{Y\subseteq I}\delta_Y(J)\left[\eta_S^{(\vartheta)}\right]^Y\left[q_S^{(\vartheta)}\right]^{I-Y} w^{(I,\vartheta)} \end{aligned} \tag{H.4}$$

Rewriting $w_\gamma(L) = \sum_{J'\in\mathcal{F}(\mathbb{B})}\delta_{J'}(L)w_\gamma(J')$ and substituting it into (H.2) yields

W. Wu et al., *Target Tracking with Random Finite Sets*,
https://doi.org/10.1007/978-981-19-9815-7

$$
\begin{aligned}
w_+^{(I,\vartheta)}(L) &= \sum_{Y\subseteq I}\delta_Y(L\cap\mathbb{L})\left[\eta_S^{(\vartheta)}\right]^Y\left[q_S^{(\vartheta)}\right]^{I-Y}w^{(I,\vartheta)}\sum_{J'\in\mathcal{F}(\mathbb{B})}\delta_{J'}(L-\mathbb{L})w_\gamma(J')\\
&= w^{(I,\vartheta)}\sum_{J'\in\mathcal{F}(\mathbb{B})}w_\gamma(J')\sum_{Y\subseteq I}\left[\eta_S^{(\vartheta)}\right]^Y\left[q_S^{(\vartheta)}\right]^{I-Y}\delta_Y(L\cap\mathbb{L})\delta_{J'}(L-\mathbb{L})\\
&= w^{(I,\vartheta)}\sum_{J'\in\mathcal{F}(\mathbb{B})}w_\gamma(J')\sum_{Y\subseteq I}\left[\eta_S^{(\vartheta)}\right]^Y\left[q_S^{(\vartheta)}\right]^{I-Y}\delta_{Y\cup J'}(L) \qquad \text{(H.5)}
\end{aligned}
$$

where the last row utilizes the relationship $L\cap\mathbb{L}=Y, L-\mathbb{L}=J'\Rightarrow L=Y\cup J'$. Then, we have

$$
\begin{aligned}
\pi_+(\mathbf{X}_+) &= \Delta(\mathbf{X}_+)\sum_{I\in\mathcal{F}(\mathbb{L})}\sum_{\vartheta\in\Xi}w^{(I,\vartheta)}\sum_{J'\in\mathcal{F}(\mathbb{B})}\sum_{Y\subseteq I}\delta_{Y\cup J'}(\mathcal{L}(\mathbf{X}_+))w_\gamma(J')\times\\
&\quad [\eta_S^{(\vartheta)}]^Y[q_S^{(\vartheta)}]^{I-Y}[p_+^{(\vartheta)}]^{\mathbf{X}_+}\\
&= \Delta(\mathbf{X}_+)\sum_{I\in\mathcal{F}(\mathbb{L})}\sum_{\vartheta\in\Xi}w^{(I,\vartheta)}\sum_{J'\in\mathcal{F}(\mathbb{B})}\sum_{Y\in\mathcal{F}(\mathbb{L})}1_I(Y)\delta_{Y\cup J'}(\mathcal{L}(\mathbf{X}_+))\times\\
&\quad w_\gamma(J')[\eta_S^{(\vartheta)}]^Y[q_S^{(\vartheta)}]^{I-Y}[p_+^{(\vartheta)}]^{\mathbf{X}_+}\\
&= \Delta(\mathbf{X}_+)\sum_{Y\in\mathcal{F}(\mathbb{L})}\sum_{J'\in\mathcal{F}(\mathbb{B})}\sum_{\vartheta\in\Xi}\left(w_\gamma(J')[\eta_S^{(\vartheta)}]^Y\sum_{I\in\mathcal{F}(\mathbb{L})}1_I(Y)[q_S^{(\vartheta)}]^{I-Y}w^{(I,\vartheta)}\right)\\
&\quad \times\ \delta_{Y\cup J'}(\mathcal{L}(\mathbf{X}_+))[p_+^{(\vartheta)}]^{\mathbf{X}_+}\\
&= \Delta(\mathbf{X}_+)\sum_{Y\in\mathcal{F}(\mathbb{L})}\sum_{J'\in\mathcal{F}(\mathbb{B})}\sum_{\vartheta\in\Xi}w_\gamma(J')w_S^{(\vartheta)}(Y)\delta_{Y\cup J'}(\mathcal{L}(\mathbf{X}_+))[p_+^{(\vartheta)}]^{\mathbf{X}_+}\\
&= \Delta(\mathbf{X}_+)\sum_{(I_+,\vartheta)\in\mathcal{F}(\mathbb{L}_+)\times\Xi}w_\gamma(I_+\cap\mathbb{B})w_S^{(\vartheta)}(I_+\cap\mathbb{L})\delta_{I_+}(\mathcal{L}(\mathbf{X}_+))[p_+^{(\vartheta)}]^{\mathbf{X}_+}\\
&= \Delta(\mathbf{X}_+)\sum_{(I_+,\vartheta)\in\mathcal{F}(\mathbb{L}_+)\times\Xi}w_+^{(I_+,\vartheta)}\delta_{I_+}(\mathcal{L}(\mathbf{X}_+))[p_+^{(\vartheta)}]^{\mathbf{X}_+} \qquad \text{(H.6)}
\end{aligned}
$$

where since $Y\subseteq\mathbb{L}$, $J'\subseteq\mathbb{B}$, and $\mathbb{L}$ is disjoint from $\mathbb{B}$, we have $Y=I_+\cap\mathbb{L}$ and $J'=I_+\cap\mathbb{B}$, and then by setting $I_+=Y\cup J'$, the penultimate row of the above formula can be obtained.

(2) Proof of Proposition 18

According to Proposition 16 and Definition 3, then we have

$$
\pi(\mathbf{X}|Z)=\Delta(\mathbf{X})\sum_{(I,\vartheta)\in\mathcal{F}(\mathbb{L})\times\Xi}\sum_{\theta\in\Theta(I)}w^{(I,\vartheta,\theta)}(\mathcal{L}(\mathbf{X})|Z)\left[p^{(\vartheta,\theta)}(\cdot|Z)\right]^{\mathbf{X}} \qquad \text{(H.7)}
$$

where

$$w^{(I,\vartheta,\theta)}(L|Z) = \frac{\delta_{\theta^{-1}(\{0:|Z|\})}(L)\delta_I(L)\left[\eta_Z^{(\vartheta,\theta)}\right]^L w^{(I,\vartheta)}}{\sum_{(I,\vartheta)\in\mathcal{F}(\mathbb{L})\times\Xi}\sum_{\theta\in\Theta(I)}\sum_{J\subseteq\mathbb{L}}\delta_{\theta^{-1}(\{0:|Z|\})}(J)\delta_I(J)\left[\eta_Z^{(\vartheta,\theta)}\right]^J w^{(I,\vartheta)}} \tag{H.8}$$

with $p^{(\vartheta,\theta)}(\boldsymbol{x},\ell|Z)$ and $\eta_Z^{(\vartheta,\theta)}(\ell)$ being given by Proposition 18.

For $w^{(I,\vartheta,\theta)}(L|Z)$, note that $\delta_{\theta^{-1}(\{0:|Z|\})}(L)\delta_I(L) = \delta_{\theta^{-1}(\{0:|Z|\})}(I)\delta_I(L)$, and the sum over $J \subseteq \mathbb{L}$ in the denominator is

$$\sum_{J\subseteq\mathbb{L}}\delta_{\theta^{-1}(\{0:|Z|\})}(J)\delta_I(J)\left[\eta_Z^{(\vartheta,\theta)}\right]^J w^{(I,\vartheta)} = \delta_{\theta^{-1}(\{0:|Z|\})}(I)\left[\eta_Z^{(\vartheta,\theta)}\right]^I w^{(I,\vartheta)} \tag{H.9}$$

Then we have

$$\begin{aligned} w^{(I,\vartheta,\theta)}(L|Z) &= \frac{\delta_{\theta^{-1}(\{0:|Z|\})}(I)\left[\eta_Z^{(\vartheta,\theta)}\right]^I w^{(I,\vartheta)}}{\sum_{(I,\vartheta)\in\mathcal{F}(\mathbb{L})\times\Xi}\sum_{\theta\in\Theta(I)}\delta_{\theta^{-1}(\{0:|Z|\})}(I)\left[\eta_Z^{(\vartheta,\theta)}\right]^I w^{(I,\vartheta)}}\delta_I(L) \\ &= w^{(I,\vartheta,\theta)}(Z)\delta_I(L) \end{aligned} \tag{H.10}$$

Appendix I
Derivation of LMB Prediction

Before the proof of Proposition 19, the following lemma is introduced first.

Lemma 6

$$\frac{(1-\varepsilon^{(\cdot)})^{\mathbb{L}}\eta_S^L}{(1-\eta_S)^L}\sum_{I\supseteq L}(1-\eta_S)^I\left(\frac{\varepsilon^{(\cdot)}}{1-\varepsilon^{(\cdot)}}\right)^I=\left(1-\varepsilon^{(\cdot)}\eta_S\right)^{\mathbb{L}}\left(\frac{\varepsilon^{(\cdot)}\eta_S}{1-\varepsilon^{(\cdot)}\eta_S}\right)^L \tag{I.1}$$

Proof of Lemma 6 Let

$$f^{(L)}(I)=\left(1-\varepsilon^{(\cdot)}\right)^{\mathbb{L}}\eta_S^L\frac{(1-\eta_S)^I}{(1-\eta_S)^L}\left(\frac{\varepsilon^{(\cdot)}}{1-\varepsilon^{(\cdot)}}\right)^I \tag{I.2}$$

then for any $\ell\in L$, we have

$$\begin{aligned}f^{(L-\{\ell\})}(I)&=(1-\varepsilon^{(\cdot)})^{\mathbb{L}}\eta_S^{L-\{\ell\}}\frac{(1-\eta_S)^I}{(1-\eta_S)^{L-\{\ell\}}}\left(\frac{\varepsilon^{(\cdot)}}{1-\varepsilon^{(\cdot)}}\right)^I\\&=(1-\varepsilon^{(\cdot)})^{\mathbb{L}}\eta_S^L\frac{1-\eta_S(\ell)}{\eta_S(\ell)}\frac{(1-\eta_S)^I}{(1-\eta_S)^L}\left(\frac{\varepsilon^{(\cdot)}}{1-\varepsilon^{(\cdot)}}\right)^I=\frac{1-\eta_S(\ell)}{\eta_S(\ell)}f^{(L)}(I)\\&=\frac{\varepsilon^{(\ell)}-\varepsilon^{(\ell)}\eta_S(\ell)}{\varepsilon^{(\ell)}\eta_S(\ell)}f^{(L)}(I)\end{aligned} \tag{I.3}$$

$$\begin{aligned}f^{(L-\{\ell\})}(I-\{\ell\})&=\frac{\varepsilon^{(\ell)}-\varepsilon^{(\ell)}\eta_S(\ell)}{\varepsilon^{(\ell)}\eta_S(\ell)}f^{(L)}(I-\{\ell\})\\&=\frac{\varepsilon^{(\ell)}-\varepsilon^{(\ell)}\eta_S(\ell)}{\varepsilon^{(\ell)}\eta_S(\ell)}(1-\varepsilon^{(\cdot)})^{\mathbb{L}}\eta_S^L\frac{(1-\eta_S)^{I-\{\ell\}}}{(1-\eta_S)^L}\left(\frac{\varepsilon^{(\cdot)}}{1-\varepsilon^{(\cdot)}}\right)^{I-\{\ell\}}\\&=\frac{\varepsilon^{(\ell)}-\varepsilon^{(\ell)}\eta_S(\ell)}{\varepsilon^{(\ell)}\eta_S(\ell)}(1-\varepsilon^{(\cdot)})^{\mathbb{L}}\eta_S^L\frac{1}{1-\eta_S(\ell)}\frac{(1-\eta_S)^I}{(1-\eta_S)^L}\times\end{aligned}$$

W. Wu et al., *Target Tracking with Random Finite Sets*,
https://doi.org/10.1007/978-981-19-9815-7

$$\left(\frac{\varepsilon^{(\cdot)}}{1-\varepsilon^{(\cdot)}}\right)^{I}\left(\frac{1-\varepsilon^{(\ell)}}{\varepsilon^{(\ell)}}\right)$$

$$=\frac{1-\varepsilon^{(\ell)}}{\varepsilon^{(\ell)}\eta_S(\ell)}f^{(L)}(I) \tag{I.4}$$

Therefore, by adding the two expressions (i.e., (I.3) and (I.4)) together, we can obtain

$$f^{(L-\{\ell\})}(I)+f^{(L-\{\ell\})}(I-\{\ell\})=\frac{1-\varepsilon^{(\ell)}\eta_S(\ell)}{\varepsilon^{(\ell)}\eta_S(\ell)}f^{(L)}(I) \tag{I.5}$$

Meanwhile, note that since the set containing $\{\ell_1,\ldots,\ell_{n-1}\}$ is composed of the set I containing $\{\ell_1,\ldots,\ell_n\}$ and the set $I-\{\ell_n\}$. For any function $g:\mathcal{F}(\mathbb{L})\to\mathbb{R}$, we have

$$\sum_{I\supseteq\{\ell_1,\ldots,\ell_{n-1}\}} g(I)=\sum_{I\supseteq\{\ell_1,\ldots,\ell_n\}}[g(I)+g(I-\{\ell_n\})] \tag{I.6}$$

Therefore, the left-hand side of (I.1) is $\sum_{I\supseteq L} f^{(L)}(I)$ and is abbreviated as $f(L)$. To prove Lemma 6, it is necessary to prove that $f(L)$ is equal to the right-hand side of (I.1) by induction, and this lemma holds for $L=\mathbb{L}$. Suppose that for some $L=\{\ell_1,\ldots,\ell_n\}$, this lemma holds, i.e.,

$$f(L)=(1-\varepsilon^{(\cdot)}\eta_S)^{\mathbb{L}}\left(\frac{\varepsilon^{(\cdot)}\eta_S}{1-\varepsilon^{(\cdot)}\eta_S}\right)^{L}$$

then we have

$$\begin{aligned}f(L-\{\ell_n\})&=\sum_{I\supseteq L-\{\ell_n\}} f^{(L-\{\ell_n\})}(I)=\sum_{I\supseteq L}\left[f^{(L-\{\ell_n\})}(I)+f^{(L-\{\ell_n\})}(I-\{\ell_n\})\right]\\&=\frac{1-\varepsilon^{(\ell_n)}\eta_S(\ell_n)}{\varepsilon^{(\ell_n)}\eta_S(\ell_n)}\sum_{I\supseteq L} f^{(L)}(I)\\&=\frac{1-\varepsilon^{(\ell_n)}\eta_S(\ell_n)}{\varepsilon^{(\ell_n)}\eta_S(\ell_n)}(1-\varepsilon^{(\cdot)}\eta_S)^{\mathbb{L}}\left(\frac{\varepsilon^{(\cdot)}\eta_S}{1-\varepsilon^{(\cdot)}\eta_S}\right)^{L}\\&=(1-\varepsilon^{(\cdot)}\eta_S)^{\mathbb{L}}\left(\frac{\varepsilon^{(\cdot)}\eta_S}{1-\varepsilon^{(\cdot)}\eta_S}\right)^{L-\{\ell_n\}}\end{aligned} \tag{I.7}$$

In the above equation, the second row and the third row are obtained according to (I.5) and (I.6), respectively. Therefore, by induction, this lemma is proved.

According to Lemma 6, the proof of Proposition 19 is as follows.

Proof Enumerate the label $\mathbb{L}=\{\ell_1,\ldots,\ell_M\}$ and rewrite the weight $w_S(L)$ in (7.4.12) as

$$
\begin{aligned}
w_S(L) &= \eta_S^L \sum_{I \supseteq L} [1-\eta_S]^{I-L} w(I) \\
&= \eta_S^L \sum_{I \supseteq L} [1-\eta_S]^{I-L} \prod_{i \in \mathbb{L}} (1-\varepsilon^{(i)}) \prod_{\ell \in I} \frac{1_{\mathbb{L}}(\ell)\varepsilon^{(\ell)}}{1-\varepsilon^{(\ell)}} \\
&= \eta_S^L \sum_{I \supseteq L} \frac{(1-\eta_S)^I}{(1-\eta_S)^L} (1-\varepsilon^{(\cdot)})^{\mathbb{L}} \left(\frac{\varepsilon^{(\cdot)}}{1-\varepsilon^{(\cdot)}} \right)^I \\
&= \sum_{I \supseteq L} (1-\varepsilon^{(\cdot)})^{\mathbb{L}} \eta_S^L \frac{(1-\eta_S)^I}{(1-\eta_S)^L} \left(\frac{\varepsilon^{(\cdot)}}{1-\varepsilon^{(\cdot)}} \right)^I \\
&= \sum_{I \supseteq L} f^{(L)}(I)
\end{aligned} \tag{I.8}
$$

where $f^{(L)}(I)$ is given by (I.2). Further, according to (I.1), we can obtain

$$
\begin{aligned}
w_S(L) &= (1-\varepsilon^{(\cdot)}\eta_S)^{\mathbb{L}} \left(\frac{\varepsilon^{(\cdot)}\eta_S}{1-\varepsilon^{(\cdot)}\eta_S} \right)^L \\
&= \prod_{\ell' \in \mathbb{L}} (1-\varepsilon^{(\ell')}\eta_S(\ell')) \prod_{\ell \in L} \frac{1_{\mathbb{L}}(\ell)\varepsilon^{(\ell)}\eta_S(\ell)}{1-\varepsilon^{(\ell)}\eta_S(\ell)}
\end{aligned} \tag{I.9}
$$

It can be seen from the above equation that the weight $w_S(L)$ has the form of the LMB weight. Since weights $w_S(I_+ \cap \mathbb{L})$ and $w_\gamma(I_+ \cap \mathbb{B})$ both have the form of the LMB weight, $w_+(I_+) = w_S(I_+ \cap \mathbb{L})w_\gamma(I_+ \cap \mathbb{B})$ also has the form of the LMB weight. As a result, the LMB component of the predicted density consists of the birth LMB component $\{\varepsilon_\gamma^{(\ell)}, p_\gamma^{(\ell)}\}_{\ell \in \mathbb{B}}$ and the surviving LMB component $\{\varepsilon_{+,S}^{(\ell)}, p_{+,S}^{(\ell)}\}_{\ell \in \mathbb{L}}$ given by (7.4.14)–(7.4.15).

Appendix J
Mδ-GLMB Approximation

(1) Proof of Proposition 21

Before proving Proposition 21, we first introduce the following lemma.

Lemma 7 *Let $f : (\mathbb{X}\times\mathbb{Y})^n \rightarrow \mathbb{R}$ be a symmetric function, then function $g : \mathbb{X}^n \rightarrow \mathbb{R}$ given by the following equation is also a symmetric function over $\mathbb{X}^n$*

$$g(\boldsymbol{x}_1, \ldots, \boldsymbol{x}_n) = \int f\big((\boldsymbol{x}_1, \boldsymbol{y}_1), \ldots, (\boldsymbol{x}_n, \boldsymbol{y}_n)\big)d(\boldsymbol{y}_1, \ldots, \boldsymbol{y}_n) \tag{J.1}$$

Proof Let σ be the permutation of $\{1, \ldots, n\}$, then we have

$$\begin{aligned}
g(\boldsymbol{x}_{\sigma(1)}, \ldots, \boldsymbol{x}_{\sigma(n)}) &= \int f\big((\boldsymbol{x}_{\sigma(1)}, \boldsymbol{y}_{\sigma(1)}), \ldots, (\boldsymbol{x}_{\sigma(n)}, \boldsymbol{y}_{\sigma(n)})\big)d(\boldsymbol{y}_{\sigma(1)}, \ldots, \boldsymbol{y}_{\sigma(n)})\\
&= \int f\big((\boldsymbol{x}_1, \boldsymbol{y}_1), \ldots, (\boldsymbol{x}_n, \boldsymbol{y}_n)\big)d(\boldsymbol{y}_{\sigma(1)}, \ldots, \boldsymbol{y}_{\sigma(n)})\\
&= \int f\big((\boldsymbol{x}_1, \boldsymbol{y}_1), \ldots, (\boldsymbol{x}_n, \boldsymbol{y}_n)\big)d(\boldsymbol{y}_1, \ldots, \boldsymbol{y}_n)\\
&= g(\boldsymbol{x}_1, \ldots, \boldsymbol{x}_n)
\end{aligned}$$

where the penultimate row utilizes the commutative property of the integration order.

Proof of Proposition 21 According to Lemma 7, since $p_{\{\ell_1,\ldots,\ell_n\}}(\boldsymbol{x}, \ell)$ is symmetric with respect to $\ell_1, \ldots, \ell_n$, $\hat{p}^{(I)}(\boldsymbol{x}, \ell)$ is actually a function of set I. The fact that (7.5.4) can be rewritten as $\hat{\pi}(\mathbf{X}) = \hat{w}(\mathcal{L}(\mathbf{X}))\hat{p}(\mathbf{X})$ is used for the proof, where $\hat{w}(L) = \hat{w}^{(L)}$ and $\hat{p}(\mathbf{X}) = \Delta(\mathbf{X})[\hat{p}^{(\mathcal{L}(\mathbf{X}))}]^{\mathbf{X}}$.

According to (3.3.41) and (3.3.44), the cardinality distribution and PHD of $\hat{\pi}$ (7.5.4) can be obtained as

$$\hat{\rho}(|\mathbf{X}|) = \sum_{L\subseteq\mathbb{L}} \delta_{|\mathbf{X}|}(|L|)\hat{w}^{(L)} \tag{J.2}$$

W. Wu et al., *Target Tracking with Random Finite Sets*,
https://doi.org/10.1007/978-981-19-9815-7

$$\hat{v}(\boldsymbol{x},\ell)=\sum_{L\subseteq\mathbb{L}}1_L(\ell)\hat{w}^{(L)}\hat{p}^{(L)}(\boldsymbol{x},\ell) \tag{J.3}$$

In order to prove that the cardinalities of $\hat{\pi}$ and π are equal, note that the cardinality distribution of any labeled RFS can be completely determined by the joint existence probability $w(\cdot)$ of labels, i.e.,

$$\begin{aligned}\rho(|\mathrm{X}|)&=\frac{1}{|\mathrm{X}|!}\int\pi(\{\mathbf{x}_1,\ldots,\mathbf{x}_{|\mathrm{X}|}\})d(\mathbf{x}_1,\ldots,\mathbf{x}_{|\mathrm{X}|})\\&=\frac{1}{|\mathrm{X}|!}\sum_{(\ell_1,\ldots,\ell_{|\mathrm{X}|})\in\mathbb{L}^{|\mathrm{X}|}}\int w(\{\ell_1,\ldots,\ell_{|\mathrm{X}|}\})\times\\&\quad p(\{(\boldsymbol{x}_1,\ell_1),\ldots,(\boldsymbol{x}_n,\ell_{|\mathrm{X}|})\})\\&\quad d(\boldsymbol{x}_1,\ldots,\boldsymbol{x}_{|\mathrm{X}|})\\&=\sum_{L\subseteq\mathbb{L}}\delta_{|\mathrm{X}|}(|L|)w(L)\end{aligned} \tag{J.4}$$

Since both $\hat{\pi}$ and π have the same joint existence probability, i.e., $\hat{w}(L)=\hat{w}^{(L)}=w(L)$, their cardinality distributions are also the same.

In order to prove that the PHD of $\hat{\pi}$ is the same as that of π, according to (J.3) and substituting (7.5.5) and (7.5.8), the PHD of $\hat{\pi}$ can be obtained as

$$\begin{aligned}\hat{v}(\boldsymbol{x},\ell)&=\sum_{n=0}^{\infty}\frac{1}{n!}\sum_{(\ell_1,\ldots,\ell_n)\in\mathbb{L}^n}\hat{w}^{(\{\ell,\ell_1,\ldots,\ell_n\})}\hat{p}^{(\{\ell,\ell_1,\ldots,\ell_n\})}(\boldsymbol{x},\ell)\\&=\sum_{n=0}^{\infty}\frac{1}{n!}\sum_{(\ell_1,\ldots,\ell_n)\in\mathbb{L}^n}w(\{\ell,\ell_1,\ldots,\ell_n\})\times\\&\quad\int p(\{(\boldsymbol{x},\ell),(\boldsymbol{x}_1,\ell_1),\ldots,(\boldsymbol{x}_n,\ell_n)\})\\&\quad d(\boldsymbol{x}_1,\ldots,\boldsymbol{x}_n)\end{aligned} \tag{J.5}$$

Note that the right-hand side of the above equation is the set integral $\int\pi(\{(\boldsymbol{x},\ell)\}\cup\mathrm{X})\delta\mathrm{X}=v(\boldsymbol{x},\ell)$. Hence, $\hat{v}(\boldsymbol{x},\ell)=v(\boldsymbol{x},\ell)$.

In order to prove that the KLD of $\hat{\pi}$ and π has a minimum, note that the KLD between π and any δ-GLMB of the form (7.5.4) is

$$\begin{aligned}\mathcal{D}_{\mathrm{KL}}(\pi||\hat{\pi})&=\int\pi(\mathrm{X})\log\frac{\pi(\mathrm{X})}{\hat{\pi}(\mathrm{X})}\delta\mathrm{X}=\int w(\mathcal{L}(\mathrm{X}))p(\mathrm{X})\log\frac{w(\mathcal{L}(\mathrm{X}))p(\mathrm{X})}{\hat{w}(\mathcal{L}(\mathrm{X}))\hat{p}(\mathrm{X})}\delta\mathrm{X}\\&=\sum_{n=0}^{\infty}\frac{1}{n!}\sum_{(\ell_1,\ldots,\ell_n)\in\mathbb{L}^n}w(\{\ell_1,\ldots,\ell_n\})\log\left(\frac{w^{(\{\ell_1,\ldots,\ell_n\})}}{\hat{w}^{(\{\ell_1,\ldots,\ell_n\})}}\right)\times\end{aligned}$$

$$\int p(\{(\boldsymbol{x}_1, \ell_1), \ldots, (\boldsymbol{x}_n, \ell_n)\}) d(\boldsymbol{x}_1, \ldots, \boldsymbol{x}_n)$$

$$+ \sum_{n=0}^{\infty} \frac{1}{n!} \sum_{(\ell_1, \ldots, \ell_n) \in \mathbb{L}^n} w(\{\ell_1, \ldots, \ell_n\})$$

$$\times \int p(\{(\boldsymbol{x}_1, \ell_1), \ldots, (\boldsymbol{x}_n, \ell_n)\}) \times$$

$$\log\left(\frac{p(\{(\boldsymbol{x}_1, \ell_1), \ldots, (\boldsymbol{x}_n, \ell_n)\})}{\prod_{i=1}^{n} \hat{p}^{(\{\ell_1, \ldots, \ell_n\})}(\boldsymbol{x}_i, \ell_i)}\right) d(\boldsymbol{x}_1, \ldots, \boldsymbol{x}_n) \tag{J.6}$$

According to (7.5.2), the integral of the joint density $p(\{(\cdot, \ell_1), \ldots, (\cdot. \ell_n)\})$ conditioned on the labels is 1, so that we have

$$\mathcal{D}_{\mathrm{KL}}(\pi || \hat{\pi}) = \mathcal{D}_{\mathrm{KL}}(w || \hat{w}) + \sum_{n=0}^{\infty} \frac{1}{n!} \sum_{(\ell_1, \ldots, \ell_n) \in \mathbb{L}^n} w(\{\ell_1, \ldots, \ell_n\}) \times$$

$$\mathcal{D}_{\mathrm{KL}}(p(\{(\cdot, \ell_1), \ldots, (\cdot, \ell_n)\}) || \prod_{i=1}^{n} \hat{p}^{(\{\ell_1, \ldots, \ell_n\})}(\cdot, \ell_i)) \tag{J.7}$$

Because $\hat{w}(I) = w(I)$, then $\mathcal{D}_{\mathrm{KL}}(w || \hat{w}) = 0$. Moreover, for each n and each $\{\ell_1, \ldots, \ell_n\}$, $\hat{p}^{(\{\ell_1, \ldots, \ell_n\})}(\cdot, \ell_i)$, $i = 1, \ldots, n$ is the marginal probability of $p(\{(\cdot, \ell_1), \ldots, (\cdot, \ell_n)\})$. Furthermore, according to [446], it can be obtained that each KLD in the sum in (J.7) is the minimum value. Therefore, $\mathcal{D}_{\mathrm{KL}}(\pi || \hat{\pi})$ is the minimum value over the δ-GLMB classes of the form (7.5.4).

(2) Proof of Proposition 23

First, rewrite the δ-GLMB density in (3.3.55) as the general form of the labeled RFS density [152], i.e., $\pi(\mathrm{X}) = w(\mathcal{L}(\mathrm{X})) p(\mathrm{X})$, where

$$\begin{aligned}
w(\{\ell_1, \ldots, \ell_n\}) &\triangleq \int_{\mathbb{X}^n} \pi(\{(\boldsymbol{x}_1, \ell_1), \ldots, (\boldsymbol{x}_n, \ell_n)\}) d(\boldsymbol{x}_1, \ldots, \boldsymbol{x}_n) \\
&= \sum_{I \in \mathcal{F}(\mathbb{L})} \delta_I(\{\ell_1, \ldots, \ell_n\}) \sum_{\vartheta \in \Xi} w^{(I, \vartheta)} \times \\
&\quad \int_{\mathbb{X}^n} p^{(\vartheta)}(\boldsymbol{x}_1, \ell_1) \ldots p^{(\vartheta)}(\boldsymbol{x}_n, \ell_n) d\boldsymbol{x}_1 \ldots d\boldsymbol{x}_n \\
&= \sum_{\vartheta \in \Xi} w^{(\{\ell_1, \ldots, \ell_n\}, \vartheta)} \sum_{I \in \mathcal{F}(\mathbb{L})} \delta_I(\{\ell_1, \ldots, \ell_n\}) \\
&= \sum_{\vartheta \in \Xi} w^{(\{\ell_1, \ldots, \ell_n\}, \vartheta)}
\end{aligned} \tag{J.8}$$

and

$$
\begin{aligned}
p(\{(\boldsymbol{x}_1,\ell_1),\ldots,(\boldsymbol{x}_n,\ell_n)\}) &\triangleq \frac{\pi(\{(\boldsymbol{x}_1,\ell_1),\ldots,(\boldsymbol{x}_n,\ell_n)\})}{w(\{\ell_1,\ldots,\ell_n\})} \\
&= \frac{\Delta(\{(\boldsymbol{x}_1,\ell_1),\ldots,(\boldsymbol{x}_n,\ell_n)\})}{w(\{\ell_1,\ldots,\ell_n\})} \times \\
&\quad \sum_{I\in\mathcal{F}(\mathbb{L})} \delta_I(\{\ell_1,\ldots,\ell_n\}) \times \\
&\quad \sum_{\vartheta\in\Xi} w^{(I,\vartheta)}[p^{(\vartheta)}]^{\{(\boldsymbol{x}_1,\ell_1),\ldots,(\boldsymbol{x}_n,\ell_n)\}}
\end{aligned} \tag{J.9}
$$

Therefore, according to Proposition 21, the Mδ-GLMB approximation parameters $\hat{w}^{(I)}$ and $\hat{p}^{(I)}$ that match the PHD and cardinality of the original GLMB can be obtained as

$$
\hat{w}^{(I)} = w(I) = \sum_{\vartheta\in\Xi} w^{(I,\vartheta)} \tag{J.10}
$$

$$
\begin{aligned}
\hat{p}^{(I)}(\boldsymbol{x},\ell) &= 1_I(\ell)p_{I-\{\ell\}}(\boldsymbol{x},\ell) \\
&= 1_I(\ell)\int p(\{(\boldsymbol{x},\ell),(\boldsymbol{x}_1,\ell_1),\ldots,(\boldsymbol{x}_n,\ell_n)\})d(\boldsymbol{x}_1,\ldots,\boldsymbol{x}_n)
\end{aligned} \tag{J.11}
$$

where $I = \{\ell,\ell_1,\ldots,\ell_n\}$ and $I - \{\ell\} = \{\ell_1,\ldots,\ell_n\}$. Substituting (J.9) into (J.11) yields

$$
\begin{aligned}
\hat{p}^{(I)}(\boldsymbol{x},\ell) &= \frac{1_I(\ell)\Delta(\{(\boldsymbol{x},\ell),(\boldsymbol{x}_1,\ell_1),\ldots,(\boldsymbol{x}_n,\ell_n)\})}{w(\{\ell,\ell_1,\ldots,\ell_n\})} \\
&\quad \times \sum_{J\in\mathcal{F}(\mathbb{L})} \delta_J(\{\ell,\ell_1,\ldots,\ell_n\}) \sum_{\vartheta\in\Xi} w^{(J,\vartheta)} \times \\
&\quad \int [p^{(\vartheta)}]^{\{(\boldsymbol{x},\ell),(\boldsymbol{x}_1,\ell_1),\ldots,(\boldsymbol{x}_n,\ell_n)\}} d(\boldsymbol{x}_1,\ldots,\boldsymbol{x}_n) \\
&= \frac{1_I(\ell)\Delta(\{(\boldsymbol{x},\ell),(\boldsymbol{x}_1,\ell_1),\ldots,(\boldsymbol{x}_n,\ell_n)\})}{w(\{\ell,\ell_1,\ldots,\ell_n\})} \times \\
&\quad \sum_{J\in\mathcal{F}(\mathbb{L})} \delta_J(\{\ell,\ell_1,\ldots,\ell_n\}) \sum_{\vartheta\in\Xi} w^{(J,\vartheta)} p^{(\vartheta)}(\boldsymbol{x},\ell)
\end{aligned} \tag{J.12}
$$

Note that only one term in the summation of the above equation over J is non-zero, so we have

$$
\hat{p}^{(I)}(\boldsymbol{x},\ell) = 1_I(\ell)\Delta(\{(\boldsymbol{x},\ell),(\boldsymbol{x}_1,\ell_1),\ldots,(\boldsymbol{x}_n,\ell_n)\}) \frac{\sum_{\vartheta\in\Xi} w^{(I,\vartheta)} p^{(\vartheta)}(\boldsymbol{x},\ell)}{\sum_{\vartheta\in\Xi} w^{(I,\vartheta)}} \tag{J.13}
$$

Therefore, according to Proposition 21, the Mδ-GLMB approximation is

$$
\begin{aligned}
\pi(\mathrm{X}) &= \Delta(\mathrm{X}) \sum_{I \in \mathcal{F}(\mathbb{L})} \delta_I(\mathcal{L}(\mathrm{X}))\hat{w}^{(I)}[\hat{p}^{(I)}]^{\mathrm{X}} \\
&= \Delta(\mathrm{X}) \sum_{I \in \mathcal{F}(\mathbb{L})} \delta_I(\mathcal{L}(\mathrm{X})) \sum_{\vartheta \in \Xi} w^{(I,\vartheta)} \left[1_I(\cdot) \frac{\sum_{\vartheta \in \Xi} w^{(I,\vartheta)} p^{(\vartheta)}(\cdot,\cdot)}{\sum_{\vartheta \in \Xi} w^{(I,\vartheta)}} \right]^{\mathrm{X}} \\
&= \Delta(\mathrm{X}) \sum_{I \in \mathcal{F}(\mathbb{L})} \delta_I(\mathcal{L}(\mathrm{X}))w^{(I)}[p^{(I)}]^{\mathrm{X}}
\end{aligned}
\tag{J.14}
$$

where $w^{(I)}$ and $p^{(I)}$ are given by (7.5.17) and (7.5.18), respectively.

Appendix K
Derivation of TBD Measurement Updated PGFl Under IIDC Prior

Proof of Corollary 6 Since X is an IIDC with PHD v and cardinality distribution ρ, its PGFl is

$$G[h] = G(\langle v, h\rangle/\langle v, 1\rangle) \tag{K.1}$$

where $G(z) = \sum_{n=0}^{\infty} \rho(n)z^n$ is the PGF of $|X|$. According to Proposition 28 and using $\langle v, g_Z\rangle = \langle vg_Z, 1\rangle$, we have

$$\begin{aligned} G[h|Z] &= \frac{G[hg_Z]}{G[g_Z]} = \frac{G(\langle vg_Z, h\rangle/\langle v, 1\rangle)}{G(\langle v, g_Z\rangle/\langle v, 1\rangle)} \\ &= \frac{\sum_{n=0}^{\infty} \rho(n)(\langle vg_Z, h\rangle/\langle v, 1\rangle)^n}{\sum_{j=0}^{\infty} \rho(j)(\langle v, g_Z\rangle/\langle v, 1\rangle)^j} \\ &= \frac{\sum_{n=0}^{\infty} \rho(n)(\langle v, g_Z\rangle/\langle v, 1\rangle)^n(\langle vg_Z, h\rangle/\langle vg_Z, 1\rangle)^n}{\sum_{j=0}^{\infty} \rho(j)(\langle v, g_Z\rangle/\langle v, 1\rangle)^j} \end{aligned} \tag{K.2}$$

In order to prove that the posterior RFS is indeed an IIDC with PHD (10.3.7) and cardinality distribution (10.3.8), one needs to prove that the posterior PGFl (K.2) has the form of $G[h|Z] = G(\langle v(\cdot|Z), h\rangle/\langle v(\cdot|Z), 1\rangle|Z)$. Note that $\langle vg_Z, h\rangle/\langle vg_Z, 1\rangle = \langle v(\cdot|Z), h\rangle/\langle v(\cdot|Z), 1\rangle$ can be obtained from (10.3.7), and substituting it into $(\langle vg_Z, h\rangle/\langle vg_Z, 1\rangle)$ in (K.2) and using (10.3.8) yields

$$\begin{aligned} G[h|Z] &= \sum_{n=0}^{\infty} \frac{\rho(n)(\langle v, g_Z\rangle/\langle v, 1\rangle)^n}{\sum_{j=0}^{\infty} \rho(j)(\langle v, g_Z\rangle/\langle v, 1\rangle)^j} \left(\frac{\langle v(\cdot|Z), h\rangle}{\langle v(\cdot|Z), 1\rangle}\right)^n \\ &= \sum_{n=0}^{\infty} \rho(n|Z)(\langle v(\cdot|Z), h\rangle/\langle v(\cdot|Z), 1\rangle)^n \\ &= G(\langle v(\cdot|Z), h\rangle/\langle v(\cdot|Z), 1\rangle|Z) \end{aligned} \tag{K.3}$$

W. Wu et al., *Target Tracking with Random Finite Sets*,
https://doi.org/10.1007/978-981-19-9815-7

Hence, the posterior is an IIDC process with PHD (10.3.7) and cardinality distribution (10.3.8).

Remark The posterior PHD (10.3.7) and cardinality distribution (10.3.8) can be obtained by differentiating the posterior PGFl $G[\cdot|Z]$ and PGF $G(\cdot|Z)$, respectively.

Appendix L
Uniqueness of Partitioning Measurement Set

Proof of Proposition 29 This proof involves two steps. First, the construction method is used to prove that there is a partition that satisfies the condition of Proposition 29. Then, the contradictory method is used to prove that this partition is unique.

(1) A partition satisfying the condition of Proposition 29 exists

Consider the algorithm in Table L.1. In this algorithm, a partition composed of single-element set corresponding to each measurement is formed first, and then the cells in the partition are combined. Obviously, the number of cells generated by the algorithm in the table decreases monotonously with iteration. Since the number of cells has a lower bound, it is easy to see that the number of iterations of the algorithm is at most $|Z| - 1$.

When this algorithm stops, the resulting partition will satisfy the condition of Proposition 29. This can be proven by the mathematical induction. In the first iteration, all cells are singletons and therefore the condition of Proposition 29 is naturally satisfied. Assume that at the beginning of the ith iteration, this partition satisfies the condition of Proposition 29. If the algorithm stops (row 7 in the table), then this partition naturally satisfies the condition of Proposition 29. If the algorithm doesn't stop, a new partition is formed through merging of the two cells. The unmerged cells remain unchanged, and, for any of these cells, any two measurements $z^{(p)}$ and $z^{(q)}$ also naturally satisfy (11.2.30) or (11.2.31).

Measurements in merged cells also satisfy (11.2.30) or (11.2.31), which is proved as follows. Suppose that at the ith iteration, the cell $W_{s_*+t_*}^{i-1}$ is formed by merging cells $W_{s_*}^{i-1}$ and $W_{t_*}^{i-1}$ (see row 8 in the table), and let $z^{(m_*)}$ and $z^{(n_*)}$ be a pair of measurements such that the cells are merged, i.e.,

$$\left(z^{(m_*)}, z^{(n_*)}\right) = \underset{\substack{z^{(m)} \in W_{s_*}^{i-1} \\ z^{(n)} \in W_{t_*}^{i-1}}}{\arg\min}\, d\left(z^{(m)}, z^{(n)}\right) \tag{L.3}$$

W. Wu et al., *Target Tracking with Random Finite Sets*,
https://doi.org/10.1007/978-981-19-9815-7

Table L.1 Solving the partition $\mathcal{P}$ satisfying the condition of Proposition 29

No.	Input: $Z = [z^{(1)}, \ldots, z^{(\lvert Z\rvert)}]$
1:	Set $\mathcal{P}_0 = \{\{z^{(1)}\}, \{z^{(2)}\}, \ldots, \{z^{(\lvert Z\rvert)}\}\}$, i.e., set $W_j^0 = \{z^{(j)}\},\ j = 1, \ldots, \lvert Z\rvert$
2:	Set $i = 1$
3:	% calculate all pairwise distances between cells of $\mathcal{P}_{i-1}$
4:	$\eta_{st}^{i-1} = \min_{z^{(m)} \in W_s^{i-1}, z^{(n)} \in W_t^{i-1}} d(z^{(m)}, z^{(n)}) \quad$ (L.1)
5:	% find out the cell corresponding to the minimum pairwise distance
6:	$(s_*, t_*) = \arg\min_{1 \le s \ne t \le \lvert\mathcal{P}_{i-1}\rvert} \eta_{st}^{i-1} \quad$ (L.2)
7:	If $\eta_{s_* t_*}^{i-1} > T_l$, this algorithm stops, because $\mathcal{P}_{i-1}$ is a partition that satisfies the condition of Proposition 29.
8:	Otherwise, a new single cell $W_{s_*+t_*}^{i-1} = W_{s_*}^{i-1} \cup W_{t_*}^{i-1}$ is formed, and cells $W_{s_*}^{i-1}$ and $W_{t_*}^{i-1}$ are deleted.
9:	Set $\mathcal{P}_i$ to be the set of cells obtained in line 8.
10:	Set $i = i + 1$ and jump to line 3.

Let $z^{(p)}$ and $z^{(q)}$ be two measurements in the merged cell $W_{s_*+t_*}^{i-1}$. If $z^{(p)}$ and $z^{(q)}$ are both within $W_{s_*}^{i-1}$ (or within $W_{t_*}^{i-1}$), they naturally satisfy (11.2.30) or (11.2.31). If $z^{(p)}$ is in $W_{s_*}^{i-1}$ while $z^{(q)}$ is in $W_{t_*}^{i-1}$ (or vice versa), under the assumption of $p \ne m_*$ and $q \ne n_*$, one of the following four relationships must hold:

① Relationship 1:

$$d\left(z^{(p)}, z^{(m_*)}\right) \le T_l \tag{L.4a}$$

$$d\left(z^{(m_*)}, z^{(n_*)}\right) \le T_l \tag{L.4b}$$

$$d\left(z^{(n_*)}, z^{(q)}\right) \le T_l \tag{L.4c}$$

② Relationship 2:

$$d\left(z^{(p)}, z^{(m_*)}\right) \le T_l \tag{L.5a}$$

$$d\left(z^{(m_*)}, z^{(n_*)}\right) \le T_l \tag{L.5b}$$

$$d\left(z^{(n_*)}, z^{(r_1)}\right) \le T_l \tag{L.5c}$$

$$d\left(z^{(r_a)}, z^{(r_{a+1})}\right) \le T_l, \quad a = 1, 2, \ldots, R - 1 \tag{L.5d}$$

$$d\left(z^{(r_R)}, z^{(q)}\right) \le T_l \tag{L.5e}$$

where $\{z^{(r_1)}, z^{(r_2)}, \ldots, z^{(r_R)}\} \subset W_{t_*}^{i-1} - \{z^{(q)}, z^{(n_*)}\}$.

③ Relationship 3:

$$d\left(z^{(p)}, z^{(r_1)}\right) \le T_l \tag{L.6a}$$

$$d\left(z^{(r_a)}, z^{(r_{a+1})}\right) \le T_l, \quad a = 1, 2, \ldots, R-1 \tag{L.6b}$$

$$d\left(z^{(r_R)}, z^{(m_*)}\right) \le T_l \tag{L.6c}$$

$$d\left(z^{(m_*)}, z^{(n_*)}\right) \le T_l \tag{L.6d}$$

$$d\left(z^{(n_*)}, z^{(q)}\right) \le T_l \tag{L.6e}$$

where $\{z^{(r_1)}, z^{(r_2)}, \cdots, z^{(r_R)}\} \subset W_{s_*}^{i-1} - \{z^{(p)}, z^{(m_*)}\}$.

④ Relationship 4:

$$d\left(z^{(p)}, z^{(r_1)}\right) \le T_l \tag{L.7a}$$

$$d\left(z^{(r_a)}, z^{(r_{a+1})}\right) \le T_l, \quad a = 1, 2, \ldots, R-1 \tag{L.7b}$$

$$d\left(z^{(r_R)}, z^{(m_*)}\right) \le T_l \tag{L.7c}$$

$$d\left(z^{(m_*)}, z^{(n_*)}\right) \le T_l \tag{L.7d}$$

$$d\left(z^{(n_*)}, z^{(c_1)}\right) \le T_l \tag{L.7e}$$

$$d\left(z^{(c_b)}, z^{(c_{b+1})}\right) \le T_l, \quad b = 1, 2, \ldots, B-1 \tag{L.7f}$$

$$d\left(z^{(c_B)}, z^{(q)}\right) \le T_l \tag{L.7g}$$

where $\{z^{(r_1)}, z^{(r_2)}, \ldots, z^{(r_R)}\} \subset W_{s_*}^{i-1} - \{z^{(p)}, z^{(m_*)}\}$ and $\{z^{(c_1)}, z^{(c_2)}, \ldots, z^{(c_B)}\} \subset W_{t_*}^{i-1} - \{z^{(q)}, z^{(n_*)}\}$.

For the case of $p = m_*$, $q \ne n_*$, and $p \ne m_*$, $q = n_*$, the above conclusion can be obtained in a similar way. But for the case of $p = m_*$, $q = n_*$, $d(z^{(p)}, z^{(q)}) \le T_l$, so $z^{(p)}$ and $z^{(q)}$ satisfy (11.2.30). This shows that the merged cell satisfies (11.2.30) or (11.2.31), and this partition satisfies the condition of Proposition 29. Therefore, the algorithm in Table L.1 will stop at a finite number of iterations, and the partition given by the algorithm satisfies the condition of Proposition 29.

(2) The uniqueness of the partition

Assuming that there are two distinct partitions ($\mathcal{P}_i$ and $\mathcal{P}_j$) that satisfy the condition of Proposition 29, then there must be (at least) one measurement $z^{(m)} \in Z$, one cell $W^i_{m_i} \in \mathcal{P}_i$, and one cell $W^j_{m_j} \in \mathcal{P}_j$ such that $z^{(m)} \in W^i_{m_i}$, $z^{(m)} \in W^j_{m_j}$, and $W^i_{m_i} \neq W^j_{m_j}$. This only requires (at least) a single measurement $z^{(n)} \in Z$ within either $W^i_{m_i}$ or $W^j_{m_j}$. Without loss of generality, assume that $z^{(n)} \in W^i_{m_i}$ and $z^{(n)} \notin W^j_{m_j}$. Since both $z^{(m)}$ and $z^{(n)}$ are in $W^i_{m_i}$, either $d(z^{(m)}, z^{(n)}) \leq T_l$ (which contradicts $z^{(n)} \notin W^j_{m_j}$), or there must be a non-void subset $\{z^{(r_1)}, z^{(r_2)}, \ldots, z^{(r_R)}\} \subset W^i_{m_i} - \{z^{(m)}, z^{(n)}\}$, such that the following condition holds: when $R = 1$, then

$$d\left(z^{(m)}, z^{(r_1)}\right) \leq T_l \tag{L.8a}$$

$$d\left(z^{(r_1)}, z^{(n)}\right) \leq T_l \tag{L.8b}$$

otherwise, when $R > 1$, then

$$d\left(z^{(m)}, z^{(r_1)}\right) \leq T_l \tag{L.8c}$$

$$d\left(z^{(r_s)}, z^{(r_{s+1})}\right) \leq T_l, \quad s = 1, 2, \ldots, R-1 \tag{L.8d}$$

$$d\left(z^{(r_R)}, z^{(n)}\right) \leq T_l \tag{L.8e}$$

However, the condition in (L.2) means that the following measurements should all be in the same cell

$$\begin{cases} \left\{z^{(m)}, z^{(r_1)}, z^{(n)}\right\}, & \text{if } R = 1 \\ \left\{z^{(m)}, z^{(r_1)}, z^{(r_2)}, \ldots, z^{(r_R)}, z^{(n)}\right\}, & \text{if } R > 1 \end{cases} \tag{L.9}$$

For $\mathcal{P}_j$, this cell is $W^j_{m_j} \ni z^{(m)}$, which conflicts with $z^{(n)} \notin W^j_{m_j}$. Therefore, the initial assumption that there are two different partitions satisfying the condition of Proposition 29 is not valid, thus completing the proof process.

Appendix M
Derivation of Measurement Likelihood for Extended Targets

The proof of Proposition 30 is as follows.

First, consider the case where there are no false alarm (i.e., all measurements are generated by targets). According to Assumptions A.23 and A.24, given the set X of extended targets, the likelihood of measurement set Y is [3]

$$g_D(Y|\mathrm{X}) = \sum_{(W_1,\ldots,W_{|\mathrm{X}|}):\uplus_{i=1}^{|\mathrm{X}|} W_i = Y} g'(W_1|\boldsymbol{x}_1, \ell_1)\ldots g'(W_{|\mathrm{X}|}|\boldsymbol{x}_{|\mathrm{X}|}, \ell_{|\mathrm{X}|}) \tag{M.1}$$

where $\uplus$ means the union operation of sets, and

$$g'(W|\boldsymbol{x}, \ell) \propto \begin{cases} p_D(\boldsymbol{x}, \ell)\tilde{g}(W|\boldsymbol{x}, \ell), & W \neq \emptyset \\ q_D(\boldsymbol{x}, \ell), & W = \emptyset \end{cases} \tag{M.2}$$

Note that a partition of any set S is defined as a mutually exclusive set of non-empty subsets of S, and its union is S. Nevertheless, in (M.1), the set $W_1, \ldots, W_{|\mathrm{X}|}$ may be either an empty set or a non-empty set and therefore they don't satisfy the definition of the partition. However, the non-empty sets in $W_1, \ldots, W_{|\mathrm{X}|}$ actually constitute a partition of Y. Therefore, by decomposing (M.1) into the product of the empty set and the non-empty set W_i, we can obtain

$$g_D(Y|\mathrm{X}) = [q_D]^{\mathrm{X}} \sum_{i=1}^{|\mathrm{X}|} \sum_{\mathcal{P}(Y)\in\mathcal{G}_i(Y)} \sum_{1\leq j_1\neq\cdots\neq j_i\leq|\mathrm{X}|} \prod_{k=1}^{i} \frac{p_D(\boldsymbol{x}_{j_k}, \ell_{j_k})\tilde{g}(\mathcal{P}_k(Y)|\boldsymbol{x}_{j_k}, \ell_{j_k})}{q_D(\boldsymbol{x}_{j_k}, \ell_{j_k})} \tag{M.3}$$

where $\mathcal{P}_k(Y)$ is the kth cell in the partition $\mathcal{P}(Y)$. Similar to the derivation in [14, p. 420], the above equation can be expressed as

W. Wu et al., *Target Tracking with Random Finite Sets*,
https://doi.org/10.1007/978-981-19-9815-7

$$g_D(Y|\mathrm{X}) = [q_D]^{\mathrm{X}} \sum_{i=1}^{|\mathrm{X}|} \sum_{\mathcal{P}(Y)\in\mathcal{G}_i(Y)} \sum_{\theta\in\Theta(\mathcal{P}(Y))} \prod_{j:\theta(j)>0} \frac{p_D(\boldsymbol{x}_j, \ell_j)\tilde{g}(\mathcal{P}_{\theta(j)}(Y)|\boldsymbol{x}_j, \ell_j)}{q_D(\boldsymbol{x}_j, \ell_j)} \tag{M.4}$$

Now, consider the case where false alarms may also exist. According to Assumption A.25, the set K of false alarms follows the distribution $g_C(K)$, set Y and set K are independent, and the total measurement set is $Z = Y \cup K$. Therefore, according to the convolution formula, Z obeys the following distribution

$$\begin{aligned} g(Z|\mathrm{X}) &= \sum_{W\subseteq Z} g_C(Z - W) g_D(W|\mathrm{X}) \\ &= \sum_{W\subseteq Z} \exp(-\langle\kappa, 1\rangle)\kappa^{Z-W}[q_D]^{\mathrm{X}} \times \\ &\quad \sum_{i=1}^{|\mathrm{X}|} \sum_{\substack{\mathcal{P}(W)\in\mathcal{G}_i(W) \\ \theta\in\Theta(\mathcal{P}(W))}} \prod_{j:\theta(j)>0} \frac{p_D(\boldsymbol{x}_j, \ell_j)\tilde{g}(\mathcal{P}_{\theta(j)}(W)|\boldsymbol{x}_j, \ell_j)}{q_D(\boldsymbol{x}_j, \ell_j)} \\ &= \exp(-\langle\kappa, 1\rangle)\kappa^{Z}[q_D]^{\mathrm{X}} \times \\ &\quad \sum_{W\subseteq Z} \sum_{i=1}^{|\mathrm{X}|} \sum_{\substack{\mathcal{P}(W)\in\mathcal{G}_i(W) \\ \theta\in\Theta(\mathcal{P}(W))}} \prod_{j:\theta(j)>0} \frac{p_D(\boldsymbol{x}_j, \ell_j)\tilde{g}(\mathcal{P}_{\theta(j)}(W)|\boldsymbol{x}_j, \ell_j)}{q_D(\boldsymbol{x}_j, \ell_j)[\kappa]^{\mathcal{P}_{\theta(j)}(W)}} \end{aligned} \tag{M.5}$$

where since $\mathcal{P}(W)$ is a partition of W, the relationship of $\kappa^W = \prod_{j:\theta(j)>0} [\kappa]^{\mathcal{P}_{\theta(j)}(W)}$ is used in the last row.

Finally, the above equation can be further simplified by treating the set $Z - W$ as an extra element attached to each $\mathcal{P}(W)$, thus converting it into a partition of Z. In this process, two summations on the partition $\mathcal{P}(W) \in \mathcal{G}_i(W)$ of $W \subseteq Z$ and $i = 1 \to |\mathrm{X}|$ can be expressed as a single summation on the partition $\mathcal{P}(Z)$ of Z and $i = 1 \to |\mathrm{X}| + 1$, i.e.,

$$g(Z|\mathrm{X}) = g_C(Z)[q_D]^{\mathrm{X}} \sum_{i=1}^{|\mathrm{X}|+1} \sum_{\substack{\mathcal{P}(Z)\in\mathcal{G}_i(Z) \\ \theta\in\Theta(\mathcal{P}(Z))}} \prod_{j:\theta(j)>0} \frac{p_D(\boldsymbol{x}_j, \ell_j)\tilde{g}(\mathcal{P}_{\theta(j)}(Z)|\boldsymbol{x}_j, \ell_j)}{q_D(\boldsymbol{x}_j, \ell_j)[\kappa]^{\mathcal{P}_{\theta(j)}(Z)}} \tag{M.6}$$

Note that when $\theta(j) > 0$, $q_D(\boldsymbol{x}_j, \ell_j)$ in the denominator cancels out the corresponding term in $[q_D]^{\mathrm{X}}$, while for each $j : \theta(j) = 0$, only the term $q_D(\boldsymbol{x}_j, \ell_j)$ is left. Therefore, (M.6) can be equivalently expressed in the form of (11.3.84). □

Appendix N
Derivation of Tracker with Merged Measurements

(1) Proof of Proposition 31

The density of surviving targets is

$$
\begin{aligned}
\pi_S(\mathrm{S}) &= \int \phi_S(\mathrm{S}|\mathrm{X})\pi(\mathrm{X})\delta\mathrm{X} \\
&= \int \Delta(\mathrm{S})\Delta(\mathrm{X})1_{\mathcal{L}(\mathrm{X})}(\mathcal{L}(\mathrm{S}))[\Phi(\mathrm{S};\cdot)]^{\mathrm{X}}\Delta(\mathrm{X})\sum_{c\in\mathbb{C}} w^{(c)}(\mathcal{L}(\mathrm{X}))p^{(c)}(\mathrm{X})\delta\mathrm{X} \\
&= \Delta(\mathrm{S})\sum_{c\in\mathbb{C}}\int \Delta(\mathrm{X})1_{\mathcal{L}(\mathrm{X})}(\mathcal{L}(\mathrm{S}))w^{(c)}(\mathcal{L}(\mathrm{X}))[\Phi(\mathrm{S};\cdot)]^{\mathrm{X}}p^{(c)}(\mathrm{X})\delta\mathrm{X} \\
&= \Delta(\mathrm{S})\sum_{c\in\mathbb{C}}\sum_{L\subseteq\mathbb{L}} 1_L(\mathcal{L}(\mathrm{S}))w^{(c)}(L)\times \\
&\quad \int\prod_{i=1}^{|L|}\Phi(\mathrm{S};\boldsymbol{x}_i,\ell_i)p^{(c)}(\{(\boldsymbol{x}_1,\ell_1),\ldots,(\boldsymbol{x}_{|L|},\ell_{|L|})\})d\boldsymbol{x}_{1:|L|} \\
&= \Delta(\mathrm{S})\sum_{c\in\mathbb{C}}\sum_{L\subseteq\mathbb{L}} 1_L(\mathcal{L}(\mathrm{S}))w^{(c)}(L)\int\prod_{i=1}^{|L|}\Phi(\mathrm{S};\boldsymbol{x}_i,\ell_i)p^{(c)}_{\ell_{1:|L|}}(\boldsymbol{x}_{1:|L|})d\boldsymbol{x}_{1:|L|} \\
&= \Delta(\mathrm{S})\sum_{c\in\mathbb{C}}\sum_{L\subseteq\mathbb{L}} 1_L(\mathcal{L}(\mathrm{S}))w^{(c)}(L)\eta_S^{(c)}(L)p^{(c)}_{S,L}(\mathrm{S})
\end{aligned}
\tag{N.1}
$$

where $\phi_S(\mathrm{S}|\mathrm{X})$, $\Phi(\mathrm{S};\boldsymbol{x},\ell)$, $p^{(c)}_{S,L}(\mathrm{S})$, and $\eta_S^{(c)}(L)$ are defined in (3.4.9), (3.4.10), (11.4.8), and (11.4.9), respectively, and the fourth line is obtained by applying Lemma 3 to $h(L) = 1_L(\mathcal{L}(\mathrm{S}))w^{(c)}(L)$ and $g(\mathrm{X}) = [\Phi(\mathrm{S};\cdot)]^{\mathrm{X}}p^{(c)}(\mathrm{X})$.

The predicted density is the product of the birth density and the survival density, i.e.,

$$\pi_+(\mathrm{X}_+) = \pi_\gamma(\mathrm{B})\pi_S(\mathrm{S})$$

W. Wu et al., *Target Tracking with Random Finite Sets*,
https://doi.org/10.1007/978-981-19-9815-7

$$
\begin{aligned}
&= \Delta(\mathrm{B})\Delta(\mathrm{S}) \sum_{c\in\mathbb{C}} \sum_{L\subseteq\mathbb{L}} 1_L(\mathcal{L}(\mathrm{S}))w_\gamma(\mathcal{L}(\mathrm{B}))w^{(c)}(L)\eta_S^{(c)}(L)[p_\gamma(\cdot)]^{\mathrm{B}} p_{S,L}^{(c)}(\mathrm{S}) \\
&= \Delta(\mathrm{X}_+) \sum_{c\in\mathbb{C}} \sum_{L\subseteq\mathbb{L}} 1_L(\mathcal{L}(\mathrm{X}_+)\cap\mathbb{L})w_\gamma(\mathcal{L}(\mathrm{X}_+)-\mathbb{L})w^{(c)}(L)\times \\
&\quad \eta_S^{(c)}(L)p_\gamma^{\mathrm{X}_+-\mathbb{X}\times\mathbb{L}} p_{S,L}^{(c)}(\mathrm{X}_+\cap\mathbb{X}\times\mathbb{L}) \\
&= \Delta(\mathrm{X}_+) \sum_{c\in\mathbb{C}} \sum_{L\subseteq\mathbb{L}} w_{+,L}^{(c)}(\mathcal{L}(\mathrm{X}_+))p_{+,L}^{(c)}(\mathrm{X}_+)
\end{aligned} \tag{N.2}
$$

where the birth density $\pi_\gamma(\cdot)$ is given in (3.4.11), $\mathrm{B} = \mathrm{X}_+ - \mathbb{X}\times\mathbb{L}$, $\mathrm{S} = \mathrm{X}_+\cap\mathbb{X}\times\mathbb{L}$, and $w_{+,L}^{(c)}$ and $p_{+,L}^{(c)}$ are given by (11.4.6) and (11.4.7), respectively. □

(2) Proof of Proposition 32

The product of the prior distribution (7.5.9) and the likelihood (11.4.4) is

$$
\begin{aligned}
\pi(\mathrm{X})g(Z|\mathrm{X}) &= \Delta(\mathrm{X})\exp(-\langle\kappa, 1\rangle)\kappa^Z \times \\
&\sum_{c\in\mathbb{C}} \sum_{\substack{\mathcal{P}(\mathrm{X})\in\mathcal{G}(\mathrm{X}) \\ \theta\in\Theta(\mathcal{P}(\mathcal{L}(\mathrm{X})))}} w^{(c)}(\mathcal{L}(\mathrm{X}))[\tilde{\varphi}_Z(\cdot;\theta)]^{\mathcal{P}(\mathrm{X})} p^{(c)}(\mathrm{X}) \\
&= \Delta(\mathrm{X})\exp(-\langle\kappa, 1\rangle)\kappa^Z \times \\
&\sum_{c\in\mathbb{C}} \sum_{\substack{\mathcal{P}(\mathrm{X})\in\mathcal{G}(\mathrm{X}) \\ \theta\in\Theta(\mathcal{P}(\mathcal{L}(\mathrm{X})))}} w^{(c)}(\mathcal{L}(\mathrm{X}))\eta_Z^{(c,\theta)}(\mathcal{P}(\mathcal{L}(\mathrm{X})))p^{(c,\theta)}(\mathcal{P}(\mathrm{X})|Z)
\end{aligned} \tag{N.3}
$$

where $p^{(c,\theta)}(\mathcal{P}(\mathrm{X})|Z)$ and $\eta_Z^{(c,\theta)}(\mathcal{P}(L))$ are defined by (11.4.13) and (11.4.14), respectively.

The integral of (N.3) with respect to the multi-target state is

$$
\begin{aligned}
\int \pi(\mathrm{X})g(Z|\mathrm{X})\delta\mathrm{X} &= \int \Delta(\mathrm{X})\exp(-\langle\kappa, 1\rangle)\kappa^Z \sum_{c\in\mathbb{C}} \sum_{\substack{\mathcal{P}(\mathrm{X})\in\mathcal{G}(\mathrm{X}) \\ \theta\in\Theta(\mathcal{P}(\mathcal{L}(\mathrm{X})))}} w^{(c)}(\mathcal{L}(\mathrm{X}))\times \\
&\quad \eta_Z^{(c,\theta)}(\mathcal{P}(\mathcal{L}(\mathrm{X})))p^{(c,\theta)}(\mathcal{P}(\mathrm{X})|Z)\delta\mathrm{X} \\
&= \exp(-\langle\kappa, 1\rangle)\kappa^Z \times \\
&\quad \sum_{c\in\mathbb{C}} \sum_{n=0}^{\infty} \frac{1}{n!} \sum_{(\ell_{1:n})\in\mathbb{L}^n} \sum_{\substack{\mathcal{P}(\{\ell_{1:n}\})\in\mathcal{G}(\{\ell_{1:n}\}) \\ \theta\in\Theta(\mathcal{P}(\{\ell_{1:n}\}))}} w^{(c)}(\{\ell_{1:n}\})\eta_Z^{(c,\theta)}(\mathcal{P}(\{\ell_{1:n}\})) \\
&\quad \times \int p^{(c,\theta)}(\mathcal{P}(\{(\boldsymbol{x}_1,\ell_1),\ldots,(\boldsymbol{x}_n,\ell_n)\})|Z)d\boldsymbol{x}_{1:n} \\
&= \exp(-\langle\kappa, 1\rangle)\kappa^Z \sum_{c\in\mathbb{C}} \sum_{n=0}^{\infty} \sum_{L\in\mathcal{F}_n(\mathbb{L})} \sum_{\substack{\mathcal{P}(L)\in\mathcal{G}(L) \\ \theta\in\Theta(\mathcal{P}(L))}} w^{(c)}(L)\eta_Z^{(c,\theta)}(\mathcal{P}(L))
\end{aligned}
$$

$$= \exp(-\langle \kappa, 1 \rangle)\kappa^Z \sum_{c \in \mathbb{C}} \sum_{L \subseteq \mathbb{L}} \sum_{\substack{\mathcal{P}(L) \in \mathcal{G}(L) \\ \theta \in \Theta(\mathcal{P}(L))}} w^{(c)}(L)\eta_Z^{(c,\theta)}(\mathcal{P}(L)) \tag{N.4}$$

Substituting (N.3) and (N.4) into the Bayesian update equation yields (11.4.11). □

(3) Derivation of Eq. (11.4.15)

The product of the prior distribution (3.3.38) and the likelihood (11.4.4) is

$$\pi(\mathrm{X})g(Z|\mathrm{X}) = \Delta(\mathrm{X}) \exp(-\langle \kappa, 1 \rangle)\kappa^Z \times \sum_{c \in \mathbb{C}} \sum_{\substack{\mathcal{P}(\mathrm{X}) \in \mathcal{G}(\mathrm{X}) \\ \theta \in \Theta(\mathcal{P}(\mathcal{L}(\mathrm{X})))}} w^{(c)}(\mathcal{L}(\mathrm{X}))[\tilde{\varphi}_Z(\cdot; \theta)]^{\mathcal{P}(\mathrm{X})}[p^{(c)}(\cdot)]^{\mathrm{X}} \tag{N.5}$$

Using (11.4.18), we have

$$[\tilde{p}^{(c)}(\cdot)]^{\mathcal{P}(\mathrm{X})} = \prod_{\mathrm{Y} \in \mathcal{P}(\mathrm{X})} \tilde{p}^{(c)}(\mathrm{Y}) = \prod_{\mathrm{Y} \in \mathcal{P}(\mathrm{X})} [p^{(c)}(\cdot)]^{\mathrm{Y}} = [p^{(c)}(\cdot)]^{\mathrm{X}} \tag{N.6}$$

Then, we can obtain

$$\begin{aligned} \pi(\mathrm{X})g(Z|\mathrm{X}) &= \Delta(\mathrm{X}) \exp(-\langle \kappa, 1 \rangle)\kappa^Z \times \\ &\sum_{c \in \mathbb{C}} \sum_{\substack{\mathcal{P}(\mathrm{X}) \in \mathcal{G}(\mathrm{X}) \\ \theta \in \Theta(\mathcal{P}(\mathcal{L}(\mathrm{X})))}} w^{(c)}(\mathcal{L}(\mathrm{X}))[\tilde{p}^{(c)}(\cdot)\tilde{\varphi}_Z(\cdot; \theta)]^{\mathcal{P}(\mathrm{X})} \\ &= \Delta(\mathrm{X}) \exp(-\langle \kappa, 1 \rangle)\kappa^Z \times \\ &\sum_{c \in \mathbb{C}} \sum_{\substack{\mathcal{P}(\mathrm{X}) \in \mathcal{G}(\mathrm{X}) \\ \theta \in \Theta(\mathcal{P}(\mathcal{L}(\mathrm{X})))}} w^{(c)}(\mathcal{L}(\mathrm{X}))[\eta_Z^{(c,\theta)}]^{\mathcal{P}(\mathcal{L}(\mathrm{X}))}[p^{(c,\theta)}(\cdot|Z)]^{\mathcal{P}(\mathrm{X})} \end{aligned} \tag{N.7}$$

Its corresponding integral is

$$\begin{aligned} \int \pi(\mathrm{X})g(Z|\mathrm{X})\delta\mathrm{X} &= \exp(-\langle \kappa, 1 \rangle)\kappa^Z \times \\ &\sum_{c \in \mathbb{C}} \sum_{\substack{\mathcal{P}(\mathrm{X}) \in \mathcal{G}(\mathrm{X}) \\ \theta \in \Theta(\mathcal{P}(\mathcal{L}(\mathrm{X})))}} \int \Delta(\mathrm{X})w^{(c)}(\mathcal{L}(\mathrm{X}))[\eta_Z^{(c,\theta)}]^{\mathcal{P}(\mathcal{L}(\mathrm{X}))}[p^{(c,\theta)}(\cdot|Z)]^{\mathcal{P}(\mathrm{X})}\delta\mathrm{X} \\ &= \exp(-\langle \kappa, 1 \rangle)\kappa^Z \sum_{c \in \mathbb{C}} \sum_{L \subseteq \mathbb{L}} \sum_{\substack{\mathcal{P}(L) \in \mathcal{G}(L) \\ \theta \in \Theta(\mathcal{P}(L))}} w^{(c)}(L)[\eta_Z^{(c,\theta)}]^{\mathcal{P}(L)} \end{aligned} \tag{N.8}$$

Substituting (N.7) and (N.8) into the Bayesian update equation yields the posterior (11.4.15). □

Appendix O
Information Fusion and Weighting Operators

The following defines the information fusion operator and weighting operator for multi-target densities (including single-target densities).

Given two multi-target densities $\pi_i(X)$ and $\pi_j(X)$, the information fusion operation is defined as

$$\pi_i(X) \oplus \pi_j(X) \triangleq \frac{\pi_i(X)\pi_j(X)}{\langle \pi_i, \pi_j \rangle} = \frac{\pi_i(X)\pi_j(X)}{\int \pi_i(X)\pi_j(X)\delta X} \tag{O.1}$$

Moreover, given positive scalar ω and multi-target density $\pi(X)$, the weighting operation is defined as

$$\omega \odot \pi(X) = \frac{\pi^{\omega}(X)}{\langle \pi^{\omega}, 1 \rangle} = \frac{\pi^{\omega}(X)}{\int \pi^{\omega}(X)\delta X} \tag{O.2}$$

According to these definitions, it is easy to know that these two operators $\oplus$ and $\odot$ have the following properties

$$\pi_i \oplus \pi_j = \pi_j \oplus \pi_i \tag{O.3}$$

$$(\pi_i \oplus \pi_j) \oplus \pi_n = \pi_i \oplus (\pi_j \oplus \pi_n) = \pi_i \oplus \pi_j \oplus \pi_n \tag{O.4}$$

$$(\alpha\beta) \odot \pi = \alpha \odot (\beta \odot \pi) \tag{O.5}$$

$$1 \odot \pi = \pi \tag{O.6}$$

W. Wu et al., *Target Tracking with Random Finite Sets*,
https://doi.org/10.1007/978-981-19-9815-7

$$\alpha \odot (\pi_i \oplus \pi_j) = (\alpha \odot \pi_i) \oplus (\alpha \odot \pi_j) \tag{O.7}$$

$$(\alpha + \beta) \odot \pi = (\alpha \odot \pi) \oplus (\beta \odot \pi) \tag{O.8}$$

where π is the multi-target density and α and β are positive scalars.

Appendix P
Derivation of Weighted KLA of Multi-target Densities

Proof of Proposition 33 Let

$$\tilde{\pi}(X) = \prod_i \pi_i^{\omega_i}(X), \quad c = \int \tilde{\pi}(X)\delta X \tag{P.1}$$

so that the fused multi-target density in (12.4.4) can be expressed as $\overline{\pi}(X) = \tilde{\pi}(X)/c$, then according to the KLA, the minimum cost is

$$\begin{aligned}
J(\pi) &= \sum_i \omega_i \mathcal{D}_{\text{KL}}(\pi || \pi_i) = \sum_i \omega_i \int \pi(X) \log \frac{\pi(X)}{\pi_i(X)} \delta X \\
&= \sum_i \omega_i \sum_{n=0} \frac{1}{n!} \int \pi(\{\boldsymbol{x}_1, \ldots, \boldsymbol{x}_n\}) \log \frac{\pi(\{\boldsymbol{x}_1, \ldots, \boldsymbol{x}_n\})}{\pi_i(\{\boldsymbol{x}_1, \ldots, \boldsymbol{x}_n\})} d\boldsymbol{x}_1 \ldots d\boldsymbol{x}_n \\
&= \sum_{n=0}^{\infty} \frac{1}{n!} \int \pi(\{\boldsymbol{x}_1, \ldots, \boldsymbol{x}_n\}) \sum_i \omega_i \log \frac{\pi(\{\boldsymbol{x}_1, \ldots, \boldsymbol{x}_n\})}{\pi_i(\{\boldsymbol{x}_1, \ldots, \boldsymbol{x}_n\})} d\boldsymbol{x}_1 \ldots d\boldsymbol{x}_n \\
&= \sum_{n=0}^{\infty} \frac{1}{n!} \int \pi(\{\boldsymbol{x}_1, \ldots, \boldsymbol{x}_n\}) \log \frac{\pi(\{\boldsymbol{x}_1, \ldots, \boldsymbol{x}_n\})}{\prod_i \pi_i^{\omega_i}(\{\boldsymbol{x}_1, \ldots, \boldsymbol{x}_n\})} d\boldsymbol{x}_1 \ldots d\boldsymbol{x}_n \\
&= \sum_{n=0}^{\infty} \frac{1}{n!} \int \pi(\{\boldsymbol{x}_1, \ldots, \boldsymbol{x}_n\}) \log \frac{\pi(\{\boldsymbol{x}_1, \ldots, \boldsymbol{x}_n\})}{\tilde{\pi}(\{\boldsymbol{x}_1, \ldots, \boldsymbol{x}_n\})} d\boldsymbol{x}_1 \ldots d\boldsymbol{x}_n \\
&= \sum_{n=0}^{\infty} \frac{1}{n!} \int \pi(\{\boldsymbol{x}_1, \ldots, \boldsymbol{x}_n\}) \log \frac{\pi(\{\boldsymbol{x}_1, \ldots, \boldsymbol{x}_n\})}{c \cdot \overline{\pi}(\{\boldsymbol{x}_1, \ldots, \boldsymbol{x}_n\})} d\boldsymbol{x}_1 \ldots d\boldsymbol{x}_n \\
&= \sum_{n=0}^{\infty} \frac{1}{n!} \int \pi(\{\boldsymbol{x}_1, \ldots, \boldsymbol{x}_n\}) \log \frac{\pi(\{\boldsymbol{x}_1, \ldots, \boldsymbol{x}_n\})}{\overline{\pi}(\{\boldsymbol{x}_1, \ldots, \boldsymbol{x}_n\})} d\boldsymbol{x}_1 \ldots d\boldsymbol{x}_n \\
&\quad - \log c \sum_{n=0}^{\infty} \frac{1}{n!} \int \pi(\{\boldsymbol{x}_1, \ldots, \boldsymbol{x}_n\}) d\boldsymbol{x}_1 \ldots d\boldsymbol{x}_n
\end{aligned}$$

W. Wu et al., *Target Tracking with Random Finite Sets*,
https://doi.org/10.1007/978-981-19-9815-7

$$
\begin{aligned}
&= \int \pi(X)\log\frac{\pi(X)}{\overline{\pi}(X)}\delta X - \log c \cdot \int \pi(X)\delta X \\
&= \mathcal{D}_{\mathrm{KL}}(\pi||\overline{\pi}) - \log c \qquad \text{(P.2)}
\end{aligned}
$$

Since the KLD is always non-negative and it is zero if and only if its two parameters coincide almost everywhere, $\overline{\pi}(X)$ defined by (12.4.4) minimizes the above $J(\pi)$. □

Appendix Q
Fusion of PHD Posteriors and Fusion of Bernoulli Posteriors

(1) Fusion of PHD posteriors

By substituting the Poisson cardinality distribution into (12.4.11) and (12.4.12), the EMD fusion of two PHD filters can be obtained according to (12.4.19).

Corollary 9 *Consider the multi-target Poisson distribution given by* (12.4.11) *and* (12.4.12), *where* $\rho_i(n)$ *and* $\rho_j(n)$ *are Poisson densities with parameters* μ_i *and* μ_j, *then the corresponding EMD is also the multi-target Poisson distribution, where the positional distribution is given by* (12.4.18) *and the cardinality distribution* $\overline{\rho}$ *is a Poisson density with the following parameter* $\overline{\mu}$, *i.e.,*

$$\overline{\mu} = \mu_i^{(1-\omega)} \mu_j^{\omega} \eta_{i,j}(\omega) \tag{Q.1}$$

where $\eta_{i,j}(\omega)$ *is given by* (12.4.16).

Proof Equation (12.4.19) holds when the multi-target Poisson distribution is the IIDC with the Poisson cardinality. After substituting the Poisson density, the EMD cardinality distribution given by (12.4.20) is

$$\begin{aligned}
\overline{\rho}(|X|) &= \frac{1}{K}\left[\mu_i^{|X|}\exp(-\mu_i)/|X|!\right]^{(1-\omega)}\left[\mu_j^{|X|}\exp(-\mu_j)/|X|!\right]^{\omega}\eta_{i,j}^{|X|}(\omega) \\
&= \frac{1}{K}\frac{\left[\mu_i^{(1-\omega)}\mu_j^{\omega}\eta_{i,j}(\omega)\right]^{|X|}}{|X|!}\exp\left[-\mu_i(1-\omega)-\mu_j\omega\right]
\end{aligned} \tag{Q.2}$$

where the denominator K is

$$\begin{aligned}
K &= \sum_{|X|=0}^{\infty}\left[\mu_i^{|X|}\exp(-\mu_i)/|X|!\right]^{(1-\omega)}\left[\mu_j^{|X|}\exp(-\mu_j)/|X|!\right]^{\omega}\eta_{i,j}^{|X|}(\omega) \\
&= \exp\left[-\mu_i(1-\omega)-\mu_j\omega\right]\sum_{|X|=0}^{\infty}\left[\mu_i^{(1-\omega)}\mu_j^{\omega}\eta_{i,j}(\omega)\right]^{|X|}/|X|!
\end{aligned}$$

W. Wu et al., *Target Tracking with Random Finite Sets*,
https://doi.org/10.1007/978-981-19-9815-7

$$= \exp\left[-\mu_i(1-\omega) - \mu_j\omega\right] \cdot \exp\left[\mu_i^{(1-\omega)}\mu_j^{\omega}\eta_{i,j}(\omega)\right] \tag{Q.3}$$

Substituting (Q.3) into (Q.2) leads to

$$\overline{\rho}(|X|) = \left[\mu_i^{(1-\omega)}\mu_j^{\omega}\eta_{i,j}(\omega)\right]^{|X|} \exp\left[-\mu_i^{(1-\omega)}\mu_j^{\omega}\eta_{i,j}(\omega)\right]/|X|! \tag{Q.4}$$

The above equation is a Poisson distribution with the parameter given by (Q.1), which indicates that the EMD cardinality distribution is in Poisson form. □

(2) Fusion of Bernoulli posteriors

For the EMD fusion of Bernoulli filters, substituting the cardinality with Bernoulli form into (12.4.19) can obtain the following corollary.

Corollary 10 *Consider the distributions of two Bernoulli RFSs given by* (12.4.11) *and* (12.4.12)*, where* $\rho_i(|X|)$ *and* $\rho_j(|X|)$ *are Bernoulli densities with parameters* ε_i *and* ε_j*, respectively, i.e.,*

$$\rho_i(|X|) = \begin{cases} 1-\varepsilon_i, & |X| = 0, \\ \varepsilon_i, & |X| = 1, \\ 0, & \text{otherwise.} \end{cases} \tag{Q.5}$$

$\rho_j(|X|)$ *is similar to the above equation. The corresponding EMD is also a Bernoulli RFS distribution, whose positional distribution is given by* (12.4.18)*, and the cardinality parameter is*

$$\overline{\varepsilon} = \frac{\varepsilon_i^{(1-\omega)}\varepsilon_j^{\omega}\eta_{i,j}(\omega)}{(1-\varepsilon_i)^{(1-\omega)}(1-\varepsilon_j)^{\omega} + \varepsilon_i^{(1-\omega)}\varepsilon_j^{\omega}\eta_{i,j}(\omega)} \tag{Q.6}$$

Proof For the Bernoulli RFS distribution, Eq. (12.4.19) holds when it is an IIDC with the Bernoulli cardinality. By substituting the Bernoulli density into (12.4.20) and calculating $\overline{\rho}(|X|)$, $|X| = 0, 1, 2, \ldots$, we can obtain

$$\overline{\rho}(|X|) = \begin{cases} (1-\varepsilon_i)^{(1-\omega)}(1-\varepsilon_j)^{\omega}/K, & |X| = 0, \\ \varepsilon_i^{(1-\omega)}\varepsilon_j^{\omega}\eta_{i,j}(\omega)/K, & |X| = 1, \\ 0, & \text{otherwise.} \end{cases} \tag{Q.7}$$

where

$$K = (1-\varepsilon_i)^{(1-\omega)}(1-\varepsilon_j)^{\omega} + \varepsilon_i^{(1-\omega)}\varepsilon_j^{\omega}\eta_{i,j}(\omega) \tag{Q.8}$$

This shows that the EMD cardinality is a Bernoulli with cardinality parameter $\overline{\varepsilon}$ (Q.6).

Appendix R
Weighted KLA of Mδ-GLMB Densities

Proof of Proposition 35 For the sake of simplicity, only two Mδ-GLMB densities are considered temporarily. At this time, according to (12.4.4), we can obtain

$$\overline{\pi}(\mathrm{X}) = \frac{1}{K}\left[\Delta(\mathrm{X}) \sum_{L \in \mathcal{F}(\mathbb{L})} \delta_L(\mathcal{L}(\mathrm{X})) w_1^{(L)} [p_1^{(L)}]^{\mathrm{X}}\right]^{\omega} \times \left[\Delta(\mathrm{X}) \sum_{L \in \mathcal{F}(\mathbb{L})} \delta_L(\mathcal{L}(\mathrm{X})) w_2^{(L)} [p_2^{(L)}]^{\mathrm{X}}\right]^{1-\omega} \tag{R.1}$$

Note that

$$\pi_i^{\omega}(\mathrm{X}) = \begin{cases} \Delta(\mathrm{X})(w_i^{(L_1)})^{\omega}[(p_i^{(L_1)})^{\omega}]^{\mathrm{X}}, \ if \ \mathcal{L}(\mathrm{X}) = L_1 \\ \vdots \qquad\qquad \vdots \\ \Delta(\mathrm{X})(w_i^{(L_n)})^{\omega}[(p_i^{(L_n)})^{\omega}]^{\mathrm{X}}, \ if \ \mathcal{L}(\mathrm{X}) = L_n \end{cases}, \quad i = 1, 2 \tag{R.2}$$

Therefore, Eq. (R.1) becomes

$$\begin{aligned} \overline{\pi}(\mathrm{X}) &= \frac{\Delta(\mathrm{X})}{K} \sum_{L \in \mathcal{F}(\mathbb{L})} \delta_L(\mathcal{L}(\mathrm{X}))(w_1^{(L)})^{\omega}(w_2^{(L)})^{1-\omega}[(p_1^{(L)})^{\omega}(p_2^{(L)})^{1-\omega}]^{\mathrm{X}} \\ &= \frac{\Delta(\mathrm{X})}{K} \sum_{L \in \mathcal{F}(\mathbb{L})} \delta_L(\mathcal{L}(\mathrm{X}))(w_1^{(L)})^{\omega}(w_2^{(L)})^{1-\omega} \\ &\quad \times \left[\int (p_1^{(L)}(\cdot,\ell))^{\omega}(p_2^{(L)}(\cdot,\ell))^{1-\omega} d\boldsymbol{x} \frac{(p_1^{(L)}(\boldsymbol{x},\ell))^{\omega}(p_2^{(L)}(\boldsymbol{x},\ell))^{1-\omega}}{\int (p_1^{(L)}(\cdot,\ell))^{\omega}(p_2^{(L)}(\cdot,\ell))^{1-\omega} d\boldsymbol{x}}\right]^{\mathrm{X}} \\ &= \frac{\Delta(\mathrm{X})}{K} \sum_{L \in \mathcal{F}(\mathbb{L})} \delta_L(\mathcal{L}(\mathrm{X}))(w_1^{(L)})^{\omega}(w_2^{(L)})^{1-\omega} \times \end{aligned}$$

W. Wu et al., *Target Tracking with Random Finite Sets*,
https://doi.org/10.1007/978-981-19-9815-7

$$\left[\int (p_1^{(L)}(\boldsymbol{x},\cdot))^{\omega}(p_2^{(L)}(\boldsymbol{x},\cdot))^{1-\omega}d\boldsymbol{x}\right]^L$$
$$\times\left[\frac{(p_1^{(L)}(\cdot,\cdot))^{\omega}(p_2^{(L)}(\cdot,\cdot))^{1-\omega}}{\int (p_1^{(L)}(\boldsymbol{x},\cdot))^{\omega}(p_2^{(L)}(\boldsymbol{x},\cdot))^{1-\omega}d\boldsymbol{x}}\right]^{\mathrm{X}}$$
$$=\frac{\Delta(\mathrm{X})}{K}\sum_{L\in\mathcal{F}(\mathbb{L})}\delta_L(\mathcal{L}(\mathrm{X}))(w_1^{(L)})^{\omega}(w_2^{(L)})^{1-\omega}\times$$
$$\left[\int (p_1^{(L)}(\boldsymbol{x},\cdot))^{\omega}(p_2^{(L)}(\boldsymbol{x},\cdot))^{1-\omega}d\boldsymbol{x}\right]^L\times$$
$$\left[(\omega\odot p_1^{(L)})\oplus((1-\omega)\odot p_2^{(L)})\right]^{\mathrm{X}} \tag{R.3}$$

According to (3.3.20), the normalized constant K can be obtained as

$$K=\int \Delta(\mathrm{X})\sum_{L\in\mathcal{F}(\mathbb{L})}\delta_L(\mathcal{L}(\mathrm{X}))(w_1^{(L)})^{\omega}(w_2^{(L)})^{1-\omega}[(p_1^{(L)})^{\omega}(p_2^{(L)})^{1-\omega}]^{\mathrm{X}}\delta\mathrm{X}$$
$$=\sum_{L\subseteq\mathbb{L}}(w_1^{(L)})^{\omega}(w_2^{(L)})^{1-\omega}\left[\int (p_1^{(L)}(\boldsymbol{x},\cdot))^{\omega}(p_2^{(L)}(\boldsymbol{x},\cdot))^{1-\omega}d\boldsymbol{x}\right]^L \tag{R.4}$$

Substituting (R.4) into (R.3) yields

$$\overline{\pi}(\mathrm{X})=\Delta(\mathrm{X})\sum_{L\in\mathcal{F}(\mathbb{L})}\delta_L(\mathcal{L}(\mathrm{X}))\overline{w}^{(L)}[\overline{p}^{(L)}]^{\mathrm{X}} \tag{R.5}$$

where

$$\overline{w}^{(L)}=\frac{(w_1^{(L)})^{\omega}(w_2^{(L)})^{1-\omega}\left[\int (p_1^{(L)}(\boldsymbol{x},\cdot))^{\omega}(p_2^{(L)}(\boldsymbol{x},\cdot))^{1-\omega}d\boldsymbol{x}\right]^L}{\sum_{J\subseteq\mathbb{L}}(w_1^{(J)})^{\omega}(w_2^{(J)})^{1-\omega}\left[\int (p_1^{(J)}(\boldsymbol{x},\cdot))^{\omega}(p_2^{(J)}(\boldsymbol{x},\cdot))^{1-\omega}d\boldsymbol{x}\right]^J} \tag{R.6}$$

$$\overline{p}^{(L)}=\left[(\omega\odot p_1^{(L)})\oplus((1-\omega)\odot p_2^{(L)})\right]^{\mathrm{X}} \tag{R.7}$$

Then, Proposition 35 can be proved by generalizing the case of two Mδ-GLMB densities to N Mδ-GLMB densities through the induction method.

Another proof of Proposition 35 is as follows.

Based on the Mδ-GLMB form of (7.5.19), the numerator of (12.4.4) is calculated first, which is

$$\begin{aligned}
\prod_{i\in\mathcal{I}}\pi_i^{\omega_i}(\mathrm{X}) &= \prod_{i\in\mathcal{I}}\left[\Delta(\mathrm{X})w_i^{(\mathcal{L}(\mathrm{X}))}[p_i^{(\mathcal{L}(\mathrm{X}))}]^{\mathrm{X}}\right]^{\omega_i}\\
&= \Delta(\mathrm{X})\prod_{i\in\mathcal{I}}\left[w_i^{(\mathcal{L}(\mathrm{X}))}\right]^{\omega_i}\prod_{i\in\mathcal{I}}\left[[p_i^{(\mathcal{L}(\mathrm{X}))}]^{\omega_i}\right]^{\mathrm{X}}\\
&= \Delta(\mathrm{X})\prod_{i\in\mathcal{I}}\left[w_i^{(\mathcal{L}(\mathrm{X}))}\right]^{\omega_i}[\eta^{(\mathcal{L}(\mathrm{X}))}]^{\mathcal{L}(\mathrm{X})}\frac{\prod_{i\in\mathcal{I}}[(p_i^{(\mathcal{L}(\mathrm{X}))})^{\omega_i}]^{\mathrm{X}}}{\left[\int\prod_{i\in\mathcal{I}}[p_i^{(\mathcal{L}(\mathrm{X}))}(\boldsymbol{x},\ell)]^{\omega_i}d\boldsymbol{x}\right]^{\mathrm{X}}}\\
&= \Delta(\mathrm{X})\prod_{i\in\mathcal{I}}\left[w_i^{(\mathcal{L}(\mathrm{X}))}\right]^{\omega_i}\left[\eta^{(\mathcal{L}(\mathrm{X}))}\right]^{\mathcal{L}(\mathrm{X})}\left[\overline{p}^{(\mathcal{L}(\mathrm{X}))}\right]^{\mathrm{X}}\\
&= \sum_{J\in\mathcal{F}(\mathbb{L})}\prod_{i\in\mathcal{I}}\left[w_i^{(J)}\right]^{\omega_i}\left[\eta^{(J)}\right]^{J}\Delta(\mathrm{X})\delta_J(\mathcal{L}(\mathrm{X}))\left[\overline{p}^{(J)}\right]^{\mathrm{X}}
\end{aligned}\tag{R.8}$$

where $\overline{p}^{(I)}$ is defined in (12.4.26), and

$$\eta^{(J)}(\ell) = \int\prod_{i\in\mathcal{I}}\left[p_i^{(J)}(\boldsymbol{x},\ell)\right]^{\omega_i}d\boldsymbol{x} \tag{R.9}$$

By integrating (R.1), applying (3.3.20), and noting that $\int\overline{p}^{(J)}(\cdot,\ell)d\boldsymbol{x} = 1$, the denominator in (12.4.4) can be obtained as

$$\begin{aligned}
\int\prod_{i\in\mathcal{I}}\pi_i^{\omega_i}(\mathrm{X})\delta\mathrm{X} &= \int\sum_{J\in\mathcal{F}(\mathbb{L})}\prod_{i\in\mathcal{I}}\left[w_i^{(J)}\right]^{\omega_i}\left[\eta^{(J)}\right]^{J}\Delta(\mathrm{X})\delta_J(\mathcal{L}(\mathrm{X}))\left[\overline{p}^{(J)}\right]^{\mathrm{X}}\delta\mathrm{X}\\
&= \sum_{J\in\mathcal{F}(\mathbb{L})}\prod_{i\in\mathcal{I}}\left[w_i^{(J)}\right]^{\omega_i}\left[\eta^{(J)}\right]^{J}\int\Delta(\mathrm{X})\delta_J(\mathcal{L}(\mathrm{X}))\left[\overline{p}^{(J)}\right]^{\mathrm{X}}\delta\mathrm{X}\\
&= \sum_{J\in\mathcal{F}(\mathbb{L})}\prod_{i\in\mathcal{I}}\left[w_i^{(J)}\right]^{\omega_i}\left[\eta^{(J)}\right]^{J}\sum_{L\in\mathcal{F}(\mathbb{L})}\delta_J(L)\\
&= \sum_{J\in\mathcal{F}(\mathbb{L})}\prod_{i\in\mathcal{I}}\left[w_i^{(J)}\right]^{\omega_i}\left[\eta^{(J)}\right]^{J}
\end{aligned}\tag{R.10}$$

Substituting (R.8) and (R.10) into (12.4.4) leads to

$$\overline{\pi}(X) = \frac{\prod_{i\in\mathcal{I}}\pi_i^{\omega_i}(\mathrm{X})}{\int\prod_{i\in\mathcal{I}}\pi_i^{\omega_i}(\mathrm{X})\delta\mathrm{X}} = \Delta(\mathrm{X})\overline{w}^{(\mathcal{L}(\mathrm{X}))}[\overline{p}^{(\mathcal{L}(\mathrm{X}))}]^{\mathrm{X}} \tag{R.11}$$

where

$$\overline{w}^{(I)} = \frac{\sum_{J \in \mathcal{F}(\mathbb{L})} \prod_{i \in \mathcal{I}} \left[w_i^{(J)} \right]^{\omega_i} \left[\eta^{(J)} \right]^J \delta_J(I)}{\sum_{J \in \mathcal{F}(\mathbb{L})} \prod_{i \in \mathcal{I}} \left[w_i^{(J)} \right]^{\omega_i} \left[\eta^{(J)} \right]^J}$$

$$= \frac{\prod_{i \in \mathcal{I}} \left[w_i^{(I)} \right]^{\omega_i} \left[\eta^{(I)} \right]^I}{\sum_{J \in \mathcal{F}(\mathbb{L})} \prod_{i \in \mathcal{I}} \left[w_i^{(J)} \right]^{\omega_i} \left[\eta^{(J)} \right]^J} \tag{R.12}$$

The above equation is (12.4.25), which is proved. □

Appendix S
Weighted KLA of LMB Densities

Before proving Proposition 36, the following lemma is first introduced.

Lemma 8 *Let* $\pi_1(\mathrm{X}) = \{(\varepsilon_1^{(\ell)}, p_1^{(\ell)})\}_{\ell\in\mathbb{L}}$ *and* $\pi_2(\mathrm{X}) = \{(\varepsilon_2^{(\ell)}, p_2^{(\ell)})\}_{\ell\in\mathbb{L}}$ *be two LMB densities on* $\mathbb{X}\times\mathbb{L}$*, and* $\omega \in (0, 1)$*, then we have*

$$K \triangleq \int \pi_1^{\omega}(\mathrm{X})\pi_2^{1-\omega}(\mathrm{X})\delta\mathrm{X} = \langle \omega \odot \pi_1, (1-\omega)\odot \pi_2\rangle \tag{S.1a}$$

$$= [\tilde{\varepsilon}^{(\cdot)} + \tilde{q}^{(\cdot)}]^{\mathbb{L}} \tag{S.1b}$$

where

$$\tilde{\varepsilon}^{(\ell)} = \int (\varepsilon_1^{(\ell)} p_1^{(\ell)}(\boldsymbol{x}))^{\omega}(\varepsilon_2^{(\ell)} p_2^{(\ell)}(\boldsymbol{x}))^{1-\omega} d\boldsymbol{x} \tag{S.2}$$

$$\tilde{q}^{(\ell)} = (1-\varepsilon_1^{(\ell)})^{\omega}(1-\varepsilon_2^{(\ell)})^{1-\omega} \tag{S.3}$$

Proof According to (3.3.20), Eq. (S.1a) becomes

$$K = [\tilde{q}^{(\cdot)}]^{\mathbb{L}} \sum_{L\subseteq\mathbb{L}} \left(\tilde{\varepsilon}^{(\cdot)}/\tilde{q}^{(\cdot)}\right)^{L} \tag{S.4}$$

Applying the following Binomial theorem [447] to the above equation

$$\sum_{L\subseteq\mathbb{L}} f^{L} = (1+f)^{\mathbb{L}} \tag{S.5}$$

leads to

$$K = \left[\tilde{q}^{(\cdot)}\right]^{\mathbb{L}}\left[1+\tilde{\varepsilon}^{(\cdot)}/\tilde{q}^{(\cdot)}\right]^{\mathbb{L}} = \left[\tilde{\varepsilon}^{(\cdot)}+\tilde{q}^{(\cdot)}\right]^{\mathbb{L}} \tag{S.6}$$

W. Wu et al., *Target Tracking with Random Finite Sets*,
https://doi.org/10.1007/978-981-19-9815-7

Proof of Proposition 36 For the sake of simplicity, only the case of two LMB densities is considered temporarily. At this time, according to (12.4.27) and Lemma 8, we have

$$\begin{aligned}
\overline{\pi}(\mathrm{X}) &= \frac{1}{K}\big[\Delta(\mathrm{X})w_1(\mathcal{L}(\mathrm{X}))p_1^{\mathrm{X}}\big]^{\omega}\big[\Delta(\mathrm{X})w_2(\mathcal{L}(\mathrm{X}))p_2^{\mathrm{X}}\big]^{1-\omega} \\
&= \frac{\Delta(\mathrm{X})}{K}\Big[(1-\varepsilon_1^{(\cdot)})^{\omega}(1-\varepsilon_2^{(\cdot)})^{1-\omega}\Big]^{\mathbb{L}} \times \\
&\left[1_{\mathbb{L}}(\cdot)\left(\frac{\varepsilon_1^{(\cdot)}}{1-\varepsilon_1^{(\cdot)}}\right)^{\omega}\left(\frac{\varepsilon_2^{(\cdot)}}{1-\varepsilon_2^{(\cdot)}}\right)^{1-\omega}\right]^{\mathcal{L}(\mathrm{X})} [p_1^{\omega}p_2^{1-\omega}]^{\mathrm{X}} \\
&= \frac{\Delta(\mathrm{X})}{K}(\tilde{q}^{(\cdot)})^{\mathbb{L}}\left[1_{\mathbb{L}}(\cdot)\left(\frac{\varepsilon_1^{(\cdot)}}{1-\varepsilon_1^{(\cdot)}}\right)^{\omega}\left(\frac{\varepsilon_2^{(\cdot)}}{1-\varepsilon_2^{(\cdot)}}\right)^{1-\omega}\int p_1^{\omega}p_2^{1-\omega}d\boldsymbol{x}\right]^{\mathcal{L}(\mathrm{X})} \\
&\quad \times [(\omega\odot p_1)\oplus((1-\omega)\odot p_2)]^{\mathrm{X}}
\end{aligned} \tag{S.7}$$

Then, using (S.2) and (S.3) yields

$$\begin{aligned}
\overline{w}(L) &= \frac{(\tilde{q}^{(\cdot)})^{\mathbb{L}}(1_{\mathbb{L}}(\cdot)\tilde{\varepsilon}^{(\cdot)}/\tilde{q}^{(\cdot)})^{L}}{(\tilde{q}^{(\cdot)}+\tilde{\varepsilon}^{(\cdot)})^{\mathbb{L}}} = \left[\frac{\tilde{q}^{(\cdot)}}{\tilde{q}^{(\cdot)}+\tilde{\varepsilon}^{(\cdot)}}\right]^{\mathbb{L}}\left[1_{\mathbb{L}}(\cdot)\frac{\tilde{\varepsilon}^{(\cdot)}}{\tilde{q}^{(\cdot)}}\right]^{L} \\
&= [1-\overline{\varepsilon}^{(\cdot)}]^{\mathbb{L}}\left[\frac{1_{\mathbb{L}}(\cdot)\overline{\varepsilon}^{(\cdot)}}{1-\overline{\varepsilon}^{(\cdot)}}\right]^{L}
\end{aligned} \tag{S.8}$$

$$\overline{p}^{(\ell)}(\boldsymbol{x}) = \Big[(\omega\odot p_1^{(\ell)})\oplus((1-\omega)\odot p_2^{(\ell)})\Big]^{\mathrm{X}} \tag{S.9}$$

where

$$\overline{\varepsilon}^{(\ell)} = \tilde{\varepsilon}^{(\ell)}/[\tilde{q}^{(\ell)}+\tilde{\varepsilon}^{(\ell)}] \tag{S.10}$$

Proposition 36 can be proved by generalizing the case of two LMB densities to N LMB densities by induction.

Another proof of Proposition 36 is as follows.

Based on the LMB form of (3.3.35), the numerator in (12.4.4) is calculated first, which is

$$\begin{aligned}
\prod_{i\in\mathcal{I}}\pi_i^{\omega_i}(\mathrm{X}) &= \prod_{i\in\mathcal{I}}\big[\Delta(\mathrm{X})w_i(\mathcal{L}(\mathrm{X}))p_i^{\mathrm{X}}\big]^{\omega_i} \\
&= \Delta(\mathrm{X})\prod_{i\in\mathcal{I}}\Big[(1-\varepsilon_i^{(\cdot)})^{\mathbb{L}-\mathcal{L}(\mathrm{X})}(1_{\mathbb{L}}(\cdot)\varepsilon_i^{(\cdot)})^{\mathcal{L}(\mathrm{X})}\Big]^{\omega_i}\prod_{i\in\mathcal{I}}[p_i^{\omega_i}]^{\mathrm{X}} \\
&= \Delta(\mathrm{X})\left[\prod_{i\in\mathcal{I}}(1-\varepsilon_i^{(\cdot)})^{\omega_i}\right]^{\mathbb{L}-\mathcal{L}(\mathrm{X})}\left[1_{\mathbb{L}}(\cdot)\prod_{i\in\mathcal{I}}(\varepsilon_i^{(\cdot)})^{\omega_i}\right]^{\mathcal{L}(\mathrm{X})}
\end{aligned}$$

$$
\begin{aligned}
&\times \eta^{\mathcal{L}(\mathrm{X})} \frac{\prod_{i\in\mathcal{I}} [p_i^{\omega_i}]^{\mathrm{X}}}{\left[\int \prod_{i\in\mathcal{I}} [p_i(\boldsymbol{x},\ell)]^{\omega_i} d\boldsymbol{x}\right]^{\mathrm{X}}} \\
&= \Delta(\mathrm{X})[\tilde{q}^{(\cdot)}]^{\mathbb{L}-\mathcal{L}(\mathrm{X})} \left[1_{\mathbb{L}}(\cdot) \left[\eta(\cdot) \prod_{i\in\mathcal{I}} (\varepsilon_i^{(\cdot)})^{\omega_i} \right]\right]^{\mathcal{L}(\mathrm{X})} \overline{p}^{\mathrm{X}} \\
&= \Delta(\mathrm{X})[\tilde{q}^{(\cdot)}]^{\mathbb{L}-\mathcal{L}(\mathrm{X})} \left[1_{\mathbb{L}}(\cdot)\tilde{\varepsilon}^{(\cdot)} \right]^{\mathcal{L}(\mathrm{X})} \overline{p}^{\mathrm{X}}
\end{aligned} \tag{S.11}
$$

where $\overline{p}(\boldsymbol{x},\ell) = \overline{p}^{(\ell)}(\boldsymbol{x})$ is defined in (12.4.29), and

$$
\eta(\ell) = \int \prod_{i\in\mathcal{I}} [p_i(\boldsymbol{x},\ell)]^{\omega_i} d\boldsymbol{x} = \int \prod_{i\in\mathcal{I}} \left[p_i^{(\ell)}(\boldsymbol{x}) \right]^{\omega_i} d\boldsymbol{x} \tag{S.12}
$$

$$
\tilde{q}^{(\ell)} = \prod_{i\in\mathcal{I}} (1 - \varepsilon_i^{(\ell)})^{\omega_i} \tag{S.13}
$$

$$
\tilde{\varepsilon}^{(\ell)} = \eta(\ell) \prod_{i\in\mathcal{I}} (\varepsilon_i^{(\ell)})^{\omega_i} \tag{S.14}
$$

By integrating (S.11), applying (3.3.20), and noting that $\int \overline{p}(\cdot,\ell) d\boldsymbol{x} = 1$, the denominator in (12.4.4) is obtained as

$$
\begin{aligned}
\int \prod_{i\in\mathcal{I}} \pi_i^{\omega_i}(\mathrm{X})\delta\mathrm{X} &= \int \Delta(\mathrm{X})[\tilde{q}^{(\cdot)}]^{\mathbb{L}-\mathcal{L}(\mathrm{X})} \left[1_{\mathbb{L}}(\cdot)\tilde{\varepsilon}^{(\cdot)} \right]^{\mathcal{L}(\mathrm{X})} \overline{p}^{\mathrm{X}} \delta\mathrm{X} \\
&= \sum_{L\in\mathcal{F}(\mathbb{L})} [\tilde{q}^{(\cdot)}]^{\mathbb{L}-L} [\tilde{\varepsilon}^{(\cdot)}]^{L} \\
&= \left[\tilde{q}^{(\cdot)} + \tilde{\varepsilon}^{(\cdot)} \right]^{\mathbb{L}}
\end{aligned} \tag{S.15}
$$

where the last step makes use of the following binomial theorem [447]

$$
\sum_{L\subseteq\mathbb{L}} g^{\mathbb{L}-L} f^{L} = [g + f]^{\mathbb{L}} \tag{S.16}
$$

Substituting (S.11) and (S.15) into (12.4.4) leads to

$$
\begin{aligned}
\overline{\pi}(X) &= \frac{\prod_{i\in\mathcal{I}} \pi_i^{\omega_i}(\mathrm{X})}{\int \prod_{i\in\mathcal{I}} \pi_i^{\omega_i}(\mathrm{X})\delta\mathrm{X}} = \frac{\Delta(\mathrm{X})[\tilde{q}^{(\cdot)}]^{\mathbb{L}-\mathcal{L}(\mathrm{X})} \left[1_{\mathbb{L}}(\cdot)\tilde{\varepsilon}^{(\cdot)} \right]^{\mathcal{L}(\mathrm{X})} \overline{p}^{\mathrm{X}}}{[\tilde{q}^{(\cdot)} + \tilde{\varepsilon}^{(\cdot)}]^{\mathbb{L}}} \\
&= \Delta(\mathrm{X}) \frac{[\tilde{q}^{(\cdot)}]^{\mathbb{L}-\mathcal{L}(\mathrm{X})} \left[1_{\mathbb{L}}(\cdot)\tilde{\varepsilon}^{(\cdot)} \right]^{\mathcal{L}(\mathrm{X})}}{[\tilde{q}^{(\cdot)} + \tilde{\varepsilon}^{(\cdot)}]^{\mathbb{L}-\mathcal{L}(\mathrm{X})} [\tilde{q}^{(\cdot)} + \tilde{\varepsilon}^{(\cdot)}]^{\mathcal{L}(\mathrm{X})}} \overline{p}^{\mathrm{X}} \\
&= \Delta(\mathrm{X}) \left(\frac{\tilde{q}^{(\cdot)}}{\tilde{q}^{(\cdot)} + \tilde{\varepsilon}^{(\cdot)}} \right)^{\mathbb{L}-\mathcal{L}(\mathrm{X})} \left(1_{\mathbb{L}}(\cdot) \frac{\tilde{\varepsilon}^{(\cdot)}}{\tilde{q}^{(\cdot)} + \tilde{\varepsilon}^{(\cdot)}} \right)^{\mathcal{L}(\mathrm{X})} \overline{p}^{\mathrm{X}} \\
&= \Delta(\mathrm{X}) \left(1 - \overline{\varepsilon}^{(\cdot)} \right)^{\mathbb{L}-\mathcal{L}(\mathrm{X})} \left(1_{\mathbb{L}}(\cdot)\overline{\varepsilon}^{(\cdot)} \right)^{\mathcal{L}(\mathrm{X})} \overline{p}^{\mathrm{X}}
\end{aligned}
$$

$$= \Delta(\mathrm{X})\overline{w}(\mathcal{L}(\mathrm{X}))\overline{p}^{\mathrm{X}} \tag{S.17}$$

where

$$\overline{w}(\mathcal{L}(\mathrm{X})) = \left(1 - \overline{\varepsilon}^{(\cdot)}\right)^{\mathbb{L}-\mathcal{L}(\mathrm{X})}\left(1_{\mathbb{L}}(\cdot)\overline{\varepsilon}^{(\cdot)}\right)^{\mathcal{L}(\mathrm{X})} \tag{S.18}$$

$$\overline{\varepsilon}^{(\cdot)} = \frac{\tilde{\varepsilon}^{(\cdot)}}{\tilde{q}^{(\cdot)} + \tilde{\varepsilon}^{(\cdot)}} \tag{S.19}$$

Equation (S.19) is equal to (12.4.28). □

References

1. A.F. Garcia-Fernandez, M.R. Morelande, J. Grajal, Bayesian sequential track formation. IEEE Trans. Signal Process. **62**(24), 6366–6379 (2014)
2. R. Mahler, "Statistics 101" for multisensor, multitarget data fusion. IEEE Mag. Aerosp. Electron. Syst. **19**(1), 53–64 (2004)
3. R. Mahler, PHD filters of higher order in target number. IEEE Trans. Aerosp. Electron. Syst. **43**(4), 1523–1543 (2007)
4. B.-T. Vo, Random finite sets in multi-object filtering, Ph.D. dissertation, School of Electrical, Electronic and Computer Engineering, The University of Western Australia, Australia (2008)
5. B.-T. Vo, B. Vo, A. Cantoni, The cardinality balanced multi-target multi-Bernoulli filter and its implementations. IEEE Trans. Signal Process. **57**(2), 409–423 (2009)
6. R. Mahler, Multitarget Bayes filtering via first-order multitarget moments. IEEE Trans. Aerosp. Electron. Syst. **39**(4), 1152–1178 (2003)
7. R. Mahler., Multitarget moments and their application to multitarget tracking, in *Proceedings of Workshop on Estimation, Tracking, and Fusion: A Tribute to Y. Bar-Shalom, Naval Postgraduate School*, Monterey CA (2001), pp. 134–166
8. A. Farina, F. Studer, *Radar Data Processing*, vols. I and II (Wiley, New York, 1985)
9. S.S. Blackman, *Multiple Target Tracking with Radar Application* (Artech House, Norwood, MA, 1986)
10. S.S. Blackman, R. Popoli, *Design and Analysis of Modern Tracking Systems* (Artech House, Norwood, MA, 1999)
11. R.L. Popp, T. Kirubarajan, K.R. Pattipati, *Multitarget/Multisensor Tracking: Applications and Advances III* (Artech House, Norwood, 2000)
12. Y. Bar-Shalom, P.K. Willett, X. Tian, *Tracking and Data Fusion: A Handbook of Algorithms* (Academic Press, Orlando. FL, 2011)
13. W. Koch, *Tracking and Sensor Data Fusion* (Springer, Berlin Heidelberg, 2014)
14. R. Mahler, *Statistical Multisource-Multitarget Information Fusion* (Artech House, Norwood, 2007)
15. R. Mahler, *Advances in Statistical Multisource-Multitarget information Fusion* (Artech House, Norwood, 2014)
16. R.E. Kalman, A new approach to linear filtering and prediction problems. Trans. ASME J. Basic Eng. **82**(1), 35–45 (1960)
17. P. Zarchan, H. Musoff. *Fundamentals of Kalman Filtering: A Practical Approach*, 3rd edn (Virginia, American Institute of Aeronautics and Astronautics, 2009)
18. G.Y. Kulikov, M.V. Kulikova, Accurate numerical implementation of the continuous-discrete extended Kalman filter. IEEE Trans. Autom. Control **59**(1), 273–279 (2014)
19. V.J. Aidala, Kalman filter behavior in bearings-only tracking applications. IEEE Trans. Aerosp. Electron. Syst. **15**(1), 29–39 (1979)

W. Wu et al., *Target Tracking with Random Finite Sets*,
https://doi.org/10.1007/978-981-19-9815-7

20. D. Lerro, Y. Bar-Shalom, Tracking with debiased consistent converted measurements versus EKF. IEEE Trans. Aerosp. Electron. Syst. **29**(3), 1015–1022 (1993)
21. M.S. Schlosser, K. Kroschel, Limits in tracking with extended Kalman filters. IEEE Trans. Aerosp. Electron. Syst. **40**(4), 1351–1359 (2004)
22. S.J. Julier, J.K. Uhlmann, Unscented filtering and nonlinear estimation. Proc. of the IEEE **92**(3), 401–422 (2004)
23. S. Sarkka, On unscented Kalman filtering for state estimation of continuous-time nonlinear systems. IEEE Trans. Autom. Control **52**(9), 1631–1641 (2007)
24. I. Arasaratnam, S. Haykin, Cubature Kalman filters. IEEE Trans. Autom. Control **54**(6), 1254–1269 (2009)
25. I. Arasaratnam, S. Haykin, T.R. Hurd, Cubature Kalman filtering for continuous-discrete systems: theory and simulations. IEEE Trans. Signal Process. **58**(10), 4977–4993 (2010)
26. M.S. Arulampalam, S. Maskell, N. Gordon et al., A tutorial on particle filters for online nonlinear/non-Gaussian bayesian tracking. IEEE Trans. Signal Process. **50**(2), 174–188 (2002)
27. O. Cappe, S.J. Godsill, E. Moulines, An overview of existing methods and recent advances in sequential Monte Carlo. Proc. IEEE **95**(5), 899–924 (2007)
28. S. Julier, J. Uhlmann, H.F. Durrant-Whyte, A new method for the nonlinear transformation of means and covariances in filters and estimators. IEEE Trans. Autom. Control **45**(3), 477–482 (2000)
29. M. Orton, W. Fitzgerald, A Bayesian approach to tracking multiple targets using sensor arrays and particle filters. IEEE Trans. Signal Process. **50**(2), 216–223 (2002)
30. C.B. Chang, R.H. Whiting, M. Athans, On the state and parameter estimation for maneuvering reentry vehicles. IEEE Trans. Autom. Control **22**(1), 99–105 (1977)
31. Y. Bar-Shalom, K. Birmiwal, Variable dimension filter for maneuvering target tracking. IEEE Trans. Aerosp. Electron. Syst. **18**(5), 611–619 (1982)
32. Y.T. Chan, J.B. Plant, J. Bottomley, A Kalman tracker with a simple input estimator. IEEE Trans. Aerosp. Electron. Syst. **18**(2), 235–241 (1982)
33. S. Robert, Estimating optimal tracking filter performance for manned maneuvering targets. IEEE Trans. Aerosp. Electron. Syst. **6**(4), 473–483 (1970)
34. S. Ghosh, S. Mukhopadhyay, Tracking reentry ballistic targets using acceleration and jerk models. IEEE Trans. Aerosp. Electron. Syst. **47**(1), 666–683 (2011)
35. X.R. Li, Y. Bar-shalom, Performance prediction of the interacting multiple model algorithm. IEEE Trans. Aerosp. Electron. Syst. **31**(4), 755–771 (1993)
36. A. Houles, Y. Bar-shalom, Multisensor tracking of a maneuvering target in clutter. IEEE Trans. Aerosp. Electron. Syst. **25**(2), 176–189 (1989)
37. H. Benoudnine, M. Keche, A. Ouamri et al., New efficient schemes for adaptive selection of the update time in the IMMJPDAF. IEEE Trans. Aerosp. Electron. Syst. **48**(1), 197–214 (2012)
38. S.S. Blackman, R.J. Dempster, M.T. Busch et al., IMM/MHT solution to radar benchmark tracking problem. IEEE Trans. Aerosp. Electron. Syst. **35**(2), 730–738 (1999)
39. M.A. Olson, Simulation of a multitarget, multisensor, track-splitting tracker for maritime surveillance. Naval Postgraduate School (1999)
40. Y. Bar-Shalom, F. Daum, J. Huang, The probabilistic data association filter. IEEE Control. Syst. **29**(6), 82–100 (2009)
41. X.R. Li, Y. Bar-Shalom, Tracking in clutter with nearest neighbor filters: analysis and performance. IEEE Trans. Aerosp. Electron. Syst. **32**(3), 995–1010 (1996)
42. S. Deb, M. Yeddanapudi, K. Pattipati et al., A generalized S-D assignment algorithm for multisensor-multitarget state estimation. IEEE Trans. Aerosp. Electron. Syst. **33**(2), 523–538 (1997)
43. D. Musicki, R. Evans, A. Stankovic, Integrated probabilistic data association [J]. IEEE Trans. Autom. Control **39**(6), 1237–1241 (1994)
44. T.E. Fortmann, Y. Bar-shalom, M. Scheffe, Multi-target tracking using joint probabilistic data association, in *19th IEEE Conference on Decision and Control including the Symposium on Adaptive Processes*, Albuquerque, USA (1980), pp. 807–812

45. D. Musicki, R. Evans, Joint integrated probabilistic data association: JIPDA. IEEE Trans. Aerosp. Electron. Syst. **40**(3), 1093–1099 (2004)
46. D. Musicki, B. Lascala, R.J. Evans, Integrated track splitting filter—efficient multi-scan single target tracking in clutter. IEEE Trans. Aerosp. Electron. Syst. **43**(4), 1409–1425 (2007)
47. D. Musicki, R.J. Evans, Multiscan multitarget tracking in clutter with integrated track splitting filter. IEEE Trans. Aerosp. Electron. Syst. **45**(4), 1432–1447 (2009)
48. J.L. Williams, R. Lau, Approximate evaluation of marginal association probabilities with belief propagation. IEEE Trans. Aerosp. Electron. Syst. **50**(4), 2942–2959 (2014)
49. F. Meyer, P. Braca, P. Willett et al., A scalable algorithm for tracking an unknown number of targets using multiple sensors. IEEE Trans. Signal Process. **65**(13), 3478–3493 (2017)
50. F.R. Kschischang, B.J. Frey, H.-A. Loeliger, Factor graphs and the sum-product algorithm. IEEE Trans. Inf. Theory **47**(2), 498–519 (2001)
51. H.-A. Loeliger, An introduction to factor graphs. IEEE Signal Process. Mag. **21**(1), 28–41 (2004)
52. S. Oh, S. Russell, S. Sastry, Markov chain monte carlo data association for multi-target tracking. IEEE Trans. Autom. Control **54**(3), 481–497 (2009)
53. C. Andrieu, A. Doucet, R. Holenstein, Particle markov chain monte carlo methods. J. Roy. Stat. Soc. **72**(3), 269–342 (2010)
54. D.B. Reid, An algorithm for tracking multiple targets. IEEE Trans. Autom. Control **24**(6), 843–854 (1979)
55. R.L. Popp, K.R. Pattipati, Y. Bar-Shalom, M-best S-D assignment algorithm with application to multitarget tracking. IEEE Trans. Aerosp. Electron. Syst. **37**(1), 22–39 (2001)
56. R.D. Palkki, A.D. Lanterman, W.D. Blair, Addressing track hypothesis coalescence in sequential k-best multiple hypothesis tracking. IEEE Trans. Aerosp. Electron. Syst. **47**(3), 1551–1563 (2011)
57. R. Mahler, "Statistics 102" for multisource-multitarget detection and tracking. IEEE J. Sel. Top. Signal Process. **7**(3), 376–389 (2013)
58. S. Singh, B.-N. Vo, A. Baddeley et al., Filters for spatial point processes. SIAM J. Control Optim. **48**(4), 2275–2295 (2009)
59. C. Ouyang, H. Ji, C. Li, Improved multi-target multi-bernoulli filter. Proc. IET Radar Sonar Nav. **6**(6), 458–464 (2012)
60. B.-T. Vo, B.-N. Vo, R. Hoseinnezhad et al., Robust multi-bernoulli filtering. IEEE J. Sel. Top. Signal Process. **7**(3), 399–409 (2013)
61. B.-T. Vo, B.-N. Vo, R. Hoseinnezhad, et al., Multi-Bernoulli filtering with unknown clutter intensity and sensor field-of-view, in *Proceeding of IEEE Conference on Information Science System* Baltimore, MD, USA, (IEEE, 2011), pp. 1–6
62. B.-N. Vo, S. Singh, A. Doucet, Sequential Monte Carlo methods for multi-target filtering with random finite sets. IEEE Trans. Aerosp. Electron. Syst. **41**(4), 1224–1245 (2005)
63. B.-N. Vo, W.-K. Ma, The Gaussian mixture probability hypothesis density filter. IEEE Trans. Signal Process. **54**(11), 4091–4104 (2006)
64. B.-T. Vo, B.-N. Vo, A. Cantoni, Analytic implementations of the cardinalized probability hypothesis density filter. IEEE Trans. Signal Process. **55**(7), 3553–3567 (2007)
65. D. Clark, B.-N. Vo, Convergence analysis of the Gaussian mixture PHD filter. IEEE Trans. Signal Process. **55**(4), 1204–1212 (2007)
66. D. Clark, J. Bell, Convergence results for the particle PHD filter. IEEE Trans. Signal Process. **54**(7), 2652–2661 (2006)
67. A.M. Johansen, S. Singh, A. Doucet et al., Convergence of the SMC implementation of the PHD filter. Methodol. Comput. Appl. Probab. **8**(2), 265–291 (2006)
68. F. Lian, C. Li, C. Han et al., Convergence analysis for the SMC-MeMBer and SMC-CBMeMBer filters. J. Appl. Math. **2012**(3), 701–708 (2012)
69. B.-T. Vo, B.-N. Vo, Labeled random finite sets and multi-object conjugate priors. IEEE Trans. Signal Process. **61**(13), 3460–3475 (2013)
70. B.-N. Vo, B.-T. Vo, D. Phung, Labeled random finite sets and the Bayes multi-target tracking filter. IEEE Trans. Signal Process. **62**(24), 6554–6567 (2014)

71. S. Reuter, B.-T. Vo, B.-N. Vo et al., The labeled multi-Bernoulli filter. IEEE Trans. Signal Process. **62**(12), 3246–3260 (2014)
72. C. Fantacci, B.-T. Vo, F. Papi, et al., The marginalized δ-GLMB filter (2015). Available online at https://arxiv.org/abs/1501.00926
73. T. Vu, B.-N. Vo, R.J. Evans, A Particle marginal metropolis-hastings multitarget tracker. IEEE Trans. Signal Process. **62**(15), 3953–3964 (2014)
74. R. Mahler, A brief survey of advances in random-set fusion, in *International Conference on Control, Automation and Informatica Sciences,* Changshu, China (2015), pp. 62–67
75. B. Ristic, M. Beard, C. Fantacci, An overview of particle methods for random finite set models. Inf. Fusion **31**(C), 110–126 (2016)
76. M. Beard, B.T. Vo, B.-N. Vo, A solution for large-scale multi-object tracking (2018), [Online]. Available: https://arxiv.org/abs/1804.06622
77. D. Clark, I.T. Ruiz, Y. Petillot et al., Particle PHD filter multiple target tracking in sonar image. IEEE Trans. Aerosp. Electron. Syst. **43**(1), 409–416 (2007)
78. R. Georgescu, P. Willett, The GM-CPHD tracker applied to real and realistic multistatic sonar data sets. IEEE J. Oceanic Eng. **37**(2), 220–235 (2012)
79. W.K. Ma, B.-N. Vo, S. Singh et al., Tracking an unknown time-varying number of speakers using TDOA measurements: a random finite set approach. IEEE Trans. Signal Process. **54**(9), 3291–3304 (2006)
80. R. Hoseinnezhad, B.-N. Vo, B.-T. Vo, Visual tracking in background subtracted image sequences via multi-bernoulli filtering. IEEE Trans. Signal Process. **61**(2), 392–397 (2013)
81. R. Hoseinnezhad, B.-N. Vo, D. Suter et al., Visual tracking of numerous targets via multi-Bernoulli filtering of image data. Pattern Recogn. **45**(10), 3625–3635 (2012)
82. J. Mullane, B.-N. Vo, M. Adams et al., A random finite set approach to Bayesian SLAM. IEEE Trans. Robotics **27**(2), 268–282 (2011)
83. H. Deusch, S. Reuter, K. Dietmayer, The labeled multi-bernoulli SLAM filter. IEEE Signal Process. Lett. **22**(10), 1561–1565 (2015)
84. G. Battistelli, L. Chisci, S. Morrocchi, et al., Traffic intensity estimation via PHD filtering, in *5th European Radar Conference on Amsterdam*, The Netherlands (2008), pp. 340–343
85. M. Ulmke, O. Erdinc, P. Willett, GMTI Tracking via the Gaussian mixture cardinalized probability hypothesis density filter. IEEE Trans. Aerosp. Electron. Syst. **46**(4), 1821–1833 (2010)
86. F. Papi, D.Y. Kim, A particle multi-target tracker for superpositional measurements using labeled random finite sets. IEEE Trans. Signal Process. **63**(6), 4348–4358 (2015)
87. K. Pikora, F. Ehlers, Analysis of the FKIE passive radar data set with GMPHD and GMCPHD, in *16th International Conference on Information Fusion,* Istanbul, Turkey (ISIF, 2013), pp. 272–279
88. S.J. Wong, B.-T. Vo, Square root Gaussian mixture PHD filter for multi-target bearings only tracking, in *Proceeding of 3rd Conference Information Technology Applications in Biomedicine,* Arlington, VA (IEEE, 2011), pp. 520–525
89. J.D. Glass, A.D. Lanterman, MIMO radar target tracking using the probability hypothesis density filter, in *Proceeding of IEEE Conference on Decision and Control,* Shanghai, China, (IEEE, 2012), pp. 1–8
90. X. Zhang, Adaptive control and reconfiguration of mobile wireless sensor networks for dynamic multi-target tracking. IEEE Trans. Autom. Control **56**(10), 2429–2444 (2011)
91. M. Uney, D. Clark, S. Julier, Distributed fusion of PHD filters via exponential mixture densities. IEEE J. Sel. Top. Signal Process. **7**(3), 521–531 (2013)
92. G. Battistelli, L. Chisci, C. Fantacci et al., Consensus CPHD filter for distributed multitarget tracking. IEEE J. Sel. Top. Signal Process. **7**(3), 508–520 (2013)
93. R. Mahler, PHD filters for nonstandard targets, II: Unresolved targets, in *12th International Conference on Information Fusion* (IEEE, 2009), pp. 922–929
94. M. Beard, B.-T. Vo, B.-N. Vo, Bayesian multi-target tracking with merged measurements using labelled random finite sets. IEEE Trans. Signal Process. **63**(6), 1433–1447 (2015)

95. K. Granstrom, C. Lundquist, U. Orguner, Extended target tracking using a gaussian-mixture PHD filter. IEEE Trans. Aerosp. Electron. Syst. **48**(4), 3268–3286 (2012)
96. A. Swain, D. Clark, The single-group PHD filter: an analytic solution, in *Proceeding of International Conference Information Fusion*, Chicago, IL, USA (IEEE, 2011), pp. 1–8
97. D. Clark, S. Godsill, Group target tracking with the Gaussian mixture probability hypothesis density filter, in *Proceeding of IEEE Conference on Decision and Control* Shanghai, China (IEEE, 2007), pp. 149–154
98. N. Whiteley, S. Singh, S. Godsill, auxiliary particle implementation of probability hypothesis density filter. IEEE Trans. Aerosp. Electron. Syst. **46**(3), 1437–1454 (2010)
99. J.H. Yoon, D.Y. Kim, K.-J. Yoon, Gaussian mixture importance sampling function for unscented SMC-PHD filter. Signal Process. **93**(9), 2664–2670 (2013)
100. M. Vihola, Rao-blackwellised particle filtering in random set multitarget tracking. IEEE Trans. Aerosp. Electron. Syst. **43**(2), 689–705 (2007)
101. R. Sithiravel, X. Chen, R. Tharmarasa et al., The spline probability hypothesis density filter. IEEE Trans. Signal Process. **61**(24), 6188–6203 (2013)
102. W. Li, Y. Jia, Nonlinear gaussian mixture PHD filter with an H∞ criterion. IEEE Trans. Aerosp. Electron. Syst. **52**(4), 2004–2016 (2016)
103. M. Yazdian-Dehkordi, Z. Azimifar, M.A. Masnadi-Shirazi, Competitive Gaussian mixture probability hypothesis density filter for multiple target tracking in the presence of ambiguity and occlusion. IET Radar Sonar Navig. **6**(4), 251–262 (2012)
104. R. Mahler, Detecting, tracking, and classifying group targets: a unified approach, in *Proceeding of SPIE*, Orlando (2001), pp. 217–228
105. W. Yang, Y. Fu, J. Long et al., Joint detection, tracking, and classification of multiple targets in clutter using the PHD filter. IEEE Trans. Aerosp. Electron. Syst. **48**(4), 3594–3609 (2012)
106. K. Panta, B.-N. Vo, S. Singh, Novel data association schemes for the probability hypothesis density filter. IEEE Trans. Aerosp. Electron. Syst. **43**(2), 556–570 (2007)
107. D.E. Clark, J. Bell, Multi-target state estimation and track continuity for the particle PHD filter. IEEE Trans. Aerosp. Electron. Syst. **43**(4), 1441–1453 (2007)
108. W. Liu, C. Han, F. Lian et al., Multitarget state extraction for the PHD filter using MCMC approach. IEEE Trans. Aerosp. Electron. Syst. **46**(2), 864–883 (2010)
109. M. Tobias, A. Lanterman, Techniques for birth-particle placement in the probability hypothesis density particle filter applied to passive radar. IET Radar Sonar Navig. **2**(5), 351–365 (2008)
110. X. Tang, P. Wei, Multi-target state extraction for the particle probability hypothesis density filter. IET Radar, Sonar Navig. **5**(8), 877–883 (2011)
111. Y. Petetin, F. Desbouvries, A mixed GM/SMC implementation of the probability hypothesis density filter, in *18th Mediterranean Conference on Control & Automation,* Marrakech, Morocco, (2012), pp. 1423–1428
112. D. Clark, B.-T. Vo, B.-N. Vo, Gaussian particle implementations of probability hypothesis density filters, in *Proceeding of 3rd Conference Information Technology Applications in Biomedicine* (IEEE, 2011), pp. 1–11
113. P. Kusha, B.-N. Vo, S, et al., Probability hypothesis density filter versus multiple hypothesis tracking, in *Proceedings of the SPIE* (2004), pp. 284–295
114. K. Panta, D.E. Clark, B.-N. Vo, Data association and track management for the gaussian mixture probability hypothesis density filter. IEEE Trans. Aerosp. Electron. Syst. **45**(3), 1003–1016 (2009)
115. L. Lin, Y. Bar-Shalom, T. Kirubarajan, Track labeling and PHD filter for multitarget tracking. IEEE Trans. Aerosp. Electron. Syst. **42**(3), 778–795 (2006)
116. F. Papi, G. Battistelli, L. Chisci, et al., Multitarget tracking via joint PHD filtering and multi-scan association, in *12th International Conference on Information Fusion,* Seattle, WA, USA (ISIF, 2009), pp. 1163–1170
117. O. Erdinc, P. Willett, Y. Bar-Shalom, Probability hypothesis density filter for multitarget multisensor tracking, in *8th International Conference on Information Fusion* (IEEE, 2005), pp. 146–153

118. Á.F. García-Fernández, B.-N. Vo, Derivation of the PHD and CPHD filters based on direct Kullback-Leibler divergence minimisation. IEEE Trans. Signal Process. **63**(21), 5812–5820 (2015)
119. D. Franken, M. Schmidt, M. Ulmke, "Spooky Action at a Distance" in the cardinalized probability hypothesis density filter. IEEE Trans. Aerosp. Electron. Syst. **45**(4), 1657–1664 (2009)
120. C. Ouyang, H.-B. Ji, Y. Tian, Improved gaussian mixture CPHD tracker for multitarget tracking. IEEE Trans. Signal Process. **49**(2), 1177–1191 (2013)
121. B.A. Jones, S. Gehly, P. Axelrad, Measurement-based birth model for a space object cardinalized probability hypothesis density filter, in *Proceeding of AIAA/AAS Astrodynamics Specialist Conference* (2014)
122. M. Lundgren, L. Svensson, L. Hammarstrand, A CPHD filter for tracking with spawning models. IEEE J. Sel. Topics Signal Process. **7**(3), 496–507 (2013)
123. D.S. Bryant, E.D. Delande, S. Gehly et al., The CPHD filter with target spawning. IEEE Trans. Signal Process. **65**(5), 1324–1338 (2017)
124. C.A. Charalambides, *Enumerative Combinatorics* (CRC Press, BocaRaton, FL, USA, 2002)
125. P. Jing, J. Zou, Y. Duan et al., Generalized CPHD filter modeling spawning targets. Signal Process. **128**, 48–56 (2016)
126. S. Challa, B.-N. Vo, X. Wang, Bayesian approaches to track existence-IPDA and random sets, in *Proceeding of 5th IEEE International Conference Information Fusion* (2002), pp. 1228–1235
127. E. Baser, M. McDonald, T. Kirubarajan et al., A joint multitarget estimator for the joint target detection and tracking filter. IEEE Trans. Signal Process. **63**(15), 3857–3871 (2015)
128. E. Baser, T. Kirubarajan, M. Efe et al., A novel joint multitarget estimator for multi-bernoulli models. IEEE Trans. Signal Process. **64**(19), 5038–5051 (2016)
129. B.-T. Vo, C.-M. See, N. Ma et al., Multi-sensor joint detection and tracking with the bernoulli filter. IEEE Trans. Aerosp. Electron. Syst. **48**(2), 1385–1402 (2012)
130. B. Ristic, B.-T. Vo, B.-N. Vo et al., A tutorial on bernoulli filters: theory, implementation and applications. IEEE Trans. Signal Process. **61**(13), 3406–3430 (2013)
131. B.-T. Vo, D. Clark, B.-N. Vo et al., Bernoulli forward-backward smoothing for joint target detection and tracking. IEEE Trans. Signal Process. **59**(9), 4473–4477 (2011)
132. E. Baser, T. Kirubarajan, M. Efe et al., Improved MeMBer filter with modeling of spurious targets. IET Radar Sonar Navig. **10**(2), 285–298 (2015)
133. J.L. Williams, An efficient, variational approximation of the best fitting multi-bernoulli filter. IEEE Trans. Signal Process. **63**(1), 258–273 (2015)
134. L. Svensson, D. Svensson, M. Guerriero et al., Set JPDA filter for multitarget tracking. IEEE Trans. Signal Process. **59**(10), 4677–4691 (2011)
135. S. Reuter, B. Wilking, J. Wiest et al., Real-time multi-object tracking using random finite sets. IEEE Trans. Aerosp. Electron. Syst. **49**(4), 2666–2678 (2013)
136. H. Zhang, Z. Jing, S. Hu, Gaussian mixture CPHD filter with gating technique. Signal Process. **89**(8), 1521–1530 (2009)
137. D. Macagnano, G.F. de Abreu, Adaptive gating for multitarget tracking with gaussian mixture filters. IEEE Trans. Signal Process. **60**(3), 1533–1538 (2012)
138. R. Mahler, Linear-complexity CPHD filters, in *Proceeding of 13th International Conference Information Fusion* Edinburgh, UK (IEEE, 2010), pp. 1–8
139. D.S. Bryant, B.-T. Vo, B.-N. Vo, et al., A generalized labeled multi-bernoulli filter with object spawning (2017). Available: https://arxiv.org/abs/1705.01614
140. B.T. Vo, B.N. Vo, A multi-scan labeled random finite set model for multi-object state estimation (2018). Available: https://arxiv.org/abs/1805.10038
141. M.H. Sepanj, Z. Azimifar, N-scan δ-generalized labeled multi-bernoulli-based approach for multi-target tracking, in *Artificial Intelligence and Signal Processing Conference (AISP)* (IEEE, Shiraz, Iran, 2017)
142. R. Mahler, A generalized labeled multi-bernoulli filter for correlated multitarget systems, in *Proceedings of SPIE 10646, Signal Processing, Sensor/Information Fusion, and Target Recognition XXVII* (2018)

143. R. Mahler, Integral-Transform derivations of exact closed-form multitarget trackers, in *19th International Conference on Information Fusion* (IEEE, Heidelberg, Germany, 2016) pp. 950–957
144. D.F. Crouse, On implementing 2D rectangular assignment algorithms. IEEE Trans. Aerosp. Electron. Syst. **52**(4), 1679–1696 (2016)
145. H.G. Hoang, B.-T. Vo, B.-N. Vo, A generalized labeled multi-bernoulli filter implementation using gibbs sampling (2015), [Online]. Available: http://arxiv.org/abs/1506.00821
146. B.-N. Vo, B.-T. Vo, H.G. Hoang, An efficient implementation of the generalized labeled multi-bernoulli filter, IEEE Trans. Signal Process. **65**(8), 1975–1987 (2017)
147. D.Y. Kim, B.-N. Vo, B.-T. Vo, Multi-object particle filter revisited, in *2016 International Conference on Control, Automation and Information Sciences (ICCAIS)* Ansan, Korea, (2016), pp. 42–47
148. M. Beard, S. Reuter, K. Granström et al., Multiple extended target tracking with labeled random finite sets. IEEE Trans. Signal Process. **64**(7), 1638–1653 (2016)
149. C. Fantacci, F. Papi, Scalable multisensor multitarget tracking using the marginalized δ-GLMB density. IEEE Signal Process. Lett. **23**(6), 863–867 (2016)
150. A.K. Gostar, R. Hoseinnezhad, A. Bab-Hadiashar, Sensor control for multi-object tracking using labeled multi-bernoulli filter, in *17th International Conference on Information Fusion* (IEEE, Salamanca, Spain, 2014), pp. 1–8
151. S. Li, W. Yi, R. Hoseinnezhad et al., Multi-object tracking for generic observation model using labeled random finite sets. IEEE Trans. Signal Process. **66**(2), 368–383 (2018)
152. F. Papi, B.-N. Vo, B.-T. Vo et al., Generalized labeled multi-bernoulli approximation of multi-object densities. IEEE Trans. Signal Process. **63**(20), 5487–5497 (2015)
153. S. Reuter, *Multi-Object Tracking Using Random Finite Sets* (Ulm, Ulm University, Diss. Zugl, 2014)
154. Ulm University. Project homepage autonomous driving (2014), [Online]. Available: https://www.uni-ulm.de/in/automatisiertes-fahren
155. P. Wang, L. Ma, K. Xue, Efficient approximation of the labeled multi-bernoulli filter for online multitarget tracking. Math. Probl. Eng. **2017**(3), 1–9 (2017)
156. A. Danzer, S. Reuter, K. Dietmayer, The adaptive labeled multi-bernoulli filter, in *19th International Conference on Information Fusion* (IEEE, Heidelberg, Germany, 2016), pp. 1531–1538
157. S. Kullback, R. Leibler, On information and sufficiency. Ann. Math. Stat. **22**(1), 79–86 (1951)
158. C.E. Shannon, A mathematical theory of communication. Bell Syst. Tech. J. **27**(3), 379–423 (1948)
159. W. Yi, S. Li, Enhanced approximation of labeled multi-object density based on correlation analysis, in *19th International Conference on Information Fusion* (IEEE, Heidelberg, Germany, 2016)
160. Z. Lu, W. Hu, T. Kirubarajan, Labeled random finite sets with moment approximation. IEEE Trans. Signal Process. **65**(13): 3384–3398 (2017)
161. Y. Boers, E. Sviestins, H. Driessen, Mixed labelling in multitarget particle filtering. IEEE Trans. Aerosp. Electron. Syst. **46**(2), 792–802 (2010)
162. M. Beard, B.-T. Vo, B.-N. Vo et al., A partially uniform target birth model for Gaussian mixture PHD/CPHD filtering. IEEE Trans. Aerosp. Electron. Syst. **49**(4), 2835–2844 (2013)
163. B. Ristic, D. Clark, B.-N. Vo et al., Adaptive target birth intensity for PHD and CPHD filters. IEEE Trans. Aerosp. Electron. Syst. **48**(2), 1656–1668 (2012)
164. B.A. Jones, cphd filter birth modeling using the probabilistic admissible region. IEEE Trans. Aerosp. Electron. Syst. **54**(3), 1456–1469 (2018)
165. I.E. Maggio, M. Taj, A. Cavallaro, Efficient multi-target visual tracking using random finite sets. IEEE Trans. Circuits Syst. Video Technol. **18**(8), 1016–1027 (2008)
166. J. Houssineau, D. Laneuville, PHD filter with diffuse spatial prior on the birth process with applications to GM-PHD filter, in *Proceedings of 13th International Conference on Information Fusion*, Edinburgh (2010), pp. 21–28

167. B. Cho, S. Park, E. Kim, A Newborn track detection and state estimation algorithm using bernoulli random finite sets. IEEE Trans. Signal Process. **64**(10), 2660–2674 (2016)
168. S. Lin, B.T. Vo, S.E. Nordholm, Measurement driven birth model for the generalized labeled multi-bernoulli filter, in *International Conference on Control, Automation and Information Sciences (ICCAIS)*, Ansan, Korea (2016), pp. 94–99
169. B. Ristic, D. Clark, B.N. Vo, Improved SMC implementation of the PHD filter, in *Proceedings of 13th International Conference on Information Fusion*, Edinburgh (2010), pp. 1–8
170. Y. Zhu, S. Zhou, H. Zou et al., Probability hypothesis density filter with adaptive estimation of target birth intensity. IET Radar Sonar Navig. **10**(5), 901–911 (2016)
171. F. Lian, C. Han, W. Liu, Estimating unknown clutter intensity for PHD filter. IEEE Trans. Aerosp. Electron. Syst. **46**(4), 2066–2078 (2010)
172. X. Chen, R. Tharmarasa, M. Pelletier et al., Integrated clutter estimation and target tracking using poison point processes. IEEE Trans. Aerosp. Electron. Syst. **48**(2), 1210–1235 (2012)
173. R. Mahler, B.-T. Vo, B.-N. Vo, CPHD filtering with unknown clutter rate and detection profile. IEEE Trans. Signal Process. **59**(8), 3497–3513 (2011)
174. M. Beard, B.-T. Vo, B.-N. Vo, Multitarget filtering with unknown clutter density using a bootstrap GMCPHD filter. IEEE Signal Process. Lett. **20**(4), 323–326 (2013)
175. C. Yuan, J. Wang, P. Lei, et al., Multi-target tracking based on multi-bernoulli filter with amplitude for unknown clutter rate. Sensors **15**(12), 30385–30402 (2015)
176. C.B. Chang, M. Athans, State estimation for dicrete system with switching parameters. IEEE Trans. Aerosp. Electron. Syst. **14**(3), 418–425 (1978)
177. A. Pasha, B.-N. Vo, H.D. Tuan, et al., Closed-form PHD filtering for linear jump Markov models, in *Proceedings of 9th International Conference on Information Fusion*, Florence, Italy (IEEE, 2006), pp. 1–8
178. S.A. Pasha, B.-N. Vo, H.D. Tuan et al., A Gaussian mixture PHD filter for jump Markov system models. IEEE Trans. Aerosp. Electron. Syst. **45**(3), 919–936 (2009)
179. B.-N. Vo, W.-K. Ma, Joint detection and tracking of multiple maneuvering targets in clutter using random finite sets, in *Proceedings of the 8th International Conference on Control, Automation, Robotics and Vision* Kunming, China (IEEE, 2004), pp. 1485–1490
180. P. Dong, Z. Jing, M. Li, et al., The variable structure multiple model GM-phd filter based on likely-model set algorithm, in *19th International Conference on Information Fusion* (IEEE, Heidelberg, Germany, 2016)
181. T.M. Wood, Interacting methods for manoeuvre handling in the GM-PHD filter. IEEE Trans. Aerosp. Electron. Syst. **47**(4), 3021–3025 (2011)
182. W. Li, Y. Jia, Gaussian mixture PHD filter for jump Markov models based on best-fitting Gaussian approximation. Signal Process. **91**(4), 1036–1042 (2011)
183. R. Georgescu, P. Willett, The multiple model CPHD tracker. IEEE Trans. Signal Process. **60**(4), 1741–1751 (2012)
184. S.H. Rezatofighi, S. Gould, B.T. Vo, et al., Multi-target tracking with time-varying clutter rate and detection profile: application to time-lapse cell microscopy sequences. IEEE Trans. Med. Imaging **34**(6), 1336–1348 (2015)
185. B. Vo, A. Pasha, H.D. Tuan, A Gaussian mixture PHD filter for nonlinear jump Markov models, in *Proceedings of the 45th IEEE Conference on Decision and Contol* (IEEE, San Diego, CA, USA, 2006), pp. 3162–3167
186. S.A. Pasha, H.D. Tuan, P. Apkarian, Nonlinear jump Markov models in multi-target tracking, in *Proceeding 48th IEEE Conference on Decision and Control*. Shanghai, China, (IEEE, 2009), pp. 5478–5483
187. R. Sithiravel, M. Mcdonald, B. Balaji et al., Multiple model spline probability hypothesis density filter. IEEE Trans. Aerosp. Electron. Syst. **52**(3), 1210–1226 (2016)
188. K. Punithakumar, T. Kirubarajan, A. Sinha, Multiple-model probability hypothesis density filter for tracking maneuvering targets. IEEE Trans. Aerosp. Electron. Syst. **44**(1), 87–98 (2008)
189. K. Punithakumar, T. Kirubarajan, A. Sinha, A multiple model probability hypothesis density filter for tracking maneuvering targets, in *Signal and Data Processing of Small Targets, Proceeding of SPIE*, (2004), pp. 113–121

190. R. Mahler, On multitarget jump-markov filters, in *15th International Conference on Information Fusion* (IEEE, Singapore, 2012), pp. 149–156
191. Y. Wei, F. Yaowen, L. Jianqian et al., Random finite sets-based joint manoeuvring target detection and tracking filter and its implementation. IET Signal Proc. **6**(7), 648–660 (2012)
192. D. Dunne, T. Kirubarajan, Multiple model multi-bernoulli filters for manoeuvering targets. IEEE Trans. Aerosp. Electron. Syst. **49**(4), 2679–2692 (2013)
193. X. Yuan, F. Lian, C.Z. Han, Multiple-model cardinality balanced multi-target multi-bernoulli filter for tracking maneuvering targets. J. Appl. Math. 1–16 (2013)
194. J.-L. Yang, H.-B. Ji, H.-W. Ge, Multi-model particle cardinality-balanced multi-target multi-bernoulli algorithm for multiple manoeuvring target tracking. IET Radar Sonar Navig. **7**(2), 101–112 (2013)
195. S. Reuter, A. Scheel, K. Dietmayer, The multiple model labeled multi-bernoulli filter, in *18th International Conference on Information Fusion* (IEEE, Washington, DC, 2015), pp. 1574–1580
196. Y. Punchihewa, B.-N. Vo, B.-T. Vo, A generalized labeled multi-bernoulli filter for maneuvering targets, in *19th International Conference on Information Fusion* (IEEE, Heidelberg, Germany, 2016)
197. W. Yi, M. Jiang, R. Hoseinnezhad, The multiple model vo-vo filter. IEEE Trans. Aerosp. Electron. Syst. **53**(2), 1045–1054 (2017)
198. S.-W. Yeom, T. Kirubarajan, Y. Bar-shalom, Track segment association, fine-step IMM and initialization with Doppler for improved track performance. IEEE Trans. Aerosp. Electron. Syst. **40**(1), 293–309 (2004)
199. X. Wang, D. Musicki, R. Ellem, Fast track confirmation for multi-target tracking with Doppler measurements, in *3rd International Conference on Intelligent Sensors, Sensor Networks and Information*, Melbourne, Qld., Australia (2007), pp. 263–268
200. X. Wang, D. Musicki, R. Ellem et al., Efficient and enhanced multi-target tracking with Doppler measurements. IEEE Trans. Aerosp. Electron. Syst. **45**(4), 1400–1417 (2009)
201. R. Georgescu, P. Willett, Predetection fusion with Doppler measurements and amplitude information. IEEE J. Ocean. Eng. **37**(1), 56–65 (2012)
202. S. Zollo, B. Ristic, On polar and versus Cartesian coordinates for target tracking, in *Proceedings of the fifth International Symposium on Signal Processing and Its Applications*, Salisbury, SA, Australia (1999), pp. 499–502
203. Mallick, M., S.Arulampalam, Comparison of nonlinear filtering algorithms in ground moving target indicator (GMTI) tracking, in *Signal and Data Processing of Small Targets*, San Diego, USA (SPIE, 2003)
204. T. Kirubarajan, Y. Bar-shalom, Tracking evasive move-stop-move targets with a GMTI radar using a VS-IMM estimator. IEEE Trans. Aerosp. Electron. Syst. **39**(3), 1098–1103 (2003)
205. J. Wang, P. He, T. Long, Use of the radial velocity measurement in target tracking. IEEE Trans. Aerosp. Electron. Syst. **39**(2), 401–413 (2003)
206. S. Zhang, Y. Bar-shalom, Track segment association for GMTI tracks of evasive move-stop-move maneuvering targets. IEEE Trans. Aerosp. Electron. Syst. **47**(3), 1899–1914 (2011)
207. Y. Bar-Shalom, Negative correlation and optimal tracking with Doppler measurements. IEEE Trans. Aerosp. Electron. Syst. **37**(3), 1117–1120 (2001)
208. W. Wu, J. Jiang, W. Liu, et al., A sequential converted measurement Kalman filter in the ECEF coordinate system for airborne Doppler radar. Aerosp. Sci. Technol. **51**, 11–17 (2016)
209. M. Mallick, B.F. La Scala, Comparison of single-point and two-point difference track initiation algorithms using position measurements, in *Proceedings of International Colloquium on Information Fusion*, Xian, China (2007)
210. D. Musicki, T.L. Song, Track initialization: prior target velocity and acceleration moments. IEEE Trans. Aerosp. Electron. Syst. **49**(1), 655–670 (2013)
211. B. Ristic, S. Arulampalam, N.J. Gordon, *Beyond the Kalman Filter: Particle Filters for Tracking Applications* (Artech House, 2004)
212. M. Longbin, S. Xiaoquan, Z. Yiyu et al., Unbiased converted measurements for tracking. IEEE Trans. Aerosp. Electron. Syst. **34**(2), 1023–1027 (1998)

213. M. Mallick, Y. Bar-Shalom, T. Kirubarajan et al., An improved single-point track initiation using GMTI measurements. IEEE Trans. Aerosp. Electron. Syst. **51**(4), 2697–2714 (2015)
214. W. Koch, On exploiting 'negative' sensor evidence for target tracking and sensor data fusion. Inf. Fusion **8**(1), 28–39 (2007)
215. W. Koch, R. Klemm, Ground target tracking with STAP radar. IEE Proc.-Radar, Sonar Navig. **148**(3), 173–185 (2001)
216. M. Mertens, W. Koch, T. Kirubarajan, Exploiting Doppler blind zone information for ground moving target tracking with bistatic airborne radar. IEEE Trans. Aerosp. Electron. Syst. **52**(3), 1408–1420 (2016)
217. N. Gordon, B. Ristic, Tracking airborne targets occasionally hidden in the blind Doppler. Digital Signal Process. **12**(12), 383–393 (2002)
218. S. Du, Z. Shi, W. Zang et al., Using interacting multiple model particle filter to track airborne targets hidden in blind Doppler. J. Zhejiang Univ. Sci. A **8**(8), 1277–1282 (2007)
219. J. Clark, P. Kountouriotis, R. Vinter, A new Gaussian mixture algorithm for GMTI tracking under a minimum detectable velocity constraint. IEEE Trans. Autom. Control **54**(12), 2745–2756 (2009)
220. M. Yu, C. Liu, B. Li, et al., An enhanced particle filtering method for GMTI radar tracking. IEEE Trans. Aerosp. Electron. Syst. **12**(12), 383-393 (2002)
221. S. Zhang, Y. Bar-Shalom, Tracking move-stop-move targets with state-dependent mode transition probabilities. IEEE Trans. Aerosp. Electron. Syst. **47**(3), 2037–2054 (2011)
222. L. Lin, Y. Bar-Shalom, T. Kirubarajan, New assignment-based data association for tracking move-stop-move targets. IEEE Trans. Aerosp. Electron. Syst. **40**(2), 714–725 (2004)
223. J.H. Yoon, D.Y. Kim, S.H. Bae et al., Joint initialization and tracking of multiple moving objects using doppler information. IEEE Trans. Signal Process. **59**(7), 3447–3452 (2011)
224. S. Subedi, Y.D. Zhang, M.G. Amin et al., Group sparsity based multi-target tracking in passive multi-static radar systems using doppler-only measurements. IEEE Trans. Signal Process. **64**(4), 3619–3634 (2016)
225. F. Papi, Multi-sensor δ-GLMB filter for multi-target tracking using doppler only measurements, in *European Intelligence and Security Informatics Conference*, Manchester, UK (2015) pp. 83–89
226. R. Kohlleppel, Ground target tracking with signal adaptive measurement error covariance matrix, in *15th International Conference on Information Fusion,* Singapore (IEEE, 2012), pp. 550–557
227. M. Mallick, V. Krishnamurthy, B.-N. Vo, *Integrated Tracking, Classification, and Sensor Management* (Wiley, 2013)
228. W. Wu, W. Liu, J. Jiang, et al., GM-PHD filter-based multi-target tracking in the presence of Doppler blind zone. Digit. Signal Process. **52**(C), 1–12 (2016)
229. W. Wu, J. Jiang, W. Liu, et al., Augmented state GM-PHD filter with registration errors for multi-target tracking by Doppler radars. Signal Process. **120**(3), 117-128 (2016)
230. S.B. Colegrove, A.W. Davis, J.K. Ayliffe, Track initiation and nearest neighbours incorporated into probabilistic data association. J. Electr. Electron. Eng. Austral. **6**(3), 191–198 (1986)
231. G. van Keuk, Multihypothesis tracking using incoherent signal-strength information. IEEE Trans. Aerosp. Electron. Syst. **32**(3), 1164–1170 (1996)
232. Y. Barniv, Dynamic programming solution for detecting dim moving targets. IEEE Trans. Aerosp. Electron. Syst. **21**(1), 144–156 (1985)
233. L.R. Moyer, J. Spak, P. Lamanna, A multi-dimensional hough transform-based track-before-detect technique for detecting weak targets in strong clutter backgrounds. IEEE Trans. Aerosp. Electron. Syst. **47**(4), 3062–3068 (2011)
234. Y. Boers, J.N. Driessen, Multitarget particle filter track before detect application. IEE Proc.-Radar Sonar Navig. **151**(6), 351–357 (2004)
235. G.W. Pulford, B.F.L. Scala, Multihypothesis Viterbi data association: algorithm development and assessment. IEEE Trans. Aerosp. Electron. Syst. **46**(2), 583–609 (2010)
236. S.M. Tonissen, Y. Bar-Shalom, Maximum likelihood track-before-detect with fluctuating target amplitude. IEEE Trans. Aerosp. Electron. Syst. **34**(3), 796–806 (1998)

237. S.J. Davey, M.G. Rutten, B. Cheung, A comparison of detection performance for several track-before-detect algorithms, in *International Conference on Information Fusion*, Cologne, Germany (2008)
238. D. Clark, B. Ristic, B.-N. Vo et al., Bayesian multi-object filtering with amplitude feature likelihood for unknown object SNR. IEEE Trans. Signal Process. **58**(1), 26–37 (2010)
239. S.P. Ebenezer, A. Papandreou-Suppappola, Generalized recursive track-before-detect with proposal partitioning for tracking varying number of multiple targets in low SNR. IEEE Trans. Signal Process. **64**(11), 2819–2834 (2016)
240. W. Yi, M.R. Morelande, L. Kong et al., An efficient multi-frame track-before-detect algorithm for multi-target tracking. IEEE J. Sel. Top. Signal Process. **7**(3), 421–434 (2013)
241. S. Buzzi, M. Lops, L. Venturino et al., Track-before-detect procedures in a multi-target environment. IEEE Trans. Aerosp. Electron. Syst. **44**(3), 1135–1150 (2008)
242. D.A. Castanon, Efficient algorithms for finding the best K paths through a trellis. IEEE Trans. Aerosp. Electron. Syst. **26**(2), 405–410 (1990)
243. E. Grossi, M. Lops, L. Venturino, A novel dynamic programming algorithm for track-before-detect in radar systems. IEEE Trans. Signal Process. **61**(10), 2608–2619 (2013)
244. E. Grossi, M. Lops, L. Venturino, A heuristic algorithm for track-before-detect with thresholded observations in radar systems. IEEE Signal Process. Lett. **20**(8), 811–814 (2013)
245. E. Grossi, M. Lops, L. Venturino, A track-before-detect algorithm with thresholded observations and closely-spaced targets. IEEE Signal Process. Lett. **20**(12), 1171–1174 (2013)
246. A. Aprile, E. Grossi, M. Lops et al., Track-before-detect for sea clutter rejection: tests with real data. IEEE Trans. Aerosp. Electron. Syst. **52**(3), 1035–1045 (2016)
247. S. Wong, B.T. Vo, F. Papi, Bernoulli forward-backward smoothing for track-before-detect. IEEE Signal Process. Lett. **21**(6), 727–731 (2014)
248. F. Papi, V. Kyovtorov, R. Giuliani et al., Bernoulli filter for track-before-detect using MIMO radar. IEEE Signal Process. Lett. **21**(9), 1145–1149 (2014)
249. B.-N. Vo, B.-T. Vo, N.-T. Pham et al., Joint detection and estimation of multiple objects from image observations. IEEE Trans. Signal Procesing **58**(10), 5129–5241 (2010)
250. J. Wong, B.-T. Vo, B.-N. Vo, et al., Multi-Bernoulli based track-before-detect with road constraints, in *Proceedings of 15th Annual Conference on Information Fusion*, Singapore (2012)
251. F. Papi, B.-T. Vo, M. Bocquel, et al., Multi-target track-before-detect using labeled random finite set, in *International Conference on Control, Automation and Information Sciences*, Vietnam (2013)
252. S. Li, W. Yi, B. Wang, et al. Labeled multi-object tracking algorithms for generic observation model, in *19th International Conference on Information Fusion* (IEEE, Heidelberg, Germany, 2016)
253. F.P. Yuthika Punchihewa, R. Hoseinnezhad, Multiple target tracking in video data using labeled random finite set, in *International Conference on Control, Automation and Information Sciences* (2014), pp. 13–18
254. T. Rathnayake, A.K. Gostar, R. Hoseinnezhad, et al., Labeled multi-bernoulli track-before-detect for multi-target tracking in video, in *18th International Conference on Information Fusion*, Washington, DC. (IEEE, 2015), pp. 1353–1358
255. M. Li, J. Li, Y. Zhou, Labeled RFS-Based track-before-detect for multiple maneuvering targets in the infrared focal plane array. Sensors **15**(12), 30839–30855 (2015)
256. A.F. Garcia-Fernandez, Track-before-detect labeled multi-bernoulli particle filter with label switching. IEEE Trans. Aerosp. Electron. Syst. **52**(5), 2123–2138 (2016)
257. M. Bocquel, Labeled random finite sets in multi-target track-before-detect (2017), Available: https://ris.utwente.nl/ws/portalfiles/portal/5091241/ASSL_1994_Pollnau.pdf
258. S. Nannuru, M. Coates, R. Mahler, Computationally-tractable approximate PHD and CPHD filters for superpositional sensors. IEEE J. Sel. Top. Signal Process. **7**(3), 410–420 (2013)
259. F. Papi, Constrained δ-GLMB filter for multi-target track-before-detect using radar measurements, in *European Intelligence and Security Informatics Conference* (IEEE, 2015), pp. 90–97

260. F. Papi, M. Bocquel, M. Podt et al., Fixed-Lag smoothing for bayes optimal knowledge exploitation in target tracking. IEEE Trans. Signal Process. **62**(12), 3143–3152 (2014)
261. K. Gilholm, D. Salmond, Spatial distribution model for tracking extended objects. IEE Proc.-Radar, Sonar Navig. **152**(5), 364–371 (2005)
262. D.-Y. Kim, B.-T. Vo, B.-N. Vo, Data Fusion in 3D vision using a RGB-D data via switching observation model and its application to people tracking, in *Proceedings of International Conference on Control Automation and Information Sciences*, Vietnam (2013)
263. K. Gilholm, S. Godsill, S. Maskell, et al., Poisson models for extended target and group tracking, in *Proceedings of SPIE, Signal and Data Processing of Small Targets*, San Diego, CA (2005), pp. 230–241
264. J.W. Koch, Bayesian approach to extended object and cluster tracking using random matrices. IEEE Trans. Aerosp. Electron. Syst. **44**(3), 1042–1059 (2008)
265. M. Feldmann, D. Fränken, J.W. Koch, Tracking of extended objects and group targets using random matrices. IEEE Trans. Signal Process. **59**(4), 1409–1420 (2011)
266. M. Wieneke, W. Koch, Probabilistic tracking of multiple extended targets using random matrices, in *Proceedings of SPIE, Signal and Data Processing of Small Targets*, Orlando, Florida, United States (2010)
267. M. Baum, U. Hanebeck, Extended object tracking with random hypersurface models. IEEE Trans. Aerosp. Electron. Syst. **50**(1), 149–159 (2014)
268. M. Baum, M. Feldmann, D. Fraenken, et al., Extended object and group tracking: a comparison of random matrices and random hypersurface models. Inf. Fusion 904–906 (2011)
269. C. Lundquist, K. Granstrom, U. Orguner, Estimating the shape of targets with a PHD filter, in *Proceedings of the 14th International Conference on Information Fusion* Chicago, IL (IEEE, 2011), pp. 1–8
270. N. Wahlstrom, E. Ozkan, Extended target tracking using gaussian processes. IEEE Trans. Signal Process. **63**(16), 4165–4178 (2015)
271. B.-T. Vo, B.-N. Vo, A. Cantoni, Bayesian filtering with random finite set observations. IEEE Trans. Signal Process. **56**(4), 1313–1326 (2008)
272. B. Ristic, J. Sherrah, Bernoulli filter for joint detection and tracking of an extended object in clutter. IET Radar, Sonar, Navig. **7**(1), 26–35 (2013)
273. R. Mahler, PHD filters for nonstandard targets, I: extended targets, in *12th International Conference on Information Fusion,* Seattle, WA (IEEE, 2009), pp. 915–921
274. K. Granstrom, U. Orguner, R. Mahler et al., Corrections on: extended target tracking using a gaussian-mixture PHD filter. IEEE Trans. Aerosp. Electron. Syst. **53**(2), 1055–1058 (2017)
275. X. Tang, X. Chen, M. McDonald, et al., A multiple-detection probability hypothesis density filter, in IEEE Trans. Signal Process. **63**(8), 2007–2019 (2015)
276. K. Granstrom, U. Orguner, A PHD filter for tracking multiple extended targets using random matrices. IEEE Trans. Signal Process. **60**(11), 5657–5671 (2012)
277. K. Granström, A. Natale, P. Braca et al., Gamma gaussian inverse wishart probability hypothesis density for extended target tracking using x-band marine radar data. IEEE Trans. Geosci. Remote Sens. **53**(12), 6617–6631 (2015)
278. A. Swain, D. Clark, Extended object filtering using spatial independent cluster processes, in *13th International Conference on Information Fusion*, Edinburgh, U.K. (IEEE, 2010), pp. 1–8
279. C. Lundquist, K. Granstrom, U. Orguner, An extended target CPHD filter and a gamma gaussian inverse wishart implementation. IEEE J. Sel. Top. Signal Process. **7**(3), 472–483 (2013)
280. K. Granstrom, M. Fatemi, L. Svensson, Poisson multi-bernoulli conjugate prior for multiple extended object estimation (2016), [Online]. Available: https://arxiv.org/abs/1605.06311
281. J.L. Williams, Marginal multi-bernoulli filters: RFS derivation of MHT, JIPDA, and association-based MeMBer. IEEE Trans. Aerosp. Electron. Syst. **51**(3), 1664–1687 (2015)
282. Karl Granstrom, Maryam Fatemi, Lennart Svensson. Gamma Gaussian inverse-Wishart Poisson multi-Bernoulli Filter for Extended Target Tracking, in *19th International Conference on Information Fusion* (IEEE, Heidelberg, Germany, 2016), pp. 893–900

283. A. Scheel, S. Reuter, K. Dietmayer, Using separable likelihoods for laser-based vehicle tracking with a labeled multi-bernoulli filter, in *19th International Conference on Information Fusion* Heidelberg, Germany (2016), pp. 1200–1207
284. T. Hirscher, A. Scheel, S. Reuter, et al., Multiple extended object tracking using gaussian processes, in *19th International Conference on Information Fusion* (IEEE, Heidelberg, Germany, 2016), pp. 868–875
285. H. Zhu, C. Han, C. Li, An extended target tracking method with random finite set observations, in *Proceedings of the 14th International Conference on Information Fusion,* Chicago, IL (IEEE, 2011), pp. 1–6
286. M. Baum, U.D. Hanebeck, Random hypersurface models for extended object tracking, in *Proceedings of the IEEE International Symposium on Signal Processing and Information Technology*, Ajman, United Arab Emirates (2009), pp. 178–183
287. K. Granstrom, M. Baum, S. Reuter, Extended object tracking: introduction, overview and applications (2016), Available: https://arxiv.org/abs/1604.00970
288. K.-C. Chang, Y. Bar-Shalom, Joint probabilistic data association for multitarget tracking with possibly unresolved measurements and maneuvers. IEEE Trans. Autom. Control **29**(7), 585–594 (1984)
289. H.A.P. Blom, E.A. Bloem, Bayesian tracking of two possibly unresolved maneuvering targets. IEEE Trans. Aerosp. Electron. Syst. **43**(2), 612–627 (2007)
290. S. Jeong, J.K. Tugnait, Tracking of two targets in clutter with possibly unresolved measurements. IEEE Trans. Aerosp. Electron. Syst. **44**(2), 748–765 (2008)
291. D. Svensson, M. Ulmke, L. Hammarstrand, Multitarget sensor resolution model and joint probabilistic data association. IEEE Trans. Aerosp. Electron. Syst. **48**(4), 3418–3434 (2012)
292. W. Koch, G. V. Keuk, Multiple hypothesis track maintenance with possibly unresolved measurements. IEEE Trans. Aerosp. Electron. Syst. **33**(3),883–892 (1997)
293. T. Kirubarajan, Y. Bar-Shalom, K.R. Pattipati, Multiassignment for tracking a large number of overlapping objects. IEEE Trans. Aerosp. Electron. Syst. **37**(1), 2–21 (2001)
294. Z. Khan, T. Balch, F. Dellaert, Multitarget tracking with split and merged measurements, in *Proceedings of IEEE Conference on Computer Vision Pattern Recognition* (2005), pp. 605–610
295. S.J. Davey, Tracking possibly unresolved targets with PMHT, in *Proceedings of Infringement Decision, Control* (2007)
296. D. Musicki, M. Morelande, Finite resolution multitarget tracking, in *Proceedings of SPIE Signal Data Process, Small Targets* (2005)
297. D. Musicki, W. Koch, Multi scan target tracking with finite resolution sensors, in *Proceedings of 11th International Conference on Information Fusion*, Cologne, Germany, (2008)
298. F. Lian, C. Han, W. Liu et al., Unified cardinalized probability hypothesis density filters for extended targets and unresolved targets. Signal Process. **92**(7), 1729–1744 (2012)
299. R. Olfati-Saber, J.A. Fax, R. Murray, Consensus and cooperation in networked multi-agent systems. Proc. IEEE **95**(1), 215–233 (2007)
300. G. Battistelli, L. Chisci, C. Fantacci et al., Consensus-based multiple-model Bayesian filtering for distributed tracking. IET Radar Sonar Navig. **9**(4), 401–410 (2015)
301. R.P.S. Mahler, Optimal/robust distributed data fusion: a unified approach, in *Proceedings of SPIE Signal Processing, Sensor Fusion, and Target Recognition IX*, Orlando, FL, USA (2000)
302. M.B. Hurley, An information-theoretic justification for covariance intersection and its generalization, in *Proceedings of FUSION Conference* (2002), pp. 505–511
303. S.J. Julier, Fusion without independence, in *IET Seminar on Target Tracking and Data Fusion: Algorithms and Applications*, Birmingham, UK (2008), pp. 3–4
304. K.C. Chang, C.-Y. Chong, S. Mori, Analytical and computational evaluation of scalable distributed fusion algorithms. IEEE Trans. Aerosp. Electron. Syst. **46**(4), 2022–2034 (2010)
305. G. Battistelli, L. Chisci, Kullback-Leibler average, consensus on probability densities, and distributed state estimation with guaranteed stability. Automatica **50**(3), 707–718 (2014)
306. R. Mahler, The multisensor PHD filter—part I: general solution via multitarget calculus, in *Proceedings of SPIE-Signal Process. Sensor Fusion Target Recognit*, vol. XVIII (2009) pp. 1–12

307. R. Mahler, The multisensor PHD filter—part II: erroneous solution via poisson magic, in *Proceedings of SPIE-Signal Process. Sensor Fusion Target Recognit*, vol. XVIII (2009)
308. X. Jian, F.-M. Huang, Z.-L. Huang, The multi-sensor PHD filter: analytic implementation via Gaussian mixture and effective binary partition, in *Proceedings of the International Conference on Information Fusion*, Istanbul, Turkey (2013)
309. E. Delande, E. Duflos, D. Heurguier, et al., Multi-target PHD filtering: proposition of extensions to the multi-sensor case. INRIA, Research Report RR-7337 (2010)
310. P. Braca, S. Marano, V. Matta et al., Asymptotic efficiency of the PHD in multi-target/multisensor estimation. IEEE J. Sel. Top. Signal Process. **7**(3), 553–564 (2013)
311. S. Nannuru, S. Blouin, M. Coates et al., Multisensor CPHD filter. IEEE Trans. Aerosp. Electron. Syst. **52**(4), 1834–1854 (2016)
312. E. Delande, E. Duflos, P. Vanheeghe, et al., Multi-Sensor PHD: construction and implementation by space partitioning, in *IEEE International Conference on Acoustics, Speech and Signal Processing* (2011), pp. 3632–3635
313. N.T. Pham, W. Huang, S.H. Ong, Multiple sensor multiple object tracking with GMPHD filter, in *10th International Conference on Information Fusion*, Quebec, Que. (IEEE, 2007), pp. 1–7
314. S. Nagappa, D.E. Clark, On the ordering of the sensors in the iterated-corrector probability hypothesis density (PHD) filter, in *Proceedings SPIE International Conference on Signal Processing, Sensor Fusion, Target Recognition*, Orlando, FL, (2011)
315. G. Battistelli, L. Chisci, S. Morrocchi et al., Robust multisensor multitarget tracker with application to passive multistatic radar tracking. IEEE Trans. Aerosp. Electron. Syst. **48**(4), 3450–3472 (2012)
316. R. Mahler, Approximate multisensor CPHD and PHD filters, in *Proceedings of the International Conference on Information Fusion*, Edinburgh, United Kingdom (2010)
317. C. Ouyang, H. Ji, Scale unbalance problem in product multisensor PHD filter. Electron. Lett. **47**(22), 1247–1249 (2011)
318. B. Ristic, A. Farina, Target tracking via multi-static doppler shifts. IET Proc. Radar Sonar Navig. **7**(5), 508–516 (2013)
319. M.B. Guldogan, Consensus Bernoulli filter for distributed detection and tracking using multi-static doppler shifts. IEEE Signal Process. Lett. **21**(6), 672–676 (2014)
320. A.-A. Saucan, M.J. Coates, M. Rabbat, A multisensor multi-bernoulli filter. IEEE Trans. Signal Process. **65**(20), 5495–5509 (2017)
321. G. Battistelli, L. Chisci, C. Fantacci, et al., Distributed fusion of multitarget densities and consensus PHD/CPHD filters, in *SPIE Signal Processing, Sensor/Information Fusion, and Target Recognition*, vol. XXIV (2015)
322. T. Li, J.M. Corchado, S. Sun, Partial consensus and conservative fusion of gaussian mixtures for distributed PHD fusion (2017). CoRR abs/1711.10783
323. T. Li. Distributed SMC-PHD fusion for partial, arithmetic average consensus (2017). CoRR abs/1712.06128
324. D. Clark, S. Julier, R. Mahler, et al., Robust multi-object sensor fusion with unknown correlations, in *Proceedings of Sensor Signal Processing for Defence*, London, UK (2010), pp 1–5
325. M. Üney, D. Clark, S. Julier, Information measures in distributed multitarget tracking, in *Proceedings of 14th International Conference on Information Fusion* (2011), pp. 1–8
326. C. Fantacci, Distributed multi-object tracking over sensor networks: a random finite set approach (2015), Available: https://arxiv.org/abs/1508.04158
327. B.-N. Vo, B.-T. Vo, An implementation of the multi-sensor generalized labeled multi-Bernoulli filter via Gibbs sampling, in *20th International Conference on Information Fusion* (IEEE, Xi'an, China, 2017)
328. B.N. Vo, B.T. Vo, Multi-Sensor multi-object tracking with the generalized labeled multi-bernoulli filter (2017), Available: https://arxiv.org/abs/1702.08849
329. C. Fantacci, B.-N. Vo, B.-T. Vo, et al., Consensus labeled random finite set filtering for distributed multi-object tracking (2015), Available: https://arxiv.org/pdf/1501.01579

330. B.L. Wang, W. Yi, S.Q. Li, et al., Distributed fusion of labeled multi-object densities via label spaces matching (2016), CoRR abs/1603.08336
331. B. Wang, W. Yi, R. Hoseinnezhad et al., Distributed fusion with multi-bernoulli filter based on generalized covariance intersection. IEEE Trans. Signal Process. **65**(1), 242–255 (2017)
332. S. Li, W. Yi, R. Hoseinnezhad et al., Robust distributed fusion with labeled random finite sets. IEEE Trans. Signal Process. **66**(2), 278–293 (2018)
333. M. Jiang, W. Yi, R. Hoseinnezhad, et al., Distributed multi-sensor fusion using generalized multi-bernoulli densities, in *19th International Conference on Information Fusion* (IEEE, Heidelberg, Germany, 2016), pp. 1332–1339
334. Z. Li, S. Chen, H. Leung, Joint data association, registration, and fusion using EM-KF. IEEE Trans. Aerosp. Electron. Syst. **46**(2), 496–507 (2010)
335. F. Lian, C. Han, W. Liu et al., Joint spatial registration and multi-target tracking using an extended probability hypothesis density filter. IET Radar Sonar Navig. **5**(4), 441–448 (2011)
336. R. Mahler, A. El-Fallah, Bayesian unified registration and tracking, in *Proceedings of SPIE* (2011)
337. B. Ristic, D. Clark, N. Gordon, Calibration of multi-target tracking algorithms using non-cooperative targets. IEEE Trans. Signal Process. **7**(3), 390–398 (2013)
338. B. Ristic, D. Clark, Particle filter for joint estimation of multi-object dynamic state and multi-sensor bias, in *Proceedings of IEEE International Conference on Acoustics, Speech, Signal Process*, Kyoto, Japan (IEEE, 2012), pp. 3877–3880
339. W. Li, Y. Jia, J. Du et al., Gaussian mixture PHD filter for multi-sensor multi-target tracking with registration errors. Signal Process. **93**(1), 86–99 (2013)
340. A.N. Bishop, Gaussian-sum-based probability hypothesis density filtering with delayed and out-of-sequence measurements, in *18th Mediterranean Conference on Control & Automation*, Marrakech, Morocco (2010), pp. 1423–1428
341. R.P.S. Mahler, T.R. Zajic, Probabilistic objective functions for sensor management, in *Proceedings of SPIE* (2004), pp. 233–244
342. H.G. Hoang, B.T. Vo, Sensor management for multi-target tracking via multi-bernoulli filtering. Automatica **50**(4), 1135–1142 (2014)
343. A.K. Gostar, R. Hoseinnezhad, A. Bab-Hadiashar, Multi-Bernoulli sensor control via minimization of expected estimation errors. IEEE Trans. Aerosp. Electron. Syst. **51**(3), 1762–1773 (2015)
344. A.K. Gostar, R. Hoseinnezhad, A. Bab-Hadiashar, Multi-Bernoulli sensor-selection for multi-target tracking with unknown clutter and detection profiles. Signal Process. 28–42 (2015)
345. T.M. Cover, J.A. Thomas, *Elements of Information Theory* (Wiley-Interscience, New York, NY, USA, 1991)
346. R.P.S. Mahler, Global posterior densities for sensor management, in *Proceedings of SPIE* (1998), pp. 252–263
347. B. Ristic, S. Arulampalam, Bernoulli particle filter with observer control for bearings-only tracking in clutter. IEEE Trans. Aerosp. Electron. Syst. **48**(3), 1–11 (2012)
348. B. Ristic, B.-N. Vo, Sensor control for multi-object state-space estimation using random finite sets. Automatica **46**(11), 1812–1818 (2010)
349. H.G. Hoang, B.-N. Vo, B.-T. Vo, et al., The cauchy–schwarz divergence for poisson point processes. IEEE Trans. Inf. Theory **61**(8), 4475–4485 (2015)
350. M. Beard, B.-T. Vo, B.-N. Vo et al., Void probabilities and cauchy-schwarz divergence for generalized labeled multi-bernoulli models. IEEE Trans. Signal Process. **65**(19), 5047–5061 (2017)
351. P. Tichavsky, C. Muravchik, A. Nehorai, Posterior Cramér-Rao bounds for discrete time nonlinear filtering. IEEE Trans. Signal Process. **46**(5), 1386–1396 (1998)
352. H.S. Tong, H. Zhang, H.D. Meng et al., A comparison of error bounds for a nonlinear tracking system with detection probability Pd < 1. Sensors **12**(12), 17390–17413 (2012)
353. M. Rezaeain, B.-N. Vo, Error bounds for joint detection and estimation of a single object with random finite set observation. IEEE Trans. Signal Process. **58**(3), 1493–1506 (2010)

354. M. Hernandez, B. Ristic, A. Farina, PCRLB for tracking in cluttered environments: measurement sequence conditioning approach. IEEE Trans. Aerosp. Electr. Syst. **42**(2), 680–704 (2006)
355. M. Hernandez, B. Ristic, A. Farina et al., A comparison of two Cramér-Rao bounds for nonlinear filtering with Pd < 1. IEEE Trans. Signal. Process. **52**(9), 2361–2370 (2004)
356. H. Tong, H. Zhang, H. Meng et al., The recursive form of error bounds for RFS state and observation with Pd < 1. IEEE Trans. Signal. Process. **61**(10), 2632–2646 (2013)
357. F. Lian, G.-H. Zhang, Z.-S. Duan, et al., Multi-target joint detection and estimation error bound for the sensor with clutter and missed detection. Sensors **16**(2), 1–18 (2016)
358. J. Hoffman, R. Mahler, Multitarget miss distance via optimal assignment. IEEE Trans. Syst. Man Cybern. **34**(3), 327–336 (2004)
359. D. Schuhmacher, B.-T. Vo, B.-N. Vo, A consistent metric for performance evaluation of multi-object filters. IEEE Trans. Signal. Process. **56**(8), 3447–3457 (2008)
360. B. Ristic, B.-N. Vo, D. Clark et al., A metric for performance evaluation of multi-target tracking algorithms. IEEE Trans. Sig. Process. **59**(7), 3452–3457 (2011)
361. E.H. Aoki, P.K. Mandal, L. Svensson, et al., Labeling uncertainty in multitarget tracking. IEEE Trans. Aerosp. Electron. Syst. **52**(3), 1006–1020 (2016)
362. X. He, R. Tharmarasa, T. Kirubarajan et al., A track quality based metric for evaluating performance of multitarget filters. IEEE Trans. Aerosp. Electron. Syst. **49**(1), 610–616 (2013)
363. M. Beard, B.T. Vo, B.-N. Vo, OSPA(2): using the OSPA metric to evaluate multi-target tracking performance, in *International Conference on Control, Automation and Information Sciences (ICCAIS)*, Chiang Mai, Thailand (2017)
364. R. Mahler, Divergence detectors for multitarget tracking algorithms, in *Proceedings SPIE Signal Processing, Sensor Fusion, and Target Recognition XXII* (2013)
365. S. Reuter, B.-T. Vo, B. Wilking, et al., Divergence detectors for the δ-generalized labeled multi-bernoulli filter, in *Workshop on Sensor Data Fusion: Trends, Solutions, Applications*, Bonn, Germany (2013)
366. B.-N. Vo, B.-T. Vo, R. Mahler, Closed form solutions to forward-backward smoothing. IEEE Trans. Signal. Process. **60**(1), 2–17 (2012)
367. B.D. Anderson, J.B. Moore, *Optimal Filtering* (Prentice-Hall, Englewood Cliffs, NJ, 1979)
368. R.P.S. Mahler, B.-T. Vo, B.-N. Vo, Forward-Backward probability hypothesis density smoothing. IEEE Trans. Aerosp. Electron. Syst. **48**(1), 707–728 (2012)
369. N. Nandakumaran, T. Kirubarajan, T. Lang et al., Multitarget tracking using probability hypothesis density smoothing. IEEE Trans. Aerosp. Electron. Syst. **47**(4), 2344–2360 (2011)
370. B.-N. Vo, B.-T. Vo, R.P.S. Mahler, Closed-Form solutions to forward-backward smoothing. IEEE Trans. Signal Process. **60**(1), 2–17 (2012)
371. S. Nagappa, E.D. Delande, D.E. Clark et al., A tractable forward-backward CPHD smoother. IEEE Trans. Aerosp. Electron. Syst. **53**(1), 201–217 (2017)
372. D. Li, C. Hou, D. Yi, Multi-bernoulli smoother for multi-target tracking. Aerosp. Sci. Technol. **48**, 234–245 (2016)
373. M. Beard, B.T. Vo, B.-N. Vo, Generalised labelled multi-bernoulli forward-backward smoothing, in *19th International Conference on Information Fusion* (IEEE, Heidelberg, Germany, 2016) pp. 688–694
374. Y.K.L. Keith, F. Inostroza, M. Adams, Relating random vector and random finite set estimation in navigation, mapping, and tracking. IEEE Trans. Signal. Process. **65**(17), 4609–4623 (2017)
375. O. Erdinc, P. Willett, Y. Bar-Shalom, The bin-occupancy filter and its connection to the PHD filters. IEEE Trans. Signal. Process. **57**(11), 4232–4246 (2009)
376. R. Mahler, Measurement-to-track association and finite-set statistics. Available: (2017), https://arxiv.org/abs/1701.07078
377. T.L. Song, D. Musicki, D.S. Kim et al., Gaussian mixtures in multi-target tracking: a look at Gaussian mixture probability hypothesis density and integrated track splitting. IET Radar Sonar Navig. **6**(5), 359–364 (2012)
378. D. Svensson, J. Wintenby, L. Svensson, Performance evaluation of MHT and GM-CPHD in a ground target tracking scenario, in *12th International Conference on Information Fusion*, Seattle, WA, USA (IEEE, 2009), pp. 6–9

379. E. Pollard, B. Pannetier, M. Rombaut, Hybrid algorithms for multitarget tracking using MHT and GM-CPHD. IEEE Trans. Aerosp. Electron. Syst. **47**(2), 832–847 (2011)
380. X.R. Li, V.P. Jilkov, Survey of maneuvering target tracking. Part I: dynamic models. IEEE Trans. Aerosp. Electron. Syst. **39**(4), 1333–1364 (2003)
381. D.L. Alspach, H.W. Sorenson, Nonlinear Bayesian estimation using Gaussian sum approximations. IEEE Trans. Autom. Control **17**(4), 439–448 (1972)
382. A. Runnalls, Kullback-Leibler approach to Gaussian mixture reduction. IEEE Trans. Aerosp. Electron. Syst. **43**(3), 989–999 (2007)
383. D. Salmond, Mixture reduction algorithms for point and extended object tracking in clutter. IEEE Trans. Aerosp. Electron. Syst. **45**(2), 667–686 (2009)
384. A. Doucet, N. de Freitas, N.J. Gordon, *Sequential Monte Carlo Methods in Practice* (Springer, New York, 2001), pp. 17–41
385. J.S. Liu, R. Chen, Sequential Monte Carlo methods for dynamical systems. J. Amer. Statist. Assoc. **93**(443), 1032–1044 (1998)
386. M. Pitt, N. Shephard, Filtering via simulation: auxiliary particle filters. J. Amer. Statist. Assoc. **94**(446), 590–599 (1999)
387. M. Bolic, P. Djuric, S. Hong, Resampling algorithms and architectures for distributed particle filters. IEEE Trans. Sig. Process **53**(7), 2442–2450 (2005)
388. J. Miguez, Analysis of parallelizable resampling algorithms for particle filtering. Elsevier Signal Process. **87**(12), 3155–3174 (2007)
389. R. Chen, J. Liu, Mixture Kalman filters. J. Roy. Stat. Soc.: Ser. B (Methodol.) **62**(3), 493–508 (2000)
390. J.H. Kotecha, P.M. Djuric, Gaussian particle filtering. IEEE Trans. Signal Process. **51**(10), 2592–2601 (2003)
391. J.H. Kotecha, P.M. Djuric, Gaussian sum particle filtering. IEEE Trans. Sig. Process. **51**(10), 2602–2612 (2003)
392. I. Goodman, R. Mahler, H. Nguyen, *Mathematics of Data Fusion* (Kluwer Academic, Norwell, MA, 1997)
393. D. Daley, D. Vere-Jones, *An Introduction to the Theory of Point Processes* (Springer, Berlin, Germany, 1988)
394. D. Stoyan, D. Kendall, J. Mecke, *Stochastic Geometry and Its Applications* (Wiley, New York, 1995)
395. J. Moller, R. Waagepetersen, *Statistical Inference and Simulation for Spatial Point Processes* (Chapman & Hall, Boston, MA, USA, 2004)
396. M.C. Stein, C.L. Winter, An additive theory of bayesian evidence accrual. Los Alamos National Laboratotries Report, LA-UR-93-3336 (1987). Available: https://fas.org/sgp/othergov/doe/lanl/dtic/ADA364591.pdf
397. W. Gilks, C. Berzuini, Following a moving target-Monte Carlo inference for dynamic Bayesian models. J. Roy. Stat. Soc. B **63**, 127–146 (2001)
398. P.J. Green, Reversible jump MCMC computation and Bayesian model determination. Biometrika **82**(4), 711–732 (1995)
399. G. Matheron, *Random Sets and Integral Geometry* (Wiley, New York, 1975)
400. O.E. Drummond, Methodologies for performance evaluation of multitarget multisensor tracking, in *Proceedings of SPIE, Signal and Data Processing of Small Targets* (1999), pp. 355–369
401. O.E. Drummond, B.E. Fridling, Ambiguities in evaluating performance of multiple target tracking algorithms, in *Proceedings of SPIE, Signal and Data Processing of Small Targets* (1992), pp. 326–337
402. N. Whiteley, S. Singh, S. Godsill, Auxiliary particle implementation of the probability hypothesis density filter. IEEE Trans. Aerosp. Electron. Syst. **46**(3), 1437–1454 (2010)
403. M. Beard, B.-T. Vo, B.-N. Vo, Multi-target filtering with unknown clutter density using a bootstrap GM-CPHD filter. IEEE Signal Process. Lett. **20**(4), 323–326 (2013)
404. D. Eppstein, Finding the k shortest paths. SIAM J. Comput. **28**(2), 652–673 (1998)
405. R. Bellman, On a routing problem. Q. Appl. Math. **16**, 87–90 (1958)

406. J. Munkres, Algorithms for the assignment and transportation problems. J. Soc. Ind. Appl. Math. **5**(1), 32–38 (1957)
407. R. Jonker, T. Volgenant, A shortest augmenting path algorithm for dense and sparse linear assignment problems. Computing **38**(11), 325–340 (1987)
408. I. Cox, M. Miller, On finding ranked assignments with application to multitarget tracking and motion correspondence. IEEE Trans. Aerosp. Electron. Syst. **32**(1), 486–489 (1995)
409. M. Miller, H. Stone, I. Cox, Optimizing Murty's ranked assignment method. IEEE Trans. Aerosp. Electron. Syst. **33**(3), 851–862 (1997)
410. C. Pedersen, L. Nielsen, K. Andersen, An algorithm for ranking assignments using reoptimization. Comput. Oper. Res. **35**(11), 3714–3726 (2008)
411. R. Lopez, P. Danes, Low-complexity IMM smoothing for Jump Markov nonlinear systems. IEEE Trans. Aerosp. Electron. Syst. **53**(3), 1261–1272 (2017)
412. Li, X-R., Jilkov, V. Survey of maneuvering target tracking. Part V: multiple-model methods. IEEE Trans. Aerosp. Electron. Syst. **41**(4), 1255–1321 (2005)
413. J.F.C. Kingman, *Poisson Processes* (Oxford University Press, London, 1993)
414. T. Kirubarajan, Y. Bar-Shalom, Kalman filter versus IMM estimator: when do we need the latter? IEEE Trans. Aerosp. Electron. Syst. **39**(4), 1452–1457 (2003)
415. S.V. Bordonaro, P. Willett, Y. Bar-Shalom, Unbiased tracking with converted measurements, in *Proceedings of IEEE Radar Conference* (2012), pp. 0741–0745
416. S.M. Kay, *Fundamentals of Statistical Signal Processing, Estimation Theory*, vol. I (Prentice Hall, 1993)
417. Y. Chen, On suboptimal detection of 3-dimensional moving targets. IEEE Trans. Aerosp. Electron. Syst. **25**(3), 343–350 (1989)
418. P.L. Chu, Optimal projection for multidimensional signal detection. IEEE Trans. Acoust. Speech Signal Process. **36**(5), 775–786 (1988)
419. S.D. Blostein, T.S. Huang, Detecting small, moving objects in image sequences using sequential hypothesis testing. IEEE Trans. Signal Process. **39**(7), 1611–1629 (1991)
420. M. Rollason, D.J. Samond, A particle filter for track-before-detect of a target with unknown amplitude, in *IEE Target Tracking: Algorithms and Applications*, Enschede, Netherlands (2002)
421. L.A. Johnson, V. Krishnamurthy, Performance analysis of a dynamic programming track before detect algorithm. IEEE Trans. Aerosp. Electron. Syst. **38**(1), 228–242 (2002)
422. A. Lepoutre, O. Rabaste, F. Le Gland, Multitarget likelihood computation for track-before-detect applications with amplitude fluctuations of type swerling 0, 1, and 3. IEEE Trans. Aerosp. Electron. Syst. **52**(3), 1089–1107 (2016)
423. K. Granstrom, C. Lundquist, U. Orguner, Tracking rectangular and elliptical extended targets using laser measurements, in *Proceedings of the International Conference on Information Fusion*, Chicago, IL (2011), pp. 592–599
424. D. Clark, K. Panta, B.-N. Vo, The GM-PHD filter multiple target tracker, in *Proceedings of the International Conference on Information Fusion*, Florence, Italy (2006)
425. M. Ulmke, O. Erdinc, P. Willett, Gaussian mixture cardinalized PHD filter for ground moving target tracking, in *Proceedings of the International Conference on Information Fusion*, Quebec, Canada (2007), pp. 1–8
426. G.-C. Rota, The number of partitions of a set. Am. Math. Mon. **71**(5), 498–504 (1964)
427. A. Gupta, D. Nagar, *Matrix Variate Distributions* (Chapman & Hall, London, 2000)
428. K. Granström, U. Orguner, On the reduction of Gaussian inverse Wishart mixtures, in *15th International Conference on Information Fusion*, Singapore (2012)
429. M.E. Liggins II, C.-Y. Chong, I. Kadar, et al., Distributed fusion architectures and algorithms for target tracking. Proc. IEEE **85**(2), 95–107 (1997)
430. L.-L. Ong, T. Bailey, H. Durrant-Whyte, Decentralised particle filtering for multiple target tracking in wireless sensor networks, in *Proceedings of 11th International Conference of Information Fusion*, Cologne, Germany (2008), pp. 1–8
431. L. Xiao, L.S. Boyd, S. Lall, A scheme for robust distributed sensor fusion based on average consensus, in *Proceedings of the 4th International Symposium on Information Processing in Sensor Networks* (2005), pp. 63–70

432. G.C. Calafiore, F. Abrate, Distributed linear estimation over sensor networks. Int. J. Control **82**(5), 868–882 (2009)
433. R. Carli, A. Chiuso, L. Schenato, et al., Distributed Kalman filtering based on consensus strategies. IEEE J. Sel. Areas Commun. **26**(4), 622–633 (2008)
434. S.S. Stankovic, M.S. Stankovic, D.M. Stipanovic, Consensus based overlapping decentralized estimation with missing observations and communication faults. Automatica **45**(6), 1397–1406 (2009)
435. M. Farina, G. Ferrari-Trecate, R. Scattolini, Distributed moving horizon estimation for linear constrained systems. IEEE Trans. Autom. Control **55**(11), 2462–2475 (2010)
436. T. Heskes, Selecting weighting factors in logarithmic opinion pools, in *Advances in Neural Information Processing Systems* (MIT Press, Cambridge, MA, 1998), pp. 266–272
437. L. Campbell, Equivalence of Gauss's principle and minimum discrimination information estimation of probabilities. Ann. Math. Statist. **41**(3), 1011–1015 (1970)
438. B. Silverman, *Density Estimation for Statistics and Data Analysis* (Chapman & Hall, London, 1986)
439. C. Fraley, A.W. Raftery, Model based clustering, discriminant analysis, and density estimation. J. Am. Statist. Assoc. **97**(458), 611–631 (2002)
440. M.C. Jones, J.S. Marron, S.J. Sheather, A brief survey of bad-width selection for density estimation. J. Am. Statist. Assoc. **91**(433), 401–407 (1996)
441. C.P. Robert, G. Casella, *Monte Carlo Statistical Methods* (Springer, New York, 2004)
442. A. Dabak, A geometry for detection theory. Ph.D. dissertation, Rice University, Houston, 1993. Available: https://scholarship.rice.edu/bitstream/handle/1911/16613/9408610.PDF?sequence=1
443. O. Hlinka, F. Hlawatsch, P.M. Djuric, Consensus-based distributed particle filtering with distributed proposal adaptation. IEEE Trans. Signal Process. **62**(12), 3029–3041 (2014)
444. M. Coates, Distributed particle filters for sensor networks, in *Proceedings of the 3rd International Symposium on Information Processing in Sensor Networks*, Berkeley, CA (2004), pp. 99–107
445. P.C. Mahalanobis, Analysis of race mixture in Bengal. J. Asiat. Soc. **23**, 301–310 (1927)
446. J.-F. Cardoso, T.-W. Lee, Dependence, correlation and Gaussianity in independent component analysis. J. Mach. Learn. Res. **4**(7–8), 1177–1203 (2004)
447. M. Abramowitz, I.A. Stegun, *Handbook of Mathematical Functions: With Formulas, Graphs, and Mathematical Tables* (National Bureau of Standards, Washington, DC, 1964). Republished by Courier Dover Publications, 2012. Available free online at http://people.math.sfu.ca/~cbm/aands/abramowitz_and_stegun.pdf

MIX
Papier aus verantwortungsvollen Quellen
Paper from responsible sources
FSC® C105338

If you have any concerns about our products,
you can contact us on
ProductSafety@springernature.com

In case Publisher is established outside the EU,
the EU authorized representative is:
Springer Nature Customer Service Center GmbH
Europaplatz 3, 69115 Heidelberg, Germany

Printed by Libri Plureos GmbH
in Hamburg, Germany